VOLUME SEVENTY ONE

ADVANCES IN BOTANICAL RESEARCH

Sea Plants

ADVANCES IN BOTANICAL RESEARCH

Series Editors

VOLUME SEVENTY ONE

ADVANCES IN BOTANICAL RESEARCH

Sea Plants

Volume Editor

NATHALIE BOURGOUGNON

Université de Bretagne-Sud, Laboratoire de Biotechnologie et Chimie Marine, EA 3884, Institut Universitaire Européen de la Mer, Vannes, France

ELSEVIER

AMSTERDAM • BOSTON • HEIDELBERG • LONDON
NEW YORK • OXFORD • PARIS • SAN DIEGO
SAN FRANCISCO • SINGAPORE • SYDNEY • TOKYO
Academic Press is an imprint of Elsevier

Academic Press is an imprint of Elsevier
32 Jamestown Road, London NW1 7BY, UK
Radarweg 29, PO Box 211, 1000 AE Amsterdam, The Netherlands
The Boulevard, Langford Lane, Kidlington, Oxford, OX5 1GB, UK
225 Wyman Street, Waltham, MA 02451, USA
525 B Street, Suite 1800, San Diego, CA 92101-4495, USA

First edition 2014

ISBN: 978-0-12-408062-1
ISSN: 0065-2296

For information on all Academic Press publications visit our website at store.elsevier.com

CONTENTS

CONTRIBUTORS

Paula Ximena Arata
Cátedra de Química de Biomoléculas, Departamento de Biología Aplicada y Alimentos, Facultad de Agronomía, Universidad de Buenos Aires, Argentina

Ravi Singh Baghel
Discipline of Marine Biotechnology and Ecology, CSIR-Central Salt and Marine Chemicals Research Institute, Bhavnagar, Gujarat, India; Academy of Scientific & Innovative Research (AcSIR-CSMCRI), CSIR-Central Salt and Marine Chemicals Research Institute, Bhavnagar, Gujarat, India

Gilles Bedoux
Université de Bretagne-Sud, Laboratoire de Biotechnologie et Chimie Marine, EA 3884, Institut Universitaire Européen de la Mer, Vannes, France

Nathalie Bourgougnon
Université de Bretagne-Sud, Laboratoire de Biotechnologie et Chimie Marine, EA 3884, Institut Universitaire Européen de la Mer, Vannes, France

Catherine Boyen
Sorbonne Universités, UPMC Univ Paris 06, UMR 8227, Integrative Biology of Marine Models, Station Biologique de Roscoff, Roscoff, France; CNRS, UMR 8227, Integrative Biology of Marine Models, Station Biologique de Roscoff, Roscoff, France

Stéphanie Brulé-Josso
Divers-cités, Laboscop, Chrysalide, Quimper, France

Anne-Sophie Burlot
Université de Bretagne-Sud, Laboratoire de Biotechnologie et Chimie Marine, EA 3884, Institut Universitaire Européen de la Mer, Vannes, France

Alejandro H. Buschmann
Centro i-mar, Universidad de Los Lagos, Puerto Montt, Chile

Jean-Paul Cadoret
Laboratoire de physiologie et biotechnologie des algues, IFREMER, Nantes, France

Carolina Camus
Consorcio BalBiofuel, Camino Chinquihue km 6, Puerto Montt, Chile

Grégory Carrier
Laboratoire de physiologie et biotechnologie des algues, IFREMER, Nantes, France

Stéphane Cérantola
Service RMN-RPE, Université de Bretagne Occidentale, UFR Sciences et Techniques, Brest, France

Marina Ciancia
Cátedra de Química de Biomoléculas, Departamento de Biología Aplicada y Alimentos, Facultad de Agronomía, Universidad de Buenos Aires, Argentina; CIHIDECAR-CONICET, Departamento de Química Orgánica, Facultad de Ciencias Exactas y Naturales, Universidad de Buenos Aires, Argentina

Jonas Collén
Sorbonne Universités, UPMC Univ Paris 06, UMR 8227, Integrative Biology of Marine Models, Station Biologique de Roscoff, Roscoff, France; CNRS, UMR 8227, Integrative Biology of Marine Models, Station Biologique de Roscoff, Roscoff, France

M. Lynn Cornish
Acadian Seaplants Limited, Dartmouth, NS, Canada

James Craigie
Aquatic and Crop Resource Development, National Research Council Canada, Halifax, NS, Canada

Alan T. Critchley
Acadian Seaplants Limited, Dartmouth, NS, Canada

Bernardo A.P. da Gama
Marine Chemical Ecology Lab, Departament of Marine Biology, Biology Institute, Universidade Federal Fluminense, Niterói, Rio de Janeiro, Brazil

Delmail David
UMR 6226 ISCR, Equipe PNSCM, Faculté des Sciences Pharmaceutiques et Biologiques, Université de Rennes 1, Rennes Cedex, France

Nikolai Davidovich
Karadag Nature Reserve of the National Academy of Sciences, Feodosiya, Ukraine

Justine Dumay
LUNAM, Université de Nantes, Nantes, France

James Ehrman
Department of Biology, Mount Allison University, Sackville, NB, Canada

Sylvain Faugeron
Centro de Conservación Marina, Facultad de Ciencias Biológicas, Pontificia Universidad Católica de Chile, Santiago, Chile

Paula Virginia Fernández
Cátedra de Química de Biomoléculas, Departamento de Biología Aplicada y Alimentos, Facultad de Agronomía, Universidad de Buenos Aires, Argentina

Elizabeth Ficko-Blean
Sorbonne Universités, UPMC Univ Paris 06, UMR 8227, Integrative Biology of Marine Models, Station Biologique de Roscoff, Roscoff, France; CNRS, UMR 8227, Integrative Biology of Marine Models, Station Biologique de Roscoff, Roscoff, France

Helen J. Fitton
Marinova Pty. Ltd Cambridge, Tasmania, Australia

Joël Fleurence
LUNAM, Université de Nantes, Nantes, France

Le Devehat Françoise
UMR 6226 ISCR, Equipe PNSCM, Faculté des Sciences Pharmaceutiques et Biologiques, Université de Rennes 1, Rennes Cedex, France

Patrick Gardet
Territoires en ProjetS, Chrysalide, Quimper, France

Vicki-Anne Gardiner
Marinova Pty. Ltd Cambridge, Tasmania, Australia

Romain Gastineau
LUNAM, FR 3473, MMS Mer, Molécules, Santé, Université du Maine, Le Mans, France

Fabienne Guérard
UMR 6539 CNRS UBO IRD IFREMER, LEMAR-IUEM-UBO, UEB, Technopôle Brest-Iroise, Plouzane, France

Vishal Gupta
Discipline of Marine Biotechnology and Ecology, CSIR-Central Salt and Marine Chemicals Research Institute, Bhavnagar, Gujarat, India

Alfonso Gutíerrez
Centro i-mar, Universidad de Los Lagos, Puerto Montt, Chile

Gert Hansen
Department of Biology, Marine Biological Section, University of Copenhagen, Copenhagen, Denmark

Gustaaf Hallegraeff
Institute for Marine and Antarctic Studies (IMAS), Hobart, Australia

Kevin Hardouin
Université de Bretagne-Sud, Laboratoire de Biotechnologie et Chimie Marine, EA 3884, Institut Universitaire Européen de la Mer, Vannes, France; OLMIX SA, Bréhan, France

Yann Hardivillier
LUNAM, FR 3473, MMS Mer, Molécules, Santé, Université du Maine, Le Mans, France

Dorothée Hermann
LUNAM, FR 3473, MMS Mer, Molécules, Santé, Université du Maine, Le Mans, France

María Carmen Hernández-González
Centro i-mar, Universidad de Los Lagos, Puerto Montt, Chile

Cécile Hervé
Sorbonne Universités, UPMC Univ Paris 06, UMR 8227, Integrative Biology of Marine Models, Station Biologique de Roscoff, Roscoff, France; CNRS, UMR 8227, Integrative Biology of Marine Models, Station Biologique de Roscoff, Roscoff, France

Thüs Holger
Natural History Museum, Cromwell Road, London, UK

Javier Infante
Consorcio BalBiofuel, Camino Chinquihue km 6, Puerto Montt, Chile

Anne-Gaëlle Jacquin
La Route des Algonautes, Laboscop, Chrysalide, Quimper, France

Boris Jacquette
LUNAM, FR 3473, MMS Mer, Molécules, Santé, Université du Maine, Le Mans, France

Camille Jégou
EA 3882 LUBEM-UBO, Quimper, France

Bhavanath Jha
Discipline of Marine Biotechnology and Ecology, CSIR-Central Salt and Marine Chemicals Research Institute, Bhavnagar, Gujarat, India

Boustie Joël
UMR 6226 ISCR, Equipe PNSCM, Faculté des Sciences Pharmaceutiques et Biologiques, Université de Rennes 1, Rennes Cedex, France

Irena Kaczmarska
Department of Biology, Mount Allison University, Sackville, NB, Canada

Sam S. Karpiniec
Marinova Pty. Ltd Cambridge, Tasmania, Australia

Stephen Kelloway
Department of Environmental Sciences, Faculty of Agriculture, Dalhousie University, Truro, NS, Canada

Stacy A. Krueger-Hadfield
College of Charleston, Grice Marine Laboratory, Charleston, SC, USA

Manoj Kumar
Discipline of Marine Biotechnology and Ecology, CSIR-Central Salt and Marine Chemicals Research Institute, Bhavnagar, Gujarat, India; Institute of Plant Sciences, Agricultural Research Organization (ARO), The Volcani Center, Bet Dagan, Israel.

Puja Kumari
Discipline of Marine Biotechnology and Ecology, CSIR-Central Salt and Marine Chemicals Research Institute, Bhavnagar, Gujarat, India

Catherine Leblanc
Sorbonne Universités, UPMC Univ Paris 06, UMR 8227, Integrative Biology of Marine Models, Station Biologique de Roscoff, Roscoff, France; CNRS, UMR 8227, Integrative Biology of Marine Models, Station Biologique de Roscoff, Roscoff, France

Vincent Leignel
LUNAM, FR 3473, MMS Mer, Molécules, Santé, Université du Maine, Le Mans, France

Klervi Le Lann
UMR 6539 CNRS UBO IRD IFREMER, LEMAR-IUEM-UBO, UEB, Technopôle Brest-Iroise, Plouzane, France

Cécile Le Guillard
LUNAM, Université de Nantes, Nantes, France

Florian Maumus
Unité de Recherche en Génomique-Info, UR 1164, INRA Centre de Versailles-Grignon, Versailles, France

Abasq Marie-Laurence
UMR 6226 ISCR, Equipe PNSCM, Faculté des Sciences Pharmaceutiques et Biologiques, Université de Rennes 1, Rennes Cedex, France

Vona Méléder
LUNAM FR 3473, MMS Mer, Molécules, Santé, Université de Nantes, Nantes, France

Gurvan Michel
Sorbonne Universités, UPMC Univ Paris 06, UMR 8227, Integrative Biology of Marine Models, Station Biologique de Roscoff, Roscoff, France; CNRS, UMR 8227, Integrative Biology of Marine Models, Station Biologique de Roscoff, Roscoff, France

Jean-Luc Mouget
LUNAM, FR 3473, MMS Mer, Molécules, Santé, Université du Maine, Le Mans, France

Michèle Morançais
LUNAM, Université de Nantes, Nantes, France

Mathilde Munier
LUNAM, Université de Nantes, Nantes, France

Pi Nyvall-Collén
Amadéite SAS, Bréhan, France

Renato C. Pereira
Marine Chemical Ecology Lab, Departament of Marine Biology, Biology Institute, Universidade Federal Fluminense, Niterói, Rio de Janeiro, Brazil

Rupert Perkins
School of Earth and Ocean Sciences, Cardiff University, Cardiff, Wales, UK

Erwan Plouguerné
Marine Chemical Ecology Lab, Departament of Marine Biology, Biology Institute, Universidade Federal Fluminense, Niterói, Rio de Janeiro, Brazil

Philippe Potin
Centre National de la Recherche Scientifique, Sorbonne Universités UPMC Univ. Paris 06, Roscoff cedex, Brittany, France

Steven Prescott
Institute of Aquaculture, University of Stirling, Stirling, UK

Balakrishnan Prithiviraj
Department of Environmental Sciences, Faculty of Agriculture, Dalhousie University, Truro, NS, Canada

C.R.K. Reddy
Discipline of Marine Biotechnology and Ecology, CSIR-Central Salt and Marine Chemicals Research Institute, Bhavnagar, Gujarat, India; Academy of Scientific & Innovative Research (AcSIR-CSMCRI), CSIR-Central Salt and Marine Chemicals Research Institute, Bhavnagar, Gujarat, India

Jan Rines
Graduate School of Oceanography University of Rhode Island, Narragansett, RI, USA

Bruno Saint-Jean
Laboratoire de physiologie et biotechnologie des algues, IFREMER, Nantes, France

Jatinder Singh Sangha
Department of Environmental Sciences, Faculty of Agriculture, Dalhousie University, Truro, NS, Canada

Valérie Stiger-Pouvreau
UMR 6539 CNRS UBO IRD IFREMER, LEMAR-IUEM-UBO, UEB, Technopôle Brest-Iroise, Plouzane, France

Wendy A. Stirk
Research Centre for Plant Growth and Development, School of Life Sciences, University of KwaZulu-Natal Pietermaritzburg Campus, Scottsville, South Africa

Damien Stringer
Marinova Pty. Ltd Cambridge, Tasmania, Australia

Rajendra Singh Thakur
Analytical Discipline and Centralized Instrument Facility, CSIR-Central Salt and Marine Chemicals Research Institute, Bhavnagar, Gujarat, India

Thierry Tonon
Sorbonne Universités, UPMC Univ Paris 06, UMR 8227, Integrative Biology of Marine Models, Station Biologique de Roscoff, Roscoff, France; CNRS, UMR 8227, Integrative Biology of Marine Models, Station Biologique de Roscoff, Roscoff, France

Johannes Van Staden
Research Centre for Plant Growth and Development, School of Life Sciences, University of KwaZulu-Natal Pietermaritzburg Campus, Scottsville, South Africa

Daniel A. Varela
Centro i-mar, Universidad de Los Lagos, Puerto Montt, Chile

Julio A. Vásquez
Facultad de Ciencias del Mar, Universidad Católica del Norte, Center for Advanced Studies in Arid Zones CEAZA, Coquimbo, Chile

Pia C. Winberg
School of Medicine, University of Wollongong, Wollongong, NSW, Australia; Venus Shell Systems Pty. Ltd, Narrawallee, NSW, Australia

Angela Wulff
Department of Biological and Environmental Sciences, University of Gothenburg, Göteborg, Sweden

Marian Yallop
School of Biological Sciences, University of Bristol, Bristol, UK

PREFACE

When I was born in 1946 the global human population was about 1.2 billion people; seaweed farming was virtually limited to Japan; seaweeds were generally held in low esteem and the main global use for seaweeds was as raw material for the manufacture of marine colloids. As of 2014 there are more than 7 billion of us populating the planet; seaweed farming has developed all over the world; the value of seaweeds for global wellbeing is increasingly recognised and a broad array of economic uses for seaweeds is being developed.

The collection of papers presented in *Advances in Botanical Research* (ABR) is testimony to the broad scope and the diverse range of research and development that is leading to advances in the state of the art of human utilisation of seaweeds. The included papers cover topics that include seaweed biogeography and productivity; cultivation techniques; pests and diseases; bioactive substances; aquaculture impacts, bioremediation and integrated multitrophic aquaculture (IMTA); seaweed resources and harvesting impacts; seaweed diversity and biogeography; ecophysiology; seaweed extracts; natural seaweed biomass; molecular biology and systematics; reproductive biology; population ecology and seaweeds as food and feed.

Papers included in ABR report the results of essential research and development necessary to advance the state of the art in seaweed production and management of ecosystem services. Such work is essential if we are to reach the point where the true potential of seaweeds as primary producers in integrated aquaculture systems can be realised. It is estimated by Food and Agriculture Organization (FAO) that within the next few years aquaculture will account for over 50% of the world's seafood consumption. Presently, 95.5% of the world seaweed supply comes from aquaculture (FAO 2012). Seaweeds have been the largest group of organisms cultured in the sea since 2004 and they comprise about half of total world mariculture production. It is a reflection of the growth in human population cited above that there is a vast global need for aquaculture. FAO forecast demand is for added annual production of 33 M tons of animal-based aquaculture production worth at least 46 B USD by 2030. Capture fisheries cannot increase production so sustainable, integrated; algae-based aquaculture is the path towards the future.

With respect to seaweed utilisation, papers included in *ABR* reflect a wide range of advances beyond the field of hydrocolloid applications. In particular there are widespread and expanding advances in knowledge and technologies in fields including the extraction and production of bioactive substances from seaweeds; seaweeds for direct use in food, feed and fertilisation applications; biotechnology based on polymers synthesised by seaweeds; ecosystem services such as nutrient biomitigation, oxygen provision, carbon sequestration and reduction of ocean acidification and the production of biofuels from algae including seaweeds.

ABR is about advances in botanical research and it is gratifying to see that the sluggish pace of advancement typical of the twentieth century seems to be picking up as we move into the twenty-first century.

Iain C. Neish

Macroalgae (seaweeds) are ecologically and economically important, providing essential ecosystem services and biomass for foods, phycocolloids, soil additives, animal feeds and nutraceuticals. Selection of these algal species is intensively cultured. In 2013, 11.3 million tons of fresh algae were produced commercially forming a $5.7 billion market. Asian countries, such as China, Korea and Japan, which have traditionally consumed seaweed for food, are responsible for three quarters of this production. In contrast, seaweed is very seldom considered as food in European countries. About 220 species of algae are consumed globally as food. Six genera (*Laminaria* (kombu, 40.1%), *Undaria* (wakame, 22.3%), *Porphyra* (nori, 12.4%), *Eucheuma/Kappaphycus* (11.6%) and *Gracilaria* (8.4%)) are responsible for 94.8% of the total algal production. The hydrocolloid industry is known to have the second largest consumption of seaweeds. The most common hydrocolloids manufactured from seaweeds are alginates, carrageenans and agars. In recent years, the application of macroalgae as animal feed supplement has grown. Alginates from seaweeds are used as a binder in feeds for farm animals. Laboratory investigations have also been carried out to evaluate macroalgae as possible alternative protein sources for farmed fish because of their high protein content and productivity.

Due to their rapid growth rates, ability to assimilate nutrients, such as nitrogen and phosphorous and to sequester other elements, such as heavy metals, they are also utilised for the bioremediation of waste waters, in particular from aquaculture systems. Algae have a number of characteristics that have a potential to significantly improve the sustainability of many industries. These include high biomass productivity, the possibility of utilising marginal, infertile land, salt water, waste streams as nutrient supply and combustion gas as CO_2 source to generate a wide range of fuel and nonfuel products. At the same time, eutrophication related to anthropogenic inputs from land has led to several green tides in shallow sandy bays. There has been research to use this biomass for animal feed, composting, biodigestion and methane production, fertiliser applications, etc., but none of these approaches proved economically feasible and capable of handling the variety of situations outside a laboratory context.

The harvesting of seaweed can exert pressure on the natural stocks of macroalgae. This pressure is made more severe as several macroalgae populations are in decline due to climate change. Seaweed harvesting can potentially cause a disruption in the natural habitats of shellfish and fish. This could be minimised and better understood with long-term monitoring. Rational management appears to be a promising solution for maintaining such biodiversity. Various research groups also highlight the valuable contribution of a 'fundamental research' approach to deal with practical management issues.

More sustainable production of seaweeds using new aquaculture technologies can help commercially develop green algae and simultaneously improve the environment. Technologies, such as IMTA have bioremedial properties which help improve and stabilise nutrient fluxes in the oceans. This development of seaweeds will depend on the successful domestication of several species.

For all these reasons, it is indeed imperative to continue research in the fields of metabolism, physiology, biology of algae, relations between plant and environment, and management, to see and understand the conditions ensuring a culture of quality, to highlight compounds of interest for food, cosmetics, agriculture, bioactives and to explore them as raw material for chemistry and bioenergy.

The authors are all recognised as experts in their respective areas. From Brazil to India, from France to South Africa or from Australia to Canada, they enrich our knowledge about the incredible world of algae. Two special guests the first with a species of microalgae and the second with of coastal lichens… a simple link to algae.

I would like to thank all the contributors for their patience, enthusiasm, hard work and commitment.

I would like also to acknowledge the valuable contributions of Elsevier team (specials thanks to Sarah Lay), Dr Erwan Ar Gall (University of Brest, France), Prof Rossitza Atanassova (University of Poitiers, France), Dr Denis Bailly (Université of Brest, France), Dr Carole Deleu (University of Rennes, France), Prof Nathalie Giglioli-Guivarc'h (University of Tours, France), Dr Claire Hellio (University of Porthsmouth, UK), Prof Jean-Michel Kornprobst (University of Nantes, France), Dr Marc Lahaye (INRA, France),

Dr Rafael Loureiro (Brazil), Prof Angèle Lengo Mambu (University of Limoges, France), Dr Marion Millot (University of Limoges, France), Prof Iain C. Neish (Satouni Co), Dr Laurent Picot (University of La Rochelle, France), Dr Florent Yvergnaux (Solabia, France) and some own authors who provided valuable inputs and support in reviewing this special edition of ABR.

I hope you will enjoy this textbook as much as we have enjoyed putting it together.

Prof Nathalie Bourgougnon

CHAPTER ONE

Selected Comments on the Role of Algae in Sustainability

Anne-Gaëlle Jacquin*[,1], Stéphanie Brulé-Josso[†], M. Lynn Cornish[‡], Alan T. Critchley[‡] and Patrick Gardet[§]

*La Route des Algonautes, Laboscop, Chrysalide, Quimper, France
[†]Divers-cités, Laboscop, Chrysalide, Quimper, France
[‡]Acadian Seaplants Limited, Dartmouth, NS, Canada
[§]Territoires en ProjetS, Chrysalide, Quimper, France
[1]Corresponding author: e-mail address: contact@routedesalgonautes.org

Contents

Abstract

Since the late nineteenth century, technical improvements in the lifestyles of societies have been accompanied by unprecedented environmental change. Scientists from various disciplines are attempting to develop a shared definition of 'sustainability', which includes elements from economics, environmental science and social sciences.

Advances in Botanical Research, Volume 71
ISSN 0065-2296
http://dx.doi.org/10.1016/B978-0-12-408062-1.00001-9

The diverse assemblage of photosynthetic organisms grouped together as 'the algae' provide important organisational roles and functional ecosystem services, which could make significant contributions to the pillars of sustainable development. Phycology could play an important role in the nascent development of sustainability science.

While many of the properties and potential contributions of the algae to global issues of sustainability are evident to the scientific community, relatively few of the opportunities are economically realised. It is unfortunate that the general 'Joe Public' perception of most algae (both micro- and macroform), particularly in the Western nations, is negative and frequently associated with environmental problems, such as green tides or toxic, harmful algal blooms. There is a pressing need to bridge scientific knowledge and public opinions in order to help spur innovations for future health and environmental benefits and for economic prosperity.

This article gives a personal and selected overview of the significant potential for advances phycology can make towards global sustainability concerns. The outcomes of the action-oriented research project 'The Algonauts Road' illustrate how applied aspects of phycology can be promoted by acting at the interface between science and society.

1.1 GENERAL ELEMENTS ABOUT SUSTAINABILITY

1.1.1 A Concept in the Making

In many ways societies have the potential to jeopardise the environmental goods and services, which will be bequeathed to future generations. Looming threats to human life on a global scale are thought to include environmental destruction, shortages of drinking water and food, water pollution (due to both biological and chemical substances), radioactivity and energy problems (Notoya, 2010). Until recently, these threats were limited both in scope and effects; but have reached a scale to encompass the entire planet. The Dutch Nobel Laureate, Crutzen (2002), suggested that society should recognise we are now entering a new geological epoch, the Anthropocene era, which started around 1780, when industrialisation began to change the geochemical fabric of the planet.

The general concept of sustainable development emerged in the second half of the twentieth century, tracing its origin to the seventeenth century. This term was officially formalised in the report *Our Common Future* (Brundtland, 1987), also called the 'Brundtland Report' from the name of the President of the World Commission on the Environment and the Development in 1987. Sustainable development was defined as: 'A development that meets the needs of the present without compromising the ability of future generations to meet their own needs.' It affirmed the interdependence of economic spheres, and social and environmental aspects.

In the Rio Declaration, signed by 178 countries at the United Nations Conference and Development in 1992, 27 general concepts firmly placed the ideology of sustainable development at the heart of global issues for the future and it laid down the groundwork for major international issues to be addressed during the following decades (Baudin, 2009). However, despite many international conventions and regulations endorsed by the political leaders of various countries, a fundamental imbalance in sustainability still prevails in the world (Ramade, 2012).

The literature on sustainability has expanded exponentially and reached vast proportions in various contexts, with nuances in the definitions and meanings according to the personal contexts and field of authors (i.e. economists, agronomists, climatologists, sociologists, etc.; Euzen, Eymard, & Gaill, 2013). The expression 'sustainability' has become so popular, that it is sometimes inadvertently abused by public authorities to advance agenda that may even oppose the original concepts of the term (Ramade, 2012). There is today, a general awareness about the lack of precision of this concept and the need to address the complexities of such a global problem with interdisciplinary collaborations (Euzen et al., 2013; Mancebo, 2013).

1.1.2 A New Paradigm to Organise Sustainability Science

To reach universal objectives of sustainability, there is a clear necessity for a truly interdisciplinary approach and an integration of knowledge from experience and from science. The archetypal convention of separating natural and social sciences in academia has resulted in an incomplete understanding of nature–society interactions and the integrated dynamics of the 'Earth System' as a whole (Clark, Crutzen, & Schellnhuber, 2004, pp. 1–28; Schellnhuber, 1999). In 1999, in a report called 'Our common journey: a transition toward sustainability', the Board on Sustainable Development from the United States, National Research Council (NRC), introduced the concept of: 'sustainability science'. Kates et al. (2001), in an international collaborative article, stated that a new field of sustainability science was emerging, which sought to understand the fundamental characteristics of the interactions between nature and society. The emergence of this new science is an important attempt to bridge the natural and social sciences in seeking creative solutions to complex challenges (Jerneck et al., 2011).

The literature on the general topic of sustainability has proliferated all around the world in recent years (Quental & Lourenço, 2012). Summaries of progress and priorities to actions have been prepared reflecting American

(Levin & Clark, 2010), European (Jäger, 2009) and Asian (Komiyama, Takeuchi, Shiroyama, & Mino, 2011) perspectives in this endeavour. A comprehensive article by Kates (2010) reviewed the primary literature of the subdomains of sustainability science. World-wide, transdisciplinary approaches are currently focussed on establishing a practical and realistic research agenda for effectively restructuring sustainability science, including action-oriented research efforts (Jerneck et al., 2011; Lang et al., 2012) and policy-oriented research (Hopkins, Bailly, & Støttrup, 2011).

1.1.3 Science in Society for Sustainability

The Rio Declaration pointed out that: 'the best way to deal with environmental issues is to ensure participation of all citizens and stakeholders concerned, at a relevant level'. Along with the emergence of interdisciplinary approaches to sustainability science, a new relationship between science and society had evolved for allowing citizens to participate in decision making. The need for a new contract between science and society, for the development of an informed, sustainable and inclusive knowledge society was vital (European Commission, 2013). This need led to an increasing body of literature in the past decade. The expression 'Science and Society' has tended to evolve around the notion of 'Science In Society', which should be an independent object of research itself (Alix, 2007).

Putting into practise 'science and society' implies the creation of new partnerships with all stakeholders, particularly on modalities of knowledge co-production, and decision making (Lang et al., 2012; Nauen, 2012). Nauen (2012), President of Mundus Maris, formerly a Senior Scientific Officer, European Commission Coordinator, asserted that in a knowledge society 'it is of the utmost importance to link sustainability research to education, life-long learning and innovation in all their forms'. May (2006) advised that education should concentrate on improving the understanding of how science works, rather than focussing on rote memorisation of definitions provided without context. This sharing of methods and enthusiasm can be achieved through the processes of collaborative knowledge production. In this proposed 'new deal' between science and society for the implementation of sustainability, the trans- and multidisciplinary, community-based, interactive, or participatory research practises have a common focus. This focus converges on research collaborations amongst scientists from different disciplines and nonacademic stakeholders, including business, government and civil society, in order to address challenges to sustainable activities and develop options for solutions (Lang et al., 2012).

For several centuries, amateurs have contributed to scientific knowledge in disciplines, such as: botany, zoology, entomology, ornithology, geology and astronomy. Information dissemination and flow was further accelerated with the spread of the Internet. Participatory approaches in research have increased in the past decade (Alix, 2007). Nevertheless, in its current form, their legitimacy poses questions in several dimensions (i.e. persons driving the processes, tools used, knowledge produced, neutrality and power asymmetries amongst stakeholders, etc.), which are poorly addressed and require reflexivity prior to implementation of all of the participatory processes (Barnaud, 2013).

The transfer and ownership of knowledge is another area of concern for science in society. For example, the Do-It-Yourself (DIY) biology movement promotes open access to resources surrounding modern molecular biology. The movement strives to democratise science and, in particular, to be a source of cheaper and simpler solutions for environmental monitoring, personal diagnosis and the use of biomaterials (Landrain, Meyer, Perez, & Sussan, 2013; Ledford, 2010; Wolinsky, 2009). According to advocates of this movement, citizens can contribute to sustainable development in an active and innovative way in parallel to political, scientific and economic institutions (Meyer, 2013a). Originating in the discipline of electronics and information technology, DIY movements have successfully created a generation of innovative products, which have become part of daily life. Despite the educational value, the perceived economic benefits of this transfer leading to innovation on new commercial products and the potential for co-production of knowledge, 'garage biology' or 'kitchen biology', poses important questions regarding the potential risks for the person and/or for environment, in which this 'tinkering' takes place. These risks, in large part, are due to inappropriate biological experimentation (Meyer, 2013b). Some security analysts worry about the prospect of possible abuse for nefarious purposes (Wolinsky, 2009).

Beyond the co-production and transfer of scientific knowledge and know-how, another challenge posed by interactions at the interface of science and society lies in the firm understanding of the relevance of information necessary to advance society towards embracing and implementation of the concepts of sustainability. This debate should be brought to open discussion in the society and not left to restricted forums of science academies. This new form of engagement already exists with science strategic councils in which industry and NGOs participate and could be more open to public debate.

Yet, in this context of sustainability and the relationships between science and society, where does phycology fit in and what contributions can this discipline of science make on a large-scale relevant to global needs? Could algae be a part (if only small but significant) of the solution for addressing the numerous and diverse challenges to sustainable development? Despite the fact that phycology still appears to be a speciality field, encompassing research in both microalgal and macroalgal spheres, the results of a wide range of studies and innovations are fundamentally related to the societal issues in question. As a first step, there is a question of how to transfer the general public mind-set from the negative connotations often associated with algae (principally in Western countries) to the view of three large and diverse taxonomic divisions with high economic, societal and environmental potentials, will require concerted efforts.

1.2 PHYCOLOGY: A MAJOR SCIENCE, WITH POTENTIAL CONTRIBUTIONS TO BE MADE FOR SUSTAINABLE DEVELOPMENT?

1.2.1 Past and Present Environmental, Economic and Social Importance of Algae

The term: 'phycology' is derived from the Greek word *phykos* which means 'seaweed' (Lee, 2008), but today it also refers to the study of 'the algae', a collective term for both microalgae and (macro) seaweeds. The word: 'algology' has also been employed to describe a science, but, it has progressively fallen into disuse. The Greek original prefix 'algos' refers to 'pain', which has led, in many languages, to a large lexicological field related to it. As a result, the term 'phycology' now tends to predominate.

Algae are ubiquitous organisms: comprising green, red, brown or even apparently blueish hues, due to the wide palette of photosynthetic and auxiliary pigments they utilise for photosynthesis and/or as compounds for screening and self-protection from various stresses, including damaging UV-A and UV-B components of sunlight.

In truth, the words 'alga/algae' are singular/plural descriptives of convenience, used in order to group together very different photosynthetic organisms, some might also describe as 'marine plants': they range from the giant kelp, forming coastal marine forests, to microscopic organisms which may colour snow red in the mountains – or which form obligate relationships with other organisms to form the symbiotic associations known as lichens and corals (Graham, Graham, & Wilcox, 2009). All these organisms are photoautotrophic entities, which can occupy almost any environmental condition where there is some light and water (even if only periodic): be it fresh, marine, brackish or terrestrial habitats. Defined as: 'thallophytes' in an ancient classification of the Plantae, members of this polyphyletic group have no true roots, leaves or stems, but in the majority of cases have chlorophyll *a* in order to carry out the primary photosynthesis.

Globally, the algae have been, and continue to be, instrumental in shaping the biosphere of the planet. Life on Earth appeared in the oceans, probably as a bacterial form. The first photosynthetic organisms are purported to have been cyanobacteria, also called 'blue-green algae' (as a result of the combination of accessory pigments). These organisms are prokaryotes, with chlorophyll *a* and specialised phycobiliproteins as components of their photosynthetic apparatus. These pigments allowed cyanobacteria to utilise sunlight, water and carbon dioxide to photosynthesise and release oxygen into the primordial, anoxic atmosphere of the Earth. Evidence of their presence from 3.7×10^9 years ago until the present day, exists in the form of stromatolites, which are calcareous biogenic structures, found in several parts of the world (Golubic, 1991). The proliferation of the first cyanobacteria is presumed to have stimulated an 'endosymbiotic event', which led to the evolution of all current plastids and eukaryotic organisms (Keeling, 2004; Margulis, 1974; McFadden & van Dooren, 2004). Molecular biological evidence suggested that chloroplasts arose from the engulfment of an oxygenic cyanobacterium by a host cell that almost certainly contained a mitochondrion (Falkowski & Raven, 2007, pp. 484; Kasting & Siefert, 2002). This event profoundly impacted the Earth's atmosphere and led to an oxygen-based world and the establishment of the ozone layer.

Furthermore, the ability to precipitate soluble bicarbonate in the form of calcium carbonate is responsible for geological structures and provided an important functional CO_2 sink. These processes had the effect, until relatively recently, to reduce the pH levels of the oceans. Such a significant biogenic development allowed for the establishment of conditions, which were favourable to the expansion of all life as we know it; first in the marine environment, and later on land (Pérez, 1997). The biological production of gases contributed to major environmental changes at the origin of the diversification and expansion of life on the planet.

Modern algae remain a vitally important group for all ecosystem functioning. It is estimated that 71% of the world is covered by ocean, the upper layers of which are inhabited by (micro and macro) algae (Wiencke & Bischof, 2012). Marine primary production is responsible for the majority of the world's atmospheric O_2 (Kasting & Siefert, 2002). Marine photosynthesis is a net source of O_2, since only a small fraction (~0.1%) of the organic matter synthesised in the oceans is actually buried into the sediments. By comparison, terrestrial photosynthesis contributes a smaller fraction to atmospheric O_2, because it is effectively in balance by the reverse processes of respiration and decay.

Harvesting sunlight to produce energy and oxygen by photosynthesis is the key biological process driving the development of heterotrophic life. Thus, primary production plays a vital role in ecosystem functioning, as it is at the basis of all trophic webs in the aquatic environment. In some coastal waters, seaweeds may form large underwater forests, which can provide habitats and nursery grounds for the survival, breeding and feeding of a vast array of marine organisms including fish, crustaceans, filter feeders and zooplankton. In addition, seaweeds cast upon the shore as 'storm toss', especially on sandy shores, play an important role in stabilisation and organic enrichment and soil development for dunes, as well as for supporting meiofaunal species (Kirkman & Kendrick, 1997).

In addition to major environmental functions, macroalgae have been used for centuries by humans and some species have continued to play large economic, and social roles (Lüning & Pang, 2003). Recent archeological discoveries in Monte Verde in Southern Chile attested to the utilisation of seaweeds by early man, likely for food, trade, or medicinal purposes, dating from around 14,000 years ago (Dillehay et al., 2008). A hypothesis has been also formulated about the role that a kelp forest – as one of the most productive habitats on Earth – could have played in the migration of maritime peoples, hunter-gatherers, from Asia to America near the end of the Pleistocene, about 16,000 years ago (Erlandson et al., 2007). Algae are also commonly utilised in traditional Chinese and aboriginal medicines, with references dating from about 5000 years ago. There is even evidence of incorporation of seaweeds in the human diet from the fourth century in Japan and the sixth century in China (McHugh, 2003). However, the West, with the possible exception of Iceland, Wales and Ireland, has been somewhat slower to adapt, and the general usage of algae for food has been relatively quite limited in much of Europe and North America, as compared to Asia, Polynesia, South America, Australia and New Zealand. Seaweeds have been traditionally used in coastal regions all over the world, primarily as an animal feedstock, or field fertiliser compost (Craigie, 2011), for the extraction of iodine (Kelly, 1961), gunpowder (Neushul, 1989), glass production and more recent technologically innovative products developed from extraction technologies (Mouritsen, 2013). Indeed there are many potential extracts to be derived from seaweeds and microalgae. In the broadest terms, these fall into the categories of: 'colloids' (for their rheological or physical properties) and 'actives', which are algal, secondary metabolites, storage compounds or pigments, with interesting biological activities, once removed and stabilised from the algal biomass. Some of these activities are at the scale of large multinational corporations, all the way to entrepreneurial microbusinesses.

The biomass required can be harvested from wild/natural stocks/beds and/or grown on an intensive basis of cultivation. Extracts from algae (seaweeds and microalgae) provide materials that can be used as commodity stabilisers (for their rheological properties), as stiffeners and thickeners, in a variety of processed food industries, and as speciality ingredients (with active principles) in cosmetics, the pharmaceutical industry and various biotechnological applications. The development of algal cultivation techniques (both extensive and intensive) and further micropropagation, allowed for the increased development of aquaculture which provided feedstock for the production of a large number of algal species and varieties with increasing economic value (Wikfors & Ohno, 2001). The Fisheries and Aquaculture Department of the Food and Agriculture Organization of the United Nations records data concerning the collection of 'aquatic plant material' from the wild, as well as aquaculture production, and publishes them in their biannual flagship report: 'The State of the World Fisheries and Aquaculture'. These data provide an overview of major production and trends, which are consolidated into an accessible source of information for guidance. According to the latest edition, more than 19 million tonnes of marine macroalgae were commercially produced in 2010, 95.5% of that was farmed and the remaining 4.5% collected from the wild.

Global production is dominated by macroalgae, grown in both marine and brackish waters (FAO, 2012). The production of algae, by volume, increased at annual rates of 9.5% in the 1990s and 7.4% in the 2000s (production was 3.8 million tonnes in 1990). The total value of farmed aquatic algae rose from US$4.4 billion in 2008, to US$5.7 billion in 2010. In contrast to fish aquaculture, the cultivation of algae is practised in fewer countries, but there are remarkable differences in the tonnages produced. Only 31 countries and territories were recorded with algal farming production in 2010, and 99.6% of the globally cultivated algal production is derived from just eight countries: viz. China (58.4%), Indonesia (20.6%), Philippines (9.5%), Republic of Korea (4.7%), Democratic People's Republic of Korea (2.3%), Japan (2.3%), Malaysia (1.1%) and the United Republic of Tanzania (0.7%) (FAO, 2012).

On the other hand, cultivation of microalgae, is still in its relative early stages (Buchholz, Krause, & Buck, 2012). After having received considerable attention during the 'oil crisis' of the 1970s, there is a current flurry of interest as a feedstock associated with biofuels, but relatively little is grown at large commercial scale (Richmond & Hu, 2013). However, microalgal cultivation currently plays a vital role in providing the correct size-class feedstuffs for rearing aquatic animals in aquaculture, and as such provides primary production trophic level of an important economy (Guedes & Malcata, 2012; Hemaiswarya, Raja, Kumar, Ganesan, & Anbazhagan, 2011; Muller-Feuga, 2000).

Despite the undeniable major ecological (as in primary production and ecosystem services provided), economical and societal importance, there are very often misconceptions and erroneously negative connotations associated with all algae, especially the seaweeds. In 1997, using specific examples from France, Perez introduced his book on seaweeds, highlighting the fact that the public can draw erroneous perceptions: 'unpleasant', 'slippery', 'nuisance', 'toxic algae', 'killer algae', 'invasive algae', are just a few of the terms in people's general vocabulary used to describe algae. Such comments came from bathers, inlanders, walkers, anglers, or reported/broadcast through newspapers and the media, amongst others. Perez mentioned that it was rare to report, in an objective manner, the positive and fundamental roles algae played in aquatic ecosystems. More than 15 years later, the perception of algae by the general public, particularly in Western countries, seems to have remained largely unchanged (Mouritsen, 2013).

Since 2009, particularly in Brittany, France, algae received a lot of negative attention from the media and general public, due to major green tides, caused by anthropogenic eutrophication resulting in huge amounts of rotting biomass being cast on shores at high tides. During the decomposition of these beach-cast green tides, hydrogen sulphide was a product of anaerobic decomposition from bacterial degradation of the algae, which was suspected to have caused a number of tragic incidents including the death of a worker involved in the removal of biomass from the shore. A horse, dogs and wild boars were also reported to have succumbed to the toxic gas. These green tides have occurred since the 1970s on the coast of Brittany, and resulted in some beaches having restricted access, at certain times of the year. Ironically the root cause of the problems was not the algae per se, but agricultural runoff, particularly intensive pig farming, which were revealed as major point sources of coastal eutrophication (Guyomarc'h & Le Foll, 2011).

As a consequence of the negative economic impacts of such pollution, there were local political and social problems: this became a point of conflict between different stakeholders and resulted in major oppositions between ecologists and farmers, which were frequently reported in national media broadcasts (Le Chêne, 2012). This issue also demonstrated the typical absence of appropriate communication between science and society, between and among all stakeholder parties. The problem of green tides – the most visible part of a global problem of coastal eutrophication – has increased in many parts of the world, both in extent and in public perception over the last three decades (Ye et al., 2011).

In May–July 2008, just prior to the Olympic Games in China, the coastline of Qingdao – the major site for the sailing events – experienced the world's largest green tide (Leliaert, Malta, Engelen, Mineur, & De Clerck, 2008), caused by the proliferation of *Ulva prolifera* (Pang et al., 2010). This phenomenon tended to convey a global, negative image and to occlude the complexity and the potential benefits of the algae, without addressing the key cause – coastal eutrophication. However, fortunately, in the last few decades, research and applications in phycology have progressed along with the emergence of the concepts of sustainable production, associated coastal development, and resulting socioeconomic benefits. Regarding their utility throughout the history of the world, algae collectively fulfilled fundamental roles and as such are well placed to provide vital contributions to the future challenges facing human development, health and societal wellness.

1.2.2 A Selected Overview of the Potential of Algae towards the Goals of Sustainability

In addition to the significant roles all algae play in the functioning of our biosphere, it is noteworthy that 'the algae' have been inexplicably absent in international debates on biodiversity (Connan, 2004; Norton, Melkonian, & Andersen, 1996) and there has been little effort, or progress globally, regarding their conservation (Brodie, Andersen, Kawachi, & Millar, 2009; Mollo & Thomas-Bourgneuf, 2009). Public discussions such as climate change, production of feedstocks for biofuels and green technologies often refer to 'plants', but the role and potential sustainability of: 'the algae', is generally known only to experts in the field (Reisser, 2010). Nonetheless, since the beginning of the twenty-first century, the literature linking algae to the concerns of sustainability has increased, particularly in the domain of bioenergy from algae. A scientometric, year-on-year, analysis showed that the number of published papers with the search terms 'algae' coupled to 'bioenergy' increased exponentially during the past three decades (see Konur, 2011; Oncel, 2013).

1.2.2.1 Bioenergy from Algae

The threat of global climate change is largely attributed to anthropogenic, greenhouse gas emissions from fossil fuel use, deforestation and increased development of farmland, but stochastic megaevents, such as volcanic eruptions cannot be discounted as major contributors to gas emissions. It is a fact that oil runs the modern world, and there is an entrenched hydrocarbon economy, which has been the cause of multiple conflicts in oil producing regions. The oil and its products are necessary in order to provide 'cheap' fuel for the internal combustion engine, which is required for transportation. In addition, various grades of oil are produced for numerous applications with wide ranging applications as 'energy' from candle wax, to plastics to much needed medicines (such as aspirin, psychotropic drugs, antihistaminic or antibiotic substances): an evaluation of the limitations and geopolitical disruption of global reserves and the perceptions of declining availability and increased prices of 'dollars per barrel of crude oil' have led to increasing efforts for developing sources of alternative, renewable energies. Amongst these are biofuels (Brennan & Owende, 2010; Sing, Isdepsky, Borowitzka, & Moheimani, 2013), including those that are produced from algal-derived feedstocks. The need for energy production (bioenergy, energy from solar light, wind, etc.) partnered with environmental conservation approaches without adding to the global CO_2 sink is required (Notoya, 2010).

Microalgae present several advantages as compared to land plants, due to their high growth rates, ability to grow on nonarable land and their appreciable content of energy-rich compounds (i.e. lipids) that can be optimised under controlled conditions, harvested and chemically converted to fuels. They can potentially produce biofuels in several forms – such as biodiesel, biogas, biohydrogen and bioethanol – and it has been shown that they are more efficient than terrestrial plants in converting sunlight to biochemical energy. Indeed some species have excellent oil productivity (Chisti, 2008; Stephenson, Moore, Terry, Zubkov, & Bibby, 2011). To fully compete with fossil fuels, major challenges remain, such as the poor returns on both financial and energy investments, which weaken the argument for the feasibility of commercial scaleup, and consequently, sustainability (Rhodes, 2010; Sing et al., 2013). Economic viability is an important pillar of sustainability. The biological reality remains firm, that algae can be used as a biomass for many types of biofuels. This has been shown to be possible for a long time, but even with many modern advances, the unsubsidised, scaleup from pilot plant to long-term viability of commercial operations remains elusive, but it will be instructive to watch the many operations currently advancing in these endeavours. Unfortunately, reduced oil supplies, caused by natural disasters and/or future global conflict and wars may be the necessary economic conditions which will tip the balance in favour of economically viable biofuels. Of course the demand for these biofuels is maintained by the attachment to current energy resources, but developments of other energy sources, such as hydrogen, for example, may reduce the demand for 'expensive' photosynthetically derived alternatives. Today, global awareness within societies has been kept alive with recurrent news items and promotional fancy slogans about biofuels. However, there is some speculation that political jargon of the 'future' of microalgal biofuels is meant only to appease certain factions interested in alternative energy sources. Since relatively little was done compared to the talk, further calls for government subsidisation for the development of microalgal biofuels is losing ground (Oncel, 2013). The announcement of the need of a fourth generation biofuel from genetically modified algae for boosting the production and the commercialisation of fuel oil (Chisti, 2013) could rekindle the public interest, but since GMO and or 'engineered', photosynthetic life forms would be involved, there is huge potential for major conflicts ahead of that initiative. Genetic modification remains a sensitive issue and more controversial debate would arise requiring increased efforts of communication at the interface between science and society. One of the constraints limiting the development of microalgal biofuel technology lies in the fundamental limitations on the physicochemical efficiencies of light harvesting of the biomass, before the inefficiencies of mass cultivation are considered. As a result of their unique carbohydrate (polysaccharide) content and the relatively fast growth rates, of some species, selected seaweeds have also been considered as potential candidates for bioethanol production and some authors have promoted their candidacy as effective raw materials for biofuel production (see Kraan, 2013; Notoya, 2010, amongst others). For extensive reviews on biofuels from macroalgae see Wei et al. (2013) and Kraan (2013).

If there is an absence of competition between food and algal biofuels for agricultural surface area, there does remain competition in terms of fertiliser consumption required for both efficient fertilised agricultural (terrestrial) and algal (aquatic) productivity. The possible, widespread implementation of algal biofuel production at the commercial scale, as a sustainable and renewable resource, implies large consumption of phosphorus, currently sourced from nonrenewable phosphate rock mineral. The sustainability of this is disputed (Rhodes, 2010). A solution to the availability of P as a fertiliser might come through the recycling of phosphorus from manure and other types of plant, animal and human waste, which could allow for the maintenance of agricultural practises, at present levels, and also to provide for increased demands due to requirements from the cultivation of algae (Rhodes, 2010). This challenge is being addressed, in part, by utilising developments from other research platforms, both with microalgae and seaweeds. These include ecological engineering tools such as phycoremediation and integrated multitrophic aquaculture (IMTA).

1.2.2.2 Environmental Remediation

The bioremediation properties of algae (micro and macro) present, as yet, untapped remedial potential against pollution from others sources, such as biosorption of heavy metals from industry (Davis, Volesky, & Mucci, 2003; Volesky & Holan, 1995) or radioactive elements (Krejci, Finney, Vogt, & Joester, 2011).

Phycoremediation is the use of algae for the removal, or biotransformation, of pollutants, including nutrients and xenobiotics from wastewater and CO_2 from effluent air streams (i.e. gas, cement plants, etc.; Olguín, 2003). However, the technologies being 'reinvented' and refined are not 'new ideas', e.g. the concept of the high rate algal ponds (HRAP) was developed by Oswald et al. (1957) in the mid-1950s, and since its inception, the technology has spread around the world. It is currently utilised for commercial algal production and/or for development of phycoremediation technology for domestic or agricultural wastewater streams, such as those found on dairy farms, piggeries, etc (Bux, 2013; Rao, Kumar, Raghavan, Subramanian, & Sivasubramanian, 2011). Until relatively recently, the use of wastewater treatment HRAPs, for algal production and biofuel conversion, has received little attention (Park, Craggs, & Shilton, 2011). The demonstration of promising performance by microalgae in a 5 ha-scale wastewater HRAP, over seasonal scales, in a temperate climate, seemed to be an economically viable way for a biofuel production to occur with minimal environmental impacts (see Park et al., 2011; Sutherland, Howard-Williams, Turnbull, Broady, & Craggs, 2013 and others for discussions).

The treatment of domestic sewage and its conversion to energy has been recently highlighted in the media with the realisation of experimental architectural projects integrating algae in the glass facade of buildings, such as the Bio Intelligent Quotient building in Hamburg (Phipps, 2013). These bioreactor facades are expected to provide services such as production of energy and biomass, insulation from sound, heat and cold and also a control of the light and the provision of shade (Kim, 2013; Phipps, 2013).

Recent hot topics include the use of algae for 'carbon capture', with the biomass being used for extraction of bioactive compounds and agricultural inputs, with bioenergy as a potential 'by-product'. For example, the goal of the NRC's Algal Carbon Conversion Flagship, is to provide Canadian industry with a cost-competitive, value-generating solution to divert CO_2 emissions into algal biomass, which can then be processed into biofuels and other marketable products. This initiative was presented as an opportunity to address environmental, energy and economic issues associated with industrial emissions. It explores how CO_2 emissions can be sustainably and profitably converted by algae into valuable products, without displacing food crop resources (see http://www.nrc-cnrc.gc.ca/eng/solutions/collaborative/algal_index.html). CO_2 sequestration by marine macroalgae has been considered in the last decade as a possible sink for anthropogenic CO_2 emissions: according to this hypothesis – particularly supported within the 'Asian Network for using algae as CO_2 sink'. Through experiments in South Korea, from 2006 – seaweed cultivation could be one measure, among others, for mitigating global warming by CO_2 capture, and oxygen production through algal photosynthesis (Chung, Beardall, Mehta, Sahoo, & Stojkovic, 2011; Chung et al., 2013). These initiatives would seem to promise to provide environmental services and even opportunities for emissions-trading in the context of the Kyoto Protocol. Further economic analysis and integrated business plans are needed in order to fix the value as well as further analysis for measuring the effective CO_2 capture, in relatively short-lived algae (as compared to entrapment for many decades in long-lived trees, for example).

Seaweeds are also considered as good candidates to provide environmental services in the aquaculture industry. Notably the concept of 'IMTA' was developed at the beginning of this century (Chopin et al., 2001) and was inspired by traditional, mixed fish, crustacean and some seaweed farming, as practised in Asia. With increased industrialisation of agriculture and aquaculture, producers moved to monoculture types of production. Such intensive systems required significant energy inputs and caused different environmental issues, such as eutrophication, due to excess nutrients, particularly nitrogen. The IMTA concept required the recycling of nutrients through various trophic levels of organisms (i.e. but not exclusively fish (salmon), bivalve molluscs (mussels), echinoderms (sea cucumbers, sea urchins) and seaweeds (kelp spp.)). Different species are cocultured and the waste from one becomes a resource for the other, i.e. seaweeds assimilate the fish/mollusc-excreted ammonia, phosphate and CO_2, converting them into biomass and providing an important ecosystem service. With this treatment, nutrient-stripped effluents can recirculate back to the fish ponds or be discharged without endangering the environment (Chopin et al., 2001). This system could also be beneficial for the health and productivity of the cultured fish (Abreu, Pereira, Yarish, Buschmann, & Sousa-Pinto, 2011). Neori et al. (2007) suggested that a balanced ecosystem approach in aquaculture would need a shift from caught fish to the farming of fish. An adjustment in the patterns of the seafood market, away from the present domination of carnivorous fish, would favour a trophically and ecologically balanced 'portfolio' with more seaweeds, molluscs, herbivorous finfish and omnivorous organisms.

Nutrient bioextraction, also called bioharvesting, consists into the practise of farming and harvesting shellfish and seaweed for the purpose of removing nitrogen and other nutrients from natural water bodies. By reducing the nutrient pollution, these cultivation systems could contribute to the restoration of coastal waters. For example, in the United States, the Long Island Sound program is piloting the use of red seaweed (*Gracilaria*) as a nutrient removal strategy. Along with the quantification of the potential for nutrient bioextraction, the programme also evaluates use conflicts and researches new markets for products from seaweeds (Kim & Yarish, 2012). These nutrient assimilation services by seaweeds could be potentially financially transformed into credits in the context of a nutrient trading programme (Stephenson & Shabman, 2013).

1.2.2.3 Traditional Applications towards Sustainability

The most traditional applications from algae remain the most economically efficient and the development of these sectors could contribute to sustainability through food production, new health services and increased employment in aquaculture. Algal biomass, harvested from wild stocks, cultivated in open waters or produced using 'biotech advances' on land, offer a potential wide range of products with commercial applications and value. These include fertilisers (Craigie, 2011), animal feeds (Evans & Critchley, 2013), phycocolloids (Bixler & Porse, 2011) and human foods (Cornish & Garbary, 2010).

Seaweeds have long been used as food and condiments in Asia, on a daily basis, with associated health benefits, as demonstrated through several epidemiological studies. However, their nutritional properties are known only in a general sense because studies on the nutrient bioavailability are scarce (Taboada, Millan, & Miguez, 2013), and challenging to perform. This is also true for most foods, but new science, and advanced data analysis systems are emerging to address this issue *in vivo*, which is where it is most meaningful. With the world population expected to rise to 9 billion inhabitants by 2050 (Alexandratos et al., 2006), high quality protein sources will be in demand. Animal proteins should be used in a more responsible manner because intensive cattle (beef and milk) production contributes significantly to greenhouse gas emissions (i.e. methane). In addition, increased meat consumption requires more arable land, notably for the production of cereals for cattle feed, and is thus a factor in deforestation. Novel protein sources are expected to enter the European feed and food market as a substitute for animal-derived proteins. Algae are considered to be one of them, along with insects, duckweed and rapeseed. The legislation though is still unclear regarding novel, seaweed-based food products (Spiegel, Noordam, & Fels-Klerx, 2013). The knowledge of traditional and tasty algal cuisines, in various regions of the world, could facilitate the development of westernised methods for including algae in human diets (see Mouritsen, 2013). Such traditional uses include the use of cochayuyo (*Durvillaea antarctica*) in Chile (Cardenas Alvarez & Munoz Arias, 2008), or the production of flour from 'llaska' (i.e. *Cladophora crispata*) by a local population of Titicaca lake border in Peru (Canales, Taquila, Pacheco, Salas, & Aguilar, 2008).

Microalgae (including the cyanobacteria) are already established as commercial sources of high-value compounds (i.e. within the current paradigm of how products are valued, these products withstand the inherent costs in the production) such as β-carotene, astaxanthin, docosahexaenoic acid, eicosahexaenoic acid, phycobilin pigments, as well as extracts for cosmetics, nutraceuticals and functional foods (Borowitzka, 2013; Sharma, Tiwari, Tripathi, & Rai, 2011; Spolaore, Joannis-Cassan, Duran, & Isambert, 2006). Amongst the cyanobacteria, the term: '*Spirulina*' (which actually refers to a large number of species of cyanobacteria) has been commercialised as 'blue-green algae' for dietary supplementation in Western countries. These 'algae' are claimed to have exceptional nutritional properties, as one of the most complete foods on Earth (since they are reported to contain all of the macro and micronutrients, including about 65% protein, B-complex vitamins, phycocyanin, chlorophyll, β-carotene, vitamin E, superoxide dismutase and numerous minerals; see Wang et al., 2008). High concentrations of polyunsaturated fatty acids and γ-linolenic acid may also be present (Otles & Pire, 2001). In view of the nutrient content, *Spirulina* is considered suitable as a functional food (Park et al., 2008). It has been used to treat malnutrition, for example, through the development of nutritional candies distributed to children in India (http://www.antenna.ch/en/programs/spirulina-programs/products). An obvious dietary advantage of macroalgae over microalgae, however, relates to the high quality fibres, both insoluble and fermentable components present in the multicellular algae and play very important roles in prebiotic effects on gastrointestinal tract health and whole-body immunity.

Holdt and Kraan (2011) reviewed the literature to that point, related to the vast potential of bioactive molecules from seaweeds. Seaweeds are known for their richness in polysaccharides, minerals, essential amino acids and certain vitamins. They also contain bioactive substances, such as sulphated polysaccharides, proteins, lipids and polyphenols that have pending applications or potential for animal and human health and wellness. Studies have demonstrated that various seaweeds, and seaweed extracts have antibacterial, antidiabetic, antiviral and antifungal properties, and exhibit inhibitory effects against tumours (Fedorov, Ermakova, Zvyagintseva, & Stonik, 2013) and HIV. They have been shown to reduce blood cholesterol levels and hypertension and all possess extensive antioxidant capacities, making them excellent candidates for functional food applications (see Holdt & Kraan, 2011 for more details).

Algae may also play important roles outside that of 'health', there are reports that their biomass also holds promise in the development of biodegradable, innovative, (bio)materials, such as specialised papers (Pei et al., 2012; Seo, Lee, Lee, & You, 2010), algal-sourced (bio)plastics (Lucas, 2012), or as sources of pigments for photo-electro-chemical, solar cells (Calogero et al., 2014). Algae, such as the green macroalga, *Cladophora*, has potential as a natural source of nanocellulose components, typically manufactured from wood pulp, for the specialised development of medical devices and applications in human healthcare (i.e. personal hygiene products, biological implants, wound and burn dressing materials, as well as scaffolds for tissue regeneration, artificial blood vessels, filters for haemodialysis devices, etc.; Hua, Carlsson, Strømme, Mihranyan, & Ferraz, 2013).

1.2.2.4 A Global Approach to the Applications and Future Use of Phycology?

A proof of concept 'algal-based, biorefinery' has been developed that relies on the efficient use of algal biomass through fractionation, resulting in multiple, individual products from one or multiple, raw materials or biomass (Ibañez & Cifuentes, 2013). This holds great promises for the production of a wide range of unique chemical components and bioenergy, both in quantity and in quality (Kraan, 2013).

Increasing pockets of interest in applied phycology, especially seaweed biology, act in concert as stimuli for further studies in relation to socioeconomics and the wider economy (Wiencke & Bishof, 2012). The role of seaweed cultivation in the employment of rural women, improvement of livelihoods and sustainable development of coastal communities has been highlighted in various parts of the world (Buchholz et al., 2012; Hurtado, Montaño, & Martinez-Goss, 2013; Padhi et al., 2011; Valderrama, Cai, Hishamunda, & Ridler, 2013). The potential development of future, large-scale ocean farming operations for seaweeds could create significant new employment opportunities in fisheries (Kraan, 2013), especially in areas where employment rates are low and people are 'poor' (in terms of cash economies).

Gathering knowledge on species, ecosystem functioning and restoration of algal beds through fundamental research is obligatory for the preservation of the environment. Natural systems never function in isolation and an integrated approach is an absolute necessity for any thorough analysis. Methods for measuring perturbations in seaweed populations are required and techniques must be adapted to each species of interest (Ugarte & Sharp, 2012; Vásquez, Piaget, & Vega, 2012). Special attention should be also paid to the potential impacts of intensive agriculture on marine and coastal environments which could lead to alteration of natural habitats, changes in hydrology, decreases in biodiversity, degradation of water quality, disturbance of coral reefs and eutrophication as a result of excess nutrient runoff (Wei, Quarterman, & Jin, 2013). The production of dimethylsulphoniopropionate by marine algae is an important factor in the formation of clouds and the temperature of the Earth's atmosphere (Charlson, Lovelock, Andreae, & Warren, 1987). Alteration of the iodine biogeochemical cycle by seaweeds may also occur, which can have a significant impact on atmospheric chemistry (Leblanc et al., 2006) and in some cases influencing coastal weather patterns related to rain (O'Dowd et al., 2004). These environmental impacts could be minor or even beneficial for fish populations, but a balance must be obtained (Wei et al., 2013) taking account of economical and ecological 'costs'; for instance, there should be far more focussed 'economic analyses' of financial and cost and benefit analysis, which will make the science of phycology more 'understandable' and practical to corporate thinkers and decision makers. Marine licensing (Kraan, 2013) and usage conflicts in coastal areas could be potential issues in the application of aquacultural technologies.

From highly localised development projects, to space exploration programmes, the ecological and economic roles of 'the algae' are vast and collectively, could contribute widely towards a more sustainable society. Nevertheless, in the current context of a globalised economy, development of sustainability is not purely limited by technology and profitability. It has a social component and this will need to be developed through a synchronised engagement with society, combining ideas from many disciplines, both technical and societal. Sustainability includes an ethical dimension that requires a paradigm shift (Becker, 2012), which would need a complete reevaluation of society's concepts of morals and values. As recommended by Nauen (2012), the integrated approaches, required for sustainability, must also encompass the human population including: 'the growing proportion of humanity living in cities and mostly disconnected from marine and coastal ecosystems'.

1.3 THE ALGONAUTS' ROAD: AN EXAMPLE OF PHYCOLOGY AT THE INTERFACE BETWEEN SCIENCE, SOCIETY AND SUSTAINABILITY

1.3.1 Presentation

The project 'The Algonauts' Road' (La Route des Algonautes) is an Action Research (Barbier, 1996; McNiff, 2013) project directed at the interface between science and society. It is based on selected international developments led by phycology, which have many potential and practical implications for the concepts of 'sustainability'. Intended for the general public, the aim is to stimulate and improve public awareness about a wide range of micro- and macroalgae, through the communication of exciting and relevant scientific knowledge.

In this project, 144 semistructured video-taped interviews of people from various settings were performed. This included researchers, industrialists, fishermen, NGOs, inventors, etc. all of whom deal daily with algae in various themes (i.e. aquaculture, bioenergy, bioremediation, value-added products, coastal management, education, etc.). These participants were defined as the 'Algonauts'. The interviews were conducted in the Algonauts' respective countries viz. India, China, South Korea, Japan, Malaysia, Philippines, Indonesia, Thailand, Australia, New Zealand, Chile, Peru, Brazil, United States, Canada, United Kingdom and France. Most of the interviews were carried out between October 2010 and June 2011, as part of a continuous tour with consecutive meetings through the above countries.

Discussions focussed on research and work topics, contexts that led to the development of an innovation – technological, social, organisational or service oriented – as well as the participants' background and his or her opinion regarding the potential role of algae in sustainability.

In the framework of social anthropology, this Action Research was undertaken with an inductive approach. The data collected constituted the material for long-term Action Research. Further, within the framework of anthropology, the project was performed in the following three phases:

Phase 1: Preliminary phase in which the key questions regarding sustainability and the possible role of phycology were constructed, and questions regarding the position of science and society were formulated.

Phase 2: Performing interviews of Algonauts and collection of data. The interview was performed from the perspective of a common, but concerned, citizen from the general public, to help understand the nature of communication challenges, which lie ahead for phycology and sustainability.

Phase 3: Simultaneous assembly and presentation of interview data, dissemination of key findings and pursuit of developments furthering the mission of science in society towards sustainability. Significant landmarks for the project include (1) a co-operative knowledge-sharing programme between Finistere, France and Chiloe, Chile, (2) increasing algae awareness in France through the organisation of exhibitions, conferences, workshops and articles in various popular media, and, (3) creation of interactive pedagogic tools for elementary schools on the topic of algae and their potential.

The Algonauts' Road project is supported by and exists in a novel institution known as a co-operative, specifically, the Cooperative of Activities and Employment (Bost, 2011). Working with people from diverse backgrounds, the cross-fertilisation of ideas in this co-operative has led to significant insights into the practises that could help and encourage interdisciplinary work. An illustrative example is the novel application of seaweed-based biomaterials in fashion and art. Crucially, these ideas have been equally nurtured by scientists and practitioners of the craft, and hint at the innovations that could happen globally. To encourage such unlikely collaborations and investigate the nature of co-production of knowledge, the Laboscop® was created by a team including the authors Anne-Gaëlle Jacquin and Stéphanie Brulé-Josso in June 2012 as a sandbox for such experiments in interdisciplinary developments in the spirit outlined in Draperi (2011, pp. 177–198) in his work on the research-action-training process in co-operative economy: through the seeking of socioeconomic and educational purposes, he highlighted that the complementarity between 'innovative actors and committed researchers' is probably founder of new practises of contemporary research-action.

1.3.2 Analysis: Some Elements for a Scientific Mediation

Nauen (2012) suggests that science 'needs to be complemented by "narratives" which connect the research results to millions of citizens in all walks of life. With this inspiration, the essence of the Algonaut's project is to understand how to report effectively scientific and technical advances to society, within the specific examples of phycology and sustainability. Beyond the "narrative", this informal approach can provide forms of "translation" between researchers and citizens. Within the paradigm of social anthropology, we refer to the work of the sociologists Callon and Latour (Callon & Latour, 1988; Callon & Latour, 1991) to illustrate some preliminary findings of the Algonauts' Road.

1.3.2.1 Scientific Mediation and Environmental Sustainability

How can scientific mediation be envisaged in the building of the science of sustainability? For example, in the Algonauts' Road, tackling the problem of green tide and the negative image often associated with algae, participants were asked for their opinions regarding this phenomenon. These are some of the responses: 'It's the famous saying, don't shoot the messenger. Algae "are telling" you that something humans are doing is the problem, so they are actually a good thing because they are telling us something is wrong' (interview of researcher from New Zealand, November 2010). 'If we would need something for proving the interesting properties of algae, green tides are the best evidence' (interview of researcher in France, October 2012). This point of view by researchers offers another perspective which differs from the most common and negative one about this phenomenon: the pollution begotten by humans demonstrates a natural process of phycoremediation which could be exploited and placed at the origin of solutions for more sustainable management of coastal areas through technical developments.

1.3.2.2 Scientific Mediation and Social Sustainability: A Collaborative Approach

A benefit of the Algonauts' Road was to give the same level of importance to the technical innovations as well as to the social and cultural environment of the persons who developed the work. A published paper does not necessarily reflect the life and professional histories at the origin of a technical innovation. In the following example, a collaboration between a company and a research institution in Australia led to the successful development of an integrated aquaculture system in response to a problem threatening the economic activity of the corporation. This involved the production of microalgae in large, shallow open ponds and the extraction of their carotenoid pigments used in food, cosmetics and pharmaceutical industries. It was sometimes hindered by a natural bloom of brine shrimps, primary consumers of microalgae that occurred in the ponds. The first attempts to eradicate these crustaceans were unsuccessful. The collaboration with the researcher – owning complementary skills and a knowledge of brine shrimp cultivation – has led to a paradigm shift and a follow-up success story with the design of a new innovative system. The system allowed for co-production of microalgae and brine shrimp, which is commercially viable, ensures efficient utilisation of waste and energy, and has led to the diversification of value-added products. Protagonists were asked about the origin of the collaboration: 'The first meeting was quite funny. I was bogged in the middle of nowhere, in my car while doing a site survey for aquaculture and just like my luck – or basically like both of our luck – Robert[1] just happened to be somewhere nearby and he managed to pull me out of from where I was stuck. Since then we started to have a very strong friendship, and that has obviously helped with the project' (interview of a researcher in Australia, February 2011). This example can be analysed from two levels.

1.3.2.2.1 Scientific Mediation and Serendipity

It appears that coincidence was a major factor for the development of this collaboration: the coincidence to be bogged down at the same place and time, coincidence to develop a friendship, coincidence to have complementary skills, coincidence to transform a nonprofessional relationship into a professional one. The serendipity in the co-operative approach is counter to the planned approach and corresponds to a management of opportunities in adding value to improbable encounters. Numerous innovations in science and business are attributed to serendipity (Roberts, 1989). Human relationship and coincidence in the process of innovation have been mentioned several times during the 'Algonauts' Road' project: when it leads to a success, participants spontaneously identify it. The comprehensive analysis of the field survey is in progress for measuring eventual differences among the visited societies.

1.3.2.2.2 Scientific Mediation and Consideration of Emotions

In the follow-up mediation actions of the project intended for the general public, participants have expressed interest and were touched by these narrative parts relative to the social and environmental context, such as in the

[1] The first name has been changed.

previous example. This contributed to the establishment or re-establishment of a link between the researcher and society. History and emotions play a fundamental role in the relationship between science and society as it communicates enthusiasm and interest for science and for the particular field. This is consistent with Nauen's recommendation (2011) about the necessity and the possibility to build a greater awareness of emotions into international relations in general, and more particularly into the way scientists interact with nonscientists. Emotions can link into the human relations and political processes underpinning international cooperation in order to increase the positive impact of science on society. In today's context, she stated that it is intimately associated with the necessity to support transitions towards greater sustainability of human societies on our planet.

1.4 CONCLUSION

Phycology as a science contributes to issues of concern in the world and has been a field of great interest and significant innovations over many of the most recent decades. This overview of applications and potential developments from micro-and macroalgae demonstrates the inherent sustainability of this varied biomass. Algae may play a feasible role in providing solutions to some of the global problems, such as environmental degradation or health and nutrition and energy sustainability. Applications of appropriate technologies involving algae could contribute significantly to some of the United Nations Millennium Development Goals, which are not expected to be successfully reached by the 2015 deadline.

With many initiatives all over the world, across different segments of society, programmes for sustainable development are gathering momentum. Climate change and various global and local issues represent challenges for phycologists who could apply their knowledge of algae and their associated applications contributing to several important solutions in the emerging sciences as applied to the objectives of sustainability. Generally, algae – both seaweeds and microalgae – receive less attention from the general public, than higher plants and animals. In recent years, artistic competition and exhibitions with an algal focus have been promoted in many parts of the world. Phycology would face two challenges in order to reach the public: on one hand, social science dedicated to the study of the relations between society and algae should be a part of the definition and the research organisation of phycology; on the other hand, science, natural and social, should develop its link with society beyond technical expertise, including participatory research, policy oriented work, communication, education, etc. This could require the development of mediation capacity by scientists or mediation professionals. The tremendous potential of algae needs to be communicated beyond a specialised scientific community to all citizens at the interface of science and society.

ACKNOWLEDGEMENTS

AG Jacquin thanks all partners who have financially supported the first part of the 'Algonauts' Road' project: French Ministry of Youth ('Willing to Act program' grant), Regional Council of Brittany (project call 'Education to Environment and to Sustainable Development'), General Council of Finistère district (consultancy mission in the context of Cooperation program between Finistère and Chiloe, Chile), Grouping of Local authorities of Cap Sizun (Educational activities with school), Marine Park of Iroise Sea (Consultancy), City Council of Plogoff (Local council), International Society for Applied Phycology and Acadian Seaplants Limited (participation to a congress). The author is grateful to all participants for support of all kinds, which have contributed to the completion of the first part of the project.

The opinions expressed in this article are meant to generate discussion, and are not necessarily those of the institutions some of the authors may represent.

REFERENCES

Abreu, M. H., Pereira, R., Yarish, C., Buschmann, A. H., & Sousa-Pinto, I. (2011). IMTA with *Gracilaria vermiculophylla*: productivity and nutrient removal performance of the seaweed in a land-based pilot scale system. *Aquaculture, 312*(1), 77–87.

Alexandratos, N., Bruinsma, J., Bödeker, G., Schmidhuber, J., Broca, S., Shetty, B., et al. (2006). *World agriculture: Towards 2030/2050.* Rome: Food and Agriculture Organization of the United Nations.

Alix, J. P. (2007). *Sciences et société en mutation: actes du colloque du 12 février 2007.* CNRS Editions.

Barbier, R. (1996). *La recherche-action.* Paris: Anthropos.

Barnaud, C. (2013). La participation, une légitimité en question. *Natures Sciences Sociétés, 21*(1), 24–34.

Baudin, M. (2009). *Le développement durable, nouvelle idéologie du XXIe siècle?* Paris: l'Harmattan.

Becker, C. U. (2012). *Sustainability ethics and sustainability research.* Springer.

Bixler, H. J., & Porse, H. (2011). A decade of change in the seaweed hydrocolloids industry. *Journal of Applied Phycology, 23*(3), 321–335.

Borowitzka, M. A. (2013). High-value products from microalgae—their development and commercialisation. *Journal of Applied Phycology*, 1–14.

Bost, E. (2011). *Aux entreprenants associés: la coopérative d'activités et d'emploi.* Éditions Repas.

Brennan, L., & Owende, P. (2010). Biofuels from microalgae—a review of technologies for production, processing, and extractions of biofuels and co-products. *Renewable and Sustainable Energy Reviews, 14*(2), 557–577.

Brodie, J., Andersen, R. A., Kawachi, M., & Millar, A. J. (2009). Endangered algal species and how to protect them. *Journal Information, 48*(5).

Brundtland, G. H. (1987). *Our common future.* New York, NY: Oxford University Press.

Buchholz, C. M., Krause, G., & Buck, B. H. (2012). Seaweed and man. In *Seaweed biology* (pp. 471–493). Berlin, Heidelberg: Springer.

Bux, F. (Ed.). (2013). *Biotechnological applications of microalgae: Biodiesel and value added products.* CRC Press Llc.

Calogero, G., Citro, I., Di Marco, G., Armeli Minicante, S., Morabito, M., & Genovese, G. (2014). Brown seaweed pigment as a dye source for photoelectrochemical solar cells. *Spectrochimica Acta Part A: Molecular and Biomolecular Spectroscopy, 117*, 702–706.

Callon, M., & Latour, B. (1988). *La science et ses réseaux*. Paris: La découverte.

Callon, M., & Latour, B. (1991). *La science telle qu'elle se fait*. Paris: La découverte.

Canales, A. G., Taquila, R. T., Pacheco, M.Y., Salas, P. C., & Aguilar, S.T. (2008). *Conservacion y manejo de la alga Llaska*. Centro Para el Desarrollo Sostenible (CEDESOS).

Cardenas Alvarez, R., & Munoz Arias, L. (2008). *Chiloé contado desde la cocina*. Santiago de Chile: Gobierno de Chile. Consejo Nacional de la Cultura y Artes.

Charlson, R. J., Lovelock, J. E., Andreae, M. O., & Warren, S. G. (1987). Oceanic phytoplankton, atmospheric sulphur, cloud albedo and climate. *Nature, 326*(6114), 655–661.

Chisti, Y. (2008). Biodiesel from microalgae beats bioethanol. *Trends in Biotechnology, 26*(3), 126–131.

Chisti, Y. (2013). Constraints to commercialization of algal fuels. *Journal of Biotechnology, 167*, 201–214.

Chopin, T., Buschmann, A. H., Halling, C., Troell, M., Kautsky, N., Neori, A., et al. (2001). Integrating seaweeds into marine aquaculture systems: a key toward sustainability. *Journal of Phycology, 37*(6), 975–986.

Chung, I. K., Beardall, J., Mehta, S., Sahoo, D., & Stojkovic, S. (2011). Using marine macroalgae for carbon sequestration: a critical appraisal. *Journal of Applied Phycology, 23*(5), 877–886.

Chung, I. K., Oak, J. H., Lee, J. A., Shin, J. A., Kim, J. G., & Park, K. S. (2013). Installing kelp forests/seaweed beds for mitigation and adaptation against global warming: Korean Project Overview. *ICES Journal of Marine Science: Journal du Conseil*.

Clark, W. C., Crutzen, P. J., & Schellnhuber, H. J. (2004). *Science for global sustainability. Earth system analysis for sustainability*. Cambridge: MIT.

Connan, S. (2004). *Etude de la diversité spécifique des macroalgues de la Pointe de Bretagne et analyse des composés phénoliques des Phéophycées dominantes* (Doctoral dissertation).

Cornish, M. L., & Garbary, D. J. (2010). Antioxidants from macroalgae: potential applications in human health and nutrition. *Algae, 25*(4), 155–171.

Craigie, J. S. (2011). Seaweed extract stimuli in plant science and agriculture. *Journal of Applied Phycology, 23*(3), 371–393.

Crutzen, P. J. (2002). Geology of mankind. *Nature, 415*(6867), 23.

Davis, T. A., Volesky, B., & Mucci, A. (2003). A review of the biochemistry of heavy metal biosorption by brown algae. *Water Research, 37*(18), 4311–4330.

Dillehay, T. D., Ramírez, C., Pino, M., Collins, M. B., Rossen, J., & Pino-Navarro, J. D. (2008). Monte Verde: seaweed, food, medicine, and the peopling of South America. *Science, 320*(5877), 784–786.

Draperi, J. F. (2011). *L'économie sociale et solidaire, une réponse à la crise?: Capitalisme, territoires et démocratie*. Dunod, 177–198.

Erlandson, J. M., Graham, M. H., Bourque, B. J., Corbett, D., Estes, J. A., & Steneck, R. S. (2007). The kelp highway hypothesis: marine ecology, the coastal migration theory, and the peopling of the Americas. *The Journal of Island and Coastal Archaeology, 2*(2), 161–174.

European Commission. (2013). *President Jose Manuel Barroso's Science and Technology Advisory Council*. http://ec.europa.eu/commission_2010-2014/president/advisory-council/documents/stac_policy_paper_no_1_290813.pdf.

Euzen, A., Eymard, L., & Gaill, F. (2013). *Le développement durable à découvert*. CNRS Editions.

Evans, F. D., & Critchley, A. T. (2013). Seaweeds for animal production use. *Journal of Applied Phycology*, 1–9.

FAO. (2012). *The state of the world of fisheries and aquaculture 2012 (SOFIA)*. Rome: FAO Fisheries and Aquaculture Department.

Falkowski, P. G., & Raven, J. A. (2007). *Aquatic photosynthesis* (2nd ed.). Princeton, New Jersey, U.S.A: Princeton University Press.

Fedorov, S. N., Ermakova, S. P., Zvyagintseva, T. N., & Stonik, V. A. (2013). Anticancer and cancer preventive properties of marine polysaccharides: some results and prospects. *Marine Drugs, 11*(12), 4876–4901.
Golubic, S. (1991). Modern stromatolites: a review. In *Calcareous algae and stromatolites* (pp. 541–561). Berlin, Heidelberg: Springer.
Guedes, A. C., & Malcata, F. X. (2012). Nutritional value and uses of microalgae in aquaculture. *Aquaculture, 10*(1516), 59–78.
Graham, L. E., Graham, J. M., & Wilcox, L. W. (2009). *Algae*. Benjamin-Cummings Publishing Company.
Guyomarc'h, J. P., & Le Foll, F. (2011). *Milieux côtiers, ressources marines et société: étude*. Région Bretagne, Conseil économique, social et environnemental régional. S. M. littoral.
Hemaiswarya, S., Raja, R., Kumar, R. R., Ganesan, V., & Anbazhagan, C. (2011). Microalgae: a sustainable feed source for aquaculture. *World Journal of Microbiology and Biotechnology, 27*(8), 1737–1746.
Holdt, S. L., & Kraan, S. (2011). Bioactive compounds in seaweed: functional food applications and legislation. *Journal of Applied Phycology, 23*(3), 543–597.
Hopkins, T. S., Bailly, D., & Støttrup, J. (2011). A systems approach framework for coastal zones. *Ecology and Society, 16*(4).
Hua, K., Carlsson, D. O., Strømme, M., Mihranyan, A., & Ferraz, N. (2013). *In vitro* cytocompatibility of *Cladophora* nanocellulose. *European Cells and Materials, 26*(S2), 13.
Hurtado, A. Q., Montaño, M. N. E., & Martinez-Goss, M. R. (2013). Commercial production of carrageenophytes in the Philippines: ensuring long-term sustainability for the industry. *Journal of Applied Phycology*, 1–10.
Ibañez, E., & Cifuentes, A. (2013). Benefits of using algae as natural sources of functional ingredients. *Journal of the Science of Food and Agriculture, 93*(4), 703–709.
Jäger, J. (2009). Sustainability science in Europe. In *DG research*. Brussels: European Commission.
Jerneck, A., Olsson, L., Ness, B., Anderberg, S., Baier, M., Clark, E., et al. (2011). Structuring sustainability science. *Sustainability Science, 6*(1), 69–82.
Kasting, J. F., & Siefert, J. L. (2002). Life and the evolution of Earth's atmosphere. *Science, 296*(5570), 1066–1068.
Kates, R. W. (2010). *Readings in sustainability science and technology*. Center for International Development at Harvard University.
Kates, R. W., Clark, W. C., Corell, R., Hall, J. M., Jaeger, C. C., Lowe, I., et al. (2001). Sustainability science. *Science, 292*(5517), 641–642.
Keeling, P. (2004). A brief history of plastids and their hosts. *Protist, 155*, 3–7.
Kelly, F. C. (1961). Iodine in medicine and pharmacy since its discovery—1811–1961. *Proceedings of the Royal Society of Medicine, 54*(10), 831.
Kim, K. H. (2013). *Beyond green: Growing algae facade*. . The visibility of research, 499.
Kim, J. K., & Yarish, C. (2012). Nutrient bioextraction by *Saccharina latissima* and *Gracilaria tikvahiae* in Long Island Sound and the Bronx river estuary. *Journal of Phycology, 48*, S13.
Kirkman, H., & Kendrick, G. A. (1997). Ecological significance and commercial harvesting of drifting and beach-cast macro-algae and seagrasses in Australia: a review. *Journal of Applied Phycology, 9*(4), 311–326.
Komiyama, H., Takeuchi, K., Shiroyama, H., & Mino, T. (2011). *Sustainability science: A multidisciplinary approach*. Tokyo: United Nations University Press.
Konur, O. (2011). The scientometric evaluation of the research on the algae and bio-energy. *Applied Energy, 88*(10), 3532–3540.
Kraan, S. (2013). Mass-cultivation of carbohydrate rich macroalgae, a possible solution for sustainable biofuel production. *Mitigation and Adaptation Strategies for Global Change, 18*(1), 27–46.
Krejci, M. R., Finney, L., Vogt, S., & Joester, D. (2011). Selective sequestration of strontium in desmid green algae by biogenic co-precipitation with barite. *ChemSusChem, 4*(4), 470–473.

Landrain, T., Meyer, M., Perez, A. M., & Sussan, R. (2013). Do-it-yourself biology: challenges and promises for an open science and technology movement. *Systems and Synthetic Biology*, 7(3), 115–126.

Lang, D. J., Wiek, A., Bergmann, M., Stauffacher, M., Martens, P., Moll, P., et al. (2012). Transdisciplinary research in sustainability science: practice, principles, and challenges. *Sustainability Science*, 7(1), 25–43.

Leblanc, C., Colin, C., Cosse, A., Delage, L., La Barre, S., Morin, P., et al. (2006). Iodine transfers in the coastal marine environment: the key role of brown algae and of their vanadium-dependent haloperoxidases. *Biochimie*, *88*(11), 1773–1785.

Le Chêne, M. (2012). Algues vertes, terrain glissant. *Ethnologie Française*, *42*(4), 657–665.

Ledford, H. (2010). Life hackers. *Nature*, *467*(7316), 650–652.

Lee, R. E. (2008). *Phycology*. Cambridge University Press.

Leliaert, F., Malta, E. J., Engelen, A. H., Mineur, F., & De Clerck, O. (2008). Quindao algal bloom culprit identified. *Marine Pollution Bulletin*, *56*(9), 1516.

Levin, S. A., & Clark, W. C. (2010). *Toward a science of sustainability: Report from toward a science of sustainability conference*. Princeton, NJ: Princeton University.

Lucas, R. (2012). WIPO Patent No. 2012095569. Geneva, Switzerland: World Intellectual Property Organization.

Lüning, K., & Pang, S. (2003). Mass cultivation of seaweeds: current aspects and approaches. *Journal of Applied Phycology*, *15*(2–3), 115–119.

Mancebo, F. (2013). *Développement durable*. Armand Colin.

Margulis, L. (1974). Origin and evolution of the eukaryotic cell. *Taxonomy*, *23*, 225–226.

May, R. M. (2006). Address of the President, Lord May of Oxford OM AC FRS, given at the Anniversary Meeting on 30 November 2005 Threats to Tomorrow's World. *Notes and Records of the Royal Society*, *60*(1), 109–130.

McFadden, G. I., & van Dooren, G. G. (2004). Evolution: red algal genome affirms a common origin of all plastids. *Current Biology*, *14*, R514–R516.

McHugh, D. J. (2003). *A guide to the seaweed industry–FAO*. Fisheries Technical Paper 441. Rome: Food and Agriculture Organization of the United Nations.

McNiff, J. (2013). *Action research: Principles and practice*. Routledge.

Meyer, M. (2013a). Bricoler la biologie: les politiques et les enjeux de la do-it-yourself biology. *Médecine/Sciences*, *29*, 36–37.

Meyer, M. (2013b). La durabilité domestiquée: les pratiques 'do-it-yourself'. In A. Euzen, L. Eymard, & F. Gaill (Eds.), *Le developpement durable à découvert* (pp. 316–317). Paris: CNRS Editions.

Mollo, P., & Thomas-Bourgneuf, M. (2009). *L'enjeu plancton: L'écologie de l'invisible*. Charles Léopold Mayer/ECLM.

Mouritsen, O. G. (2013). *Seaweeds: Edible, available, and sustainable*. University of Chicago Press.

Muller-Feuga, A. (2000). The role of microalgae in aquaculture: situation and trends. *Journal of Applied Phycology*, *12*(3–5), 527–534.

National Research Council (US). Policy Division. Board on Sustainable Development. (1999). *Our common journey: A transition toward sustainability*. National Academies Press.

Nauen, C. E. (2011). Negotiating a new deal between science and society: reflections on the importance of cognition and emotions in international scientific cooperation and possible implications for enabling sustainable societies. In *Psychological and political strategies for peace negotiation* (pp. 131–156). New York: Springer.

Nauen, C. (2012). Possible transitions towards rebuilding healthy fisheries: can international science and technology cooperation help? In I. Ekeland, D. Fessler, J. M. Lasry, & D. Lautier (Eds.), *The ocean as a global system* (pp. 45–66). Portland: Eska Publishing.

Neori, A., Troell, M., Chopin, T., Yarish, C., Critchley, A., & Buschmann, A. H. (2007). The need for a balanced ecosystem approach to blue revolution aquaculture. *Environment: Science and Policy for Sustainable Development*, *49*(3), 36–43.

Neushul, P. (1989). Seaweed for war: California's World War I kelp industry. *Technology and Culture, 30*(3), 561–583.

Norton, T. A., Melkonian, M., & Andersen, R. A. (1996). Algal biodiversity. *Phycologia, 35*(4), 308–326.

Notoya, M. (2010). Production of biofuel by macroalgae with preservation of marine resources and environment. In *Seaweeds and their role in globally changing environments* (pp. 217–228). Netherlands: Springer.

O'Dowd, C. D., Facchini, M. C., Cavalli, F., Ceburnis, D., Mircea, M., Decesari, S., et al. (2004). Biogenically driven organic contribution to marine aerosol. *Nature, 431*(7009), 676–680.

Olguín, E. J. (2003). Phycoremediation: key issues for cost-effective nutrient removal processes. *Biotechnology Advances, 22*(1), 81–91.

Oncel, S. S. (2013). Microalgae for a macroenergy world. *Renewable and Sustainable Energy Reviews, 26*, 241–264.

Oswald, W. J., Gotaas, H. B., Golueke, C. G., Kellen, W. R., Gloyna, E. F., & Hermann, E. R. (1957). Algae in waste treatment. *Sewage and Industrial Wastes, 29*(4), 437–457.

Otles, S., & Pire, R. (2001). Fatty acid composition of chlorella and spirulina microalgae species. *Journal of AOAC International, 84*, 1708–1714.

Padhi, S., Swain, P. K., Behura, S. K., Baidya, S., Behera, S. K., & Panigrahy, M. R. (2011). Cultivation of *Gracilaria verrucosa* (Huds) Papenfuss in Chilika Lake for livelihood generation in coastal areas of Orissa State. *Journal of Applied Phycology, 23*(2), 151–155.

Pang, S. J., Liu, F., Shan, T. F., Xu, N., Zhang, Z. H., Gao, S. Q., et al. (2010). Tracking the algal origin of the *Ulva* bloom in the Yellow Sea by a combination of molecular, morphological and physiological analyses. *Marine Environmental Research, 69*(4), 207–215.

Park, H. J., Lee, Y. J., Ryu, H. K., Kim, M. H., Chung, H. W., & Kim, W. Y. (2008). A randomized double-blind, placebo-controlled study to establish the effects of spirulina in elderly Koreans. *Annals Nutrition and Metabolism, 52*(4), 322–328.

Park, J. B. K., Craggs, R. J., & Shilton, A. N. (2011). Wastewater treatment high rate algal ponds for biofuel production. *Bioresource Technology, 102*(1), 35–42.

Pei, J., Lin, A., Zhang, F., Zhu, D., Li, J., & Wang, G. (2012). Using agar extraction waste of *Gracilaria lemaneiformis* in the papermaking industry. *Journal of Applied Phycology*, 1–7.

Pérez, R. (1997). *Ces algues qui nous entourent.* Editions IFREMER.

Phipps, S. (2013). Slime render. *Construction Research and Innovation, 4*(2), 12–13.

Quental, N., & Lourenço, J. M. (2012). References, authors, journals and scientific disciplines underlying the sustainable development literature: a citation analysis. *Scientometrics, 90*(2), 361–381.

Ramade, F. (2012). Éléments d'écologie. In *Écologie appliquée: action de l'Homme sur la biosphère* (7e éd.). Paris: Dunod.

Rao, P., Kumar, R. R., Raghavan, B. G., Subramanian, V. V., & Sivasubramanian, V. (2011). Application of phycoremediation technology in the treatment of wastewater from a leather-processing chemical manufacturing facility. *Water SA, 37*(1).

Reisser, W. (2010). The future is green: on the biotechnological potential of green algae. In Seaweeds and their Role. In *Globally changing environments* (pp. 373–383). Netherlands: Springer.

Rhodes, C. J. (2010). Biofuel from algae: salvation from peak oil? In *Seaweeds and their role in globally changing environments* (pp. 229–248). Netherlands: Springer.

Richmond, A., & Hu, Q. (2013). *Handbook of microalgal culture: Applied phycology and biotechnology* (2nd ed.). Wiley-Blackwell.

Roberts, R. M. (1989). *Serendipity: Accidental discoveries in science.* Wiley Science Editions.

Schellnhuber, H. J. (1999). 'Earth system' analysis and the second Copernican revolution. *Nature, 402*, C19–C23.

Seo, Y. B., Lee, Y. W., Lee, C. H., & You, H. C. (2010). Red algae and their use in papermaking. *Bioresource Technology, 101*(7), 2549–2553.

Sharma, N. K., Tiwari, S. P., Tripathi, K., & Rai, A. K. (2011). Sustainability and cyanobacteria (blue-green algae): facts and challenges. *Journal of Applied Phycology*, *23*(6), 1059–1081.

Sing, S. F., Isdepsky, A., Borowitzka, M. A., & Moheimani, N. R. (2013). Production of biofuels from microalgae. *Mitigation and Adaptation Strategies for Global Change*, *18*(1), 47–72.

Spiegel, M., Noordam, M. Y., & Fels-Klerx, H. J. (2013). Safety of novel protein sources (insects, microalgae, seaweed, duckweed, and rapeseed) and legislative aspects for their application in food and feed production. *Comprehensive Reviews in Food Science and Food Safety*, *12*(6), 662–678.

Spolaore, P., Joannis-Cassan, C., Duran, E., & Isambert, A. (2006). Commercial applications of microalgae. *Journal of Bioscience and Bioengineering*, *101*(2), 87–96.

Stephenson, P. G., Moore, C. M., Terry, M. J., Zubkov, M. V., & Bibby, T. S. (2011). Improving photosynthesis for algal biofuels: toward a green revolution. *Trends in Biotechnology*, *29*(12), 615–623.

Stephenson, K., & Shabman, L. A. (2013). The use of nutrient assimilation services in performance-based water quality incentive programs. In *Annual meeting, February 2–5, 2013, Orlando, Florida (No. 143093)*. Southern Agricultural Economics Association.

Sutherland, D. L., Howard-Williams, C., Turnbull, M. H., Broady, P. A., & Craggs, R. J. (2013). Seasonal variation in light utilisation, biomass production and nutrient removal by wastewater microalgae in a full-scale high-rate algal pond. *Journal of Applied Phycology*, 1–13.

Taboada, C., Millan, R., & Miguez, I. (2013). Evaluation of marine algae *Undaria pinnatifida* and *Porphyra purpurea* as a food supplement: composition, nutritional value and effect of intake on intestinal, hepatic and renal enzyme activities in rats. *Journal of the Science of Food and Agriculture*, *93*(8), 1863–1868.

Ugarte, R., & Sharp, G. (2012). Management and production of the brown algae *Ascophyllum nodosum* in the Canadian maritimes. *Journal of Applied Phycology*, *24*(3), 409–416.

Valderrama, D., Cai, J., Hishamunda, N., & Ridler, N. (Eds.). (2013). *Social and economic dimensions of carrageenan seaweed farming* (p. 204). Rome: FAO. (Fisheries and Aquaculture Technical Paper No. 580).

Vásquez, J. A., Piaget, N., & Vega, J. A. (2012). The *Lessonia nigrescens* fishery in northern Chile: 'how you harvest is more important than how much you harvest'. *Journal of Applied Phycology*, *24*(3), 417–426.

Volesky, B., & Holan, Z. R. (1995). Biosorption of heavy metals. *Biotechnology Progress*, *11*(3), 235–250.

Wang, J., Wang, Y., Wang, Z., Li, L., Qin, J., Lai, W., et al. (2008). Vitamin A equivalence of spirulina β-carotene in Chinese adults as assessed by using a stable-isotope reference method. *American Journal of Clinical Nutrition*, *87*(6), 1730–1737.

Wei, N., Quarterman, J., & Jin, Y. S. (2013). Marine macroalgae: an untapped resource for producing fuels and chemicals. *Trends in Biotechnology*, *31*(2), 70–77.

Wiencke, C., & Bischof, K. (2012). In *Seaweed biology: Novel insights into ecophysiology, ecology and utilization* (Vol. 219) Springer.

Wikfors, G. H., & Ohno, M. (2001). Impact of algal research in aquaculture. *Journal of Phycology*, *37*(6), 968–974.

Wolinsky, H. (2009). Kitchen biology. The rise of do-it-yourself biology democratizes science, but is it dangerous to public health and the environment? *EMBO Reports*, *10*(7), 683.

Ye, N. H., Zhang, X. W., Mao, Y. Z., Liang, C. W., Xu, D., Zou, J., et al. (2011). 'Green tides' are overwhelming the coastline of our blue planet: taking the world's largest example. *Ecological Research*, *26*(3), 477–485.

CHAPTER TWO

Seaweed Metabolomics: A New Facet of Functional Genomics

Vishal Gupta*, Rajendra Singh Thakur†, Ravi Singh Baghel*,‡, C.R.K. Reddy*,‡,1 and Bhavanath Jha*

*Discipline of Marine Biotechnology and Ecology, CSIR-Central Salt and Marine Chemicals Research Institute, Bhavnagar, Gujarat, India
†Analytical Discipline and Centralized Instrument Facility, CSIR-Central Salt and Marine Chemicals Research Institute, Bhavnagar, Gujarat, India
‡Academy of Scientific & Innovative Research (AcSIR-CSMCRI), CSIR-Central Salt and Marine Chemicals Research Institute, Bhavnagar, Gujarat, India
[1]Corresponding author: e-mail address: crk@csmcri.org.

Contents

Abstract

Metabolomics is one of the emerging areas of functional genomics and provides newer insights in systems biology. The integration of metabolome with transcriptome analysis facilitates our understanding of gene functionality and its regulation in various metabolic pathways. Marine organisms have a variety of unique biological processes and adaptations suitable for their successful propagation. Marine macroalgae, known as 'seaweeds', are one of the economically important renewable resources of the oceans and have characteristic morphological and physiological processes quite different from terrestrial plants. Seaweeds being attached forms in intertidal region undergo periodic diverse chronic stresses arising from variations in desiccation, irradiance, ultraviolet radiation, salinity, tidal currents and others from anthropogenic activities. Despite the advancement in transcriptomics for seaweeds in recent times, the genetic regulations controlling various biochemical pathways are still in its inception and largely remain unexplored. The study of metabolomics provides a snapshot of cell's catalytic and regulatory processes and also deciphers metabolic responses involved in plant and environment interactions. While summarizing the recent advancements made in analytical technology platforms, a comprehensive review of metabolomics was prepared and discussed from the context of functional genomics, systems biology and biotechnology to gain newer insights into various regulatory networks functioning in the seaweeds.

Advances in Botanical Research, Volume 71
ISSN 0065-2296
http://dx.doi.org/10.1016/B978-0-12-408062-1.00002-0

2.1 INTRODUCTION

Marine macroalgae are the large photoautotrophic multicellular benthic plants (occasionally free floating) occurring in marine environments and consist of as many as 25,000 species worldwide with considerable morphological and functional diversity (Holdt & Kraan, 2011). The continuous exploration of seaweeds for various chemicals of immense commercial value has significantly expanded their utility in agrichemicals, cosmetics, biomaterials, bioenergy and their long standing conventional utilisation as food, feed and hydrocolloids (Bixler & Porse, 2011; Holdt & Kraan, 2011). This has resulted spurt in seaweed production from 3.8 million tons in 1990 to whopping 15.8 million tons fresh in 2008 ranking them as one of the major mariculture crops with an annual market value over US$7.4 thousand millions. (FAO, 2010). Seaweeds represent a unique marine environment that is distinct from terrestrial one and experience various chronic stresses arising from fluctuations in various environmental factors such as desiccation, salinity, radiation, temperature and nutrients (Liu & Pang, 2010; Ross & Van Alstyne, 2007). Recently, transcriptome analysis has been employed to gain deeper insight into their evolution and adaptation to hyper-variable environmental conditions (Coelho, Simon, Ahmed, Cock, & Partensky, 2013; Collen, Guisle-Marsollier, Leger, & Boyen, 2007; Dittami et al., 2011; Dittami, Michel, Collen, Boyen, & Tonon, 2010; Dittami et al., 2009; Gravot et al., 2010; Heinrich, Valentin, Frickenhaus, John, & Wiencke, 2012; Pearson et al., 2010). The release of whole genome sequence data of a brown alga *Ectocarpus siliculosus* (Dillwyn) Lyngbye and a red alga *Chondrus crispus* (Irish moss) revealed genome features that have been evolved in this group of organisms for their successful propagation and proliferation in coastal environment. For example, the *Ectocarpus* genome explored the presence of a complex photosynthetic system facilitating its propagation even in highly variable light conditions, flavonoid pathway genes homologous to plants synthesizing high phenolic contents protecting the alga from ultra-violet radiations and also an uncommon halide metabolism has been deciphered based on the presence of 21 putative dehalogenases and 2 haloalkane dehalogenases. Moreover, the genes and gene families associated with the development of multicellularity and evolution of the brown algal lineage have been identified (Cock et al., 2010). Similarly, the genome sequence of red alga has also elucidated metabolic adaptations pertaining to halogen metabolism, synthesis of oxylipins, microRNA and transcription factors for the development of multicellularity (Collen et al., 2013). This study has also revealed unique metabolic features that are otherwise part of bacterial and fungal metabolism (cellulose synthesis and cell-wall remodelling) and are absent in genome of brown alga. A most recent study by Konotchick et al. (2013) conceptualises a depth-dependent physiology of seaweed by transcriptome analysis.

Though whole genome sequence studies explored some uncommon genomic features related to primary and secondary metabolism in seaweeds, the large part of the genomic information remained unannotated. Integration of data sets from functional- and comparative genomics with proteomics and metabolomics can generate a more accurate and holistic view of genes of unknown function (Xu, Ismail, & Ronald, 2014). A few studies have recently been carried out to understand the metabolic processes in response to salinity and oxidative stress in a brown alga linking with transcriptome data (Dittami et al., 2011; Gravot et al., 2010; Konotchick et al., 2013). These studies have revealed a few unknown pathways functioning under specific stress condition, and warranted for correlation of metabolites-genomic regulatory networks to disclose the specialised mechanisms, which are quite distinct from terrestrial counterparts.

Metabolomics is another addition to 'omic' techniques and presents the information of biological relevance as it reflects the immediate biochemical consequences of genomic and transcriptomic activity. In recent times metabolomics is gaining prominence in the area of integrated systems biology, and is emerging as an essential analytical tool in the post-genomic era (Blow, 2008; Hall 2006; Saito & Matsuda, 2010). Since the term 'metabolome' was coined in 1998, metabolomics has now become an ancillary high-resolution biochemical phenotyping tool to advance our understanding of primary and secondary metabolism. It has made unprecedented success in assessing responses to environmental stress, biomarker analysis, chemotaxonomy, comparing mutants and different growth stages, drug discovery, studying global effects of genetic manipulation, and natural product discovery (Higashi & Saito, 2013; Kim, Choi, & Verpoorte, 2011; Muranaka & Saito, 2013; Putri et al., 2013). The extensive metabolites information generated from various analytical platforms at different laboratories necessitated to frame international standards for metabolomics studies, most importantly on sample collection, precision in data interpretation and analysis. Sumner et al. (2007) proposed minimum reporting standards for chemical analysis. Subsequently, Nicholson and Lindon (2008) in Nature's Q&A news reported the facts and factual about metabolomics. Roessner and Bacic (2009) published a technical feature in Australian Biochemist on 'Dos and Don'ts of Plant Metabolomics'.

The discovery of genes by networking metabolite information with functional genomics information has led to the development of softwares enabling to analyse the huge array of data generated from various analytical platforms, attributing their roles in metabolic pathway networks and fluxome (Table 2.1). This has also led to constitute various metabolite libraries (Table 2.2). The most encouraging example of biological component

Table 2.1 Detailed Description of Software Packages Used for Data Processing, Metabolite Signature Matching in Instrument Specific Manner as well as for Metabolic Pathway Mapping

Program	Web Site	Data Type
COLMAR	http://spinportal.magnet.fsu.edu/	NMR data
FiD	http://www.cs.helsinki.fi/group/sysfys/software/fragid/	Tandem mass spectrometry
MeltDB	https://meltdb.cebitec.uni-bielefeld.de/cgi-bin/login.cgi	GC-MS
MetaboloAnalyst	http://www.metaboanalyst.ca/MetaboAnalyst/faces/Home.jsp	NMR/MS
MetaboMiner	http://wishart.biology.ualberta.ca/metabominer/	NMR
MolFind	http://metabolomics.pharm.uconn.edu/Software.html	HPLC/MS
Fiehn Lab	http://fiehnlab.ucdavis.edu/staff/kind/Metabolomics/Structure_Elucidation/	NMR/MS
OpenMS	http://open-ms.sourceforge.net/	LC-MS
Peak alignment	http://fiehnlab.ucdavis.edu/staff/kind/Metabolomics/Peak_Alignment/	NMR/MS
SetupX	http://fiehnlab.ucdavis.edu/projects/binbase_setupx/	GC-MS
Seven Golden Rules	http://fiehnlab.ucdavis.edu/projects/Seven_Golden_Rules	MS
XCMS	http://metlin.scripps.edu/xcms/	LC-MS
PRIMe	http://prime.psc.riken.jp/	NMR/MS
Metabolic Pathway Databases		
KEGG	http://www.genome.jp/kegg/	
MetaCyc	http://metacyc.org/	
BioCyc	http://biocyc.org/	
Reactome	http://www.reactome.org/	
BRENDA	http://www.brenda-enzymes.org/	
BioPath	http://www.molecular-networks.com/databases/biopath	
Biochemical Pathways	http://web.expasy.org/pathways/	
PathCase	http://nashua.cwru.edu/pathwaysweb/About.aspx#sb	
BioSystems	http://www.ncbi.nlm.nih.gov/Structure/biosystems/docs/biosystems_about.html	
Plant Metabolic Network	http://www.plantcyc.org/	

networking or co-expression analysis studies is the model plant *Arabidopsis* wherein transcriptome and metabolome analysis have successfully discovered the function of genes involved in various metabolic pathways (Quanbeck et al., 2012; Tohge et al., 2005). Such systems biology approaches have recently been gaining importance in medicinal plant research. A specialised metabolic pathway in plants is involved with synchronised

Table 2.2 Description of Databases Often Employed for Metabolites Identification for Metabolomics Studies

Database	Web Site	Data Type
The Human Metabolome Database	http://www.hmdb.ca/	Various spectral information
SetupX and BinBase	http://fiehnlab.ucdavis.edu/projects/binbase_setupx	GC-MS data
MetaboLights Database	http://www.ebi.ac.uk/metabolights/index	Various experimental studies
The BioMagResBank	http://www.bmrb.wisc.edu/metabolomics/	NMR spectra
The Madison Metabolomics Consortium Database	http://mmcd.nmrfam.wisc.edu/	MS and NMR data
MassBank	http://www.massbank.jp/	MS search
Golm Metabolome Database	http://gmd.mpimp-golm.mpg.de/	GC-MS search
METLIN	http://metlin.scripps.edu/index.php	MS/MS search
Fiehn Lib	http://fiehnlab.ucdavis.edu/Metabolite-Library-2007/	GC-MS search
Birmingham Metabolite Library Nuclear Magnetic Resonance database	http://www.bml-nmr.org/	NMR search
mzCloud	https://mzcloud.org/	MS search
PRIMe Standard Spectrum Search	http://prime.psc.riken.jp/?action=standard_index	NMR and MS search
PRIMe MS2T	http://prime.psc.riken.jp/lcms/ms2tview/ms2tview.html	LC-ESI-Q-TOF/MS
HIFI	http://spectra.psc.riken.jp/menta.cgi/netcdf/index	LC-FTICR-MS
SpinAssign	http://prime.psc.riken.jp/?action=nmr_search	NMR search
AtMetExpress	http://prime.psc.riken.jp/lcms/AtMetExpress/	LC-MS

action of various genes. Decoding the information of those genes, gene regulatory networks and the association of gene-metabolite are of paramount importance to design metabolic engineering and synthetic biology approaches to have an organism producing high value chemicals in environmentally safe manner and with low input.

Metabolite profiling in seaweeds has been carried out for decades but restricted to targeted identification of one group of metabolites particularly lipids and their derivatives (Bouarab et al., 2004; Goulitquer, Potin, & Tonon, 2012; Kumari et al. 2013a, b). Mycosporine-like amino acids and halogenated compounds are yet another targeted metabolite families often used to study the stress responses in seaweeds (Kundel et al., 2012; La Barre, Potin, Leblanc, & Delage, 2010; Yuan, Westcott, Hu, & Kitts, 2009). The metabolomic information generated for seaweeds is at its infancy in comparison to the reference libraries of metabolites constructed for terrestrial plants and can be visualised from the publications released in last one decade in comparison to the same in terrestrial plants (Figure 2.1). Recently, an attempt has been made to constitute a reference library for seaweed metabolites (Davis & Vasanthi, 2011).

This review discusses the significance of co-expression studies integrating metabolomics information with genomics to derive a comprehensive view of biological system and specialised regulatory networks. An overview of the advances made on analytical platforms for metabolite analysis is also summarised. Further, the metabolomics is presented as a new facet to functional genomics approaches in seaweeds in order to draw a holistic view of their physiological regulations and responses arising due to highly dynamic marine environment.

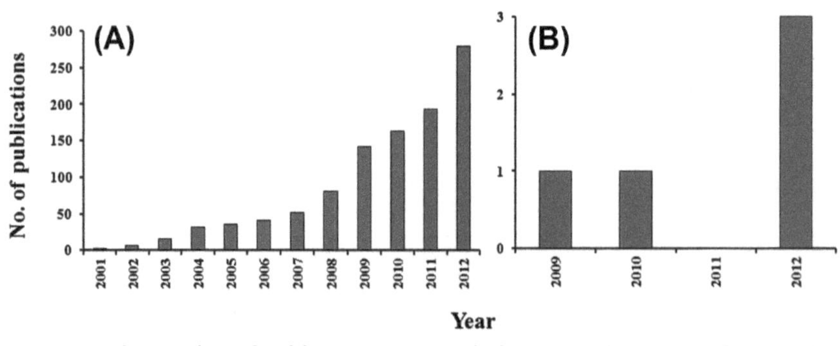

Figure 2.1 The number of publications on metabolomics studies on (A) plants and (B) seaweeds as obtained by searching web of knowledge with the key words 'metabolomics SAME plant' and 'metabolomics SAME seaweed' for former and later, respectively.

2.2 SAMPLE COLLECTION

The vast array of metabolites in the biological system coupled with their dynamic variations resulting from subtle environmental fluxes pose an incredible challenge to analyst while identifying metabolites. Metabolite profile varies with the diurnal fluctuation, availability of nutrients, variation in photosynthesis and respiration depending on the light (Davis, Fiehn, & Durnford, 2013; Dunn & Ellis, 2005; Ho, Tang, Hoefel, & Vigneswaran, 2012; Lee, Perdian, Song, Yeung, & Nikolau, 2012). Therefore, it is very crucial to fix the time and condition for plant collection before metabolite extraction. Utmost care has to be taken in such a way that all sampling should be done at the same conditions. Metabolites have dynamic exchange rates from one form to another thereby warranting for fixing the state of plant immediately after harvest and flash freezing is the best way. Freeze-drying of the harvested sample is another crucial step before metabolite extraction, as difference in water content will lead to biased quantification of metabolites. Later, sample extraction, which is to be optimised with respect to the sample type such as whole plant, specific cell type as well as according to the analytical platform to be used for analysis, is followed.

2.3 ANALYTICAL PLATFORMS FOR METABOLITE SEPARATION AND IDENTIFICATION

Because of the vast metabolite diversity, there have been 100,000 known and 200,000 unknown metabolites profiled through employing a wide range of analytical tools and techniques (Saito & Matsuda, 2010). The analytical platforms for metabolite analysis fractionate them based on their (1) different masses, (2) mass/charge ratio, (3) ionisation. Therefore, a range of analytical platforms are used for metabolite profiling and mainly includes gas chromatography–mass spectrometry (GC-MS), liquid chromatography–mass spectrometry (LC-MS), capillary electrophoresis–mass spectrometry (CE-MS), nuclear magnetic resonance (NMR) spectroscopy, Fourier Transform infrared and Raman spectroscopy. Among these, MS-based and NMR spectroscopic approaches are most popular. Most recently, matrix-assisted laser desorption/ionisation (MALDI) has also been introduced as another analytical platform for metabolite profiling (Ye et al., 2013). MS-based approaches are the most preferred techniques for metabolite profiling because of their moderate to high sensitivity, high reproducibility and superior separation potential based on electron ionisation and charge to mass ratio.

GC-MS method has been the most popular and preferred analytical platform for metabolite profiling because it is rapid, reproducible and less affected with matrix effect. The only condition for GC-MS-based analysis is to vaporise the samples by derivatization mainly by silylation reaction. This step is considered to be the most crucial step and required to be optimised in terms of the volume ratios of pyridine × methoxyamine to N-Methyl-N-(trimethylsilyl)trifluoroacetamide (MSTFA), preferred is 1: 9 but the ratio of 1:2 or 1:1 have also been reported (Fiehn et al., 2008; Hill & Roessner, 2013; Lisec, Schauer, Kopka, Willmitzer, & Fernie, 2006). The utmost care has to be taken to avoid the moisture after derivatization of samples, which otherwise lead to abundance of ions m/z 221 and m/z 281 for in GC-MS spectra that is attributed to polysiloxanes arising from hydrolysis (Fiehn et al., 2008). After derivatization, most polar metabolites or metabolites with highly polar functional groups such as sugars, amino acids, amines and fatty acids can easily be identified by GC-MS. One limitation with trimethylsilyl derivatization is the low-ion abundance, therefore, required to be optimised for derivatization duration and in some cases the absence of the molecular ion ($M^{+\cdot}$), and in many cases an $M\text{-}15^{+\cdot}$ fragment ion. Precision in the mass of the molecular ion is important for the determination of molecular formula aiding to metabolite identification (Dunn et al. 2013). Another cautionary note applies to data analysis after obtaining the GC-MS spectrum. The GC-MS spectrum is highly complex and contains numerous overlapping peaks because of the co-elution of low abundant metabolites peaks. This necessitate for deconvolution of the spectrum. While deconvolution of the spectrum, defining thresholds for peak finding is most difficult task as it leads to either false negative or false positive signals for metabolites, if set to very high and very low values respectively (Fiehn et al., 2008). In addition to software available from instrument manufacturers, softwares freely available for deconvolution are Automated Mass Spectral Deconvolution and Identification System (http://chemdata.nist.gov/mass-spc/amdis/), and Metabolite Detector (http://md.tu-bs.de/). The metabolite libraries available for metabolite identification are NIST08, as generalised chemical library (http://chemdata.nist.gov/mass-spc/ms-search/), specifically for metabolites are the Golm Metabolome Database (Kopka et al., 2005) and FiehnLib (Kind et al., 2009).

LC-MS is another emerging platform for metabolomics studies preferably for semipolar metabolites and includes soft ionisation techniques such as electrospray ionisation, atmospheric pressure chemical ionisation, atmospheric pressure photoionisation, and fast atom bombardment. The major advantages offered by this technique are no derivatization required and metabolites can be effectively extracted with aqueous alcohol solution and can directly be analysed mainly on reverse phase column. The hydrophilic interaction chromatography (HILIC) is the recent advancement in metabolomics studies, which gives more sensitivity and reproducibility in comparison to reverse phase (Gika et al. 2013). The LC-MS holds promising potentials of reproducibility in m/z values but significant variations may be observed in retention time and peak intensities. It is cautionary to run standard quality control (QC) sample every time with the sample. As retention times are prone to drift significantly, peak identification for LC-MS is based on centroiding over the m/z range generally performed on extracted ion chromatograms (EIC) (Tautenhahn, Bottcher, & Neumann, 2008). The instrument manufacturer software packages are enabled with direct acquisition of centroided data. EIC is 2D distribution intensity signals vs retention time over a small m/z interval. The other algorithms making their way for improving the peak picking include Mexican Hat wavelet analysis (Tautenhahn et al., 2008), a bi-Gaussian mixture model (Yu & Peng, 2010). Ion annotation is another critical factor for data analysis in LC-MS-based metabolomics studies because one metabolite shows many peaks in LC-MS with different m/z values. The reference mass spectral library and software packages for peak assignments for LC-MS data is under progress. A few software tools available publically for preprocessing and analysis of LC-MS data include Markerlynx, Mzmine, XCMS and MetAlign. The preprocessed results can be exported for further statistical analysis. LC therefore allows separation of compounds of wide range of polarity with little efforts comparative to GC based analysis. Advancement in LC techniques is ultra performance LC, which speeds up the run time and provides better chromatographic resolution (Flamini et al. 2013).

CE-MS has been shown as another technique with potentials for untargeted metabolite profiling for metabolomics studies. This is a complementary technique to LC, relatively inexpensive as it does not require expensive LC columns and also requires very small amount of samples and solvents. The separation of metabolites is based on the same principal of charge to mass ratio as in LC. Because of the small sample size required this technique suffers from poor sensitivity. The improvement in hardware configuration i.e. replacement of stainless steel needle for sample spraying with platinum needle resulted in better results for anionic compounds in negative ionisation mode with improved sensitivity to about 63-fold and limit of detection (LODs) between 0.03 and 0.87 μM (Soga et al., 2009). The ionisation efficiency was further enhanced by the development of background electrolyte (BGE) and sheath liquid composition. Replacing ammonium acetate with triethylamine in BGE and sheath liquid was found to be effective for enhancing the metabolite ionisation as well as improved the coverage (Kok, de Jong, & Somsen, 2011). Moreover, Ramautar et al. (2012) developed a sheath less CE-MS method for the detection of polar and charged metabolites in human urine even at a scale of subnanomolar. Even then the application of CE-MS in metabolomic studies is limited compared to other MS-based approaches mainly because of lower sensitivity, resolution and coverage, and also less experience of researchers. The potential of this technique has been described in various reviews (Barbas, Moraes, & Villasenor, 2011; Monton & Soga, 2007; Ramautar, Demirci, & de Jong, 2006; Ramautar, Somsen, & de Jong, 2013).

MALDI-MS is another powerful tool to investigate the de novo imaging and dynamics of metabolites by direct analysis of the tissue (Kaspar, Peukert, Svatos, Matros, & Mock, 2011; Lee et al., 2012; Stoeckli, Chaurand, Hallahan, & Caprioli, 2001). Though matrix generates small interfering ionised peaks on desorption/ionisation process, matrix-free laser desorption/ionisation was developed and shown to be applicable to image UV-absorbing metabolites from plant tissues (Holscher et al., 2009). The metabolite imaging application was further improved with the development of various matrix-free surface-modified/functionalised techniques including metal and carbon materials (Cha et al., 2008; Jun et al., 2010; Northen et al., 2007; Woo, Northen, Yanes, & Siuzdak, 2008). More recently, Ye et al. (2013) showed the efficiency of MALDI matrix [1,8–bis (dimethyl-amino) naphthalene, DMAN] over conventional matrix 2,5–dihydroxybenzoic acid (DHB) revealing the better resolution and coverage with the identification of large array of organic acids, amino acids, sugars, lipids, flavonoids and their conjugates.

NMR Spectroscopy is another approach employed extensively for various metabolomics studies. Spectra of nuclei of atoms having odd atomic number (^{1}H) or odd mass number (^{13}C) is acquired by application of radio frequency excitation under strong magnetic field. An NMR spectrum consists of peaks characterised by their position, multiplicity, and width and all these parameters are governed by local electronic (or chemical) environment. Recently, NMR spectroscopy has shown extensive utilisation in plant metabolomics mainly for discrimination of species, ecotypes and mutants

(Broyart et al., 2010; Kim, Choi, Erkelens, Lefeber, & Verpoorte, 2005; Ward, Harris, Lewis, & Beale, 2003), transgenic plants from their wild type siblings (Barros et al., 2010; Charlton et al., 2004; Choi, Choi, et al., 2004), infected plants from healthy ones (Choi, Tapias, et al., 2004) and also led to understand the metabolic transitions during disease (Ward et al., 2010). It also allows detection and quantification of metabolites in a complex mixture without prior knowledge of sample composition even under *in vivo* state (Oikawa & Saito, 2012). According to Fernie, Obata, Allen, Araujo, and Bowler (2012) "NMR represents the gold standard in structure identification." The usefulness of NMR spectroscopy to both species management and monitoring of agricultural product quality (Ryan & Robards, 2006) led to compile a reference library of metabolites for terrestrial plants. Extracts prepared for NMR based metabolomics have delivered comparable spectra for multiple laboratories when standardised with respect to solvent type, pH and runtime temperature (Ward et al., 2010). The application of magic angle spinning is getting popularity to evaluate the physiological state of intact cells or tissues (Dunn & Ellis, 2005). Although NMR is less sensitive than MS, NMR offers the advantage of simpler and reliable extract preparation and measurements in nondestructive and nontargeted manner. The natural abundance of ^{1}H, ^{31}P, ^{19}F, ^{13}C and ^{15}N atoms is the limiting factor to the sensitivity of NMR spectroscopy. The recent developments in hardware configurations along with the availability of coil cooling with liquid helium or liquid nitrogen (to reduce the magnitude of noise generated by thermal motion of electrons) delivers sensitivity enhancement and considerably shorten the experimental time.

Another feature of NMR spectroscopy is the ease of comparative quantification of metabolites as the spectral peak intensity obtained is directly proportional to the quantity of metabolite present. The position of NMR peaks are determined by the chemical environment around those sites and thus the technique is better suited for mixtures containing different variety of compounds i.e. a mixture of amino acids, sugars, aromatics, fatty acid etc. With this, NMR spectroscopy can be an effective tool for chemical fingerprinting and fluxome studies. The assignment of peaks to respective compounds is a daunting task in NMR spectroscopy owing to often observed peak congestion. The assignment is facilitated by dispersing the 1D spectra to 2D experiments where correlations between chemically bonded pairs are acquired. The most commonly used 2D experiments are TOCSY (total correlation spectroscopy; ^{1}H-^{1}H correlation), HSQC (hetero-nuclear single quantum correlation; ^{1}H-^{13}C correlation) and HMBC (hetero-nuclear multiple bonds correlation; ^{1}H-^{13}C correlation). TOCSY involves application of two pulses (preferably of 90° flip angle) wherein the duration between them is incremented to generate second dimension along with efficient mixing using a 17-pulse train designed by Levitt (2001). HSQC experiments involve excitation of proton magnetisation followed by Insensitive Nuclei Enhanced by Polarization Transfer (INEPT) transfer to ^{13}C followed by evolution period to generate indirect dimension. Signal is often detected on ^{1}H channel after INEPT transfer from ^{13}C channel. J-Resolved spectroscopy is additional useful 2D experiment that delivers spectra free from J-coupling in the indirect dimension and is better suited for quantitative studies.

Among various solvents employed for extraction of metabolites for NMR-based studies, perchloric acid was shown to be the most suitable for extraction of polar metabolites but found to be unsuitable for acid-labile compounds like fructose-2,6-bio-phosphate, Nicotinamide adenine dinucleotide phosphate (NADPH) and Deoxyguanosine triphosphate (dGTP) (Kruger, Troncoso-Ponce, & Ratcliffe, 2008). A common extraction method employed for GC-MS analysis i.e. aqueous $CHCl_3$/methanol suffers from the inferior extraction efficiency for number of metabolites, e.g. amino acids and nucleotides in comparison to acid-based extraction. Two-phase extraction system including the aqueous methanol in combination with chloroform is also employed to yield polar and nonpolar metabolites (Choi, Choi, et al., 2004; Choi, Tapias, et al., 2004). The additional separation step while using two-phase extraction system is a limiting factor towards high-throughput analysis. Recently, Kim, Choi, and Verpoorte (2010) showed an extraction system with aqueous methanol as most suitable for metabolite profiling. The cautionary remark while extraction of metabolite is to stabilise pH of the extraction system whose fluctuation leads to broadening of peaks in ^{1}H spectrum. Phosphate buffers of pH 6 (Kim et al., 2010) and oxalate buffer of pH 4 (Pereira et al., 2005; Son et al., 2009) are some of the most commonly used buffering systems.

Authors have made an attempt to study metabolomics of seaweeds using NMR spectroscopy. This was initiated with the simplification of sample preparation together with NMR spectral acquisition. The aqueous extract obtained from the physical crushing of the alga representing the native state was analysed. While recording the spectra from the aqueous medium, pre-saturation pulse, a weak radio frequency pulse of long duration, was applied on H_2O peak to saturate the transition leading to its suppression for both 1D as well as 2D experiments (Figure 2.2). During this experiment a drop of D_2O containing Trimethylsilyl propanoic acid (TSP) was added for controlling the drift of magnetic field as well as an internal reference respectively. Application

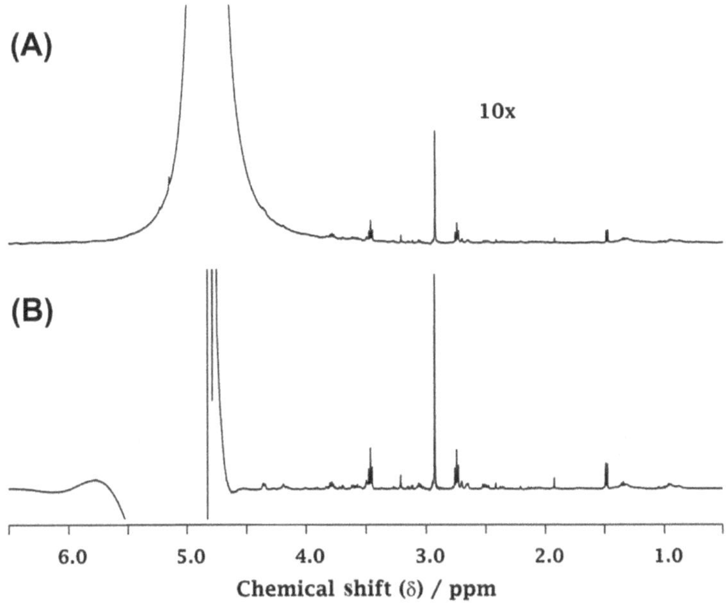

Figure 2.2 ^{1}H NMR spectra acquired from aqueous extract of *Ulva lactuca* after (A) single pulse excitation (B) pre-saturation block applied before single pulse excitation. Sensitivity gain by a factor of 10 was noticed however water peak sometimes distorted.

of pre-saturation block can distort baseline and magnetisation transfer via chemical exchange mediated saturation transfer and can affect quantitation of sugar residues in the seaweed biomass richer in such compounds.

Freeze-drying of samples is recommended for metabolite quantitation studies but in our study spectral information generated from the freeze-dried samples was found different from the spectra recorded from aqueous extract. Some of the spectral peaks were found missing from the former analysis (Figure 2.3). The peak information includes the signature for acetate and coniferaldehyde, which led to know some of the unexplored specialised features in algal species. The existence of coniferaldehyde indicated the existence of lignin precursors in seaweeds, which was then confirmed with additional experiment of derivatization followed by reductive cleavage method (Lu & Ralph, 1997). The chemical signature of acetate provided inferences towards the specialised regulatory mechanisms involved in mitigation of the effect of gaseous flux arising as a result of tidal rhythms (Gupta, Thakur, Reddy, & Jha, 2013). NMR spectra of algal species investigated by the authors seem to be quite distinct from the spectra reported in literature for terrestrial plants, thus indicating a difference in their metabolic complexity related to their distinct habitat.

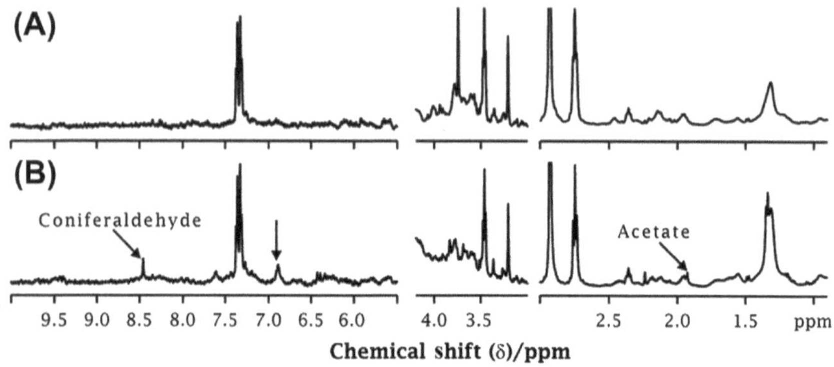

Figure 2.3 ^{1}H NMR spectra of *Ulva lactuca* (A) after freeze-drying and (B) direct aqueous extract. Arrows indicated the peaks found missing after freeze-drying. *(Source: Gupta et al., 2013).*

2.4 INTEGRATION OF METABOLOMICS DATA WITH FUNCTIONAL GENOMICS

Networking of multicomponent information generated from different 'omic' technologies led to a number of gene discoveries towards specialised mechanism in different model and nonmodel plants (Higashi & Saito, 2013). The most encouraging example of gene co-expression networking with metabolomics is the identification of genes responsible for anthocyanin biosynthesis (Tohge et al., 2005). A few more examples of gene-metabolite co-expression analysis are the identification of glycosyltransferase genes (Yonekura-Sakakibara, Fukushima, & Saito, 2013), putative bile acid transporter family protein involved in the transport of glucosinolate-related metabolites (Gigolashvili et al. 2009; Sawada et al. 2009) and genes involved in circadian clock network (Kerwin et al., 2011). The co-expression networking has now been progressed to single cell level and the example of the key findings extracted out from such studies are the role of D-enantiomer of ornithine as a selective regulator to L-arginine biosynthesis and urea cycle in tobacco cells (Gholami et al. 2013). Metabolic features of different root cell types in *Arabidopsis* (Moussaieff et al., 2013) and salt responsive network profile in rice suspension cell culture (Liu et al. 2013) are a few recent examples. Recent reviews by Higashi and Saito (2013), Seaver, Henry, and Hanson (2012), and Schilmiller, Pichersky, and Last (2012) provided exhaustive information on functional genomics of terrestrial plants by gene-metabolomics co-expression studies.

Integration of metabolite information with other functional genomics is based upon guilt-by-association principle (Saito, Hirai, & Yonekura-Sakakibara, 2008). This principle states that in a certain biological process at a given point of time a set of genes or metabolites are co-ordinately regulated reflecting a common regulatory system. One can designate the function of unknown gene or group of genes co-expressing in a biological function or during specific environmental stress. For example the stress induced by sulphur deficiency in *Arabidopsis* allowed predicting the genes coding for sulphotransferases (Hirai et al., 2005), two MYB transcription factors (Hirai et al., 2007), chain elongation enzymes (Sawada et al., 2009), and leucine biosynthesis and phenylpropanoid pathway as elicitors to adapt to the new environment (Caldana et al., 2011). Identification of cold stress-regulated raffinose by a transcription factor DREB, and dehydration-induced ABA-dependent responses (Urano et al., 2009), specialised metabolism particularly of flavonoids against exposure to UV radiations (Kusano et al., 2011), genes responsible for modifications of flavonoids (Tohge & Fernie, 2010; Yonekura-Sakakibara et al., 2008, 2007), elicitors for aliphatic glucosinolate biosynthesis (Hirai et al., 2007), isoprenoid pathway (Wille et al., 2004), brassinosteroid-regulated genes (Lisso, Steinhauser, Altmann, Kopka, & Mussig, 2005) and many more are the examples of integrative systems biology approach. Similar to model plants, co-expression analysis studies revealed secondary metabolite biosynthesis pathways and their regulatory genes and gene clusters in non-model plants such as cyanogenic glycosides in cassava, sorghum and *Lotus japonicas* (Takos et al., 2011; Takos & Rook, 2012); root of cannabinoid biosynthetic genes in *Cannabis sativa* (van Bakel et al., 2011); identification of diphosphate synthase from the tomato trichome (Schilmiller et al., 2009); tissue specificity of triterpene saponins including pharmacological glycyrrhizin in *Glycyrrhiza* (Seki et al., 2008, 2011).

A few attempts of co-expression analysis have also been made in seaweeds to understand various physiological adaptations inherited by this group of organisms but restricted to only one genus *Ectocarpus*. The study with *Ectocarpus* aimed at understanding the responses to salinity shown the active function of γ-aminobutyric acid synthesised through a salt stress-induced putrescine-degradation pathway (Dittami et al., 2011). Another study by the same research group on same alga showed impregnation of genomic alterations at metabolite level to stabilise the transition of evolutionary colonisation of alga from fresh water to marine habitat (Dittami et al., 2012). Even then much more need to be explored for seaweeds because of the existence of uncommon physiological state and behaviour of seaweeds in comparison to their terrestrial counterparts. For example, the cell wall biosynthesis in seaweeds is a complex phenomenon as it composed of hetero-polysaccharide agar and carrageenan in seaweeds belonging to red, alginates in brown and Ulvan in seaweed species of Ulvophyceae. A close understanding of biosynthesis mechanism of these hydrocolloids may help to initiate genetic manipulation studies to improve the seaweed for these traits. Likewise, the oxylipin biosynthetic machinery in brown seaweeds showed gene cluster homologous to both plant and animal. The sulphinic acid compounds and their derivatives reported by Gupta et al. (2013) in seaweeds are of great interest and their biosynthetic pathways need to be explored for exploring their role in free radical detoxification. A systems biology approach combining both gene and metabolite expression holds the potential to uncover adaptive mechanisms not revealed by whole genome sequencing.

In conventional approach, co-expression study involves simultaneous analysis of metabolomics and transcriptomics of an organism at a given point of time under specific physiological or environmental condition. A co-expression match between mRNA (gene) and targeted metabolite determines their homologous expression. Reverse genetic approach knocking out the gene then leads to identify responsible metabolites and its function. This leads to a discovery of relationship between gene regulatory networks with specialised metabolic pathway opening new avenues for metabolic engineering for the production of targeted specialised metabolites in seaweeds. Alternatively, modern synthetic biology approaches can reconstruct the specialised metabolic pathways *in vitro* for production of large amounts of desired metabolites.

ACKNOWLEDGEMENTS

Authors are thankful to CSIR-CSMCRI for providing analytical instrument facility for initializing seaweed metabolomics studies. Mr Vishal thanks CSIR, New Delhi for awarding Senior Research Fellowship.

REFERENCES

van Bakel, H., Stout, J. M., Cote, A. G., Tallon, C. M., Sharpe, A. G., Hughes, T. R., et al. (2011). The draft genome and transcriptome of *Cannabis sativa*. *Genome Biology*, *12*.

Barbas, C., Moraes, E. P., & Villasenor, A. (2011). Capillary electrophoresis as a metabolomics tool for non-targeted fingerprinting of biological samples. *Journal of Pharmaceutical and Biomedical Analysis*, *55*, 823–831.

Barros, E., Lezar, S., Anttonen, M. J., van Dijk, J. P., Roehlig, R. M., Kok, E. J., et al. (2010). Comparison of two GM maize varieties with a near-isogenic non-GM variety using transcriptomics, proteomics and metabolomics. *Plant Biotechnology Journal*, *8*, 436–451.

Bixler, H. J., & Porse, H. (2011). A decade of change in the seaweed hydrocolloids industry. *Journal of Applied Phycology*, *23*, 321–335.
Blow, N. (2008). Biochemistry's new look. *Nature*, *455*, 697–700.
Bouarab, K., Adas, F., Gaquerel, E., Kloareg, B., Salaun, J. P., & Potin, P. (2004). The innate immunity of a marine red alga involves oxylipins from both the eicosanoid and octadecanoid pathways. *Plant Physiology*, *135*, 1838–1848.
Broyart, C., Fontaine, J.-X., Molinie, R., Cailleu, D., Terce-Laforgue, T., Dubois, F., et al. (2010). Metabolic profiling of maize mutants deficient for two glutamine synthetase isoenzymes using H-1-NMR-based metabolomics. *Phytochemical Analysis*, *21*, 102–109.
Caldana, C., Degenkolbe, T., Cuadros-Inostroza, A., Klie, S., Sulpice, R., Leisse, A., et al. (2011). High-density kinetic analysis of the metabolomic and transcriptomic response of *Arabidopsis* to eight environmental conditions. *Plant Journal*, *67*, 869–884.
Charlton, A., Allnutt, T., Holmes, S., Chisholm, J., Bean, S., Ellis, N., et al. (2004). NMR profiling of transgenic peas. *Plant Biotechnology Journal*, *2*, 27–35.
Cha, S. W., Zhang, H., Ilarslan, H. I., Wurtele, E. S., Brachova, L., Nikolau, B. J., et al. (2008). Direct profiling and imaging of plant metabolites in intact tissues by using colloidal graphite-assisted laser desorption ionization mass spectrometry. *Plant Journal*, *55*, 348–360.
Choi, H. K., Choi, Y. H., Verberne, M., Lefeber, A. W. M., Erkelens, C., & Verpoorte, R. (2004). Metabolic fingerprinting of wild type and transgenic tobacco plants by H-1 NMR and multivariate analysis technique. *Phytochemistry*, *65*, 857–864.
Choi, Y. H., Tapias, E. C., Kim, H. K., Lefeber, A. W. M., Erkelens, C., Verhoeven, J. T. J., et al. (2004). Metabolic discrimination of *Catharanthus roseus* leaves infected by phytoplasma using H-1-NMR spectroscopy and multivariate data analysis. *Plant Physiology*, *135*, 2398–2410.
Cock, J. M., Sterck, L., Rouze, P., Scornet, D., Allen, A. E., Amoutzias, G., et al. (2010). The *Ectocarpus* genome and the independent evolution of multicellularity in brown algae. *Nature*, *465*, 617–621.
Coelho, S. M., Simon, N., Ahmed, S., Cock, J. M., & Partensky, F. (2013). Ecological and evolutionary genomics of marine photosynthetic organisms. *Molecular Ecology*, *22*, 867–907.
Collen, J., Guisle-Marsollier, I., Leger, J. J., & Boyen, C. (2007). Response of the transcriptome of the intertidal red seaweed *Chondrus crispus* to controlled and natural stresses. *New Phytologist*, *176*, 45–55.
Collen, J., Porcel, B., Carre, W., Ball, S. G., Chaparro, C., Tonon, T., et al. (2013). Genome structure and metabolic features in the red seaweed *Chondrus crispus* shed light on evolution of the Archaeplastida. *Proceedings of the National Academy of Sciences of the United States of America*, *110*, 5247–5252.
Davis, M. C., Fiehn, O., & Durnford, D. G. (2013). Metabolic acclimation to excess light intensity in *Chlamydomonas reinhardtii*. *Plant Cell and Environment*, *36*, 1391–1405.
Davis, G. D. J., & Vasanthi, A. H. R. (2011). Seaweed metabolite database (SWMD): a database of natural compounds from marine algae. *Bioinformation*, *5*, 361–364.
Dittami, S. M., Gravot, A., Goulitquer, S., Rousvoal, S., Peters, A. F., Bouchereau, A., et al. (2012). Towards deciphering dynamic changes and evolutionary mechanisms involved in the adaptation to low salinities in *Ectocarpus* (brown algae). *Plant Journal*, *71*, 366–377.
Dittami, S. M., Gravot, A., Renault, D., Goulitquer, S., Eggert, A., Bouchereau, A., et al. (2011). Integrative analysis of metabolite and transcript abundance during the short-term response to saline and oxidative stress in the brown alga *Ectocarpus siliculosus*. *Plant Cell and Environment*, *34*, 629–642.
Dittami, S. M., Michel, G., Collen, J., Boyen, C., & Tonon, T. (2010). Chlorophyll-binding proteins revisited - a multigenic family of light-harvesting and stress proteins from a brown algal perspective. *BMC Evolutionary Biology*, *10*.

Dittami, S. M., Scornet, D., Petit, J. L., Segurens, B., Da Silva, C., Corre, E., et al. (2009). Global expression analysis of the brown alga *Ectocarpus siliculosus* (Phaeophyceae) reveals large-scale reprogramming of the transcriptome in response to abiotic stress. *Genome Biology, 10.*

Dunn, W. B., & Ellis, D. I. (2005). Metabolomics: current analytical platforms and methodologies. *TrAC Trends in Analytical Chemistry, 24*, 285–294.

Dunn, W. B., Erban, A., Weber, R. J. M., Creek, D. J., Brown, M., Breitling, R., et al. (2013). Mass appeal: metabolite identification in mass spectrometry-focused untargeted metabolomics. *Metabolomics, 9*, 44–66.

Fernie, A. R., Obata, T., Allen, A. E., Araujo, W. L., & Bowler, C. (2012). Leveraging metabolomics for functional investigations in sequenced marine diatoms. *Trends in Plant Science, 17*, 395–403.

Fiehn, O., Wohlgemuth, G., Scholz, M., Kind, T., Lee, D.Y., Lu, Y., et al. (2008). Quality control for plant metabolomics: reporting MSI-compliant studies. *Plant Journal, 53*, 691–704.

Flamini, R., de Rosso, M., de Marchi, F., Dalla Vedova, A., Panighel, A., Gardiman, M., et al. (2013). An innovative approach to grape metabolomics: stilbene profiling by suspect screening analysis. *Metabolomics, 9*, 1243–1253.

Food and Agriculture Organisation of the United Nations.2010). *The state of world fisheries and aquaculture.* Rome: FAO.

Gholami, M., Fakhari, A. R., & Ghanati, F. (2013). Selective Regulation of Nicotine and Polyamines Biosynthesis in Tobacco Cells by Enantiomers of Ornithine. *Chirality, 25*, 22–27.

Gigolashvili, T., Yatusevich, R., Rollwitz, I., Humphry, M., Gershenzon, J., & Flugge, U. I. (2009). The plastidic bile acid transporter 5 is required for the biosynthesis of methionine-derived glucosinolates in *Arabidopsis thaliana*. *Plant Cell, 21*, 1813–1829.

Gika, H. G., Theodoridis, G. A., Plumb, R. S., & Wilson, I. D. (2013). Current practice of liquid chromatography-mass spectrometry in metabolomics and metabonomics. *Journal of Pharmaceutical and Biomedical Analysis, 87*, 12–25.

Goulitquer, S., Potin, P., & Tonon, T. (2012). Mass spectrometry-based metabolomics to elucidate functions in marine organisms and ecosystems. *Marine Drugs, 10*, 849–880.

Gravot, A., Dittami, S. M., Rousvoal, S., Lugan, R., Eggert, A., Collen, J., et al. (2010). Diurnal oscillations of metabolite abundances and gene analysis provide new insights into central metabolic processes of the brown alga *Ectocarpus siliculosus*. *New Phytologist, 188*, 98–110.

Gupta, V., Thakur, R. S., Reddy, C. R. K., & Jha, B. (2013). Central metabolic processes of marine macrophytic algae revealed from NMR based metabolome analysis. *Rsc Advances, 3*, 7037–7047.

Hall, R. D. (2006). Plant metabolomics: from holistic hope, to hype, to hot topic. *New Phytologist, 169*, 453–468.

Heinrich, S., Valentin, K., Frickenhaus, S., John, U., & Wiencke, C. (2012). Transcriptomic analysis of acclimation to temperature and light stress in *Saccharina latissima* (Phaeophyceae). *PLos One*, 7.

Higashi, Y., & Saito, K. (2013). Network analysis for gene discovery in plant-specialized metabolism. *Plant Cell and Environment, 36*, 1597–1606.

Hill, C. B., & Roessner, U. (2013). Metabolic profiling of plants by GC-MS. In W. Weckwarth, & G. Kahl (Eds.), *The handbook of plant metabolomics.* Germany: Wiley-VCH Verlag GmbH & Co.

Hirai, M.Y., Klein, M., Fujikawa, Y., Yano, M., Goodenowe, D. B., Yamazaki, Y., et al. (2005). Elucidation of gene-to-gene and metabolite-to-gene networks in *Arabidopsis* by integration of metabolomics and transcriptomics. *Journal of Biological Chemistry, 280*, 25590–25595.

Hirai, M.Y., Sugiyama, K., Sawada, Y., Tohge, T., Obayashi, T., Suzuki, A., et al. (2007). Omics-based identification of *Arabidopsis* Myb transcription factors regulating aliphatic glucosinolate biosynthesis. *Proceedings of the National Academy of Sciences of the United States of America, 104*, 6478–6483.

Holdt, S. L., & Kraan, S. (2011). Bioactive compounds in seaweed: functional food applications and legislation. *Journal of Applied Phycology*, *23*, 543–597.

Holscher, D., Shroff, R., Knop, K., Gottschaldt, M., Crecelius, A., Schneider, B., et al. (2009). Matrix-free UV-laser desorption/ionization (LDI) mass spectrometric imaging at the single-cell level: distribution of secondary metabolites of *Arabidopsis thaliana* and *Hypericum* species. *Plant Journal*, *60*, 907–918.

Ho, L., Tang, T., Hoefel, D., & Vigneswaran, B. (2012). Determination of rate constants and half-lives for the simultaneous biodegradation of several cyanobacterial metabolites in Australian source waters. *Water Research*, *46*, 5735–5746.

Jun, J. H., Song, Z. H., Liu, Z. J., Nikolau, B. J., Yeung, E. S., & Lee, Y. J. (2010). High-spatial and high-mass resolution imaging of surface metabolites of *Arabidopsis thaliana* by laser desorption-ionization mass spectrometry using colloidal silver. *Analytical Chemistry*, *82*, 3255–3265.

Kaspar, S., Peukert, M., Svatos, A., Matros, A., & Mock, H. P. (2011). MALDI-imaging mass spectrometry - an emerging technique in plant biology. *Proteomics*, *11*, 1840–1850.

Kerwin, R. E., Jimenez-Gomez, J. M., Fulop, D., Harmer, S. L., Maloof, J. N., & Kliebenstein, D. J. (2011). Network quantitative trait loci mapping of circadian clock outputs identifies metabolic pathway-to-clock linkages in *Arabidopsis*. *Plant Cell*, *23*, 471–485.

Kim, H. K., Choi, Y. H., Erkelens, C., Lefeber, A. W. M., & Verpoorte, R. (2005). Metabolic fingerprinting of *Ephedra* species using H-1-NMR spectroscopy and principal component analysis. *Chemical & Pharmaceutical Bulletin*, *53*, 105–109.

Kim, H. K., Choi, Y. H., & Verpoorte, R. (2010). NMR-based metabolomic analysis of plants. *Nature Protocols*, *5*, 536–549.

Kim, H. K., Choi, Y. H., & Verpoorte, R. (2011). NMR-based plant metabolomics: where do we stand, where do we go? *Trends in Biotechnology*, *29*, 267–275.

Kind, T., Wohlgemuth, G., Lee, D. Y., Lu, Y., Palazoglu, M., Shahbaz, S., et al. (2009). FiehnLib: mass spectral and retention Index libraries for metabolomics based on quadrupole and time-of-flight gas chromatography/mass spectrometry. *Analytical Chemistry*, *81*, 10038–10048.

Konotchick, T., Dupont, C. L., Valas, R. E., Badger, J. H., & Allen, A. E. (2013). Transcriptomic analysis of metabolic function in the giant kelp, Macrocystis pyrifera, across depth and season. *New Phytologist*, *198*, 398–407.

Kok, M. G. M., de Jong, G. J., & Somsen, G. W. (2011). Sensitivity enhancement in capillary electrophoresis-mass spectrometry of anionic metabolites using a triethylamine-containing background electrolyte and sheath liquid. *Electrophoresis*, *32*, 3016–3024.

Kopka, J., Schauer, N., Krueger, S., Birkemeyer, C., Usadel, B., Bergmuller, E., et al. (2005). GMD@CSB.DB: the Golm Metabolome Database. *Bioinformatics*, *21*, 1635–1638.

Kruger, N. J., Troncoso-Ponce, M. A., & Ratcliffe, R. G. (2008). H-1 NMR metabolite fingerprinting and metabolomic analysis of perchloric acid extracts from plant tissues. *Nature Protocols*, *3*, 1001–1012.

Kumari, P., Kumar, M., Reddy, C. R. K., & Jha, B. (2013a). Algal lipids, fatty acids and sterols. In H. Dominguez (Ed.), *Functional ingredients from algae for foods and nutraceuticals*. Cambridge, U.K: Woodhead Publishing Ltd.

Kumari, P., Bijo, A. J., Mantri, V. A., Reddy, C. R. K., & Jha, B. (2013b). Fatty acid profiling of tropical marine macroalgae: an analysis from chemotaxonomic and nutritional perspectives. *Phytochemistry*, *86*, 44–56.

Kundel, M., Thorenz, U. R., Petersen, J. H., Huang, R. J., Bings, N. H., & Hoffmann, T. (2012). Application of mass spectrometric techniques for the trace analysis of short-lived iodine-containing volatiles emitted by seaweed. *Analytical and Bioanalytical Chemistry*, *402*, 3345–3357.

Kusano, M., Tohge, T., Fukushima, A., Kobayashi, M., Hayashi, N., Otsuki, H., et al. (2011). Metabolomics reveals comprehensive reprogramming involving two independent metabolic responses of *Arabidopsis* to UV-B light. *Plant Journal*, *67*, 354–369.

La Barre, S., Potin, P., Leblanc, C., & Delage, L. (2010). The halogenated metabolism of brown algae (Phaeophyta), its biological importance and its environmental significance. *Marine Drugs*, *8*, 988–1010.
Lee, Y. J., Perdian, D. C., Song, Z. H., Yeung, E. S., & Nikolau, B. J. (2012). Use of mass spectrometry for imaging metabolites in plants. *Plant Journal*, *70*, 81–95.
Levitt, M. H. (2001). *Spin dynamics: Basics of nuclear magnetic resonance*. Chichester, England: John Wiley and Sons Ltd.
Lisec, J., Schauer, N., Kopka, J., Willmitzer, L., & Fernie, A. R. (2006). Gas chromatography mass spectrometry-based metabolite profiling in plants. *Nature Protocols*, *1*, 387–396.
Lisso, J., Steinhauser, D., Altmann, T., Kopka, J., & Mussig, C. (2005). Identification of brassinosteroid-related genes by means of transcript co-response analyses. *Nucleic Acids Research*, *33*, 2685–2696.
Liu, F., & Pang, S. J. (2010). Stress tolerance and antioxidant enzymatic activities in the metabolisms of the reactive oxygen species in two intertidal red algae *Grateloupia turuturu* and *Palmaria palmata*. *Journal of Experimental Marine Biology and Ecology*, *382*, 82–87.
Liu, D., Ford, K. L., Roessner, U., Natera, S., Cassin, A. M., Patterson, J. H., & Bacic, A. (2013). Rice suspension cultured cells are evaluated as a model system to study salt responsive networks in plants using a combined proteomic and metabolomic profiling approach. *Proteomics*, *13*, 2046–2062.
Lu, F. C., & Ralph, J. (1997). Derivatization followed by reductive cleavage (DFRC method), a new method for lignin analysis: protocol for analysis of DFRC monomers. *Journal of Agricultural and Food Chemistry*, *45*, 2590–2592.
Monton, M. R. N., & Soga, T. (2007). Metabolome analysis by capillary electrophoresis-mass spectrometry. *Journal of Chromatography A*, *1168*, 237–246.
Moussaieff, A., Rogachev, I., Brodsky, L., Malitsky, S., Toal, T. W., Belcher, H., et al. (2013). High-resolution metabolic mapping of cell types in plant roots. *Proceedings of the National Academy of Sciences of the United States of America*, *110*, E1232–E1241.
Muranaka, T., & Saito, K. (2013). Phytochemical genomics on the way. *Plant and Cell Physiology*, *54*, 645–646.
Nicholson, J. K., & Lindon, J. C. (2008). Systems biology - Metabonomics. *Nature*, *455*, 1054–1056.
Northen, T. R., Yanes, O., Northen, M. T., Marrinucci, D., Uritboonthai, W., Apon, J., et al. (2007). Clathrate nanostructures for mass spectrometry. *Nature*, *449*. 1033–U3.
Oikawa, A., & Saito, K. (2012). Metabolite analyses of single cells. *Plant Journal*, *70*, 30–38.
Pearson, G. A., Hoarau, G., Lago-Leston, A., Coyer, J. A., Kube, M., Reinhardt, R., et al. (2010). An expressed sequence tag analysis of the intertidal brown seaweeds *Fucus serratus* (L.) and *F. vesiculosus* (L.) (Heterokontophyta, Phaeophyceae) in response to abiotic stressors. *Marine Biotechnology*, *12*, 195–213.
Pereira, G. E., Gaudillere, J. P., Van Leeuwen, C., Hilbert, G., Lavialle, O., Maucourt, M., et al. (2005). H-1 NMR and chemometrics to characterize mature grape berries in four wine-growing areas in Bordeaux, France. *Journal of Agricultural and Food Chemistry*, *53*, 6382–6389.
Putri, S. P., Nakayama, Y., Matsuda, F., Uchikata, T., Kobayashi, S., Matsubara, A., et al. (2013). Current metabolomics: practical applications. *Journal of Bioscience and Bioengineering*, *115*, 579–589.
Quanbeck, S. M., Brachova, L., Campbell, A. A., Guan, X., Perera, A. He, K., et al. (2012). Metabolomics as a hypothesis-generating functional genomic tool for the annotation of *Arabidopsis thaliana* genes of "unknown function". *Frontiers in Plant Science*, *3*, 15.
Ramautar, R., Demirci, A., & de Jong, G. J. (2006). Capillary electrophoresis in metabolomics. *Trac Trends in Analytical Chemistry*, *25*, 455–466.
Ramautar, R., Busnal, J. M., Deelder, A. M., & Mayboroda, O. A. (2012). Enhancing the coverage of the urinary metabolome by sheathless capillary electrophoresis-mass spectrometry. *Analytical Chemistry*, *84*, 885–892.

Ramautar, R., Somsen, G. W., & de Jong, G. J. (2013). CE-MS for metabolomics: developments and applications in the period 2010-2012. *Electrophoresis*, *34*, 86–98.

Roessner, U., & Bacic, A. (2009). Metabolomics in plant research. *Australian Biochemist, 40*.

Ross, C., & Van Alstyne, K. L. (2007). Intraspecific variation in stress-induced hydrogen peroxide scavenging by the ulvoid macroalga *Ulva lactuca*. *Journal of Phycology*, *43*, 466–474.

Ryan, D., & Robards, K. (2006). Metabolomics: the greatest omics of them all? *Analytical Chemistry*, *78*, 7954–7958.

Saito, K., Hirai, M.Y., & Yonekura-Sakakibara, K. (2008). Decoding genes with coexpression networks and metabolomics - 'majority report by precogs'. *Trends in Plant Science*, *13*, 36–43.

Saito, K., & Matsuda, F. (2010). Metabolomics for functional genomics, systems biology, and biotechnology. *Annual Review of Plant Biology*, *61*(61), 463–489.

Sawada, Y., Toyooka, K., Kuwahara, A., Sakata, A., Nagano, M., Saito, K., et al. (2009). *Arabidopsis* bile acid: sodium symporter family protein 5 is involved in methionine-derived glucosinolate biosynthesis. *Plant and Cell Physiology*, *50*, 1579–1586.

Schilmiller, A. L., Pichersky, E., & Last, R. L. (2012). Taming the hydra of specialized metabolism: how systems biology and comparative approaches are revolutionizing plant biochemistry. *Current Opinion in Plant Biology*, *15*, 338–344.

Schilmiller, A. L., Schauvinhold, I., Larson, M., Xu, R., Charbonneau, A. L., Schmidt, A., et al. (2009). Monoterpenes in the glandular trichomes of tomato are synthesized from a neryl diphosphate precursor rather than geranyl diphosphate. *Proceedings of the National Academy of Sciences of the United States of America*, *106*, 10865–10870.

Seaver, S. M. D., Henry, C. S., & Hanson, A. D. (2012). Frontiers in metabolic reconstruction and modeling of plant genomes. *Journal of Experimental Botany*, *63*, 2247–2258.

Seki, H., Ohyama, K., Sawai, S., Mizutani, M., Ohnishi, T., Sudo, H., et al. (2008). Licorice beta-amyrin 11-oxidase, a cytochrome P450 with a key role in the biosynthesis of the triterpene sweetener glycyrrhizin. *Proceedings of the National Academy of Sciences of the United States of America*, *105*, 14204–14209.

Seki, H., Sawai, S., Ohyama, K., Mizutani, M., Ohnishi, T., Sudo, H., et al. (2011). Triterpene functional genomics in Licorice for identification of CYP72A154 involved in the biosynthesis of glycyrrhizin. *Plant Cell*, *23*, 4112–4123.

Soga, T., Igarashi, K., Ito, C., Mizobuchi, K., Zimmermann, H. P., & Tomita, M. (2009). Metabolomic profiling of anionic metabolites by capillary electrophoresis mass spectrometry. *Analytical Chemistry*, *81*, 6165–6174.

Son, H. S., Hwang, G. S., Kim, K. M., Ahn, H. J., Park, W. M., Van Den Berg, F., et al. (2009). Metabolomic studies on geographical grapes and their Wines using H-1 NMR analysis coupled with multivariate statistics. *Journal of Agricultural and Food Chemistry*, *57*, 1481–1490.

Stoeckli, M., Chaurand, P., Hallahan, D. E., & Caprioli, R. M. (2001). Imaging mass spectrometry: a new technology for the analysis of protein expression in mammalian tissues. *Nature Medicine*, 7, 493–496.

Sumner, L. W., Amberg, A., Barrett, D., Beale, M. H., Beger, R., Daykin, C. A., et al. (2007). Proposed minimum reporting standards for chemical analysis. *Metabolomics*, *3*, 211–221.

Takos, A. M., Knudsen, C., Lai, D., Kannangara, R., Mikkelsen, L., Motawia, M. S., et al. (2011). Genomic clustering of cyanogenic glucoside biosynthetic genes aids their identification in *Lotus japonicus* and suggests the repeated evolution of this chemical defence pathway. *Plant Journal*, *68*, 273–286.

Takos, A. M., & Rook, F. (2012). Why biosynthetic genes for chemical defense compounds cluster. *Trends in Plant Science*, *17*, 383–388.

Tautenhahn, R., Bottcher, C., & Neumann, S. (2008). Highly sensitive feature detection for high resolution LC/MS. *BMC Bioinformatics*, *9*.

Tohge, T., & Fernie, A. R. (2010). Combining genetic diversity, informatics and metabolomics to facilitate annotation of plant gene function. *Nature Protocols*, *5*, 1210–1227.

Tohge, T., Nishiyama, Y., Hirai, M. Y., Yano, M., Nakajima, J., Awazuhara, M., et al. (2005). Functional genomics by integrated analysis of metabolome and transcriptome of *Arabidopsis* plants over-expressing an MYB transcription factor. *Plant Journal, 42*, 218–235.

Urano, K., Maruyama, K., Ogata, Y., Morishita, Y., Takeda, M., Sakurai, N., et al. (2009). Characterization of the ABA-regulated global responses to dehydration in *Arabidopsis* by metabolomics. *Plant Journal, 57*, 1065–1078.

Ward, J. L., Baker, J. M., Miller, S. J., Deborde, C., Maucourt, M., Biais, B., et al. (2010). An inter-laboratory comparison demonstrates that H-1-NMR metabolite fingerprinting is a robust technique for collaborative plant metabolomic data collection. *Metabolomics, 6*, 263–273.

Ward, J. L., Harris, C., Lewis, J., & Beale, M. H. (2003). Assessment of (1)H NMR spectroscopy and multivariate analysis as a technique for metabolite fingerprinting of *Arabidopsis thaliana*. *Phytochemistry, 62*, 949–957.

Wille, A., Zimmermann, P., Vranova, E., Furholz, A., Laule, O., Bleuler, S., et al. (2004). Sparse graphical Gaussian modeling of the isoprenoid gene network in *Arabidopsis thaliana*. *Genome Biology, 5*.

Woo, H. K., Northen, T. R., Yanes, O., & Siuzdak, G. (2008). Nanostructure-initiator mass spectrometry: a protocol for preparing and applying NIMS surfaces for high-sensitivity mass analysis. *Nature Protocols, 3*, 1341–1349.

Xu, K., Ismail, A. M., & Ronald, P. (2014). Flood tolerance mediated by the rice Sub1A transcription factor. In A. Jenks, & P. M. Hasegawa (Eds.), *Plant abiotic stress*. John Wiley & Sons.

Ye, H., Gemperline, E., Venkateshwaran, M., Chen, R. B., Delaux, P. M., Howes-Podoll, M., et al. (2013). MALDI mass spectrometry-assisted molecular imaging of metabolites during nitrogen fixation in the *Medicago truncatula-Sinorhizobium meliloti* symbiosis. *Plant Journal, 75*, 130–145.

Yonekura-Sakakibara, K., Fukushima, A., & Saito, K. (2013). Transcriptome data modeling for targeted plant metabolic engineering. *Current Opinion in Biotechnology, 24*, 285–290.

Yonekura-Sakakibara, K., Tohge, T., Matsuda, F., Nakabayashi, R., Takayama, H., Niida, R., et al. (2008). Comprehensive flavonol profiling and transcriptome coexpression analysis leading to decoding gene-metabolite correlations in *Arabidopsis*. *Plant Cell, 20*, 2160–2176.

Yonekura-Sakakibara, K., Tohge, T., Niida, R., & Saito, K. (2007). Identification of a flavonol 7-O-rhamnosyltransferase gene determining flavonoid pattern in *Arabidopsis* by transcriptome coexpression analysis and reverse genetics. *Journal of Biological Chemistry, 282*, 14932–14941.

Yuan, Y. V., Westcott, N. D., Hu, C., & Kitts, D. D. (2009). Mycosporine-like amino acid composition of the edible red alga, *Palmaria palmata* (dulse) harvested from the west and east coasts of Grand Manan Island, New Brunswick. *Food Chemistry, 112*, 321–328.

Yu, T. W., & Peng, H. S. (2010). Quantification and deconvolution of asymmetric LC-MS peaks using the bi-Gaussian mixture model and statistical model selection. *BMC Bioinformatics, 11*.

CHAPTER THREE

Chondrus crispus – A Present and Historical Model Organism for Red Seaweeds

Jonas Collén*,†,1, M. Lynn Cornish‡, James Craigie§, Elizabeth Ficko-Blean*,†, Cécile Hervé*,†, Stacy A. Krueger-Hadfield¶, Catherine Leblanc*,†, Gurvan Michel*,†, Philippe Potin#, Thierry Tonon*,† and Catherine Boyen*,†

*Sorbonne Universités, UPMC Univ Paris 06, UMR 8227, Integrative Biology of Marine Models, Station Biologique de Roscoff, Roscoff, France

†CNRS, UMR 8227, Integrative Biology of Marine Models, Station Biologique de Roscoff, Roscoff, France

‡Acadian Seaplants Limited, Dartmouth, NS, Canada

§Aquatic and Crop Resource Development, National Research Council Canada, Halifax, NS, Canada

¶College of Charleston, Grice Marine Laboratory, Charleston, SC, USA

#Centre National de la Recherche Scientifique, Sorbonne Universités UPMC Univ. Paris 06, Roscoff cedex, Brittany, France

[1]Corresponding author: e-mail address: jonas.collen@sb-roscoff.fr

Contents

Abstract

Chondrus crispus, or Irish moss, is a common edible red seaweed that can be found on rocky shores in the Northern Atlantic. The cell wall contains carrageenan and *C. crispus* is the original source of this commercially used thickener. Because of the ecological and economic importance of this red alga a relatively important research literature exists and one of the recent achievements in *C. crispus* research is the sequencing of its genome. In this chapter we review some of the literature with the aim to promote

Advances in Botanical Research, Volume 71
ISSN 0065-2296
http://dx.doi.org/10.1016/B978-0-12-408062-1.00003-2

C. crispus as a model organism for florideophyte red seaweeds. We consider subjects like commercial and historical uses, ecology, genetics, population structure, mating systems, physiology, cell wall biology and genomics.

3.1 INTRODUCTION

The red algae are eukaryotic organisms without flagella and centrioles with floridean starch as their energy reserve, using chlorophyll *a* and with phycobiliproteins as accessory pigments, giving them their distinctive colour. There are around 6000 species of red algae described (Guiry, 2013) and most are multicellular, macroscopic, marine and have sexual reproduction. The red algae are one of the three groups forming the Archaeplastida, together with the glaucophytes and the green lineage, and are thought to have emerged as an independent lineage ~1500 million years ago (Yoon, Hackett, Ciniglia, Pinto, & Bhattacharya, 2004; Yoon, Müller, Sheath, Ott, & Bhattacharya, 2006).

Red seaweeds are important components of the inter- and subtidal flora and are commonly found where suitable substrates are available from the polar areas to the tropics and even though most species are marine, they also exist in brackish and freshwater (Woelkerling, 1990). Red algae are also economically important; over 9 million tonnes are cultivated annually (FAO, 2012), with a value of over 2000 million USD. Despite the importance of red algae, the number of researchers that dedicate themselves to their study is still deficient. We believe that one of the reasons for this lack of visibility of this interesting group of organisms is the absence of a model species.

Chondrus crispus is a red seaweed, also called Irish moss, belonging to the Florideophytes, a group of multicellular red algae comprising of 95% of extant species and most of the species with an ecological importance (Woelkerling, 1990). It exhibits dichotomous branching and can be up to 15 cm long (Figure 3.1). The colour varies from pale yellow, through green purplish red to almost black; in the gametophyte a blue iridescence can often be found. The morphology is very variable and numerous forms have been described (Chopin et al., 1996). It commonly occurs on rocky shores and other hard substrata, inter- and subtidally, in the North Atlantic (Provan & Maggs, 2012) and is a source of carrageenan. Because of the ecologic and economic interest in this species an important scientific literature exists compared to other seaweeds. In this chapter, we try to review some of the literature and promote its use as a model for florideophyte red algae. Some arguments for using *C. crispus* as a model are shown in Table 3.1.

Figure 3.1 *Chondrus crispus* Stackhouse Gigartinales and the family Gigartinaceae. *Photo by Jonas Collén.*

Table 3.1 Principal Arguments for Choosing ***Chondrus crispus*** as a Representative of the Red Macroalgae and of the Florideophytes

- Typical red seaweed – allows for generalisations to other red algae
- Well-defined species – simplifying ecological and physiological studies
- Complex multicellularity – allows for comparisons and basic studies
- Large morphological plasticity – allows for studies on correlation between morphology, environmental factors and genotype
- Well studied red alga – background knowledge available
- Common, ecologically important species – increases the importance of the research
- Relatively small genome (105 Mbp) – facilitates sequencing and genomic studies
- Easily cultivated in the laboratory – gives the possibility of controlled studies
- Three different life cycle phases, accessible in the laboratory and in the field – allows for studies on effects of ploidy
- Cell wall composed of carrageenan, with different types being produced in different life cycle stages
- Related to economically important species – can give insight into the biology of, for example, *Eucheuma* and *Kappaphycus* species
- Of commercial interest – increases the interest in applied research

3.2 HISTORY

Chondrus crispus or Irish moss has been collected and utilised for human activities for hundreds of years (Allen & Hatfield, 2004; Mouritsen, 2013; Thorbjarnarson, 1939), but it was only in the late eighteenth century that the alga was assigned a scientific name. It initially appeared as *Fucus filiformis* Hudson 1762 as well as *Fucus crispus* Linnaeus 1767 (Guiry, 2013; Taylor & Chen, 1973). Stackhouse, in 1797, determined that this taxon should be removed from the overly broad genus *Fucus* and entered Irish moss under the binomial *C. crispus* (Papenfuss, 1950). The species has long been recognised as being exceedingly polymorphic (Chopin et al., 1996), a fact that has resulted in a complicated synonomy (Newton, Devonald, & Jones, 1957). As early as 1797, Goodenough and Woodward, in their description of species for the Linnean Society added the comment regarding *F. crispus* that 'No plant can be supposed to vary more than this,' and to corroborate, Lamouroux in 1813 listed 8 varieties of the taxon, with 15 still undecided (Taylor & Chen, 1973).

Not surprisingly, the earliest references regarding the utilisation of *C. crispus* come from people living along the north eastern coastline of the Atlantic Ocean, where the alga has been collected for centuries as food and medicine (Allen & Hatfield, 2004). The properties of its major cell wall polysaccharides were noticed early by these people, and this characteristic ultimately resulted in the present day industrial utilisation of Irish moss. The first formal recognition of the gelling properties of boiled *F. crispus* were discovered by Turner (1809), and this mucilaginous matter was named carrageenin by Pereira (1840), p. 564. The gelatinous, hot water-soluble mucilage of *C. crispus* was first isolated by Schmidt (1844). Coincidentally, in the same year, Forchhammer reported on the high sulphur content of the ash from *C. crispus* (Buggeln & Craigie, 1973).

Numerous common or vernacular names for *C. crispus* appear in many languages, i.e. Breton, English, French, Gaelic, German, Icelandic, Portuguese, Scandinavian, Spanish, Japanese and others (Chopin, 1986). This fact further attests to the longstanding and widespread use of the alga in human activities. Names such as carageen, carrageen, carrragheen, carraigin, carrageen moss, carrageen rock moss and Irish moss, etc. refer to the alga *C. crispus*. The etymology of carrageen remains uncertain although the word can be dated to 1829 when it may have been introduced as a commercial marketing term; its derivation from a town place in County Waterford, Ireland has been rejected (Bliss, 1985; Mitchell & Guiry, 1983).

The domestic use of Irish moss was documented at least as early as 1809 (Bliss, 1985; Mitchell & Guiry, 1983). An 1829 reference to 'carrageen' or Irish moss alluded to its putative therapeutic properties (Mitchell & Guiry, 1983), and a traditional treatment in Ireland for chest and lung ailments used Irish moss boiled in water or milk, strained and drunk hot (Allen & Hatfield, 2004). The alga also was used to treat kidney ailments and burns. These medicinal applications were so valued that a recipe for preparing a demulcent from *C. crispus* 'for diseases of debility' was included in an early *Materia Medica* (Frazer, 1864). The relaxing effects of the extracted carrageenans on mucous membranes provided relief in tuberculosis, whooping cough, pneumonia, quinsy and for various gastrointestinal complaints. In his report regarding the potential for trade and development of new materials for commerce in Europe, carrageen or Irish rock moss was appraised as a feasible industrial commodity (Simmonds, 1854).

Irish immigrants apparently carried the knowledge of the folk uses of Irish moss to the Americans very early in the nineteenth century, many of who settled in Boston, Massachusetts. Their requirements for carrageen were met by importing the dried seaweed, which in the early 1830s sold for US$1–$2 per pound (Humm, 1951; Smith, 1905). Dr J.V. C. Smith, a former mayor of Boston, pointed out to citizens, in 1835, that the rocky Massachusetts coastline supported an abundance of Irish moss. Irish fishers, Daniel Ward and Miles O'Brien of Scituate, Massachusetts, appear to have initiated the first commercial harvesting of seaweed in the USA between 1848 and 1850 (www.stmaryscituate.org/aboutus_history.html).

Events relating to World War II provided a catalyst for the rejuvenation of the Irish moss industry. In particular, interdiction of the Asian sources of agar needed for bacteriology and penicillin production focussed attention on alternative domestic resources in both Europe and America. Relatively low biomass of agarophytes along North Atlantic coastlines meant carrageenophytes must be evaluated (Marshall, Newton, & Orr, 1949, 184 pp.). A process was developed and an agar substitute from Irish moss, termed 'British agar', was produced. It turned out to be quite suitable for bacteriological purposes, being superior in clarity to conventional agar.

By 1952, Irish moss extractives were second in volume of production (~1.75 million lbs/annum) to alginates in the USA, with food, drug and cosmetic and industrial applications consuming 50%, 40% and 10% of the moss extract respectively (Jertson, 1952). The first carload of Irish moss shipped from Canada was from Nova Scotia (10,000 lb in 1940) with a

Newfoundland harvest of 28,795 lb in 1941. Exports from Prince Edward Island began in 1941with the shipping of 208,000 lb of dry moss (Humm, 1951). Irish moss harvesting in the Canadian Maritimes increased from 1.49 million lbs in 1943 to 87.05 million lbs in 1968 with the most rapid increase occurring after 1965 (Snaith, MacFarlane, & Johnston, 1969).

The annual standing crop of *C. crispus* is often variable and the labour associated with hand harvesting is arduous, and eventually commercial harvesting in Europe gave way to smaller, localised enterprises (Guiry & Hession, 1998). In North America, drag-rake harvesting became common practice by the mid-1950s, but the dramatic increase in Irish moss landings in the Canadian Maritimes during the late 1960s (Ffrench, 1971; Pringle & Mathieson, 1987; Snaith et al., 1969) was considered unsustainable. Implementing modern farming technologies was then suggested as a means of supplementing the natural harvest (Neish, 1968; Neish & Fox, 1971). In 1972, the first pilot-scale cultivation trials were conducted in Nova Scotia at Meteghan and in New Brunswick at Point Sapin. Two commercial firms, Genu Products Canada Limited and Marine Colloids Inc., of Rockland, Maine (Craigie & Shacklock, 1989; Mathieson, 1982) were involved. Early stage *C. crispus* cultivation facilities appeared in France in 1976, and together, these clearly demonstrated proof of concept for growing Irish moss in tanks on land (Braud, 2006; Craigie, 1990; Craigie & Shacklock, 1989; Mathieson, 1982). One, Acadian Seaplants Limited, Charlesville, is in commercial operation today.

Cultivated *Eucheuma* spp. from the Philippines began to enter the carrageenophyte market in significant quantities in the mid-1970s (Ricohermoso & Deveau, 1979) resulting in a dramatic reduction in demand for the more expensive Canadian Irish moss, which, until 1975, had supplied 75% of the world's raw material for carrageenan production (Pringle & Mathieson, 1987). The worldwide shortage of carrageenophytes in the 1970s led to similar collaborations between science and industry in France and focus on cultivating *C. crispus* in tanks intensified (Braud, 2006; Braud & Delépine, 1981). Cultivation of Irish moss in the sea also was considered independently in Canada and in France by SATIA, but the cost of such farmed seaweed precluded further development (Briand, 1991; Chopin, Sharp, Belyea, Semple, & Jones, 1999; DeRoeck-Holtzhauer, 1991). Although cultivated Irish moss was now uncompetitive as a carrageenan source, the Marine Colloids Inc. technology and infrastructure for cultivating this species was acquired in 1981 by Acadian Seaplants Ltd and by the early 1990s Irish moss was used directly as a food.

3.3 ECOLOGY

Chondrus crispus is considered as a cold water and euryhaline species that inhabits the intertidal zone and is usually most abundant at 4–7 m below mean low water (MLW) but may extend from−18 m to more than +1 m and has been dredged from depths of −38 m MLW (Mathieson & Burns, 1975). The lower limits of vertical distribution are probably determined by a variety of factors such as wave action, water transparency, availability of solid substrata and competition for space. *Chondrus crispus* can be found growing on rocks in lower intertidal and shallow subtidal zones or in pools in the mid-intertidal zone in exposed locations. The most extensive populations occur on massive horizontal boulders on semi-exposed and open coastal sites, with reduced populations on smaller rocks and sand or sediment-covered rocks. Occasionally, *C. crispus* occur as detached healthy floating populations in the sublittoral zone (Prince & Kingsbury 1973a). The extensive distribution in estuarine habitats suggests a broad tolerance to temperature, salinity and light (Mathieson & Burns, 1971; Mathieson & Prince, 1973). On European coasts only a few plants occur as fixed forms at more than 20 m depth, while on the Canadian coasts, there are extensive settlements at this depth, indicating some differences in the potential of the species in these two regions (Kopp & Perez, 1978).

Chondrus crispus can be found in various forms that range from sporadic patches of thalli arising from a single holdfast to areas where the substratum is covered with a dense, uniform canopy (with numerous stipes per cm^2) on intertidal and shallow subtidal rocky shores (Johnson, 2001). In comparison with solitary thalli, dense canopies are known to offer protection against flow-induced mechanical forces (Johnson, 2001) and thus increasing productivity and growth. It can also form dense patches beneath *Fucus* or beneath canopy-forming kelps such as *Laminaria digitata* (Schaal, Riera, & Leroux, 2010).

Chondrus crispus is widely distributed in the north western and north eastern Atlantic and adjacent waters, such as the North Sea. In the western Atlantic *C. crispus* is found from Labrador to New Jersey and in the Eastern Atlantic from approximately 69° N in Norway and on the southern coasts of Iceland to Portugal (MacFarlane, 1968; Provan & Maggs, 2012). Some sources mention records of *C. crispus* from California to Japan, however, any distribution outside the Northern Atlantic needs to be verified and probably results from confusion with other species of the same genus in the Pacific Ocean (Hu, Guiry, Critchley, & Duan, 2010).

In the Atlantic only one species of *Chondrus* is known – *C. crispus*, but other species exist in the Pacific Ocean: *Chondrus ocellatus* Holmes, *Chondrus giganteus* Yendo, *Chondrus nipponicus* Yendo, *Chondrus yendoi* Yamada & Mikami, *Chondrus pinnulatus* (Harvey) Okamura, *Chondrus armatus* (Harvey) Okamura, *Chondrus verrucosus* Mikami and *Chondrus retortus* Matsumoto & Shimada. *Chondrus giganteus* can also be found in Mediterranean as an introduced species (Hu et al., 2007; Matsumoto & Shimada, 2013).

The current geographical distribution of *C. crispus* is primarily determined by temperatures and consequently it may be impacted by global change and increasing water temperatures. For instance, Lima, Ribeiro, Queiroz, Hawkins, and Santos (2007) reported a northward shift of *C. crispus* from the southern limit of Portuguese populations of 180 km since 1971. In addition, a study using temperature-controlled water baths, Lüning, Guiry, and Masuda (1986) showed that most European strains of *C. crispus* ranging from northern Norway and Iceland to northern Spain survived a uniform upper temperature limit of 28 °C with a few European strains from the North Sea and English Channel area being able to sustain a 29 °C limit. Temperature has also a drastic effect on the reproduction of the species, Prince and Kingsbury (1973b) showed that the optimum temperature for germination and growth of spores (carpospores and tetraspores) was 21 °C. At temperatures above this threshold, spores died or developed abnormally after a few days of exposure. Above 30 °C even brief shock treatments result in spore mortality. At the other extreme, cultures maintained at 4 °C displayed inhibition of germination and growth.

Natural populations of *C. crispus* can survive freezing temperatures, but repeated freezing exposures can significantly limit productivity and biomass and thus influence the competitive ability of the species. Dudgeon, Davison, and Vadas (1990) showed that chronic freezing temperatures combined with high irradiance induced bleaching and fragmentation of fronds. Holdfasts survived the exposure and were capable of regenerating. The authors postulated that *C. crispus* combats freezing stress by phenotypic acclimation through maintenance of photosynthesis rather than by genetic adaptation.

3.4 LIFE CYCLE

The sexual life history of *C. crispus* was described by Darbishire in 1902. Nonmotile, short-lived male gametes (spermatia) are released and fertilise an egg (carpogonium) retained on the female gametophyte. The zygote is mitotically amplified within the cystocarp, a structure also

Figure 3.2 The life cycle of *Chondrus crispus*. A nonmotile spermatium fertilises the carpogonium which is retained on the female gametophytic thallus (*syngamy*). Within the cystocarp, the zygote is mitotically amplified potentially liberating thousands of genetically identical diploid carpospores following a single fertilisation event. The diploid carpospores produce the diploid free-living tetrasporophyte. *Meiosis* occurs in the tetrasporophyte, releasing haploid tetraspores which form the free-living female and male gametophytes. (See the colour plate.) *Images: S.A. Krueger-Hadfield.*

retained on the female thallus. The resulting diploid carpospores are an additional dispersive stage in comparison to other algal life cycles and germinate into the free-living tetrasporophyte. The tetrasporophyte produces haploid tetraspores via meiosis, which germinate into gametophytes (Figure 3.2).

Chen and McLachlan (1972) observed an increase in the number of cystocarps produced when male and female gametophytes obtained from cultured tetraspores were crossed in aerated culture chambers. Moreover, cystocarps failed to develop in the absence of male gametophytes, providing presumptive evidence of fertilisation (Chen & McLachlan, 1972). Krueger-Hadfield, Roze, Destombe, Correa, and Valero (2014) were the first to demonstrate cross fertilisation using genetic markers in the field and in the laboratory.

Male gametophytic fronds tend to be narrower, bearing spermatangial sori near the apices, which appear as pink or white bands (Darbishire, 1902;

Taylor & Chen, 1973; Tveter-Gallgher, Mathieson, & Cheney, 1980). The white bands on reproductive males are distinct and located 3–10 mm below the apex, whereas, actively growing apices are pale tan or white only at the apex (Krueger-Hadfield, 2011). After spermatial discharge, the thalli disintegrate at the point immediately below the sori (Chen & McLachlan, 1972).

Female gametophytic fronds are slightly wider at the apices and, when fertile, bear distinctive cystocarpic tissue (Chen & McLachlan, 1972). Fredericq, Brodie, and Hommersand (1992) found female gametophytes bearing cystocarps, which contained small patches of elongate surface cells near the apex which appeared to function as spermatangial initial cells. However, paternity analyses in natural populations of *C. crispus* revealed the presence of male alleles in cystocarpic DNA, with the exception of 11 cystocarps (<2%) that only exhibited one allele at all microsatellite loci (Krueger-Hadfield et al., 2014).

Tetrasporophytic fronds have broad apices similar to female gametophytes. When fertile, the apices are covered in ovate tetrasporangial sori, which appear reddish-orange in colour and bulge slightly from both surfaces of the frond (Taylor & Chen, 1973). Once the sori have released the tetraspores, the apical tissue appears a lighter colour than the remainder of the frond.

3.5 GENETICS, POPULATION STRUCTURE AND THE MATING SYSTEM

Mendelian genetic analyses were an excellent tool for studying inheritance patterns before molecular genetic tools were readily available. van der Meer (1987) reviewed the use of colour mutants to detect fertilisation, distinguish between self- and cross fertilisation and to distinguish between sexual and asexual processes. Pigmentation mutants of *C. crispus* were found to either follow strict maternal inheritance or classic Mendelian transmission ratios, for example, cell masses in old tetrasporangial sori found in culture were the product of an asexual process in somatic tetrasporangial tissue (van der Meer, 1987). None of the studies using molecular genetic markers (e.g. Cheney & Mathieson, 1979; Chopin et al., 1996; Donaldson, Chopin, & Saunders, 2000) were sufficiently powerful in order to explore the mating system and gene flow in detail. This is all the more surprising as the majority of populations are gametophyte biased, male gametophytes were thought to be rare (but see Krueger-Hadfield, Roze, Mauger, & Valero, 2013; Tveter-Gallagher et al., 1980) and genet identification is virtually impossible when holdfasts form contiguous mats on the available substrata and give rise to numerous upright fronds (Bhattacharya, 1985). Moreover, coalescence of neighbouring genets may occur in the field, resulting in chimeric holdfasts that consist of more than one genotype (Krueger-Hadfield, 2011; Santelices et al., 1999; Scrosati & Mudge, 2004; Tveter-Gallagher & Mathieson, 1980).

Using different markers, such as isozymes (Cheney & Mathieson, 1979), restriction fragment length polymorphism (Chopin et al., 1996), amplified fragment length polymorphism (Donaldson et al., 2000) and inter-simple sequence repeat (Wang et al., 2008), *C. crispus* has been found to be highly genetically variable over its range distribution. These molecular markers were analysed at large geographic scales establishing that *C. crispus* on both sides of the North Atlantic are conspecific (Chopin et al., 1996) as well as indicating limited dispersal capacities (Donaldson et al., 2000; Wang et al., 2008; Provan & Maggs, 2012). There were, however, discrepancies among the studies regarding the relationship between gene flow and geographic distance (Donaldson et al., 2000; Hu et al., 2010; Provan, Glendinning, Kelly, & Maggs, 2013; Wang et al., 2008).

Krueger-Hadfield, Collén, Daguin–Thiébaut, and Valero (2011) were the first study to directly assess genetic structure and the mating system in *C. crispus* using codominant microsatellite loci. The authors found very few repeated multilocus genotypes, no difference in allele frequencies between haploid and diploid phases and no linkage disequilibrium, suggesting sexual reproduction was the prevailing reproductive mode. Moreover, there was no detectable cystocarpic effect as the tetrasporophytes sampled had unique multilocus genotypes. However, it was not possible to distinguish between high levels of inbreeding or spatial substructuring in driving the significant pattern of heterozygote deficiency. In order to effectively address this issue, Krueger-Hadfield et al., (2013) explored hierarchical population structure at high and low shore populations and demonstrated heterozygote deficiencies were driven by high levels of inbreeding. Second, high and low shore individuals were found to belong to genetically differentiated populations, despite being separated by less than a few meters in vertical tidal height and less than 30 m in horizontal topographical distance. Third, gene flow was restricted within the high shore habitat due to daily tidal cycles where the high shore population could be exposed for twice the amount of time as the low shore population. Within-shore genetic differentiation reduced genetic diversity and increased levels of inbreeding indicated the high shore as a marginal environment. The results from Krueger-Hadfield et al. (2011, 2013) suggested the same mechanisms as those occurring over a species' distributional range may be at work at fine scales within the intertidal shorescape with core and marginal population dynamics.

Yet, indirect estimates of the mating system (i.e. F_{is}) can be strongly influenced by spatial substructuring or biparental inbreeding (i.e. mating between close relatives). Paternity analyses have been used to describe the mating system of animals and plants (e.g. Zipperle, Coyer, Reise, Stam, & Olsen, 2011), but have rarely been applied to red seaweeds (but see, Engel, Wattierf, Destombe, & Valero, 1999). Moreover, paternity analyses enable the detection of apomixis or intragametophytic selfing (i.e. the fusion of gametes produced by the same gametophyte, resulting in homozygosity at all loci, Klekowski, 1973), which would otherwise be difficult to detect by only analysing population structure. Krueger-Hadfield et al. (2014) documented higher levels of relatedness between males siring carpogonia on a single female gametophyte than when compared to males in the entire population. Moreover, the paternity analyses verified hypotheses of restricted spermatial and clumped spore dispersal generating high levels of inbreeding as proposed in Krueger-Hadfield et al. (2013).

In order to understand the impacts of the intertidal shorescape on gene flow and the mating system in *C. crispus*, the first hurdle is the identification of a genet. The majority of studies have utilised different sampling techniques in which the genet (holdfast including all upright fronds; *sensu* Kautsky & Ehn, 1993) or the ramet (frond; e.g. Bhattacharya, 1985) was studied. Krueger-Hadfield (2011) sampled each frond from two morphologically distinct gametophytic holdfasts grown in the laboratory from tetraspores. At least six different genotypes were detected among the two holdfasts, suggesting holdfasts in the field may be composed of more than one genotype. Moreover, in natural populations, holdfasts of mixed ploidy are common (Plumb, 1999; S. A. Krueger-Hadfield, pers. obs.). By understanding genet composition and distribution within the shore, it will be possible to determine spermatial and spore dispersal distance and describe gene flow in more detail.

3.6 ECOLOGICAL AND BIOCHEMICAL ASPECTS OF BIOTIC INTERACTIONS

Field observations of biotic interactions with smaller organisms are rather limited for *C. crispus* (Andrews, 1976; Krueger-Hadfield, 2011), but it can be heavily fouled by epibionts, such as ascidians and bryozoans (Wahl & Mark, 1999) and this is strongly dependent of interactions with snails

(Stachowicz & Whitlatch, 2005). No adelphoparasite or unpigmented red algal parasite is known in *C. crispus* (Goff, 1982). However, as early as 1889, the ascomycete *Lautitia danica* was described growing on the cystocarps of *C. crispus* by Rostrup (1889) in Denmark and later more detailed studies were provided (Jones, 1898; Rosenvinge, 1906). The fungus was present all year round in cystocarps but also in tetrasporangia (Stanley, 1992; Wilson & Knoyle, 1961). In natural populations, *C. crispus* is also plagued with a variety of green (*Acrochaete* spp., *Phaeophila dendroides*) and brown (*Streblonema* spp.) algal endophytes, as well as with the oomycete *Petersenia pollagaster* (reviewed in Correa, 1996).

Surprisingly, knowledge on *C. crispus*-associated bacteria is limited as a description of surface fouling (Sieburth & Tootle, 1981). This limited knowledge may be related to the fact that the species is known to shed outer layers of cell wall and cuticle glycoproteins to clean its surface from epibionts (Craigie, Correa, & Gordon, 1992; Correa & McLachlan, 1988).

A better knowledge during the 1980s and 1990s corresponded to the establishment of large-scale cultivation farms of *C. crispus* with increased frequency of pathologies (Craigie & Correa, 1996). These authors described the etiology of a number of infection diseases in cultivated *C. crispus* from Canada. Grazer impacts were also shown to be heavy in these land-based cultivation systems at early stages. Schacklock and Croft (1981) described important damage and differential effects caused by isopod, amphipod and gastropod grazers in maricultured *C. crispus*. The main frequent epiphytes in tank cultures of *C. crispus* are *Ulva* spp. and members of the brown algae Ectocarpales and, interestingly, it was proposed to introduce some amphipod grazers in culture tanks to keep the thallus clean of epiphytes (Schacklock & Doyle, 1983).

One of the most important diseases resulting in tissue degradation was reported to be caused by the oomycete *Petersenia pollagaster* (Craigie & Shacklock, 1989; Molina, Hughes, & Craigie, 1988). The parasite selectively destroys frond apices. But studies of biotic interactions culminated with the in-depth studies of endophytism by small filamentous green algae in natural and cultivated populations of *C. crispus* in Atlantic Canada (Correa & McLachlan, 1991, 1992, 1994; Correa, Nielsen, & Grund, 1988; Nielsen & McLachlan, 1986; Correa et al., 1987). Of particular interest is its association with *Acrochaete operculata*, as it displays a high degree of host-phase specificity (Correa & McLachlan, 1991). The filaments of the green alga are able to completely invade the medullary tissues of the sporophytic generation of the red alga, whereas they do not penetrate beyond the outer cell layers of the gametophytic plants (Correa & McLachlan, 1991, 1994). In consequence, it causes cellular damage only to the sporophytes of *C. crispus* (Rhodophyta) and leads to secondary bacterial infections by facultative pathogens from the *Cytophaga/Flavobacterium* group (Correa & McLachlan, 1992; Craigie & Correa, 1996). In addition, these sporophytes are more palatable for isopod grazers (Correa & McLachlan, 1992). This evidence suggested that host specificity in *A. operculata* is determined by cell wall composition of the hosts, likely the carrageenan fraction. In contrast, *Acrochaete heteroclada* was not host specific, infecting all offered hosts, including carrageenophytes and agarophytes (Correa & McLachlan, 1991). This hypothesis of host recognition through the perception of cell wall galactans was later tested and validated by Bouarab, Potin, Correa, and Kloareg (1999). The life cycle phases of *C. crispus* differ by the degree of sulphation of their cell wall carrageenans. The preincubation of *A. operculata* zoospores with κ-type oligocarrageenans, only present in the gametophyte cell walls, was shown to significantly decrease the penetration of *A. operculata* into *C. crispus* sporophytic tissues, whereas the elicitation by sporophyte-specific λ oligocarrageenans increases its virulence against *C. crispus* gametophytes (Bouarab et al., 1999).

In *C. crispus*, the investigations of metabolic regulations during this specific interaction have contributed to a better knowledge of the immunity responses of red algae. (For reviews Bouarab, Kloareg, Potin, & Correa, 2001; Bouarab, Potin, Weinberger, Correa, & Kloareg, 2001; Cosse, Leblanc, & Potin, 2008; Potin, 2008; Weinberger & Potin, 2010). *Chondrus crispus* gametophytes respond to cell-free extract of *A. operculata*, by the emission of reactive oxygen species essential in gametophyte resistance and through the activation of a NADPH oxidase homolog (Bouarab et al., 1999). This enzyme is likely to be encoded by the homolog of respiratory burst oxidase gp91*phox*, later characterised in *C. crispus* (Hervé, Tonon, Collén, Corre, & Boyen, 2006). The induced resistance of gametophytes was shown to be dependent of the downstream activation of signalling cascades involving compounds derived from the oxidative metabolism of polyunsaturated fatty acids (PUFAs) (Bouarab et al., 2004). It was the first demonstration that the oxylipin pathways are activated during immunity responses of red algae (Bouarab et al., 2004). In this context, the methyl ester of the plant stress hormone jasmonic acid was revealed to promote the liberation of both C20 and C18 PUFAs-derivative oxylipins, such as prostaglandins, and to increase transcription of defence-related genes (Gaquerel et al., 2007; Hervé et al., 2006). Contact of *A. operculata* with κ-oligocarrageenans also enhances secretion of L-asparagine, which in turn induces a release of H_2O_2

by *C. crispus* (Weinberger, Pohnert, Kloareg, & Potin, 2002). This reaction is apparently catalysed by an apoplastic L-amino acid oxidoreductase and could prevent settlement of *A. operculata* zoospores (Weinberger et al., 2005). The gametophytes also synthesise ultraviolet absorbing compounds around the sites of *A. operculata* zoospore penetration. This reaction suggests the involvement of the phenylpropanoid metabolism, as also shown by the activation of shikimate dehydrogenase and phenylalanine ammonialyase (Bouarab et al., 2004). The emission of volatile halocarbon compounds is another chemical defence strategy used by *C. crispus*, as shown in other red algae (for reviews Potin, Bouarab, Salaün, Pohnert, & Kloareg, 2002; Weinberger & Potin, 2010).

Biotic interactions are not solely negatively affecting *C. crispus*. Positive interactions were shown to occur in raceway tanks between the diatom *Odontella aurita* and *C. crispus* as well as the edible red alga *Palmaria palmata*. Allelopathic compounds or other exudates emitted by the two species favour diatom aquaculture and prevent the contamination of these cultures by other microalgal species that often bloom in pond cultivation (Braud, 1998).

3.7 PHOTOSYNTHESIS

Chondrus crispus has a relatively important research history concerning its photosynthesis due to its abundance and its suitability for measurements of physiological parameters. This literature dates back to at least 1885 (Rattray, 1886) where it was noted that photosynthesis of *C. crispus* was relatively low in November. Intertidal algae are exposed to varying light regimes with seasons, weather patterns, diurnal changes and tidal cycles and need to be able to acclimate or adapt in order to react to the patterns of immersion and emersion and associated changes in light, temperature, pH, CO_2-content and osmolarity. A number of studies have highlighted the importance of these environmental parameters on the photosynthesis of *C. crispus*.

Light and temperature are key factors regulating photosynthesis. Maximum photosynthesis of *C. crispus* is typically found around 200 μmol photons/m^2/s (Brechignac & Andre, 1984b; Johansson & Snoeijs, 2002; Mathieson & Norall, 1975) with photoinhibition occurring at higher light intensities (Mathieson & Burns, 1971). Algae collected subtidally had maximum photosynthetic rate at lower light intensities and tetrasporophytes exhibited higher photosynthetic rates than female gametophytes (Mathieson & Norall, 1975). Optimum temperatures for photosynthesis have been reported between 20 and 25 °C (Kübler & Davison, 1993; Mathieson & Burns, 1971). This range is higher than is usually indicated as optimal for growth, 6–18 °C (Bidwell, McLachlan, & Lloyd, 1985). The efficiency of photosynthesis has been shown to decrease at noon and increase in the late afternoon (Sagert, Forster, Feuerpfeil, & Schubert, 1997).

One aspect of light and photosynthesis that has attracted much attention is the effect of UV-radiation. There is important literature on effects on thalli (e.g. Bischof et al., 2007; Kräbs & Wiencke, 2005; Roleda, Wiencke, Hanelt, & Bischof, 2007; and references therein). In intertidal algae living in northern temperate and polar regions, freezing is an extreme form of temperature stress and affects *C. crispus* in the northern part of its range. Twelve hours of freezing at −20 °C caused a decrease in photosynthesis by 58% and a decrease in respiration by 32% with a concomitant efflux of amino acids, indicating membrane damage (Davison, Dudgeon, & Ruan, 1989). *Chondrus crispus* showed full recovery of photosynthesis, 48 h after 3 h exposure to −20 °C, whereas, after 6 h at −20 °C, full recovery was not seen (Dudgeon, Davison, & Vadas, 1989). The freezing causes tissue freezing (Davison et al., 1989; Kanwisher, 1957). The effects of freezing can be reduced by acclimation; daily freezing at −5 °C reduced the photosynthesis initially with 75%, but after acclimation to subsequent freezing, photosynthesis decreased only by 40% (Dudgeon et al., 1990). The acclimation to freezing also reduced the release of amino acids after exposure.

In order to achieve efficient photosynthesis, internal concentrations of CO_2 need to be sufficient to reduce photorespiration. Carbon concentration mechanisms have been demonstrated in many seaweeds (Raven & Hurd, 2012) including *C. crispus* (Bréchignac et al., 1985a, Brechignac, Andre, & Gerbaud, 1986). Thallus and protoplasts of *C. crispus* use extracellular and intracellular carbonic anhydrase to increase the availability of CO_2 for Rubisco (Smith & Bidwell, 1987, 1989a) while the direct use of HCO_3^- remains unclear. Brechignac et al. (1986) suggested it to be important, while Smith & Bidwell (1989b) showed different results. *Chondrus crispus* does not seem to be limited by content of inorganic carbon since increased growth with increased CO_2 was only seen at high temperatures (Sarker, Bartsch, Olischläger, Gutow, & Wiencke, 2013). It should also be noted that *C. crispus* lacks pyrenoids, chloroplast Rubisco-containing bodies, which have a role in carbon concentrating mechanisms (Badger et al., 1998).

Photosynthetic organisms need systems to control the efficiency of their photosynthetic apparatus as well as balancing the energy between photosystem 1 and 2 (PS1 and PS2). In *C. crispus,* this is achieved by a spill-over from

PS1 to PS2, which is a direct funnelling of energy from PS1 to PS2 without going through the reaction centre of PS1. This process is controlled by the redox status of the plastoquinone pool (Kowalczyk et al., 2013).

In addition to oxygen production, during daytime, oxygen uptake during photosynthesis has been studied in *C. crispus* in some detail (Brechignac & Andre, 1984, 1985b; Brechignac & Furbank, 1987). The oxygen uptake represents 37% of the net photosynthesis and was always higher than dark respiration. During CO_2 limiting conditions, Rubisco oxygenase activity represented less than half of oxygen uptake, pseudocyclic photophosporylation less than 20% and other reactions, such as respiration, the remainder. These studies showed that the uptake of oxygen in the light was relatively insensitive to CO_2 concentrations, but sensitive to increased oxygen concentrations.

3.8 GENES AND GENOMES

A crucial element for the establishment of a biological model organism is an important and pertinent knowledge of the genes and genomes of the organism. This enables the use of high throughput molecular methods and for global comparisons with other organisms. Presently, the knowledge of genes and genomes in *C. crispus* is more important than in other red macroalgae allowing for its efficient use in scientific research. To our knowledge the first genome sequence from *C. crispus*, of the 18S rRNA gene dates to 1992 (Bird, Murphy, Rice, & Ragan, 1992), even though the amino acid sequence of flavodoxin was known earlier, analysed through enzymatic digestion of the protein (Wakabayashi, Kimura, Fukuyama, Matsubara, & Rogers, 1989).

The mitochondrial genes and genome of *C. crispus* have been sequenced and studied (Boyen, Leblanc, Bonnard, Rienenberger, & Kloareg, 1994; Boyen, Leblanc, Kloareg, & Loiseaux-de Goer, 1994; Leblanc, Boyen, et al., 1995; Leblanc, Kloareg, Loiseaux-de Goër, & Boyen, 1995; Leblanc et al., 1997; Richard, Bonnard, Grienenberger, Kloareg, & Boyen, 1998; Richard, Kloareg, & Boyen, 1999; Viehmann, Richard, Boyen, & Zetsche, 1996). The mitochondrial genome is circular and has a relatively small size (26 kb), and is thus very different from angiosperms, despite the fact that the genes are phylogenetically close. It codes for 51 genes, has high coding density (95.2% coding) with only one intron and contains ten overlapping regions. The genome is transcribed into two large primary transcripts that are further matured via multiple processing steps involving tRNA genes.

The plastid genome of *C. crispus* was sequenced as a part of the genome project. The relatively large of size 180,086 bp is typical for most red algae. The genome codes for 240 unique genes of which 204 are protein coding and three genes are of unknown function (Janouškovec et al., 2013). One unprecedented feature found in *C. crispus* and other florideophyte red algae is a group II intron in a tRNA-Met gene that encodes a maturase.

The nuclear genome of *C. crispus* is 105 Mbp, is estimated to code for 9606 genes (Collén et al., 2013). The genes are characterised by their compact nature, caused by the very low content of introns considering the genome size and small untranslated regions (UTR) (Figure 3.3). 88% of genes have no introns and the existing introns are typically short with an average length of only 182 bp. Taken together this means that the genome is by dominated intergenetic DNA, with a high degree of repetition. The intergenetic DNA is organised in regions with low gene density interspersed with clusters of genes with short distances between protein coding genes and with very high gene density. The low intron content, small UTR and clustered genes are generally associated with compact small genomes. This resemblance with compact genomes is also seen by the relatively low number of genes, the small genome families and with no indication of genome duplication.

Prior to the genome project a study was performed in order to gain increased knowledge of genes in red algae and in particular genes important for cell wall synthesis (Collén et al., 2006b). The study compared the transcriptome of protoplasts (cells devoid of cell wall after treatment with cell wall digesting enzymes) and nonstressed thallus plants. The approach yielded 2291 nonredundant sequences with a large repertoire of stress genes as well as many genes of unknown function. The collection of expressed

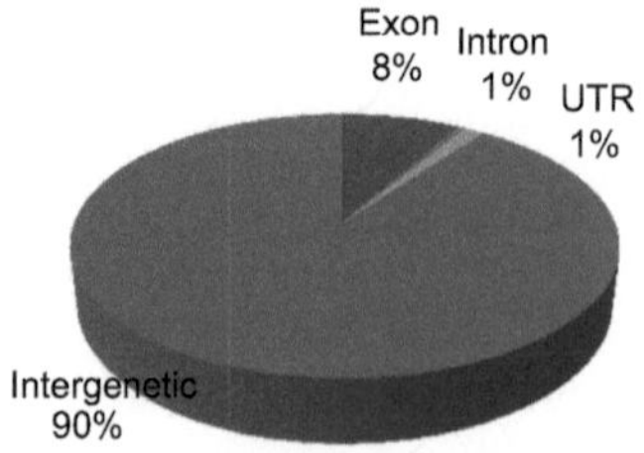

Figure 3.3 Genome composition of *Chondrus crispus*. The intergenetic part of the genome is the nonprotein coding parts between genes and the UTR are the 5′ and 3′ nontranslated part of the genes. UTR, untranslated regions.

sequence tags (ESTs) were also used to construct cDNA microarrays that were subsequently used for physiological studies (Collén et al., 2006a; Collén, Guisle–Marsollier, Léger, & Boyen, 2007). To support the annotation in the genome project another 300,000 ESTs were sequenced representing 8212 contigs (Collén et al., 2013).

3.9 PERSPECTIVES ON PRIMARY METABOLISM

Most of the primary amino acid metabolic related genes were found in single copy in the *C. crispus* genome, with the exception of glutamine synthase, aspartate transaminase and urea transporters (Collén et al., 2013). *Chondrus crispus* lacks the glutamate synthase gene, thus glutamate would necessarily be synthesised by glutamate dehydrogenase and not through the cleavage of glutamine by glutamate synthase. The genome appears to contain no ureases or arginases, important for the urea cycle, though it does have genes coding for urea transporters. It is likely that *C. crispus* can metabolise urea as a gene encoding a fusion protein of urea carboxylase and allophanate hydrolase was identified and degradation of urea by such fusion proteins has previously been shown (Genbauffe & Cooper, 1991). Due to the lack of arginases it is unlikely that proline is synthesised from arginine; however, the genome contains genes for a pyrroline-5-carboxylate reductase and a 1-pyrroline-5-carboxylate synthase suggesting that *C. crispus* is capable of synthesising proline from glutamate. For UV protection *C. crispus* synthesises the UV-absorbing mycosporine-like amino acid shinorine (Kräbs, Watanabe, & Wiencke, 2004). Other amino acids identified include gigartinine, L-citrullinyl-L-arginine, ornithine and citrulline (Laycock & Craigie, 1977; Young & Smith, 1958).

The *C. crispus* genome codes for all enzymes involved in plastidial fatty acid synthesis and their transport to the cytosol. Furthermore, it has at least one gene for each of the enzymes involved in the microsomal fatty acid elongase complex. It produces PUFAs and is notable for its high levels of arachidonic acid (Fleurence, Gutbier, Mabeau, & Leray, 1994; van Ginneken, Helsper, de Visser, van Keulen, & Brandenburg, 2011; Lamberto & Ackman, 1994). It is unlikely that fatty acid desaturation occurs in the plastid as the genome lacks an ACP desaturase although several potential desaturases have been annotated, comprising a stearyol-CoA desaturase, a $\Delta 12$ and a $\Delta 15$-desaturase and two genes for PUFA desaturases. Sphingolipids containing inositol that have been reported in other red algae (Khotimchenko, Klochkova, & Vaskovsky, 1990) and *C. crispus* shows the presence of genes required for the production of 4-hydroxysphinganine, sphingosine-1-phosphate and ceramides, together with several genes for ceramide glycosyltransferases. A full complement of genes for the synthesis of thylakoid membrane lipids and phospholipids is contained within the genome. In addition, fatty acids may be stored in triacylglycerols through the acyl-CoA dependent or independent pathway. Free C18 and C20 fatty acids may be further processed into oxylipins.

Unlike plants and green algae, which store their insoluble starch granules in their chloroplasts, the carbon energy of red algae is stored in the cytosol as floridean starch (Viola, Nyvall, & Pedersen, 2001). The glycosyl transferase (GT) families 5 and 13, involved in starch synthesis and glycoside hydrolase (GH) families 13, 14, 35 and 77, involved in starch recycling, are encoded within the *C. crispus* genome. While this organism has all the required proteins for starch synthesis and breakdown, their corresponding genes are unexpectedly low in redundancy, demonstrating a system simpler than that found in the green lineage. Four genes have been identified for synthesis and one for degradation of trehalose, a nonreducing disaccharide. In plants, trehalose is important in starch and sucrose regulation; however, sucrose metabolism is lacking in *C. crispus* and thus trehalose may have a different biological role. Another low molecular weight carbohydrate is the photo-assimilate mannosylglycerate (digeneaside; 2-D-glycerate-α-D-mannopyranoside) found in some red algae, though it has not been detected in *C. crispus*. However, the genome holds a gene coding for a mannosylglycerate synthase (GT78), which was likely acquired in red algae by horizontal gene transfer from a thermophilic marine bacterium. An alternative hypothesis is that the GT78 in *C. crispus* may be involved in the synthesis of floridoside (2-0-D-glycerol-a-o-galactopyranoside) or isofloridoside (1-0-D-glycerol-a-D-galactopyranoside), two heterosides known to be accumulated in this alga (Kremer, 1980). One of the best characterised enzymes produced by *C. crispus* corresponds to a hexose oxidase which may be involved in algal defence and which oxidises hexose sugars to their corresponding lactones and aldobionic acids (see Rand, Qvist, Walter, & Poulsen, 2006 and references therein).

The integration of information available for *C. crispus*, in particular genome annotation, will be used to reconstruct metabolic networks; however, more metabolic profiling is certainly needed to complete existing observations.

3.10 CELL WALL BIOLOGY

Red algal cell walls are a composite material often made of cellulose microfibrils embedded in a matrix composed of polysaccharides, and to a lesser extent, proteins and aromatic substances. In *C. crispus* the matrix components are highly sulphated galactans known as carrageenans. They are made up of linear chains of disaccharidic motifs of D-galactose residues with alternating α(1–3) and β(1–4) linkages and classified according to the number and position of sulphate esters and by the occurrence of a 3,6-anhydro bridge in the α-linked residue. The κ-, ι- and λ-carrageenans are respectively substituted by one, two or three sulphate ester groups per digalactose repeating unit (Figure 3.4). *Chondrus crispus* synthesis all three of these carrageenans although variations are observed depending on the life stage. Carrageenans in gametophytes are a mixture of κ- and ι–carrageenans (~70 and ~20%), with the presence of low amount of biosynthetic precursor motifs (~8% μ-carrageenan and 2% ν-carrageenan). The tetrasporophyte synthesise only λ-carrageenans (Chopin, Bodeau-Bellion, Floc'h, Guittet, & Lallemand, 1987; Tasende, Cid, & Fraga, 2012). Carrageenans naturally occur as heteropolymers composed of a sequence of several moieties, and the use of dedicated enzymes from marine bacteria has lately served to address the specific distribution patterns of the carrabiose units within the chains (Guibet et al., 2008; Jouanneau, Boulenguer, Mazoyer, & Helbert, 2010).

As for land plants, the amorphous matrix polysaccharides are expected to be interconnected to other polysaccharides or crystalline polysaccharides and thus contribute to mechanical resistance, cell expansion, and cell wall cohesion (Carpita & McCann, 2000). However, very little is known of the nature of these polysaccharides in red algae. They represent a small portion of the wall (1–8% of the dry weight). Cellulose fibrils are the most commonly found polymer, but additional polymers, such as crystalline β(1–4)-D-mannans, β(1–4) or β(1–3)-D-xylans, as sulphated α(1–3)-D-mannans or sulphated β(1–3/1–4)-D-glucans, have also been reported in various red algal species (Popper et al., 2011). Their existence in *C. crispus* still needs to be established.

The current knowledge on cell wall metabolism in red seaweed is very limited. The genome sequence of *C. crispus* allows some predictions on the cell wall metabolic pathways (Collén et al., 2013). The carrageenan

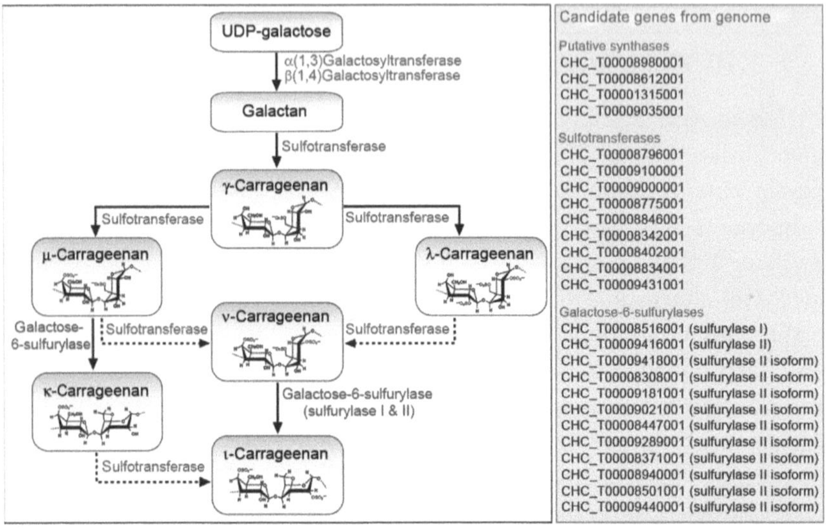

Figure 3.4 Putative carrageenan biosynthesis pathway in *Chondrus crispus* with the corresponding candidate genes retrieved from the genome sequence. Only the conversions of μ- to κ-, and ν- to ι-carrageenan, were enzymatically demonstrated in *C. crispus*.

biosynthesis pathway is yet not fully understood and most of the steps stay therefore highly speculative. It was proposed that carrageenans are first synthesised in the Golgi apparatus as a linear and neutral backbone of galactan by galactosyl transferases that catalyse the polymerisation of galactose residues (Figure 3.4). The alternation of α(1–3) and β(1–4) linkages would suggest the activity of two types of glycosyltransferases. Four genes homologous to chondroitin synthases were identified in *C. crispus*. These enzymes are involved in the biosynthesis of the sulphated glycosaminoglycans in animals and therefore the red algal homologue is a good candidate for the biosynthesis of carrageenans. Following polymerisation, the initial sulphation is likely to occur in the Golgi apparatus by sulphotransferases. In *C. crispus*, nine genes similar to carbohydrate sulphotransferases were identified. Their animal homologues are involved in the biosynthesis of glycosaminoglycans, suggesting again that genes are likely involved in the carrageenan sulphation. The sulphated galactan obtained is thereafter transported into the wall where galactose-6-sulphurylases remove the C6-sulphate from precursors to form the 3,6-anhydro bridges. This final step of carrageenan synthesis was demonstrated in *C. crispus* gametophytes with the characterisation of two distinct galactose-6-sulphurylases, named sulphurylase I and II, which convert ν- into ι-carrageenan (Genicot-Joncour et al., 2009). An enzymatic activity catalysing the conversion of μ- to κ-carrageenan has also been demonstrated in both life cycle phases of *C. crispus* (Wong & Craigie, 1978), but the corresponding gene is unknown. The *C. crispus* genomic analysis retrieved two genes identical to the previously cloned genes of sulphurylases I and II, but also ten additional paralogous genes of sulphurylase II (Figure 3.4). Among the nine sulphotransferases identified in the *C. crispus* genome, some of them might also be present into the wall for further modifications of the carrageenan chains. Finally, while sulphatases were supposed to be involved in the carrageenan remodelling of the growing wall, no homologue has been identified in the *C. crispus* genome. This result suggests that no modification in its sulphation pattern occur in carrageenan after its biosynthesis, or that new families of sulphatases may exist in the genome. By contrast, three glycosyl hydrolases related to κ-carrageenases from marine bacteria (Michel, Chantalat, Duee, et al., 2001; Michel, Chantalat, Fanchon, et al., 2001) were identified in the genome sequence and are likely involved in cell wall remodelling.

Like in land plants, the cellulose microfibrils in red algae are synthesised by clusters of cellulose synthases (CESAs) in terminal complexes (TCs), which move in the plasma membrane. However, the morphologies of the TCs and the cellulose microfibrils they produce differ from those of land plants: the TCs are organised in linear rows, randomly distributed in the plasma membrane and the resulting microfibrils show a flat-ribbon or a rectangular-parallelepiped morphology (Tsekos, 1999; Tsekos, Reiss, & Schnepf, 1993). At least three different CESA proteins are required to form a functional rosette TC in plants (Mutwil, Debolt, & Persson, 2008), while the minimum number of recruited CESAs in functional TCs in red algae is unknown. In *C. crispus*, the genome sequence analysis retrieved two CESA genes, similar to those already described in the red algae *Porphyra* sp. and *Griffithsia monilis*. However these red algal CESAs have a distinct origin to those found in land plants and their acquisition probably predated the primary endosymbiosis event. A third CESA gene is found in *C. crispus*, unrelated to the previous, and probably of an ancestral bacterium origin. Such differences within the Archaeplastida lineage might explain the structural variation in the TC organisation and the cellulose microfibrils in red algae compared to land plants. Finally, *Chondrus* also possesses GH5, GH6 and GH45 cellulases, which are absent in plants and that might be involved in remodeling of the cellulose fibrils.

Land plants require a rigid structure to support them against the pull of gravity, making crystalline cellulose an important part of the wall. By contrast marine algae need more flexible structures to accommodate the varying stresses of tidal and wave action and a cell wall mostly composed of carrageenans might fulfill this role. However the physiological significance of the variations in carrageenan composition in relation to mechanical, hydration or electrochemical regulations is still matter of debate (Kloareg & Quatrano, 1988). The *Chondrus* gametophyte, with strong gelling κ-type carrageenans, has been shown to be mechanically superior to the tetrasporophyte, with nongelling λ-type carrageenans (Carrington, Grace, & Chopin, 2001), however, this is of no impact on the distribution of the life history phases in natural populations of *C. crispus* (Garbary, Tompkins, White, Corey, & Kim, 2011). Regarding the hydration properties, if one assumes that sulphated carrageenans protect against desiccation by enhancing water absorption, we would expect the littoral populations of *C. crispus* to have increased carrageenan contents compare to the sublittoral ones. Again such assumption has yet failed to be clearly established (Fuller & Mathieson, 1972).

3.11 COMMERCIAL USES

The majority of fresh seaweeds, are not very palatable due to their texture, so they must be processed in some manner to improve their mouthfeel. In the eighteenth century, the most common method of household processing involved cooking in some way, thus *C. crispus* was recognised very early for its special thickening characteristics. Other rhodophytes, such as *Palmaria* and *Porphyra*, became the seaweeds of choice for direct consumption and as nutritional ingredients in foods, while *C. crispus* became valuable in the food and pharmaceutical industries. Less attention, therefore, was given to its potential health and nutritional benefits as a natural food.

The Irish moss based carrageenan industry in North America also developed rapidly in the early 1940s due to the pioneering work of Jacques Wolf and Company, Krim-Ko and Kraft Foods (Lewis, Stanley, & Guist, 1988). The Krim-Ko operation at Scituate, Massachusetts, changed hands to become Seaplant Corporation, while the Kraft Food operation of South Portland, Maine, was taken over by Stauffer Chemical Company in 1970 and was later closed. The Algin Corporation of America, Rockland, Maine, also began processing Irish moss soon after WWII to become a major carrageenan producer by the mid-1950s. In addition, it licensed carrageenan processing technology to SATIA of France (Lewis et al., 1988). Marine

Colloids Inc., of Rockland, Maine, was formed by an amalgamation in 1959 of the Algin and Seaplant Corporations, and by 1977 it, in turn, became a Division of FMC Corporation to remain as the sole manufacturer of carrageenan in the USA.

Commercial applications for Irish moss extract depend upon a detailed knowledge of the chemical and physicochemical properties of these phycocolloids. In brief, key events included the resolution of hot water extracts from *C. crispus* into soluble and gelling fractions based upon their sensitivity to potassium ions (Smith & Cook, 1953). The authors coined the terms 'λ-carrageenin' and 'κ-carrageenin' for these respective fractions. The proportion of the κ-/λ- carrageenan fractions controls the performance of a given extract in milk reactivity and other valued functions of the phycocolloids (Witt, 1985; Stanley, 1987). This ratio issue was resolved when it was established that the sporophyte generation of *C. crispus* produced only λ-carrageenan devoid of 3,6-anhydrogalactose, while the κ-family of carrageenans characterised both the gametophyte phases of the alga (McCandless, Craigie, & Walter, 1973). The thorough historical review by Stortz (2005) discusses the pioneering studies by Rees (1972) and his colleagues.

In the early 1970s *C. crispus* from Canada provided 75% of the world's production of carrageenan but, by 1992, *C. crispus* represented only 3.8% of the global harvest of carrageenophytes with Canada providing a mere 12% (Chopin, 1998; Pringle & Mathieson, 1987). By the late 1990s, *C. crispus* from Brittany, Vendée and Normandy represented only 9% of the carrageenophytes supplying the French carrageenan industry (Kaas, 1998). In Ireland the bulk of the *C. crispus* harvest is sold locally where it is used primarily in cooking and as a natural cold remedy (Guiry & Hession, 1998).

Acadian Seaplants Limited, in conjunction with the National Research Council of Canada and other government partners, developed a unique and innovative food product from cultivated *C. crispus*. It is utilised in a rehydrated form primarily in the *kaiso* salad market of Japan, with some usage directed to soups and garnishes due to its unique colours, shape and consistent quality. However, the global dietary seaweed market is currently dominated by cultivated species such as *Saccharina japonica* (kombu), *Undaria pinnatifida* (wakame) and *Pyropia* (*Porphyra*) spp. (Ohno & Largo, 1998).

Research shows there are many therapeutic characteristics associated with seaweeds in general and a number of species from the dominant three genera have been investigated. *Chondrus crispus* possesses characteristics typical of most seaweeds, such as antioxidant capacities and mineral content (Cornish & Garbary, 2010; Ruperez, 2002; Sangha et al., 2013), but it also provides specialised fibre in the form of both soluble and insoluble or fermentable components (Rupérez & Toledano, 2003). These are important in gut health as fibre type has been demonstrated to affect subjective appetite, acute energy intake, long-term energy intake and body weight, due in large part as a function of the various physicochemical properties of the dietary fibres (Wanders et al., 2011). Other research studies specific to *C. crispus* have demonstrated potential for therapeutic benefits in terms of anticoagulant activity (Lee, Athukorala, Lee, & Jeon, 2008), lipid induced nitric oxide inhibition (Banskota et al., 2013) and enhanced immune response (Liu, Hafting, Critchley, Banskota, & Prithiviraj, 2013).

Irish moss extracts provide valuable ingredients for cosmetic applications, in part as a result of their special hydrocolloid properties (Witt, 1985) and they also possess components with sunscreen and antiageing potential (Franklin, Kräbs, & Kuhlenkamp, 2001; Kräbs, Bischof, Hanelt, Karsten, & Wiencke, 2002; Kräbs et al., 2004; Sinha, Singh, & Häder, 2007). Tank culture permits close control of crop traceability and management. The optimisation of production rates is possible and opportunities for the appearance and isolation of phenotypic mutants are enhanced (van der Meer, 1981). Some of these may possess unique characteristics of commercial value and colour mutations enriched in blue pigment is an example (Cornish, O'Leary, & Garbary, 2013).

The long-term utilisation of Irish moss as a food and its well studied physiology make *C. crispus* a useful and versatile commercial product. Tank culture provides opportunities to exploit this alga as a vehicle for enhanced nutrition or increased functionality. Maximising control over the culture environment for *C. crispus* presents possibilities for the enhancement of useful secondary metabolite compounds and antioxidant capacity by upregulating its stress genes (Collén et al., 2007; Collén et al., 2006a; Yakovleva & Titlyanov, 2001). Cultivated *C. crispus* is now well positioned to become a global commercial success, again.

ACKNOWLEDGEMENTS

The various parts were written by different authors accordingly: Introduction (JC), History (MLC, JSC), Ecology (CB), Life cycle (SAK-H), Genetics, population structure and the mating system (SAK-H), Ecological and biochemical aspects of biotic interactions (CL, PP), Photosynthesis (JC), Genes and genomes (JC), Perspectives on primary metabolism (TT/EF-B) Cell-wall biology (CH, GM) and Commercial uses (MLC, JSC).

This work is a contribution within the project IDEALG supported by the French Government stimuli program entitled "Investissements d'avenir, Biotechnologies-Bioressources" (ANR-10-BTBR-04-02) and EFB and SAK-H were supported by Conseil Regional de Bretagne and CNRS.

REFERENCES

Allen, D. E., & Hatfield, G. (2004). *Medicinal plants in folk tradition—An ethnobotany of Britain & Ireland*. Portland, OR and Cambridge, UK:Timber Press.

Andrews, J. H. (1976). The pathology of marine algae. *Biological Review, 51*, 211–253.

Badger, M. R., Andrews, T. J., Whitney, S. M., Ludwig, M., Yellowlees, D. C., Leggat, W., et al. (1998). The diversity and coevolution of Rubisco, plastids, pyrenoids, and chloroplast-based CO_2-concentrating mechanisms in algae. *Canadian Journal of Botany, 76*, 1052–1071.

Banskota, A. H., Stefanova, R., Sperker, S., Lall, S., Craigie, J. S., & Hafting, J. T. (2013). Lipids isolated from the cultivated red alga *Chondrus crispus* inhibit nitric oxide production. *Journal of Applied Phycology*, 1–7.

Bhattacharya, D. (1985). The demography of fronds of *Chondrus crispus* Stackhouse. *Journal of Experimental Marine Biology and Ecology, 91*, 217–231.

Bidwell, R. G. S., McLachlan, J., & Lloyd, N. D. H. (1985). Tank cultivation of Irish moss, *Chondrus crispus* Stackh. *Botanica Marina, 28*, 87–98.

Bird, C. J., Murphy, C. A., Rice, E. L., & Ragan, M. A. (1992). The 18S rRNA gene sequences of four commercially important seaweeds. *Journal of Applied Phycology, 4*, 379–384.

Bischof, K., Gomez, I., Molis, M., Hanelt, D., Karsten, U., Lüder, U., et al. (2007). Ultraviolet radiation shapes seaweed communities. *Reviews in Environmental Science and Biotechnology, 5*, 141–166.

Bliss, A. (1985). Carrageen. *Notes and Queries, 32*, 382–384.

Bouarab, K., Adas, F., Gaquerel, E., Kloareg, B., Salaün, J.-P., & Potin, P. (2004). The innate immunity of a marine red alga involves oxylipins from both the eicosanoid and octadecanoid pathways. *Plant Physiology, 135*, 1838–1848.

Bouarab, K., Kloareg, B., Potin, P., & Correa, J. A. (2001). Ecological and biochemical aspects in algal infectious diseases. *Cahier de Biologie Marine, 42*, 91–100.

Bouarab, K., Potin, P., Correa, J., & Kloareg, B. (1999). Sulfated oligosaccharides mediate the interaction between a marine red alga and its green algal pathogenic endophyte. *The Plant Cell, 11*, 1635–1650.

Bouarab, K., Potin, P., Weinberger, F., Correa, J., & Kloareg, B. (2001). The *Chondrus crispus-Acrochaete operculata* host-pathogen association, a novel model in glycobiology and applied phycopathology. *Journal of Applied Phycology, 13*, 185–193.

Boyen, C., Leblanc, C., Bonnard, G., Grienenberger, J. M., & Kloareg, B. (1994). Nucleotide sequence of the cox3 gene from *Chondrus crispus*: evidence that UGA encodes tryptophan and evolutionary implications. *Nucleic Acids Research, 22*, 1400–1403.

Boyen, C., Leblanc, C., Kloareg, B., & Loiseaux-de Goer, S. (1994). Physical map and gene organization of the mitochondrial genome of *Chondrus crispus* (Rhodophyta, Gigartinales). *Plant Molecular Biology, 26*, 691–697.

Braud, J. P. (1998). Simultaneous culture in pilot tanks of the macroalga *Chondrus crispus* (Gigartinaceae) and the microalga *Odontella aurita* (Eupodiscaceae) producing EPA. In Y. Le Gal & A. Muller-Feuga (Vol. Eds.), *Marine Microorganisms for Industry. Actes de colloque IFREMER 21*, QUAE editions.

Braud, J. P. (2006). Continuous seaweed tank culture in France: from *Chondrus crispus* to co-culture of macroalge and the diatom *Odontella aurita*. In A. T. Critchley, M. Ohno, & D. B. Largo (Eds.), *Seaweed resources of the world*. Yokosuka: Japan International Cooperation Agency.

Braud, J. P., & Delépine, R. (1981). Growth response of *Chondrus crispus* (Rhodophyta, Gigartinales) to light and temperature in laboratory and outdoor tanks culture. *Proceedings International Seaweed Symposium, 10*, 553–558.

Brechignac, F., & Andre, M. (1984). Oxygen uptake and photosynthesis of the red macroalga, *Chondrus crispus*, in seawater. Effects of light and CO_2 concentration. *Plant Physiology, 75*, 919–923.

Brechignac, F., & Andre, M. (1985a). Continuous measurements of the free dissolved CO_2 concentration during photosynthesis of marine plants evidence for HCO_3^- use in *Chondrus crispus*. *Plant Physiology*, *78*, 551–554.

Brechignac, F., & Andre, M. (1985b). Oxygen uptake and photosynthesis of the red macroalga, *Chondrus crispus*, in seawater. Effects of oxygen concentration. *Plant Physiology*, *78*, 545–550.

Brechignac, F., Andre, M., & Gerbaud, A. (1986). Preferential photosynthetic uptake of exogenous HCO_3^- in the marine macroalga *Chondrus crispus*. *Plant Physiology*, *80*, 1059–1062.

Brechignac, F., & Furbank, R. T. (1987). On the nature of the oxygen uptake in the light by *Chondrus crispus*. Effects of inhibitors, temperature and light intensity. *Photosynthesis Research*, *11*, 45–59.

Briand, X. (1991). Seaweed harvesting in Europe. In M. D. Guiry, & G. Blunden (Eds.), *Seaweed resources in Europe*. Chichester: John Wiley & Sons, Ltd.

Buggeln, R. G., & Craigie, J. S. (1973). The physiology and biochemistry of *Chondrus crispus* Stackhouse. In M. J. Harvey, & J. McLachlan (Eds.), *Chondrus crispus, proceedings Nova Scotian institute of science* (Vol. 27(pp. 81–102)..

Carpita, N. C., & McCann, M. C. (2000). The cell wall. In B. Buchanan, W. Gruissem, & R. Jones (Eds.), *Biochemistry and molecular biology of plants*. Rockville, MD: American Society of Plant Physiologists.

Carrington, E., Grace, S. P., & Chopin, T. (2001). Life history phases and the biomechanical properties of the red alga *Chondrus crispus* (Rhodophyta). *Journal of Phycology*, *37*, 699–704.

Chen, L. C., & McLachlan, J. (1972). The life history of *Chondrus crispus* in culture. *Canadian Journal of Botany*, *50*, 1055–1060.

Cheney, D. P., & Mathieson, A. C. (1979). Population differentiation in *Chondrus crispus*: preliminary results. *Isozyme Bulletin*, *12*, 57.

Chopin, T. (1986). *The red alga Chondrus crispus Stackhouse (Irish moss) and carrageenans: A review. Canadian technical report of fisheries and aquatic sciences, 1514*. Ottawa, Canada: Ministère des approvisionnements et services.

Chopin, T. (1998). The seaweed resources of Eastern Canada. In A. T. Critchley, & M. Ohno (Eds.), *Seaweed resources of the world* (pp. 273–302). Yokosuka: Japan International Cooperation Agency.

Chopin, T., Bird, C. J., Murphy, C. A., Osborne, J. A., Patwary, M. U., & Floc'h, J. Y. (1996). A molecular investigation of polymorphism in the North Atlantic red alga *Chondrus crispus* (Gigartinales). *Phycological Research*, *44*, 69–80.

Chopin, T., Bodeau-Bellion, C., Floc'h, J. Y., Guittet, E., & Lallemand, J. Y. (1987). Seasonal study of carrageenan structures from female gametophytes of *Chondrus crispus* Stackhouse (Rhodophyta). *Hydrobiologia*, *151–152*, 535–539.

Chopin, T., Sharp, G., Belyea, E., Semple, R., & Jones, D. (1999). Open-water aquaculture of the red alga *Chondrus crispus* in Prince Edward Island, Canada. *Hydrobiologia*, *398/399*, 417–425.

Collén, J., Guisle–Marsollier, I., Léger, J. J., & Boyen, C. (2007). Response of the transcriptome of the intertidal red seaweed *Chondrus crispus* to controlled and natural stresses. *New Phytologist*, *176*, 45–55.

Collén, J., Hervé, C., Guisie-Marsollier, I., Léger, J. J., & Boyen, C. (2006a). Expression profiling of *Chondrus crispus* (Rhodophyta) after exposure to methyl jasmonate. *Journal of Experimental Botany*, *57*, 3869–3881.

Collén, J., Porcel, B., Carré, W., Ball, S. G., Chaparro, C., Tonon, T., et al. (2013). Genome structure and metabolic features in the red seaweed *Chondrus crispus* shed light on evolution of the Archaeplastida. *Proceedings of the National Academy of Sciences*, *110*, 5247–5252.

Collén, J., Roeder, V., Rousvoal, S., Collin, O., Kloareg, B., & Boyen, C. (2006b). An expressed sequence tag analysis of thallus and regenerating protoplasts of *Chondrus crispus* (Gigartinales, Rhodophyceae). *Journal of Phycology*, *42*, 104–112.

Cornish, M. L., & Garbary, D. J. (2010). Antioxidants from macroalgae: potential applications in human health and nutrition. *Algae, 25*, 155–171.
Cornish, M. L., O'Leary, S. J. B., & Garbary, D. J. (2013). Phycobilisome composition in *Chondrus crispus* (Gigartinales, Rhodophyta) from a wild type strain and its vegetatively derived green mutant. *Algae, 28*, 121–129.
Correa, J. A. (1996). Infectious diseases of marine algae: current knowledge and approaches. In F. E. Round & D. J. Chapman (Vol. Eds.), *Progress in Phycological research* (pp. 149–180). Bristol, UK: Biopress Ltd.
Correa, J. A., & McLachlan, J. (1988). Variation in the fine structure of the cuticle of *Chondrus crispus* Stackhouse (Rhodophyta). *British Phycological Journal, 23*, 285–286.
Correa, J. A., & McLachlan, J. L. (1991). Endophytic algae of *Chondrus crispus* (Rhodophyta). III. Host specificity. *Journal of Phycology, 27*, 448–459.
Correa, J. A., & McLachlan, J. L. (1992). Endophytic algae of *Chondrus crispus* (Rhodophyta). IV. Detrimental effects on the host following infections by *Acrochaete operculata* and *A. heteroclada* (Chlorophyta). *Marine Ecological Progress Series, 81*, 73–87.
Correa, J. A., & McLachlan, J. L. (1994). Endophytic algae of *Chondrus crispus* (Rhodophyta). V. Fine structure of the infection by *Acrochaete operculata* (Chlorophyta). *European Journal of Phycology, 29*, 33–47.
Correa, J. A., Nielsen, R., & Grund, D. W. (1988). Endophytic algae of *Chondrus crispus* (Rhodophyta). II. *Acrochaete heteroclada* sp. nov., *A. operculata* sp. nov., and *Phaeophila dendroides* (Chlorophyta). *Journal of Phycology, 24*, 528–539.
Correa, J., Nielsen, R., Grund, D. W., & McLachlan, J. (1987). Endophytic algae of Irish moss (*Chondrus crispus* Stackh.). *Proceedings of the International Seaweed Symposium, 12*, 223–228.
Cosse, A., Leblanc, C., & Potin, P. (2008). Dynamic defense of marine macroalgae against pathogens: from early activated to gene-regulated responses. *Advance in Botanical Research, 46*, 221–266.
Craigie, J. S. (1990). Irish moss cultivation: some reflections. In C. Yarish, C. A. Penniman, & P. Van Patten (Eds.), *Economically important marine plants of the Atlantic*. Groton, CT: Connecticut Sea Grant College Program.
Craigie, J. S., & Correa, J. (1996). Etiology of infectious diseases in cultivated *Chondrus crispus* (Gigartinales, Rhodophyta). *Proceeding of International Seaweed Symposium, 15*, 97–104.
Craigie, J. S., Correa, J. A., & Gordon, M. E. (1992). Cuticles from *Chondrus crispus* (Rhodophyta). *Journal of Phycology, 28*, 777–786.
Craigie, J. S., & Shacklock, P. F. (1989). Culture of Irish moss. In A. D. Boghen (Ed.), *Cold-water aquaculture in Atlantic Canada*(1st ed.) (pp. 243–270). (2nd ed.) (1995), pp. 363–390.
Darbishire, O.V. (1902). *Chondrus* . (Vol. 9)Williams & Norgate.
Davison, I. R., Dudgeon, S., & Ruan, H. M. (1989). Effect of freezing on seaweed photosynthesis. *Marine Ecology Progress Series, 58*, 123–131.
DeRoeck-Holtzhauer, Y. (1991). Uses of seaweeds in cosmetics. In M. D. Guiry, & G. Blunden (Eds.), *Seaweed resources in Europe* (pp. 83–94). Chichester: John Wiley & Sons, Ltd.
Donaldson, S. L., Chopin, T., & Saunders, G. W. (2000). An assessment of the AFLP method for investigating population structure in the red alga *Chondrus crispus* Stackhouse (Gigartinales, Florideophyceae). *Journal of Applied Phycology, 12*, 25–35.
Dudgeon, S. R., Davison, I. R., & Vadas, R. L. (1989). Effect of freezing on photosynthesis of intertidal macroalgae: relative tolerance of *Chondrus crispus* and *Mastocarpus stellatus* (Rhodophyta). *Marine Biology, 101*, 107–114.
Dudgeon, S. R., Davison, I. R., & Vadas, R. L. (1990). Freezing tolerance in the intertidal red algae *Chondrus crispus* and *Mastocarpus stellatus*: relative importance of acclimation and adaptation. *Marine Biology, 106*, 427–436.
Engel, C. R., Wattierf, R., Destombe, C., & Valero, M. (1999). Performance of non-motile male gametes in the sea: analysis of paternity and fertilization success in a natural population of a red seaweed, *Gracilaria gracilis*. *Proceedings of the Royal Society of London. Series B: Biological Sciences, 266*, 1879–1886.

FAO. (2012). *The state of world fisheries and aquaculture.* Rome: Food and Agriculture Organisation p. 230.
Ffrench, R. A. (1971). *A current appraisal of the Irish moss industry: Canadian Atlantic.* Ottawa, Canada: Department of Fisheries and Forestry.
Fleurence, J., Gutbier, G., Mabeau, S., & Leray, C. (1994). Fatty-acids from 11 marine macroalgae of the French Brittany coast. *Journal of Applied Phycology, 6,* 527–532.
Franklin, L. A., Kräbs, G., & Kuhlenkamp, R. (2001). Blue light and UV-A radiation control the synthesis of mycosporine-like amino acids in *Chondrus crispus* (Florideophyceae). *Journal of Phycology, 37,* 257–270.
Frazer, W. (1864). *Elements of Materia Medica: Containing the chemistry and natural history of drugs, their effects, doses, and adulterations, with observations on all the new remedies recently introduced into practice, and on the preparations of the British pharmacopœia* (2nd ed.). London: John Churchill and Sons.
Fredericq, S., Brodie, J., & Hommersand, M. H. (1992). Developmental morphology of *Chondrus crispus* (Gigartinaceae, Rhodophyta). *Phycologia, 31,* 542–563.
Fuller, S. W., & Mathieson, A. C. (1972). Ecological studies of economic red algae. IV. Variations of carrageenan concentration and properties in *Chondrus crispus* Stackhouse. *Journal of Experimental Marine Biology and Ecology, 10,* 49–58.
Gaquerel, E., Hervé, C., Labrière, C., Boyen, C., Potin, P., & Salaün, J. P. (2007). Evidence for oxylipin synthesis and induction of a new polyunsaturated fatty acid hydroxylase activity in *Chondrus crispus* in response to methyljasmonate. *Biochimica et Biophysica Acta (BBA)-Molecular and Cell Biology of Lipids,* 1771, 565–575.
Garbary, D. J., Tompkins, E., White, K., Corey, P., & Kim, J. K. (2011). Temporal and spatial variation in the distribution of life history phases of *Chondrus crispus* (Gigartinales, Rhodophyta). *Algae, 26,* 61–71.
Genbauffe, F. S., & Cooper, T. G. (1991). The urea amidolyase (DUR1,2) gene of *Saccharomyces cerevisiae*. *Mitochondrial DNA, 2,* 19–32.
Genicot-Joncour, S., Poinas, A., Richard, O., Potin, P., Rudolph, B., Kloareg, B., et al. (2009). The cyclization of the 3,6-anhydro-galactose ring of iota-carrageenan is catalyzed by two D-galactose-2,6-sulfurylases in the red alga *Chondrus crispus*. *Plant Physiology, 151,* 1609–1616.
Goff, L. J. (1982). The biology of parasitic red algae. *Phycology Research, 1,* 289–370.
Goodenough, S., & Woodward, T. J. (1797). XIX. Observations on the British Fuci, with particular descriptions of each species. *Transactions of the Linnean Society of London, 3,* 84–235.
Guibet, M., Boulenguer, P., Mazoyer, J., Kervarec, N., Antonopoulos, A., Lafosse, M., et al. (2008). Composition and distribution of carrabiose moieties in hybrid kappa-/iota-carrageenans using carrageenases. *Biomacromolecules, 9,* 408–415.
Guiry, M. D., & Hession, C. C. (1998). The seaweed resources of Ireland. In A. T. Critchley, & M. Ohno (Eds.), *Seaweed resources of the world* (pp. 210–216). Yokosuka: JIC.
Guiry, W. (2013). In M. D. Guiry, & G. M. Guiry (Eds.), *AlgaeBase.* Galway: World-wide electronic publication, National University of Ireland. http://www.algaebase.org.
Hervé, C., Tonon, T., Collén, J., Corre, E., & Boyen, C. (2006). NADPH oxidases in Eukaryotes: red algae provide new hints!. *Current Genetics, 49,* 190–204.
Hu, Z., Critchley, A. T., Gao, T., Zeng, X., Morrell, S. L., & Duan, D. (2007). Delineation of *Chondrus* (Gigartinales, Florideophyceae) in China and the origin of *C. crispus* inferred from molecular data. *Marine Biology Research, 3,* 145–154.
Hu, Z., Guiry, M. D., Critchley, A. T., & Duan, D. (2010). Phylogeographic patterns indicate transatlantic migration from Europe to North America in the red seaweed *Chondrus crispus* (Gigartinales, Rhodophyta). *Journal of Phycology, 46,* 889–900.
Humm, H. J. (1951). The red algae of economic importance: agar and related phycocolloids. In D. K. Tressler, & J. McW. Lemon (Eds.), *Marine products of commerce* (pp. 47–93). New York: Reinhold Publishing Corp.

Janouškovec, J., Liu, S. L., Martone, P.T., Carré, W., Leblanc, C., Collén, J., et al. (2013). Evolution of red algal plastid genomes: ancient architectures, introns, horizontal gene transfer, and taxonomic utility of plastid markers. *PloS One, 8*, e59001.

Jertson, E. C. (1952). The potential utilization of seaweed extractives in the United States. *First International Seaweed Symposium*, 70–71.

Johansson, G., & Snoeijs, P. (2002). Macroalgal photosynthetic responses to light in relation to thallus morphology and depth zonation. *Marine Ecology Progress Series, 244*, 63–72.

Johnson, A. S. (2001). Drag, drafting, and mechanical interactions in canopies of the red alga *Chondrus crispus. The Biological Bulletin, 201*, 126–135.

Jones, H. L. (1898). A new species of pyrenomycete parasitic on an alga. *Laboratory Bulletin Oberlin College, 9*, 3–4.

Jouanneau, D., Boulenguer, P., Mazoyer, J., & Helbert, W. (2010). Complete assignment of ^{1}H and ^{13}C NMR spectra of standard neo-ι-carrabiose oligosaccharides. *Carbohydrate Research, 345*, 547–551.

Kaas, R. (1998). The seaweed resources of France. In A. T. Critchley, M. Ohno, & D. B. Largo (Eds.), *Seaweed resources of the world*. Yokosuka: Japan International Cooperation Agency.

Kanwisher, J. (1957). Freezing and drying in intertidal algae. *Biological Bulletin, 113*, 275–285.

Kautsky, L., & Ehn, M. (1993). Phenotypic plasticity in different life stages of *Chondrus crispus* in response to depth, light and wave exposure. In J. C. Aldrich (Ed.), *Proceedings of the twenty-seventh European marine biology symposium* (pp. 81–91). Ashford, Ireland: Japaga Publishers.

Khotimchenko, S. V., Klochkova, N. G., & Vaskovsky, V. E. (1990). Polar lipids of marine macrophytic algae as chemotaxonomic markers. *Biochemical Systematics and Ecology, 18*, 93–101.

Klekowski, E. J., Jr. (1973). Genetic load in *Osmunda regalis* populations. *American Journal of Botany, 60*, 146–154.

Kloareg, B., & Quatrano, R. S. (1988). Structure of the cell walls of marine algae and ecophysiological functions of the matrix polysaccharides. *Oceanography and Marine Biology: An Annual Review, 26*, 259–315.

Kopp, J., & Perez, R. (1978). Contribution a l'etude de l'algue rouge *Chondrus crispus* Stack. Relation entre la croissance, la potentialite sexuelle, la quantite et la composition des carraghenanes. *Revue des Travaux de l'Institut des Pêches Maritimes, 42*, 291–324.

Kowalczyk, N., Rappaport, F., Boyen, C., Wollman, F. A., Collén, J., & Joliot, P. (2013). Photosynthesis in *Chondrus crispus*: the contribution of energy spill-over in the regulation of excitonic flux. *Biochimica et Biophysica Acta (BBA), Bioenergetics, 827*, 834–842.

Kräbs, G., Bischof, K., Hanelt, D., Karsten, U., & Wiencke, C. (2002). Wavelength-dependent induction of UV-absorbing mycosporine-like amino acids in the red alga *Chondrus crispus* under natural solar radiation. *Journal of Experimental Marine Biology and Ecology, 268*, 69–82.

Kräbs, G., Watanabe, M., & Wiencke, C. (2004). A monochromatic action spectrum for the photoinduction of the UV-absorbing mycosporine-like amino acid shinorine in the red alga *Chondrus crispus. Photochemistry and Photobiology, 79*, 515–519.

Kräbs, G., & Wiencke, C. (2005). Photosynthesis, photosynthetic pigments and mycosporine-like amino acids after exposure of the marine red alga *Chondrus crispus* (Gigartinales, Rhodophyta) to different light qualities. *Phycologia, 44*.

Kremer, B. P. (1980). Taxonomic implications of algal photoassimilate patterns. *British Phycological Journal, 15*, 399–409.

Krueger-Hadfield, S.A. (2011). *Structure des populations chez l'algue rouge haploid-diploïde Chondrus crispus: système de reproduction, différérenciation génétique et epidémiologie* (Ph.D. thesis). Paris: Université de Pierre et Marie Curie; Santiago: Pontificia Universidad Católica de Chile.

Krueger-Hadfield, S. A., Collén, J., Daguin-Thiébaut, C., & Valero, M. (2011). Genetic population structure and mating system in *Chondrus crispus* (Rhodophyta). *Journal of Phycology, 47*, 440–450.

Krueger-Hadfield, S. A., Roze, D., Destombe, C., Correa, J., & Valero, M. (2014). O father, where art thou? Paternity analyses in a natural population of the red seaweed *Chondrus crispus. Heredity (in press)*

Krueger-Hadfield, S. A., Roze, D., Mauger, S., & Valero, M. (2013). Intergametophytic selfing and microgeographic genetic structure shape populations of the intertidal red seaweed *Chondrus crispus. Molecular Ecology, 22*, 3242–3260.

Kübler, J. E., & Davison, I. R. (1993). High-temperature tolerance of photosynthesis in the red alga *Chondrus crispus. Marine Biology, 117*, 327–335.

Lamberto, M., & Ackman, R. G. (1994). Confirmation by gas chromatography/mass spectrometry of two unusual trans-3-monoethylenic fatty acids from the Nova Scotian seaweeds *Palmaria palmata* and *Chondrus crispus. Lipids, 29*, 441–444.

Laycock, M. V., & Craigie, J. S. (1977). The occurrence and seasonal variation of gigartinine and L-citrullinyl-L-arginine in *Chondrus crispus* Stackh. *Canadian Journal of Biochemistry, 55*, 27–30.

Leblanc, C., Boyen, C., Richard, O., Bonnard, G., Grienenberger, J. M., & Kloareg, B. (1995). Complete sequence of the mitochondrial DNA of the Rhodophyte *Chondrus crispus* (Gigartinales). Gene content and genome organization. *Journal of Molecular Biology, 250*, 484–495.

Leblanc, C., Kloareg, B., Loiseaux-de Goër, S., & Boyen, C. (1995). DNA sequence, structure, and phylogenetic relationship of the mitochondrial small-subunit rRNA from the red alga *Chondrus crispus* (Gigartinales, Rhodophytes). *Journal of Molecular Evolution, 41*, 196–202.

Leblanc, C., Richard, O., Kloareg, B., Viehmann, S., Zetsche, K., & Boyen, C. (1997). Origin and evolution of mitochondria: what have we learnt from red algae? *Current Genetics, 31*, 193–207.

Lee, S.-H., Athukorala, Y., Lee, J.-S., & Jeon, Y.-J. (2008). Simple separation of anticoagulant sulfated galactan from marine red algae. *Journal of Applied Phycology, 20*, 1053–1059.

Lewis, J. G., Stanley, N. F., & Guist, G. G. (1988). Commercial production and applications of algal hydrocolloids. In C. A. Lembi, & J. R. Waaland (Eds.), *Algae and human affairs* (pp. 205–236). Cambridge: Cambridge University Press.

Lima, F. P., Ribeiro, P. A., Queiroz, N., Hawkins, S. J., & Santos, A. M. (2007). Do distributional shifts of northern and southern species of algae match the warming pattern? *Global Change Biology, 13*, 2592–2604.

Liu, J., Hafting, J., Critchley, A. T., Banskota, A. H., & Prithiviraj, B. (2013). Chemical components of the cultivated red seaweed *Chondrus crispus* enhance immune response of *Caenorhabditis elegans* to *Pseudomonas aeruginosa* through pmk-1, daf-2/daf-16, and skn-1 pathways. *Applied and Environmental Microbiology, 79*, 7343–7350.

Lüning, K., Guiry, M. D., & Masuda, M. (1986). Upper temperature tolerance of North Atlantic and North Pacific geographical isolates of *Chondrus species* (Rhodophyta). *Helgoländer Meeresuntersuchungen, 41*, 297–306.

MacFarlane, C. I. (1968). *Chondrus crispus Stackhouse, a synopsis FAO report.* Halifax: Nova Scotia Research Foundation Bulletin.

Marshall, S. M., Newton, L., & Orr, A. P. (1949). *A study of certain British seaweeds and their utilization in the preparation of agar.* Ministry of Health and Ministry of Supply, HM Stationery Office.

Mathieson, A. C. (1982). Seaweed cultivation: a review. In C. J. Sindermann (Ed.), *Proceedings of the 6th US-Japan meeting on aquaculture*. Santa Barbara, CA: National Marine Fisheries Service.

Mathieson, A. C., & Burns, R. L. (1971). Ecological studies of economic red algae. I. Photosynthesis and respiration of *Chondrus crispus* Stackhouse and *Gigartina stellata* (Stackhouse) Batters. *Journal of Experimental Marine Biology and Ecology*, 7, 197–206.

Mathieson, A. C., & Burns, R. L. (1975). Ecological studies of economic red algae. V. Growth and reproduction of natural and harvested populations of *Chondrus crispus* Stackhouse in New Hampshire. *Journal of Experimental Marine Biology and Ecology, 17*, 137–156.
Mathieson, A. C., & Norall, T. L. (1975). Photosynthetic studies of *Chondrus crispus*. *Marine Biology, 33*, 207–213.
Mathieson, A. C., & Prince, J. S. (1973). Ecology of *Chondrus crispus* Stackhouse. *Proceedings of the Nova Scotian Institute of Science, 27*, 53–79.
Matsumoto, K., & Shimada, S. (2013). Taxonomic reassessment of *Chondrus verrucosus* (Rhodophyta, Gigartinales), with a description of *Chondrus retortus* sp. nov. *Phycological Research, 61*, 299–309.
McCandless, E. L., Craigie, J. S., & Walter, J. (1973). Carrageenans in the gametophytic and sporophytic stages of *Chondrus crispus*. *Planta, 112*, 201–212.
Michel, G., Chantalat, L., Duee, E., Barbeyron, T., Henrissat, B., Kloareg, B., et al. (2001). The kappa-carrageenase of *P. carrageenovora* features a tunnel-shaped active site: a novel insight in the evolution of clan-B glycoside hydrolases. *Structure, 9*, 513–525.
Michel, G., Chantalat, L., Fanchon, E., Henrissat, B., Kloareg, B., & Dideberg, O. (2001). The ι-carrageenase of *Alteromonas fortis*. *Journal of Biological Chemistry, 276*, 40202–40209.
Mitchell, M. E., & Guiry, M. D. (1983). Carrageen: a local habitation or a name? *Journal of Ethnopharmacology, 9*, 347–351.
Molina, F. I., Hughes, G. C., & Craigie, J. S. (1988). Temperature influenced infection rates in the *Chondrus crispus-Petersenia pollygaster* pathosystem: a regression analysis. *Marine Biology, 97*, 431–433.
Mouritsen, O. G. (2013). *Seaweeds edible, available & sustainable*. London: University of Chicago Press, Ltd.
Mutwil, M., Debolt, S., & Persson, S. (2008). Cellulose synthesis: a complex complex. *Current Opinion in Plant Biology, 11*, 252–257.
Neish, A. C. (1968). Marine agriculture – the prospects are good for crops from the sea. *Science Forum, 1*, 15–16.
Neish, A. C., & Fox, C. H. (1971). Greenhouse experiments on the vegetative propagation of *Chondrus crispus* (Irish moss). In *National Research Council of Canada, Atlantic regional laboratory technology report 12*.
Newton, L., Devonald, V. G., & Jones, N. R. (1957). Recherches sur *Chondrus crispus* (L.) Stackh et quelques-unes de ses soi-disant variétiés. *Écologie des Algues Marines, Colloques Internationaux, 81*, 121–139.
Nielsen, R., & McLachlan, J. (1986). Investigations of the marine algae of Nova Scotia. XVI. The occurrence of small green algae. *Canadian Journal of Botany, 64*, 808–814.
Ohno, M., & Largo, D. B. (1998). The seaweed resources of Japan. In A. T. Critchley, & M. Ohno (Eds.), *Seaweed resources of the world* (pp. 1–14). Yokosuka: Japan International Cooperation Agency.
Papenfuss, G. F. (1950). Review of the genera of algae described by Stackhouse. *Hydrobiologia, 2*, 181–208.
Pereira, J. (1840). *The elements of Materia Medica*. London: Longman. Part 2.
Plumb, J. (1999). *Population dynamics and endophytic flora of Chondrus crispus (Rhodophyta): a temporal study* (Ph.D. thesis). Bristol, UK: University of the West of England, 201 pp.
Popper, Z. A., Michel, G., Herve, C., Domozych, D. S., Willats, W. G., Tuohy, M. G., et al. (2011). Evolution and diversity of plant cell walls: from algae to flowering plants. *Annual Review of Plant Biology, 62*, 567–590.
Potin, P. (2008). Oxidative burst and related responses in biotic interactions of algae. In C. D. Amsler (Vol. Ed.), *Algal chemical ecology* (pp. 245–272). Berlin: Springer-Verlag.
Potin, P., Bouarab, K., Salaün, J.-P., Pohnert, G., & Kloareg, B. (2002). Biotic interactions of marine algae. *Current Opinion in Plant Biology, 5*, 308–317.

Prince, J. S., & Kingsbury, J. M. (1973a). The ecology of *Chondrus crispus* at Plymouth, Massachusetts. II. Field studies. *American Journal of Botany, 60*, 964–975.
Prince, J. S., & Kingsbury, J. M. (1973b). The ecology of *Chondrus crispus* at Plymouth, Massachusetts. III. Effect of elevated temperature on growth and survival. *The Biological Bulletin, 145*, 580–588.
Pringle, J. D., & Mathieson, A. C. (1987). *Chondrus crispus.* (sic!). In M. S. Doty, J. F. Caddy, & B. Santelices (Eds.), *Case study of seven commercial seaweed resources.* FAO Fisheries Technology Papers.
Provan, J., Glendinning, K., Kelly, R., & Maggs, C. A. (2013). Levels and patterns of population genetic diversity in the red seaweed *Chondrus crispus* (Florideophyceae): a direct comparison of single nucleotide polymorphisms and microsatellites. *Biological Journal of the Linnean Society, 108*, 251–262.
Provan, J., & Maggs, C. A. (2012). Unique genetic variation at a species' rear edge is under threat from global climate change. *Proceedings of the Royal Society B: Biological Sciences, 279*, 39–47.
Rand, T., Qvist, K. B., Walter, C. P., & Poulsen, C. H. (2006). Characterization of the flavin association in hexose oxidase from *Chondrus crispus. FEBS Journal, 273*, 2693–2703.
Rattray, J. (1886). Preliminary note on the evolution of oxygen by sea-weeds. *Transactions of the Botanical Society of Edinburgh, 16*, 244–258.
Raven, J.A., & Hurd, C.L. (2012). Ecophysiology of photosynthesis in macroalgae. *Photosynthesis Research, 113*, 105–125.
Rees, D. A. (1972). Shapely polysaccharides. The eighth Colworth medal lecture. *Biochemical Journal, 126*, 257–273.
Richard, O., Bonnard, G., Grienenberger, J. M., Kloareg, B., & Boyen, C. (1998). Transcription initiation and RNA processing in the mitochondria of the red alga *Chondrus crispus*: convergence in the evolution of transcription mechanisms in mitochondria. *Journal of Molecular Biology, 283*, 549–557.
Richard, O., Kloareg, B., & Boyen, C. (1999). mRNA expression in mitochondria of the red alga *Chondrus crispus* requires a unique RNA-processing mechanism, internal cleavage of upstream tRNAs at pyrimidine 48. *Journal of Molecular Biology, 288*, 579–584.
Ricohermoso, M. A., & Deveau, L. E. (1979). Review of commercial propagation of *Eucheuma* (Florideophyceae) clones in the Philippines. *Proceedings International Seaweed Symposium, 9*, 525–531.
Roleda, M.Y., Wiencke, C., Hanelt, D., & Bischof, K. (2007). Sensitivity of the early life stages of macroalgae from the Northern Hemisphere to ultraviolet radiation. *Photochemistry and Photobiology, 83*, 851–862.
Rosenvinge, L. K. (1906). Meddelelser fra den Botaniske Forening 1. Leptosphaeria marina Rostr. *Botanisk Tidsskrift, 27*, 33–35.
Rostrup, F. G. E. (1889). Mykologiske meddelelser. *Botanisk Tidsskrift, 17*, 228–237.
Ruperez, P. (2002). Mineral content of edible marine seaweeds. *Food Chemistry, 79*, 23–26.
Rupérez, P., & Toledano, G. (2003). Indigestible fraction of edible marine seaweeds. *Journal of the Science of Food and Agriculture, 83*, 1267–1272.
Sagert, S., Forster, R. M., Feuerpfeil, P., & Schubert, H. (1997). Daily course of photosynthesis and photoinhibition in *Chondrus crispus* (Rhodophyta) from different shore levels. *European Journal of Phycology, 32*, 363–371.
Sangha, J. S., Fan, D., Banskota, A. H., Stefanova, R., Khan, W., Hafting, J., et al. (2013). Bioactive components of the edible strain of red alga, *Chondrus crispus*, enhance oxidative stress tolerance in *Caenorhabditis elegans. Journal of Functional Food, 5*, 1180–1190.
Santelices, B., Correa, J. A., Aedo, D., Flores, V., Hormazábal, M., & Sánchez, P. (1999). Convergent biological processes in coalescing Rhodophyta. *Journal of Phycology, 35*, 1127–1149.
Sarker, M.Y., Bartsch, I., Olischläger, M., Gutow, L., & Wiencke, C. (2013). Combined effects of CO_2, temperature, irradiance and time on the physiological performance of *Chondrus crispus* (Rhodophyta). *Botanica Marina, 56*, 63–74.

Schaal, G., Riera, P., & Leroux, C. (2010). Trophic ecology in a Northern Brittany (Batz Island, France) kelp (*Laminaria digitata*) forest, as investigated through stable isotopes and chemical assays. *Journal of Sea Research, 63*, 24–35.
Schacklock, P. F., & Croft, G. B. (1981). Effect of grazers on *Chondrus crispus* in culture. *Aquaculture, 22*, 331–342.
Schacklock, P. F., & Doyle, R. W. (1983). Control of epiphytes in seaweed cultures using grazers. *Aquaculture, 31*, 141–151.
Schmidt, C. (1844). Über Pflanzenschleim und Bassorin. *Justus Liebigs Annalen der Chemie, 51*, 29–62.
Scrosati, R., & Mudge, B. (2004). Persistence of gametophyte predominance in *Chondrus crispus* (Rhodophyta, Gigartinaceae) from Nova Scotia after 12 years. *Hydrobiologia, 519*, 215–218.
Sieburth, J. N., & Tootle, J. L. (1981). Seasonality of microbial fouling on *Ascophyllum nodosum* (L.) Lejol., *Fucus vesiculosus* L., *Polysiphonia lanosa* (L.) Tandy and *Chondrus crispus* Stackh. *Journal of Phycology, 17*, 57–64.
Simmonds, P. L. (1854). On some undeveloped and unappreciated articles of raw produce from different parts of the world. *Journal of the Society of Arts, 2*, 33–42.
Sinha, R. P., Singh, S. P., & Häder, D.-P. (2007). Database on mycosporines and mycosporine-like amino acids (MAAs) in fungi, cyanobacteria, macroalgae, phytoplankton and animals. *Journal of Photochemistry and Photobiology, 89*, 29–35.
Smith, H. McC. (1905). The seaweed industries of Japan: the utilization of seaweeds in the United States. *Bulletin US Bureau of Fisheries, 24*, 169–181.
Smith, R. G., & Bidwell, R. G. S. (1987). Carbonic anhydrase-dependent inorganic carbon uptake by the red macroalga, *Chondrus crispus*. *Plant Physiology, 83*, 735–738.
Smith, R. G., & Bidwell, R. G. S. (1989a). Inorganic carbon uptake by photosynthetically active protoplasts of the red macroalga *Chondrus crispus*. *Marine Biology, 102*, 1–4.
Smith, R. G., & Bidwell, R. G. S. (1989b). Mechanism of photosynthetic carbon dioxide uptake by the red macroalga, *Chondrus crispus*. *Plant Physiology, 89*, 93–99.
Smith, D. B., & Cook, W. H. (1953). Fractionation of carrageenin. *Archives of Biochemistry and Biophysics, 45*, 232–233.
Snaith, W. E., MacFarlane, C. I., & Johnston, H. D. (1969). *Some notes on the marine plants industry in Canada*. Ottawa: Department of Fisheries and Forestry.
Stachowicz, J. J., & Whitlatch, R. B. (2005). Multiple mutualists provide complementary benefits to their seaweed host. *Ecology, 86*, 2418–2427.
Stanley, N. (1987). Production and properties and uses of carrageenan. In D. J. McHugh (Ed.), *Production and utilization of products from commercial seaweeds*. Rome: Fisheries Technology Papers, FAO United Nations.
Stanley, S. J. (1992). Observations on the seasonal occurrence of marine endophytic and parasitic fungi. *Canadian Journal of Botany, 70*, 2089–2096.
Stortz, C. A. (2005). Carrageenans: structural and conformational studies. In K. J. Yarima (Ed.), *Handbook of carbohydrate engineering* (pp. 211–245). Boca Raton, FL: CRC Press.
Tasende, M. G., Cid, M., & Fraga, M. I. (2012). Spatial and temporal variations of *Chondrus crispus* (Gigartinaceae, Rhodophyta) carrageenan content in natural populations from Galicia (NW Spain). *Journal of Applied Phycology, 24*, 941–951.
Taylor, A. R. A., & Chen, L. C. -M. (1973). The biology of *Chondrus crispus* Stackhouse: systematics, morphology and life history. In M. J. Harvey, & J. McLachlan (Eds.), *Chondrus crispus* (Vol. 27(pp. 1–21). Proceedings of the Nova Scotian Institute of Science.
Thorbjarnarson, T. (1939). Nyting fjφrugrasa (utilization of Irish moss). *Ægir, 9*, 193.
Tsekos, I. (1999). The sites of cellulose synthesis in algae: diversity and evolution of cellulose-synthesizing enzyme complexes. *Journal of Phycology, 35*, 635–655.
Tsekos, I., Reiss, H.-D., & Schnepf, E. (1993). Cell-wall structure and supramolecular organization of the plasma membrane of marine red algae visualized by freeze-fracture. *Acta Botanica Neerlandica, 42*, 119–132.

Turner, D. (1809). *Fuci, sive plantarum fucorum generi a botanicis ascriptarum icons, descriptions et historia*. London: John and Arthur Arch.

Tveter-Gallagher, E., & Mathieson, A. C. (1980). An electron microscopy study of sporeling coalescence in the red alga *Chondrus crispus*. *Scanning Electron Microscopy*, *3*, 571–579.

Tveter-Gallgher, E., Mathieson, A. C., & Cheney, D. P. (1980). Ecology and developmental morphology of male plants of *Chondrus crispus* (Gigartinales, Rhodophyta). *Journal of Phycology*, *16*, 257–264.

van der Meer, J. P. (1981). The inheritance of spontaneous pigment mutations in *Chondrus crispus* Stackh. (Rhodophyceae). *Proceedings of the Nova Scotian Institute of Science*, *31*, 187–192.

van der Meer, J. P. (1987). Marine algal genetics and genomes. *Hydrobiologia*, *151/152*, 49–56.

van Ginneken, V. J., Helsper, J. P., de Visser, W., van Keulen, H., & Brandenburg, W. A. (2011). Polyunsaturated fatty acids in various macroalgal species from North Atlantic and tropical seas. *Lipids in Health and Disease*, *10*, 104.

Viehmann, S., Richard, O., Boyen, C., & Zetsche, K. (1996). Genes for two subunits of succinate dehydrogenase form a cluster on the mitochondrial genome of Rhodophyta. *Current Genetics*, *29*, 199–201.

Viola, R., Nyvall, P., & Pedersen, M. (2001). The unique features of starch metabolism in red algae. *Proceedings of the Royal Society B: Biological Sciences*, *268*, 1417–1422.

Wahl, M., & Mark, O. (1999). The predominantly facultative nature of epibiosis: experimental and observational evidence. *Marine Ecology Progress Series*, *187*, 59–66.

Wakabayashi, S., Kimura, T., Fukuyama, K., Matsubara, H., & Rogers, L. J. (1989). The amino acid sequence of a flavodoxin from the eukaryotic red alga *Chondrus crispus*. *Biochemical Journal*, *263*, 981–984.

Wanders, A. J., van den Borne, J. J. G.C., de Graff, C., Hulshof, T., Jonathan, M. C., Kristensen, M., et al. (2011). Effects of dietary fibre on subjective appetite, energy intake and body weight: a systematic review of randomized controlled trials. *Obesity Reviews*, *12*, 724–739.

Wang, X., Zhao, F., Hu, Z., Critchley, A. T., Morrell, S. L., & Duan, D. (2008). Inter-simple sequence repeat (ISSR) analysis of genetic variation of *Chondrus crispus* populations from North Atlantic. *Aquatic Botany*, *88*, 154–159.

Weinberger, F., Pohnert, G., Berndt, M. L., Bouarab, K., Kloareg, B., & Potin, P. (2005). Apoplastic oxidation of L-asparagine is involved in the control of the green algal endophyte *Acrochaete operculata* Correa and Nielsen by the red seaweed *Chondrus crispus* Stackhouse. *Journal of Experimental Botany*, *56*, 1317–1326.

Weinberger, F., Pohnert, G., Kloareg, B., & Potin, P. (2002). A signal released by an endophytic attacker acts as a substrate for a rapid defensive reaction of the red alga *Chondrus crispus*. *Chembiochem*, *3*, 1260–1263.

Weinberger, F., & Potin, P. (2010). Red algal defenses in the genomics age. In J. Seckbach & D. J. Chapman (Vol. Eds.), *Red algae in the genomic age* (pp. 457–477). Dordrecht: Springer.

Wilson, I. M., & Knoyle, J. M. (1961). Three species of *Didymosphaeria* on marine algae: *D. danica* (Berlese) comb.nov., *D. pelvetiana* Suth. and *D. fucicola*. *Transactions of the British Mycological Society*, *44*, 55–71.

Witt, H. J. (1985). Carrageenan. Nature's most versatile hydrocolloid. In R. R. Colwell, E. R. Pariser, & A. J. Sinsley (Eds.), *Biotechnology of marine polysaccharides* (pp. 345–360). Washington, DC: Hemisphere Publishing Corp.

Woelkerling, W. J. (1990). An introduction. In K. M. Cole, & R. G. Sheath (Eds.), *Biology of the red algae* (pp. 1–6). Cambridge: Cambridge University Press.

Wong, K. F., & Craigie, J. S. (1978). Sulfohydrolase activity and carrageenan biosynthesis in *Chondrus crispus* (Rhodophyceae). *Plant Physiology*, *61*, 663–666.

Yakovleva, I. M., & Titlyanov, E. A. (2001). Effect of high visible and UV irradiance on subtidal *Chondrus crispus*: stress, photoinhibition and protective mechanisms. *Aquatic Botany*, *71*, 47–61.

Yoon, H. S., Hackett, J. D., Ciniglia, C., Pinto, G., & Bhattacharya, D. (2004). A molecular timeline for the origin of photosynthetic eukaryotes. *Molecular Biology and Evolution*, *21*, 809–818.
Yoon, H. S., Müller, K. M., Sheath, R. G., Ott, F. D., & Bhattacharya, D. (2006). Defining the major lineages of red algae (Rhodophyta). *Journal of Phycology*, *42*, 482–492.
Young, E. G., & Smith, D. G. (1958). Amino acids, peptides, and proteins of Irish moss, *Chondrus crispus*. *Journal of Biological Chemistry*, *233*, 406–410.
Zipperle, A. M., Coyer, J. A., Reise, K., Stam, W. T., & Olsen, J. L. (2011). An evaluation of small-scale genetic diversity and the mating system in *Zostera noltii* on an intertidal sandflat in the Wadden Sea. *Annals of Botany*, *107*, 127–134.

CHAPTER FOUR

Salinity and Desiccation Induced Oxidative Stress Acclimation in Seaweeds

Manoj Kumar*,†, Puja Kumari*, C.R.K. Reddy*,‡,1 and Bhavanath Jha*

*Discipline of Marine Biotechnology and Ecology, CSIR-Central Salt and Marine Chemicals Research Institute, Bhavnagar, Gujarat, India

†Institute of Plant Sciences, Agricultural Research Organization (ARO), The Volcani Center, Bet Dagan, Israel

‡Academy of Scientific & Innovative Research (AcSIR-CSMCRI), CSIR-Central Salt and Marine Chemicals Research Institute, Bhavnagar, Gujarat, India

[1]Corresponding author: e-mail address: crk@csmcri.org

Contents

Abstract

Marine macroalgae, commonly known as seaweeds, are assemblage of diverse groups of phototrophic marine plants and form the base of the marine trophic pyramid. Rocky intertidal zones are the most dynamic and comprises of highly stressful habitats for marine life including seaweeds. They often experience severe environmental stress as a result of periodic exposure to a wide range of ambient conditions including intense radiation, high temperature, desiccation and salinity with turning tides. The relative abundance, survivability and distribution of seaweeds in such environments are principally determined by their tolerance abilities to diverse environmental stresses. Any adverse effect on seaweeds as a result of fluctuating environmental conditions can directly or indirectly lead to

Advances in Botanical Research, Volume 71
ISSN 0065-2296
http://dx.doi.org/10.1016/B978-0-12-408062-1.00004-4

perturbations at higher trophic levels and eventually affect the integrity and sustainability of aquatic ecosystems. The recent proteome, transcriptome, metabolome and other biochemical analysis of seaweeds under oxidative stress have suggested the involvement of mannitol, proline, abscisic acid, polyamines, polyunsaturated fatty acids, oxylipins and fatty acid desaturases among others defending the seaweeds from diverse environmental stress. Both salinity and desiccation stresses are comparable in the context of a reduction of cellular water potential but differ in physiological process of ions uptake and their ratio determines the acclimation potential of seaweeds. In this chapter, we describe various tolerance and adaptive strategies of seaweeds in response to salinity fluctuations and desiccation induced oxidative stress at both biochemical and molecular levels enabling them to endure successfully for extended periods of stresses. Further, the new opportunities that became available from whole genome sequences of the brown alga *Ectocarpus siliculosus* and the red alga *Chondrus crispus*, in gaining newer insights into the cellular mechanisms of stress tolerance at molecular level in seaweeds is also discussed.

4.1 INTRODUCTION

Seaweeds are the primary producers forming the first trophic level in a food chain of aquatic ecosystem (Eklund & Kautsky, 2003). Most of seaweeds are sessile organisms that preferentially grow attached to hard substrata such as rocks, gravel, or occasionally as epiphyte on salt marshy plants (pneumatophores of mangroves) and epizoic on mussel colonies. The intertidal zone of tropical seashores where most of seaweeds occur is marked by the upper and lower limits of the tide; it is exposed at low tide and immersed at high tide, and thus exhibits zonation patterns in both flora and fauna (Myndee, 2010). The vertical distribution of organisms in this realm coincides with horizontal tidal levels, forming three discrete zones namely upper littoral, mid littoral and lower littoral. These zones experience highly dynamic environment and undergo rapid changes in physical factors with turning tides, in addition to the climatic changes owing to seasonal meteorological variations (Davison & Pearson, 1996). Thereby, seaweeds in the intertidal zone often experience severe stress as a result of periodic exposure to a wide range of direct environmental conditions such as salinity fluctuations, desiccation, intense radiations and high temperatures with each turning tide. Further, the harshness of these conditions in the intertidal zone intensifies as the height above sea level increases. The impact of all these stresses tends to be similar because they all directly or indirectly exert considerable pressure on osmotic balance of cells of seaweeds leading to perturbation of various physiological functions at cellular level and eventually affecting the productivity of aquatic ecosystems (Davison & Pearson, 1996). Adaptation and acclimation to such environmental stress are therefore of particular importance in intertidal benthic organisms such as seaweeds and other biota. Despite the publication of a great number of articles on their abilities to tolerate such extreme and regularly fluctuating environmental conditions resulting from the tidal cycles, there is still a debate on the exact mechanisms of stress tolerance.

Seaweeds distribution and relative abundance along the lower limits of intertidal zone are mainly controlled by biotic factors such as predation and intraspecific and interspecific competition, while the upper limit distribution is mostly determined by species specific width of tolerance to abiotic stress factors such as salinity and desiccation (Kirst, 1990 and references therein). Under salt stress conditions, seaweed cells are still in full contact to water with reduced water potential, whereas desiccation leads to strong cellular dehydration. Therefore, salinity stress is often defined as a 'physiological drought' (Kirst, 1990). In the open oceans salinity is around 33–35 S_A (S_A is absolute salinity (i.e. mass fraction of salt in sea water, a newly introduced standard to measure salinity), http://www.teos-10.org (Wright, Pawlowicz, McDougall, Feistel, & Marion, 2010)) and is the most common, however it gradually decreases from the tropics towards the polar seas. In near coast water salinity can also vary from 10 to 70 S_A, due to precipitations and/or fresh water influxes resulting in low salinity (hyposalinity stress), or evaporation of water may raise salinity (hypersalinity stress). Therefore, the seaweeds of higher intertidal zone must be tolerant to both salinity extremes and should be able to survive or resist to both loss of water and ions during their exposure to hypersalinity and hyposalinity, respectively.

Desiccation tolerance is the ability of an organism to dry to equilibrium with the ambient air and then recover and return to normal metabolism on remoistening (Proctor & Pence, 2002). The environmental exposure during low tide condition leads to nutrient limitation specially nitrogen and demands the intertidal seaweeds to prepare early for the desiccation followed by rehydration and associated cellular damage (Burritt, Larkindale, & Hurd, 2002). This constant state of readiness requires a great deal of energy budget and could be a contributing factor for slow growth rates of algae dwelling at the upper littoral zone as compared to those at lower littoral zone. The desiccation tolerance of marine macroalgae is considered to be a major factor responsible for vertical zonation patterns in the intertidal zone. More precisely, it is the extent to which the photosynthetic apparatus can recover from water loss upon re-immersion that clearly distinguishes desiccation-tolerant species (inhabiting in upper intertidal zone) from desiccation-sensitive species (inhabiting in lower intertidal zone). Thus, it is the ability to withstand desiccation stress (fast recovery during rehydration) and not that to avoid desiccation (water retaining ability), is the key factor determining their vertical distribution (Ji, Gao, & Tanaka, 2005).

Both hypersalinity and desiccation are considered to be closely related and reflect two different forms of water deprivation. In general high salinity or hyperosmolarity is desiccating, whereas low salinity and/or hypo-osmotic shock promote hydration. During desiccation, cellular ionic concentration increases, but the ion ratios remain constant. In contrast, during salinity stress, algal cells may not only increase their ionic concentrations, but also undergo changes in ion ratios owing to selective uptake. Salinity fluctuation and/or desiccation induced stress in algal cells result in cell membrane leakage of ions and electrolytes, pH changes, solute crystallisation and protein denaturation triggering alteration in series of physiological processes together with the accumulation of reactive oxygen species (ROS) (Bischof & Rautenberger, 2012, Collen & Davison, 1999 a,b; Karsten, 2012; Kumar & Reddy, 2012, pp. 1–160). In all aerobic organisms, the concentration of ROS is tightly controlled by ROS-scavenging pathways that metabolise ROS. However, an imbalance in generation and metabolism of ROS disrupts redox homeostasis of cell and brings it in to the state of 'oxidative stress' leading to inhibition and destruction of photosynthetic apparatus, DNA, proteins and cell membranes. Despite their destructive activity, ROS act at subtle level as signalling molecules in a variety of cellular processes including development and tolerance to environmental stresses. Because of the multifunctional roles of ROS, it is necessary for the cells to control the level of ROS tightly to avoid any oxidative injury and not to eliminate them completely. During acclimation to altered environmental conditions particularly in salinity fluctuations and desiccation stress, which are associated with the formation of ROS, seaweeds express a battery of antioxidant enzymes together with maintaining higher level of nonenzymatic antioxidants and metabolites (including proline, polyamines (PAs), low molecular weight (LMW) carbohydrates, polyols, oxylipins and polyunsaturated fatty acids (PUFAs)). These compounds contribute to control the cellular ROS levels, maintain cellular redox status, and homeostasis is of major focus in this chapter. Further, accumulation of UV absorbing compounds, dynamic distribution of photosynthetic pigments, transient inhibition of macroalgal photosynthesis and even morphological/structural changes, which have been suggested as some of successful protection strategies, are also discussed. Most importantly, the lack of genomic information in studying the stress response is gradually diminished with the influx of large expressed sequence tag (EST) projects on red, brown and green alga. Moreover, the whole genome sequences availability for the emerging brown algal model *Ectocarpus siliculosus* (Cock et al., 2010), the red algae *Chondrus crispus* (Collen, Porcel, Carre,

Ball, Chaparro, et al., 2013) and *Porphyridium purpureum* (Bhattacharya et al., 2013) provides excellent opportunities to identify genes with potential roles in the environmental stress tolerance and other stress responses. The gain of this approach into studies on the mechanisms behind the physiological and ecological stress responses at whole transcriptome level are also discussed in this chapter.

4.2 ALTERATIONS IN PHYSIOLOGICAL PERFORMANCE OF SEAWEEDS UNDER SALINITY AND DESICCATION STRESS

4.2.1 Effect on Growth, Photosynthesis, Respiration, Morphology and Ultrastructural Changes

Changes in salinity may affect organisms in three ways: (1) osmotic stress with direct impact on the cellular water potential; (2) ion (salt) stress caused by the inevitable uptake or loss of ions, which simultaneously is a part of the acclimation, and toxic effect of Na^+ on cellular structure and (3) change in the cellular ionic ratios due to the selective ion permeability of the membrane (Kirst, 1990). On the contrary, desiccation leads to increase in cellular ionic concentrations, but the ion ratios remain constant. Both salinity and desiccation stress affect the internal osmotic potential, eventually result in impairment in ecophysiological performances such as rate of survival, growth, photosynthesis and respiration.

The different environmental factors especially light, desiccation and salinity (osmotic stress) have been studied in defining the species-specific width of tolerance and has been considered as critical factors governing the vertical limits of diverse algal species in intertidal zone. It has been shown that eulittoral or supralittoral species have stronger tolerance to salinity stress compared to sublittoral species. *Ulva* together with *Ulothrix*, *Cladophora*, *Prasiola* and *Acrosiphonia,* are euryhaline algae inhabiting in eulittoral zone, and have shown to be less affected by salinity fluctuations than the sublittoral red and brown algal species. This capacity is due to their flexible thin cell membrane allowing cells to swell with the influx of water in low salinities, and they undergo deplasmolysis in higher salinities without causing injury to membranes (Russell, 1987). Karsten, Wiencke, and Kirst (1991) demonstrated 100% mortality in sublittoral brown algae such as *Phaeurus antarcticus* in contrast to >75% cell survivability in eulittoral green algal cells when subjected to triple sea water concentration. Growth under hyposalinity conditions (in estuaries) may be governed by the availability of certain inorganic ions such as Ca^{2+} which increase the lower tolerance limits of seaweeds while playing a crucial role in cell signalling, as structural component of seaweed cell walls and membranes, or other cations which balance organic anions in the plant vacuole (Verret, Wheeler, Taylor, Farnham, & Brownlee, 2010). However, growth of seaweeds is usually reduced in hypersaline water, because of both cumulative enzyme effects and reduced turgor pressure that inhibit cell division. Thus, euryhaline feature of eulittoral green seaweeds may be considered as important environmental factor in determining their vertical zonation on the shore (Russell, 1987). Recent studies, on the growth rates of *Ulva pertusa* under nonlimiting nutrient and light conditions within a salinity range from 5 to 40 S_A, demonstrated that the growth rates were maximum for salinity 15–25 S_A, and decreased considerably in both low- and high-extreme salinities (Choi, Kang, Kim, & Kim, 2010). Similarly, other *Ulva* species have been studied with different tolerance capacities to low- and high-salinity regimes such as *Ulva lactuca* in culture was more tolerant of hypersalinity than hyposalinity (Murthy, Sharma, & Rao, 1988); *Ulva rotundata* grew well in salinities 25–33 S_A (De Casabianca, 1989) and *Ulva rigida* bloomed in salinity ≥20 S_A (Fillit, 1995). Further, Ichihara, Miyaji, and Shimada (2012) compared the low salinity tolerance (5 S_A) of closely related *Ulva* sp. *Ulva linza* (intertidal species) and *Ulva prolifera* (brackish water species), and concluded that brackish water species are the most tolerant intertidal species. Recently, Yu, Lim, and Phang (2013) demonstrated the stunted growth of euryhaline red algae *Gracilaria edulis* in salinity regimes of 5–40 S_A, in contrast to *Gracilaria tenuistipitata* growth, which was not limited by salinity. Further, tetrasporophytes of both these algae have showed higher growth rates at low salinities (15 S_A) suggesting their suitable cultivation conditions in estuarine region with brackish waters. Kumar, Kumari, Gupta, Reddy, and Jha (2010) also demonstrated the higher growth of *Gracilaria corticata* in salinities ranging from 25 to 35 S_A. Other *Gracilaria* species have also been examined for their optimal growth in saline waters such as *Gracilaria cornea* best grew in 30 S_A salinity, *Gracilaria vermiculophylla* (20 S_A), *Gracilaria verrucosa*, *Gracilaria chorda* and *Gracilaria salicornia* (25–30 S_A) and *Gracilaria domingensis* (40 S_A) (Yu et al., 2013 and references therein).

In addition, the nutrient uptake by algae under salinity stress has been suggested as a key player in determining the stress tolerance limit and growth. It appeared that salinity has a greater impact on phosphate uptake than on nitrate uptake. Choi et al. (2010) demonstrated that uptake rates for phosphate in *U. pertusa* increased more rapidly than for nitrate at lowered salinities (5–20 S_A), resulting in decreased N:P ratio with higher salinities.

Similarly, the recovery of nitrate uptake after desiccation has been suggested critical in determining the zonation of algae on the shore. Desiccation appeared to enhance the short-term uptake of nitrate, ammonium and phosphate in some eulittoral species (Thomas, Turpin, & Harrison, 1987; Hurd & Dring, 1991). Recently, Kim, Kraemer, and Yarish (2009) demonstrated that both moderately and severely desiccated tissues of eulittoral *Porphyra umbilicalis* had very similar nitrate uptake rate in comparison with continuously submerged tissues, whereas the uptake rate in sublittoral *Porphyra amplissima* significantly decreased as desiccation progressed. Further, nitrogen release during emersion by *P. umbilicalis* into the environment has suggested a novel role of global N cycle influencing other physiological functions such as nitrate uptake, nitrate reductase and glutamine synthetase activity (Kim, Kraemer, & Yarish, 2013). Tolerance to desiccation instead of avoiding water loss is critical in determining the survivability and vertical distribution of intertidal macroalgae (Davison & Pearson, 1996). On the contrary, Dromgoole (1980) examined more than a dozen intertidal macroalgae and found no clear relationship between the rate of dehydration of species and their intertidal positions. Later, Murthy and Sharma (1989) suggested that higher activities of enzymes such as nitrate reductase, amylase, invertase and peroxidase in *U. lactuca* (growing in the upper littoral zone) are related with adaptation mechanism to extended periods of emergence. In addition, Burritt et al. (2002) concluded that high intertidal populations have a greater ability to reduce the build-up of hydrogen peroxide thus limiting lipid peroxidation and subsequent membrane and protein damage compared to subtidal populations conforming to their higher tolerance to desiccation.

Photosynthesis and respiration are the most important physiological aspects since their direct relation to the potential of a photosynthetic organism to grow and compete for light and other resources with other individuals and species. A relatively salt and desiccation insensitive photosynthesis and respiration seems to be a prerequisite for the successful occupation of the eulittoral habitat and may ensure long-term survival and reproduction under large amplitudes of salinity and desiccation stress in combination with other environmental factors. The photosynthetic apparatus is very sensitive to injury during desiccation and maintenance of it in a recoverable condition throughout the desiccation process or quick repair upon re-immersion is necessary for restoring photosynthetic activity. The criterion of photosynthetic recovery has been used for evaluating the adaptation of intertidal macroalgae after emersion. Dring and Brown (1982) investigated the recovery in photosynthetic performance of five brown algal species collected from different intertidal heights and found that the high intertidal species *Pelvetia canaliculata* and *Fucus spiralis* recovered from 80% to 90% water loss while the subtidal species *Laminaria digitata* could not fully recover from the water loss beyond 60%. Similarly, Abe, Kurashima, Yokohama, and Tanaka (2001) worked on 18 intertidal seaweeds and found that algae from uppermost intertidal zone *Porphyra dentata*, could fully recover the photosynthetic activity after being desiccated at a water potential of −158 MPa, when compared to those collected from lower intertidal zones. Ji and Tanaka (2002) concluded that decline in photosynthetic rate following desiccation is not correlated with height on the shore, but with the lower rates of water loss during desiccation. Later, Zou and Gao (2002) also observed similar phenomenon with significant decrease in net photosynthesis, dark respiration, photosynthetic efficiency and apparent carboxylating efficiency of *Porphyra haitanensis* following desiccation exposure during low tide conditions. Ji et al. (2005) studied the photosynthetic and respiratory performance of 12 seaweeds (collected from different intertidal regimes) following 2 h of desiccation to assess their vertical distribution. They concluded that it was the ability to withstand desiccation stress (fast recovery during rehydration) but not that to avoid desiccation (water retaining ability) for their vertical distribution. Chlorophyll fluorescence measurement using Pulse amplitude modulated (PAM) fluorometry has recently been recognised as an independent method for assessing algal physiology in the aquatic environment. Kim and Garbary (2007) studied the photosynthetic responses using PAM fluorometry in *Codium fragile* exposed to desiccation for 5 h, and found that thalli lost 20% of their mass after desiccation, but still showed high levels of photosynthetic activity. Nelson, Olson, Imhoff, and Nelson (2010) examined the role of aerial exposure in determining the elevation patterns in ulvoid algae, *U. lactuca* and *Ulvaria obscura*. Among these species *U. lactuca* found to be highly efficient in maintaining the photosynthetic integrity measured by PAM fluorometry, when both the algae were exposed to a constant time. Thus, *U. lactuca* was physiologically superior in the intertidal zone since it desiccates slowly than *U. obscura*. While photosynthetic activity during dehydration can easily be followed using chlorophyll fluorescence techniques such as PAM, this approach is not applicable for measuring respiration. Most respiration studies on green algae were done in the fluid phase using Clark-type electrodes or optodes, but these techniques are not suitable for use in desiccation stress, and hence only a few investigations have examined respiration during dehydration in marine algae. The best method to follow respiratory activity during dehydration is the infrared gas analyser (IRGA), which measures the release of carbon dioxide in the dark, under controlled atmospheric conditions. So far, IRGA investigations have been mainly applied to lichens and their green algal photobionts (Holzinger & Karsten, 2013).

Many studies have indicated that the photochemical reactions are sensitive indicators to study the physiological state of seaweeds during desiccation and rehydration. The energy transfer to the reaction centres of photosystem II (PSII) and photosystem I (PSI) results in electron transfer to ferredoxin and then reduction of $NADP^+$ to NADPH. The limitation of photosynthetic carbon fixation during dehydration decreases the utilisation of NADPH with a decline in the $NADP^+$ level, a major acceptor of electrons in PSI. Depletion of $NADP^+$ accelerates the transport of electrons from PSI to molecular oxygen and the generation of ROS. The damage to the oxygen-evolving complex of PSII and to the PSII reaction centres has commonly observed in desiccated thalli (Gao et al., 2011; Xia, Li, & Zou, 2004). The primary photosynthetic mechanism during desiccation is affected specifically at the electron transport stage between PSI and PSII. The sensitive site in *Porphyra* and *Ulva* species is most likely between plastoquinone and P 700. In *Porphyra perforata* there seem to be at least three sites in the photosynthetic apparatus that are inhibited by high salinity, namely the photoactivation of electron flow on the reducing side of PSI, the electron flow on the water side of PSII and the transfer of light energy between the pigment complexes. These authors concluded that a free electron flow at all three sites is essential to avoid photodamage through chronic photoinhibition due to accumulation of highly ROS. Recently, Gao et al. (2011) studied the changes in photosynthetic electron transport that occur during desiccation and rehydration in *Ulva* sp. and concluded that PSI-driven cyclic electron flow provides desiccation tolerance instead of linear electron flow (driven by PSII), which got eliminated during desiccation. Later, the same group (Gao et al., 2013) suggested that an increase in amylase activity during desiccation leads to starch degradation and causes the accumulation of NADPH which provide electrons for $P700^+$ and favour the operation of cyclic electron flow via PSI during severe desiccation and its recovery during rehydration. Similarly, salt stress (48–128 S_A) resulted in a significant decrease in photosynthetic O_2 evolution indicating that PSII is the main target which affects most by elevated salinity due to inactivation of reaction centre and inhibition of the electron transport at the acceptor side of PSII (Xia et al., 2004). Further, to this line of research Gylle, Nygard, Svan,

Pocock, and Ekelund (2013) evaluated the differences in relative amount of either D1 (Chlorophyll a binding core protein in PSII), PsaA (Chlorophyll a binding core protein in PSI) and Rubisco in *Fucus* sp. subjected to hyposalinity (10 S_A). Depletion in D1 protein and relatively greater amount of PsaA together with Rubisco resulted in higher ratio of D1/PsaA in hyposaline conditions. The explanation for these changes may be that algae demands more ATP, and greater amount of PsaA may increase the possibility of greater flow of cyclic electron transport around PSI to serve a higher rate of CO_2 concentrating mechanisms and/or a higher rate of CO_2 fixation by Rubisco. Zou and Gao (2002) showed that the photosynthetic and carboxylating efficiency decreased with desiccation in *P. haitanensis* during emersion, suggesting a negative impact of desiccation on both photochemical and carboxylation reactions.

Desiccation in marine macroalgae primarily leads to shrinkage process. Cytological alterations caused by desiccation or salinity fluctuations in algal cells can be best seen while staining the mitochondria or actin-linked cytoskeleton using confocal microscopy. Only a few publications have dealt with morphological and ultrastructural changes in marine algae as a response to desiccation, which primarily leads to shrinkage process. For example, Contreras-Porcia, Thomas, Flores, and Correa (2011) noticed dark purple, tightly folded, stiff and brittle fronds of desiccated *Pyropia columbina* in contrast to greenish red, expanded and translucent fully hydrated fronds. In addition, desiccated fronds were characterised by a condensed and opaque cytoplasm accompanied by a darker homogeneous pigmentation compared to hydrated cells, wherein plastids and vacuoles were clearly recognisable. At ultrastructural level desiccated cells displayed considerable folding of plasma membrane, blurred thylakoidal membranes together with the accumulation of electron dense bodies inside the chloroplast. Recently, Holzinger and Karsten (2013) described the accumulation of plastoglobules in desiccated green algae *Klebsormidium crenulatum*. Plastoglobules are lipoproteins in subcompartments of chloroplast that are permanently linked with thylakoid membranes and contain biosynthetic and metabolic enzymes, which may regulate the melting of thylakoid membrane to plastoglobule while contributing in light protection during desiccation.

4.3 ROS GENERATION AND ANTIOXIDANT SYSTEMS

The fluctuating environmental conditions in the intertidal zone trigger a series of physiological processes and accumulation of ROS (Bischof & Rautenberger, 2012; Collen & Davison, 1999a; Karsten, 2012; Kirst, 1990;

Kumar & Reddy, 2012, pp. 1–160). However, an imbalance in generation and metabolism of ROS leads to a variety of physiological challenges by disrupting redox homeostasis of cell, which is collectively known as 'oxidative stress' exacerbating cellular damages. ROS are generally produced as a by-product of normal aerobic metabolism, involving largely the membrane-linked electron transport processes, redox cascades and metabolisms, whose production are aggravated under the influence of unfavourable environmental cues. The cell organelles such as chloroplast, mitochondria and peroxisomes are vital organelles involved in highly oxidising metabolic activities with intense electron flow rate and are the major source of ROS origin (Table 4.1) (Bhattacharjee, 2005). It has been reported that 1–2% of O_2 consumed by plants is diverted to produce ROS in various subcellular loci (Gill & Tuteja, 2010). ROS are produced directly by excitation of O_2 and subsequent formation of singlet oxygen, or by the transfer of one, two or three electrons to O_2 or by protonation, which results in the formation of superoxide radicals ($O_2^{\cdot -}$), hydrogen peroxide (H_2O_2) or hydroxyl radicals ($OH\cdot$) (Bischof & Rautenberger, 2012). Furthermore, $O_2^{\cdot -}$ can be protonated to form the perhydroxyl radical ($HO_2\cdot$) and can also react with another very influential signalling free radical species i.e. nitric oxide radical ($NO\cdot$) to form more toxic peroxynitrite (OON^-) radical. Although some of these reactive species may function as important signal molecules

Table 4.1 Principal Sites for Reactive Oxygen Species Production in a Plant/Algal Cell

Location in Cell	Site of Production
Chloroplast	Photosystem I and II linked electron transport chains (ETC), ferrodoxin, iron–sulphur cluster, quinone A and B
Mitochondria	Complex I: NADPH dehydrogenase complex
	Complex II: reverse electron flow to complex I
	Complex III: ubiquinone-cytochrome region
	Enzymes: acotinase, galactonolactone dehydrogenase
Endoplasmic reticulum	NAD(P)H-dependent electron transport involving $CytP_{450}$
Plasma membrane	Several oxidoreductases, NADPH oxidases
Peroxisome	Matrix: xanthine oxidase
	Membrane: ETC flavoprotein, NADH, Cyt b
	Metabolic process: fatty acid oxidation, flavin oxidase, glycolate oxidase, polyamines degradation, disproportionation of $O_2^{\cdot -}$ radicals
Apoplast	Oxalate oxidase, amine oxidase and other oxidases
Cell wall	Cell wall linked peroxidases and diamine oxidases

that modify gene expression and modulate the activity of specific defence proteins (Mittler, 2002), at high concentration all ROS can extremely be harmful. The harmful effects include damaging of the nucleic acids, oxidation of proteins (by specific interaction with the respective amino acids involved) and lipid peroxidation in cell membranes. In this context, seaweeds express a battery of enzymes and nonenzymatic antioxidants such as ascorbate (AsA), glutathione (GSH), tocopherol, carotenoids, flavonoids (Table 4.2), which are of immense importance and play a crucial role in controlling cellular ROS levels, redox status and cellular repair, thereby minimising the risk of detrimental effects resulting from ROS-mediated oxidative stress.

Superoxide dismutase (SOD) is a metallo-enzyme and is the most effective intracellular enzymatic antioxidant that is ubiquitous in all aerobic organisms and in all subcellular compartments prone to ROS-mediated oxidative stress. SOD provides the first line of defence against the toxic effects of elevated levels of ROS and catalyses the conversion of $O_2^{\cdot -}$ to H_2O_2 (Mittler, 2002). Thus, SOD acts to keep $O_2^{\cdot -}$ concentrations low and hence decreases the risk of OH• formation via metal catalysed Haber–Weiss type reaction. This enzymatic driven reaction is 10,000-fold faster than the

Table 4.2 Enzymatic and Nonenzymatic Reactive Oxygen Species Scavengers, Their Targets and Reaction Catalyse

Nonenzymatic Antioxidants	Targets	Enzymatic Antioxidants	Reaction Catalysed
Ascorbate	$^1O_2$, $O_2^{\cdot -}$, H_2O_2, $OH^{\cdot}$	Superoxide dismutase	$2O_2^{\cdot -} + 2H^+ \rightarrow 2H_2O_2 + O_2$
Glutathione	Peroxide radicals	Catalase	$H_2O_2 \rightarrow H_2O + ½O_2$
α-Tocopherol	$ROO^{\cdot}$	Ascorbate peroxidase	$H_2O_2 + AsA \rightarrow 2H_2O + DHA$
Carotenoids	$^1O_2$, $ROO^{\cdot}$	Glutathione peroxidase	$H_2O_2 + GSH \rightarrow H_2O + GSSG$
Flavonoids	$OH^{\cdot}$,ONOOH, HOCl	Glutathione reductase	$GSSG + NADPH \rightarrow 2GSH + NADP^+$
Phytochelatins	Metal chelator	Mono-dehydroascorbate reductase	$MDHA + NADPH \rightarrow AsA + NADP^+$
Metallothioneins	$OH^{\cdot}$, metal chelator	Dehydroascorbate reductase	$DHA + 2GSH \rightarrow AsA + GSSG$

AsA, ascorbate; DHA, dehydroascorbate; GSH, glutathione; GSSG, disulphide bond of oxidised glutathione, MDHA, mono-dehydroascorbate.

spontaneous dismutation. SODs are classified by their metal cofactors in three types: (1) copper/zinc (Cu/Zn-SOD); (2) manganese (MnSOD) and (3) iron (FeSOD), which are localised in different cellular compartments (Mittler, 2002). H_2O_2 formed from $O_2^{\cdot-}$ or by the divalent reduction of oxygen (e.g. in photorespiration and β-oxidation of fatty acids) is broken down by several site-specific mechanisms. In chloroplasts, H_2O_2 is broken down mainly by ascorbate peroxidase (APX) using AsA as a substrate that results in the formation of mono-dehydroascorbate (MDHA) or dehydroascorbate (DHA) (Table 4.2). The APX family consists of at least four different isoforms including thylakoid, glyoxisome membrane forms, chloroplast stromal soluble form and cytosolic form (Gill & Tuteja, 2010). AsA can also be regenerated from MDHA and DHA using reduced ferredoxin and GSH. Another enzymatic antioxidant-glutathione reductase (GR) is a flavoprotein oxidoreductase, found in both prokaryotes and eukaryotes. It is a key enzyme of AsA-GSH cycle and plays an important role in defence system against ROS by sustaining the reduced status of GSH. GR is localised predominantly in chloroplast, but also has been found in mitochondria and cytosol (Romero-Puertas et al., 2006). GR catalyses the NADPH-dependent reduction of disulphide bond of oxidised GSH and thus maintains higher reduced GSH pool in the cell. GSH is a molecule involved in many metabolic pathways and anti-oxidative processes in algae and acts as a substrate for glutathione transferases (GSTs) in the cell. Another enzymatic antioxidant catalase (CAT) is a tetrameric heme containing enzyme with potential to directly dismutate H_2O_2 to H_2O and O_2. This enzyme has been studied extensively in higher plants and seems to be localised in peroxisomes, mitochondria and cytosol. Glutathione peroxidase (GPX) also catalyses the breakdown of H_2O_2, organic and lipid hydroperoxides to H_2O and O_2 while using GSH as reducing agent. GPXs are generally localised in cytosol, chloroplast, mitochondria and endoplasmic reticulum. Monodehydroascorbate reductase (MDHAR) is flavin adenine dinucleotide (FAD) enzyme that is present as chloroplastic and cytosolic isoenzymes, exhibits high specificity for MDHA as electron acceptor and uses NADPH as the electron donor. MDHAR reduces the MDHA to produce two molecule of AsA. If NADPH is limited, MDHA would be spontaneously disproportionate to AsA and DHA and then DHA will be reduced to generate AsA by dehydroascorbate reductase using GSH as an electron donor (Table 4.2). Apart from this, cell membranes can also be protected from oxidative damage by carotenoids and tocopherols that quench $^1O_2$ and react with lipid radicals, thereby working as chain-reaction-terminating agents (Kumar & Reddy, 2012, pp. 1–160).

4.4 ACCLIMATION STRATEGIES TO SALINITY AND DESICCATION STRESS IN SEAWEEDS

4.4.1 ROS Detoxification

In all aerobic organisms, the concentration of ROS is tightly controlled by ROS-scavenging pathways that metabolise ROS. Seaweeds possess an efficient anti-oxidative defence that protects the cell from oxidative damage. ROS detoxification is executed by redox-sensing and signalling pathways that largely regulate and control spatiotemporal titre of ROS by modifying the production and scavenging mechanisms of ROS. Phenotypic plasticity in antioxidant defence system in response to a variety of stress inducers has been documented in many higher plants (Apel & Hirt, 2004). There are many evidences wherein antioxidant enzyme activities in marine algae have been altered in response to changes in environmental conditions (Collen & Davison, 1999 a,b; Lu, Sung, & Lee, 2006; Sung, Hsu, Wu, & Lee, 2009; Kumar & Reddy, 2012, pp. 1–160 and references therein). In general, subtidal macroalgal species have shown more sensitive to environmental fluctuations and thus generate more ROS compared to high intertidal species. Higher level of antioxidant enzyme activities in intertidal *U. lactuca* (experiencing more harsh environmental conditions such as salinity variations and desiccation) compared to subtidal *U. lactuca* suggests that intertidal individuals detoxify H_2O_2 more efficiently and are better adapted physiologically in coping with environmental stresses than subtidal individuals (Ross & Van Alstyne, 2007). Similar trends have been noted in interspecific red algal comparisons wherein *Mastocarpus stellatus* scavenged exogenous H_2O_2 more rapidly than *C. crispus*, which occurs lower in the intertidal zone than *M. stellatus* (Collen & Davison, 1999b). The up-regulation of antioxidant enzymes including Mn/FeSOD, APX, GR and CAT in short-term hypersalinity (90 S_A) (Sung et al., 2009) and APX, CAT in long-term hypersalinity (Lu et al., 2006) suggested that SOD and GR are critical for *Ulva fasciata* against oxidative stress. Further, up-regulation of FeSOD gene expression and its activity in hypersalinity were mediated specifically by H_2O_2, but MnSOD, APX, GR and CAT gene expression and corresponding enzyme activity were regulated by factors other than H_2O_2 suggesting a dynamic nature of algal regulatory system in response to short-term hypersalinity that were likely changed after the long-term hypersalinity treatment (Sung et al.,

2009). Enhanced activity of either individual or combined SOD, APX and GR enzymes have also been observed in invasive red alga *Grateloupia turuturu* compared to *Palmaria palmate* (Liu & Pang, 2010) and in *G. corticata* (Kumar et al., 2010) under salinity fluctuations. In *U. prolifera* low salinities (10 S_A) led to a major increase in activities of CAT, SOD and GR coupled with a substantial increase in the contents of AsA, GSH and β-carotenoid. On the contrary, thalli exposed to hypersaline conditions (60 S_A) rapidly exhibited higher activities of CAT, SOD, GR and APX together with the accumulation of GSH but did not affect the content of AsA, α-tocopherol and β-carotenoid, suggesting that reactive oxygen scavenging enzymes played an important role in *U. prolifera* for adapting to the hypersaline conditions (Luo & Liu, 2011). Similar to these findings, the greater ability to prevent or reduce the production of ROS in desiccation exposed *Stictosiphonia arbuscula* (Burritt et al., 2002), *G. corticata* (Kumar et al., 2011) and *P. columbina* (Contreras-Porcia et al., 2011) has been demonstrated just not due to the larger antioxidant pool (AsA + GSH), but rather increased activity of the antioxidant enzymes required to regenerate AsA and GSH. Further, Kumar et al. (2011) observed specific responses of APX isoforms, APX-4 and APX-5 which were apparent under prolonged desiccation exposure (4 h) reflecting their greater ability for H_2O_2 detoxification, while at the same time inhibition of APX-3 evidenced its sensitivity for H_2O_2. The higher activity of APX compared to other enzymes (GPX and GR) involved in H_2O_2 detoxification in their study could be attributed to its easy availability throughout the cell and higher substrate affinity in the presence of AsA as a reductant (Kumar & Reddy, 2012, pp. 1–160). Increased GR and CAT activities in addition to elevated carotenoid contents have been suggested the likely sources for environmental stress tolerance, specifically the scavenging and removal of damaging ROS formed during desiccation in *P. umbilicalis* (Sampath-Wiley, Neefus, & Jahnke, 2008). The up-regulation of *PyCAT* (CAT in *Pyropia*) further strengthened the role of CAT in minimising the effect of oxidative damage in *Pyropia yezoensis* during desiccation (Li et al., 2012). In addition, contraction of the thylakoid lumen coupled with the increase in violaxanthin de-epoxidases and higher proton concentration in the lumen has been suggested for the accumulation of antheraxanthin and zeaxanthin at the expense of violaxanthin, and acting as nonphotochemical quenchers under desiccation stress in *U. pertusa* (Xie, Gao, Gu, Pan, & Wang, 2013). Up-regulation of carotene biosynthesis-related protein in *U. fasciata* exposed to hypersalinity and high light conditions suggested its role in dissipation of excessive energy, thereby acting as nonphotochemical quencher (Hsu & Lee, 2012). Also, iodine accumulation in the apoplast of brown algae such as *L. digitata* has been suggested as an advantage for nonenzymatic detoxification of ROS while interacting with O_3, $^1O_2$ and $O_2^{\cdot-}$ ions, making iodine best antioxidant available in apoplast (Kupper et al., 2008).

4.4.2 Involvements of PAs, Proline, Mycosporine-like Amino Acid, Dehydrins, LMW Carbohydrates and Other Metabolites

In response to different stresses, plants accumulate large contents of different types of compatible solutes, which are LMW, hydrophilic organic compounds and usually nontoxic at high cellular concentrations. These solutes include PAs, amino acids like proline, carbohydrates such as sucrose, polyols, trehalose and quaternary ammonium compounds such as glycine betaine, alanine betaine, proline betaine and pipecolate betaine contributing to cellular osmotic adjustment, ROS detoxification, protection of membrane integrity and enzymes/protein stabilisation (Hayat et al., 2012). PAs, a group of aliphatic amines important for both plant and algal development, are present as diamine putrescine (Put), triamine spermidine (Spd), tetramine spermine (Spm) and thermospermine in free and conjugated forms (bound to small molecules such as hydroxycinnamic acid, or to larger molecules such as proteins or nucleic acids). In the pathway of PA metabolism, arginine decarboxylase or ornithine decarboxylase catalyses the decarboxylation of L-arginine/ornithine to form Put. Enzymes namely Spd synthase and Spm synthase catalyse the synthesis of Spd and Spm from Put and Spd, respectively; while diamine oxidases and polyamine oxidases catalyse the decomposition of PAs. Further, Put can also be synthesised from methionine pathway with the involvement of methionine adenosyltransferase (MAT) and decarboxylase. Compared to higher plants relatively little is known about the role of PAs in algae, although Put, Spd and Spm, as well as some less common PAs, have been detected in macroalgae. In macroalgae, PAs functionality has been addressed for their involvement in cell division, maturation of reproductive structures and callus induction, but their functions as stress busters against salinity and desiccation stress and also as metal chelator to protect seaweeds from oxidative stress has reported recently (Garcia-Jimenez, Just, Delgado, & Robaina, 2007; Kumar, Bijo, Baghel, Reddy, & Jha, 2012; Kumar et al., 2011, 2010, Kumar & Reddy, 2012, pp. 1–160). Endogenous and/or exogenous PAs have been shown to modulate stress triggered ROS homeostasis and oxidative damage (mediated by molandialdehyde1) by enhancing the antioxidant enzyme activities including SOD, CAT, APX and GR and pools of nonenzymatic antioxidants (Kumar et al., 2012, 2010). A moderate hyposaline shock in *Grateloupia doryphora* caused an increase in the free fraction of Put, Spd and Spm, mainly due to a decrease in transglutaminase activity, together with an apparent increase in the L-arginine-dependent PAs synthesis (Garcia-Jimenez et al., 2007). Interestingly, the accumulation of Spm together with partial inhibition of spermine synthase during desiccation exposure invariably suggested the synthesis of Spm via methionine instead of ornithine (Kumar & Reddy, 2012, pp. 1–160). PAs have been suggested to form binary or tertiary complexes with cations (at physiological pH) and also with anions such as phospholipids polar head that impede auto-oxidation of Fe^{2+} and phospholipids and subsequently reduce generation of ROS (Shi & Chan, 2013). In addition, stabilisation of PSII proteins of thylakoid membranes (D1, D2, ctyb6/f) via covalent binding of PAs with proteins catalysed by transglutaminases and/or by electrostatic interaction owing to their polycationic nature has also been suggested (Shu, Guo, & Yuan, 2012). Further, Li et al. (2013) noticed the induction of protein disulphide isomerase (*UfPDI*) by hypersalinity in *U. lactuca* and reported that *UfPDI* expression get induced by Spd or Spm in contrast to Put before exposure to hypersaline conditions and continued up-regulated after hypersalinity exposure contributing to salinity tolerance and restoring the growth rate. On the contrary, Put accumulation in extreme hyposaline condition (5 S_A) and H_2O_2 generation from its decomposition has been suggested a causal factor of hyposaline injury in *U. fasciata* (Lee, 1998). The elevated expression of MAT under hypersalinity and desiccation further suggested the role of PAs to some extent in alleviating the oxidative stress conditions (Qiao, Li, Li, Tong, & Hou, 2013). Recently, Spm was found to regulate the stabilisation of DNA methylation by reducing cytosine demethylation to alleviate the cadmium induced stress in *Gracilaria dura* (Kumar et al., 2012).

PAs and proline both shares a common substrate ornithine for their biosynthesis. Therefore, the exogenous application of PAs might result in more substrate for proline biosynthesis, especially under stress conditions. Enzyme namely Δ′-pyrroline-5-carboxylate synthase (*P5CS1*) catalysing the proline synthesis, was identified as S-nitrosylation target in Spd-treated plants, with the strong induction of *P5CS1* expression by Spd application provided evidence for nitric oxide (NO)-dependent functional link between PAs and proline in response to abiotic stress (Tanou et al., 2013). NO has been proposed long back to play a role in the adaptation of the intertidal green macroalga *U. lactuca* to desiccation stress (Murthy, Rao, & Reddy, 1986). Recently, NO production in *U. fasciata* upon exposure to high light conditions and the subsequent up-regulation of methionine sulphoxide reductase expression has been suggested for acquisition of full tolerance to high light stress (Hsu & Lee, 2012). An increased activity of Δ′-pyrroline-5-carboxylate reductase (*P5CR*) together with a decrease in Ca^{2+} ions concentration and the blockage of calmodulin action resulted in elevated level of proline in *Ulva* sp. and *E. siliculosus* under hypersalinity condition suggested proline functions as an osmolyte for osmotic adjustment, in stabilising subcellular structures (membranes and proteins), scavenging free radicals, as a nitrogen-storage compound, as an energy source after the release of stress and buffering cellular redox potential under stress (Lee & Liu, 1999; Dittami et al., 2011). Further, Zhang et al. (2008) reported that copper induced proline synthesis is associated with NO generation, which subsequently resulted in higher activity and transcript level of *P5CS* and *P5CR* in *Chlamydomonas reinhardtii*. All these findings together suggest a possible link among NO-PAs-Proline and thus need a detailed investigation in this regard in fresh and/or marine macroalgae subjected to salinity and desiccation induced oxidative stress conditions.

Mycosporine-like amino acids (MAAs) including shinorine, porphyra-334, palythine, asterina-330 and the antioxidant micosporine glycine are synthesised de novo by the shikimic acid pathway and are produced by algae to shield them from UV irradiations during emersions. Mycosporine-glycine can quench $^1O_2$ whereas other MAAs can quench O^{2-}. MAAs could be induced in the lower intertidal *Porphyra leucosticta* compared to *P. umbilicalis,* which generally exposed to extreme desiccation already contains high concentrations of MAA (Blouin, Brodie, Grossman, Xu, & Brawley, 2011). A significant accumulation of UV absorbing compounds in *P. haitanensis* during desiccation stress has also been shown to have a beneficial advantage to compete in the intertidal zone (Jiang, Gao, & Helbling, 2008). In addition, dehydrin proteins have been discovered in a wide range of photosynthetic organisms, including vascular plants, ferns, cyanobacteria and algae, wherein their existence was often correlated to a state of relative desiccation tolerance. All dehydrins are localised near the plasma membranes and have one or more copies of a conserved lysine-rich sequence usually near the C-terminus, thus belong to a large group of Late embryogenesis abundant (LEA) proteins family. Their function has been suggested to protect macromolecules and membranes, as well as to prevent the accumulation of macromolecules during reduced water conditions. The constitutive expression of dehydrin like proteins in *P. umbilicalis* but not in *P. yezoensis* has been suggested in confirming the protection required for survival under desiccation conditions in intertidal regimes (Wong, 2009). Further, accumulation of LMW carbohydrates such as polyols (mannitol, sorbitol, dulcitol, heptitol volemitol and hexitol altritol), heterosides (floridosides, digeneaside and digalactosylglycerol) in various red and brown algal species under natural environmental adverse conditions such as hypersalinity and emersions have strongly suggested their involvement in the process of osmotic acclimation (Karsten, 2012 and references therein). Recently, the functionality of mannitol cycle with salinity induced gene expression and up-regulation of mannitol-1-P dehydrogenase in the brown alga *E. siliculosus* evidenced the regulation of organic osmolytes such as mannitol by salinity fluctuations (Rousvoa et al., 2011).

4.4.3 Ion Homeostasis

Both hypersalinity and desiccation stress in seaweeds affect the internal osmotic potential and thus the acclimation responses of seaweeds to these conditions are comparable, which in general involve the regulation of turgor pressure to a constant value by adjusting internal osmolyte concentrations accordingly (Kirst, 1990). The osmotic acclimation of the cytoplasm and the other cell compartments, particularly the vacuole, is achieved by a variety of feedback mechanisms acting on active ion transport, permeability of the membranes (pores/channels) and turnover rates of the organic metabolite pools. In general, the regulation of inorganic ion concentrations to conquer homeostatic conditions, ions are actively extruded under hypersaline conditions but must be imported under hyposaline stress. The main ions involved in osmotic acclimation are K^+, Na^+, Cl^-, Ca^{2+} and to a lesser extent sulphate, nitrate or phosphate. External Na^+ negatively impacts intracellular K^+ influx together with accumulation of Ca^{2+}, which apparently mediates signal for stress adaptation responses. Due to the similarity in physicochemical properties between Na^+ and K^+ (i.e. ionic radius and ion hydration energy), the former competes with K^+ for major binding sites in key metabolic processes in the cytoplasm, such as enzymatic reactions, protein synthesis and ribosome functions. Thus, uptake systems for K^+ have difficulties discriminating between both ions, and high external Na^+ amounts may result in K^+ deficiency resulting in the inhibition of K^+-dependent metabolic processes (Karsten, 2012). Therefore, all organisms tend to ensure a defined, usually high K^+/Na^+ ratio in the cytoplasm. Higher Na^+/K^+ and Na^+/Ca^{2+} ratios in *G. corticata* under hypersalinity (55 S_A) have been demonstrated with bleaching symptoms coupled with impairment in ion selectivity and integrity of the cell membrane. However, their substantial tolerance at salinity 45 S_A was attributed with the compartmentalisation of

Na^+ into the vacuole together with higher cytoplasmic Ca^{2+}, which may act in signalling mechanism of salinity induced proline accumulation for osmotic adjustment (Lee & Liu, 1999; Kumar et al., 2010).

Ion homeostasis in saline environments is dependent on transmembrane transport proteins that mediate ion fluxes, including H^+-ATPases and H^+-pyrophosphatases, Ca^{2+}-ATPases, secondary active transporters and channels. During osmotic adjustment, the changes in and control of ion composition are likely achieved by regulating the activity of above specific transport systems across plasma membrane and tonoplast of seaweeds. Ion concentrations in algae are mainly regulated by ion-selective carriers driven by the membrane potential. In addition, facilitated diffusion via ion-selective channels may play a cordial role during rapid changes and recovery of ionic composition (Kirst, 1990). Recently, the whole genome sequence of the red alga *P. purpureum* has been accomplished wherein 3.4% of total genes found encoding solute transporters, channels and pumps with the existence of putative sodium–potassium ATPases (Na^+/K^+-ATPase, namely *PyKPA1* and *PyKPA2*) which are absent in land plants (Bhattacharya et al., 2013). *PyKPA1* were expressed preferentially in sporophytes, whereas *PyKPA2* was specifically expressed in gametophytes and have been suggested to play an important role in ion homeostasis by importing K^+ into cells and Na^+ out of the cells at the expense of ATP during salt and cold stress in marine macroalgae (Uji, Hirata, Mikami, Mizuta, & Saga, 2012). Furthermore, Na^+/K^+-ATPase could provide the driving force for secondary active Na^+-coupled solute transporters such as Na^+/HCO_3^- symporter, sodium-dependent phosphate transport system, sodium/glucose cotransporter, which for the first time have been discovered in algae. The presence of these transporters suggests their involvement in transporting osmotically active solutes, such as trehalose or mannosylglycerate (Bhattacharya et al., 2013). Also, the up-regulation of two Na^+/H^+ antiporter genes (namely *PySOS1* and *PyNhaD*) in *P. yezoensis* by light irradiation suggested that coordinated activity between them is important in pH and Na^+ homeostasis under light conditions and may also be functionally active under salt stress (Uji, Monma, Mizuta, & Saga, 2012). Further, higher concentration of intracellular K^+ than in the external medium clearly points to an active uptake of this essential cation from the seaweed exterior via K^+-ATPase transporter. Most interestingly, the beauty of this complex K^+ transport system is that it requires but does not transport Na^+ (source Karsten, 2012). Such a direct activation by the presence of Na^+ would nicely explain how the K^+ transport activity is rapidly enhanced under salt stress in algae. Furthermore, down-regulation of H^+-exporting ATPase in *Gracilaria changii* under hyposaline condition suggested a tolerance strategy by reducing the loss of ions from the cytoplasm (Teo, Ho, Teoh, Rahim, & Phang, 2009). Although K^+, Na^+ and Cl^- represent the major inorganic ions involved in osmotic acclimation, some seaweeds such as *L. digitata* use NO_3^-, *Acrosiphonia arcta* and *U. rigida* use SO_4^{2-} for osmotic requirements (source Karsten, 2012).

4.4.4 Involvement of PUFAs, Oxylipins and Fatty Acid Desaturases

Lipids play crucial roles in maintaining the integrity and fluidity of membranes in seaweeds mainly by modulating the composition of PUFAs thereby altering the level of unsaturation under various abiotic stresses including salinity and desiccation (Kumar et al., 2011, 2010). An enhanced relative proportion of oleic acid (C18:1, *n*-9) and linoleic acid (C18:2, *n*-6, LA) by 1.3- to 1.6-fold with a parallel decrease in palmitoleic acid (C16:1, *n*-9) was observed in *G. corticata* at hypersalinities, >45 S_A (Kumar et al., 2010). However, it was not the case in hyposalinity (15 S_A) due to the presence of more rigid membranes during initial exposure, which prevented the ion loss to certain extent and thus increased membrane fluidity brought by higher PUFAs level was not needed. Further these authors suggested alterations in fatty acids in hypersalinity could be attributed with induced activation of Δ9 FAD enzyme responsible for converting stearic to oleic acid. Higher PUFAs could be an adaptive strategy for maintaining the requisite greater membrane fluidity, to stabilise the protein complexes of PSII and to control membrane physicochemical properties such as the increased activity of Na^+/H^+ antiporter system of the plasma membrane to cope with hypersalinity stress (Kumar et al. 2010). Moreover, Dittami et al. (2011) reported a shift from *n*-3 to *n*-6 PUFAs at higher salinities with the induction of two Δ12 and two Δ15 desaturases (one each microsomal and chloroplastic) in *E. siliculosus.*

Similarly under desiccation, enhanced levels of *n*-6 PUFAs mainly, arachidonic acid (C20:4, *n*-6, AA) and dihomogammalinolenic acid (C20:3 *n*-6, DGLA) were found after 2 h of drought exposure in *G. corticata* (Kumar et al., 2011). This increase in PUFAs was accompanied with a decrease in palmitic acid (C16:0), which decreased to a minimum value of 19.88% total fatty acids (TFA) at 2 h and corresponded to a significant reduction of 43% ($p < 0.01$) over the control (34.83% TFA). Also, with the advancement of desiccation the ratio of unsaturated to saturated fatty acids (UFA/SFA) varied with a maximum of 3.72 (2 h) and minimum 1.37 (4 h) compared with the control 1.68. The significant accumulation of unsaturated fatty acids (AA, DGLA and oleic acid) at the expense of dominant saturated fatty acids (C16:0, 18:0) and monounsaturated fatty acid (oleic acid) indicated the induction of desaturases during the short-term exposure (2 h). A possible explanation for this change could be the desiccation induced partial induction of elongases, which catalyses the conversion of LA (18:2, *n*-6) to eicosadienoic acid C20:2(*n*-6), which served as substrate for desaturation reactions involving Δ8 and Δ5 desaturases resulting in the accumulation of more PUFAs namely DGLA and AA, respectively (Kumar et al., 2011). Further, oxidised fatty acids, termed oxylipins, are an important class of signalling molecule in plants, especially related to stress responses and innate immunity, and are formed enzymatically (via lipoxygenase (LOX)) or nonenzymatically via the action of ROS. There are few reports available in seaweed biology wherein oxylipins have been shown to accumulate in response to wound, pathogen infection, heavy metal, desiccation and other stresses (Contreras, Mella, Moenne, & Correa, 2009; Kumar et al., 2011, 2010; Weinberger et al., 2011). LOX enzyme that generates singlet oxygen and superoxide anions while incorporating molecular oxygen into arachidonic, linoleic and linolenic fatty acids, to form lipid hydroperoxides has been shown induced with two new arachidonic-dependent LOX isoforms (LOX-2 of 85 kDa; LOX-3 of 65 kDa) during desiccation exposure (3–4 h) of *G. corticata* (Kumar et al., 2011), resulting in excess ROS accumulation. However, the specificity of products was not analysed and thus the specific oxylipin (fatty acid hydroperoxide) levels could not be ascertained. Contreras-Porcia et al. (2011) also reported the ROS-mediated production of fatty acid hydroperoxides in *Porphyra* sp. under desiccation stress. Earlier, Contreras et al. (2009) also described the induction of an arachidonic acid-dependent LOX activity and its role in lipoperoxides production in *Lessonia nigrescens* and *Scytosiphon lomentaria* under copper induced oxidative stress conditions.

4.4.5 Functional Genomics – A Stepping Stone in Understanding the Stress Tolerance Mechanisms

Molecular control mechanisms for abiotic stress tolerance are based on the activation and regulation of specific stress related genes. These genes are involved in the whole sequence of stress responses, such as signalling, transcriptional control, protection of membranes and proteins and free-radical and toxic-compound scavenging. Recently, research in understanding the molecular mechanisms of stress responses has started to bear fruit with the availability of whole genome sequence of algal models such as brown alga *E. siliculosus* (Cock et al., 2010), red alga *C. crispus* (Collen et al., 2013) and *P. purpureum* (Bhattacharya et al., 2013) enabling a new suite of approaches to study the mechanisms involved in their stress tolerance under adverse environmental conditions. Additional tools of functional genomics that have been developed or may be developed shortly for such studies in these and other seaweeds include RNA-seq data, a large EST collection, whole-genome tiling arrays, sRNAs sequences, EST-based expression microarrays, a genetic map, bioinformatic and proteomic tools. The use of microarray analysis which facilitates multiparallel, high-throughput and comprehensive functional genomics approach allowed the identification of several putative genes overexpressed under various stress conditions in red alga *C. crispus* (Collen, Guisle-Marsollier, Leger, & Boyen, 2007), in *G. changii* (Teo et al., 2009). For example, hyperosmotic stress resulted in up-regulation of expression of many proteins primarily involves carbohydrate and protein metabolism in one or both organism (Table 4.3). Interestingly, whole genomic sequence of *C. crispus* revealed no candidate gene for allene oxide synthase, allene oxide cyclase or jasmonic acid carboxyl methyltransferase, which indicates that methyl jasmonate and oxylipin synthesis in *Chondrus* may be carried out by enzymes other than the ones characterised so far (Collen et al., 2013). Further, Dittami et al. (2009) accomplished the global transcriptomic study of *E. siliculosus* subjected to hyposalinity/hypersalinity and came up with the identification of several new genes and pathways with a putative function in the stress response. In general, genes involving in primary metabolic process such as protein, carbohydrate, fatty acid and pigment synthesis together with GSTs and CAT were found to be down-regulated. However, several other genes encoding gluconolactase, xylulokinase, phosphoglycerate kinase, isocitrate lyase (glyoxylate cycle), glutaredoxin, arginase (urea cycle), fatty acid catabolism (oxylipin synthesis), heat shock proteins (HSP70, HSP90), soluble NSF attachment protein receptors (SNARs)-related genes (vesicle trafficking) and genes for sugar transporters and ubiquitination were up-regulated, thereby paved the way for more detailed investigations of the mechanisms underlying the stress tolerance of marine macroalgae (Table 4.3). Similar group (Dittami et al., 2011) later integrated these transcriptomic profiles with metabolic profiles of *E. siliculosus* under similar conditions and concluded that mannitol and proline accumulation but not urea and trehalose under hypersaline conditions are most responsible for osmoregulation under stress. Also, hypersalinity induced a moderate increase in the level of γ-aminobutyric acid, an

Table 4.3 Potential Marker Genes with Putative Functions Regulated Under Hyposalinity and Hypersalinity Stress Conditions in *Gracilaria changii*, *Chondrus crispus* and *Ectocarpus siliculosus*

	Hyposalinity		Hypersalinity	
Organism	**Up-Regulated**	**Down-Regulated**	**Up-Regulated**	**Down-Regulated**
Gracilaria changii	Vacuolar ATP synthase subunit H Cysteine conjugate beta-lyase Vanadium-dependent bromoperoxidase Fructose-bisphosphate aldolase Pyruvate phosphate dikinase Glutamine amido transferase Putative haemolysin III HLY-III Adult-specific rigid cuticular protein Heat shock protein (HSP) 90 Ubiquitin-conjugating enzyme E2	Water channel protein MipI H^+-transporting two-sector ATPase High-affinity phosphate transporter Putative ATP-binding cassette (ABC) transporter-like protein ATP-binding cassette, sub-family B Transmembrane protein 34 Light-harvesting complex I polypeptide	Water channel protein MipI ABC transporter Vanadium-dependent bromoperoxidase Light-harvesting protein Phycobilisome linker polypeptide Cytosolic ascorbate peroxidase Ascorbate peroxidase Phosphoglycerate kinase Phosphatidylinositol transfer protein Hydroxymethyltransferase	H^+-exporting ATPase Light-harvesting protein Phycobilisome linker polypeptide Fructose-bisphosphate aldolase 6-phosphofructo-2-kinase Phosphoenolpyruvate carboxykinase Serine acetyltransferase Putative decarboxylase Glutathione S-transferase-like protein
Chondrus crispus	Guanosine triphosphate (GTP)-binding protein DnaJ protein ATPase Aspartate aminotransferase Eosinophil peroxidase Sulphate adenyltransferase	ADP-ribosylation factor 1 Amino-acid oxidase CDH1-D Cell division protein Elongation factor Eosinophil peroxidase Glutathione peroxidase Photomorphogenic proteins pyrrolidone peptidase Ribosomal protein L23 Sugar hydrolases V-bromoperoxidase	ABC nontransporter protein Alternative oxidase Aldolase Zinc finger protein Flavohaemoglobin Glucose-6-phosphate dehydrogenase Glutaredoxin HSP90 Ferritin Dodecenoyl-CoA isomerase	Amino-acid oxidase Cell division protein Elongation factor NADH dehydrogenase Phosphatidylinositol transferase Protein disulfide isomerase Pyrrolidone peptidase Ribosomal protein L23 Sugar hydrolases

Ectocarpus siliculosus	Arginase Pyrroline-5-carboxylate reductase 3-Methyl-2-oxobutanoate dehydrogenase Dihydrolipoyl transacylase	Chloroplastic Δ15 desaturase Glutamate dehydrogenase Glutamate synthase (ferredoxin) Asparagine synthase Homoserine kinase Dihydroxy-acid dehydratase Leucyl-tRNA synthetase Indole-3-glycerol phosphate synthase Phenylalanine – tRNA ligase Phosphoglycolate phosphatase Glycolate oxidase Peroxisomal alanine aminotransferase 1-Pyrroline-5-carboxylate synthase Glycine cleavage system T-protein	Chloroplastic Δ12-desaturase 3-Methyl-2-oxobutanoate dehydrogenase Dihydrolipoyl transacylase Arginine – tRNA ligase Amine oxidase 1-Pyrroline-5-carboxylate dehydrogenase Ornithine – oxo-acid transaminase Glycine cleavage system P-protein Glycolate oxidase Peroxisomal alanine aminotransferase	Tryptophan – tRNA ligase Arginine – tRNA ligase Argininosuccinate synthase Argininosuccinate lyase Valine – tRNA ligase Isopropylmalate synthase Aspartate transaminase

amino acid known to accumulate in response to stress in plants. This compatible solute, whose synthesis was suggested from PAs catabolism rather than glutamate decarboxylation, may later contributed to stress protection by regulating pH, acting as osmoregulator or as a signalling molecule.

De novo transcriptome analysis gained insights into physiological and metabolic characteristics of *Sargassum thunbergii* wherein for the first time two pathways of guanosine diphosphate-mannose synthesis were identified together with the genes encoding the three putative enzymes responsible for mannitol cycle. Moreover, many putative genes encoding the different kinds of HSPs and a total of 119 putative genes encoding kinds of ROS scavengers were also identified in the transcriptome, indicating strong resistance or tolerance ability of *S. thunbergii* to various stresses (Liu, Sun, Wang, Liang, & Wang, 2013). Recent metabolic labelling and genome sequencing data in *U. prolifera* evidenced the coexistence of C_3 and C_4 photosynthetic pathways with induced expression of *rbcL* and pyruvate orthophosphate dikinase (*PPDK*) and other C_4 cycle associated enzymes in salinity stress but only *PPDK* in desiccation stress suggesting *PPDK*-mediated C_4-type carbon metabolism in *U. prolifera* under environmental stress condition (Xu, Fan, Zhang, Xu, Mou, et al., 2012). Transcriptome and microarray database of *U. linza* (Zhang et al., 2012) and *C. crispus* (Collen et al., 2007) has also evidenced the regulation of early light-induced proteins with higher expression under hyposaliny, high light and thermal stress; reduced expression in severe desiccation and unchanged under hypersalinity stress. It has been hypothesised that they play a protective role either by transiently binding the excited free chlorophyll molecules as transient pigment carriers and/or by binding xanthophyll pigments to dissipate the excess absorbed light energy thus acting as photoprotective agent. Overexpression of many ribosomal protein genes and two ubiquitin ribosomal protein fusion genes in EST analysis of desiccation tolerant (*Fucus vesiculosus*) and susceptible (*Fucus serratus*) suggested that ribosome function and/or biogenesis are important during cycles of rapid desiccation and rehydration in the intertidal zone (Pearson et al., 2010). Further, global gene expression and EST sequences analysis of *Furcellaria lumbricalis* subjected to hypersalinity stress revealed up-regulation of NAD-*myo*-inositol dehydrogenase, dihydroorotase, aldehyde reductase, NADH-dependent mannose-6-phosphate reductase, phospho-mannomutase and Na^+/H^+ exchanger proteins (Kostamo et al., 2011). In context to ROS attenuation by the antioxidant system during the hydration–desiccation cycle, several antioxidant enzymes and others including a peroxiredoxin, arachidonate 5-lipoxygenase, GSTs and several cytochrome P450 and HSPs (in hydrated thalli), whereas thioredoxin, CAT and low variants of HSPs70 and 90 (in desiccated thalli) were found up-regulated in *P. columbina* analysed by suppression subtractive hybridisation and ESTs (Contreras-Porcia et al., 2013). A well-regulated signalling network has been characterised extensively in plants and yeast with a central role played by cytoplasmic phosphoproteins called mitogen-activated protein kinases (MAPKs). A recent study of Parages, Heinrich, Wiencke, and Jimnez (2013) indicated that MAPK-like proteins in all three red, green and brown algal phyta were highly phosphorylated and were most prominent in rhodophytes, followed by chlorophytes and phaeophytes in response to desiccation stress, where phosphorylation of p38-like MAPK was always preceded that of JNK-like (c-Jun amino-terminal kinases) MAPK. Further, up-regulation of protein phosphatase type 2C in *P. yezoensis* revealed its role in mitigating desiccation induced oxidative stress by phosphorylating some specific proteins while interacting with MAPK (Zhang et al., 2012). Protein phosphorylation and dephosphorylation is believed to be an essential part of signalling events leading to acquisition of drought/desiccation tolerance, and changes in the phosphorylated status of a number of proteins. Furthermore, accumulation of abscisic acid (ABA) in *P. columbina* during desiccation suggested its involvement in transcription of regulatory networks of desiccation stress signals and gene expression involving sugars, proline and PAs (Contreras-Porcia et al., 2013). ABA induces protein kinases of the SNF-related serine/threonine-protein kinase (SnRK) family to mediate number of downstream events in higher plants, however ABA signalling via MAPK cascades need to be explored in seaweeds.

4.5 CONCLUSION AND FUTURE PERSPECTIVE

ROS generation and their subsequent signalling are key regulators in algal cell physiology and cellular responses to environment. At high concentration ROS exacerbates cellular and macromolecular damage; but at sublethal level ROS can act as ubiquitous 'signalling molecule' in normal plant cell functioning and initiating defence genes and adaptive responses. Although, activity of many enzymes and metabolites are known to be affected by salinity fluctuations and desiccation induced oxidative stress, the molecular mechanism underlying such adaptation is still in its rudimentary state, and further progress has been hindered by the lack of available genomic information in seaweeds. However, with the influx of transcriptome data and whole genome sequences for few red and brown algal genomes together with EST data, a rapid progress has been made in recent years wherein many ROS-responsive genes coupled with alterations in diverse metabolic pathways has been recognised. While, this information enabling a new suite of approaches in algal stress physiology, still there are many uncertainties and gaps in our knowledge of ROS formation and their functionality under environmental stress conditions in seaweeds. Future work is needed to investigate the 'omic' profiling with the integration of genomics, proteomics and metabolomics tools to elucidate and understand diverse biochemical networks involved in cellular responses to oxidative stress. Improved understanding of these will be helpful in constructing engineered algal cells and/or whole plants with in-built capacity of enhanced levels of tolerance to ROS using biotechnological approaches.

ACKNOWLEDGEMENTS

We apologise to all colleagues whose original works could not be cited in the manuscript due to space limitations. The first author (M.K.) gratefully acknowledges the Council of Scientific and Industrial Research (CSIR), New Delhi, India, and Agriculture Research Organization (ARO), Volcani Center, Israel for funding support.

REFERENCES

Abe, S., Kurashima, A., Yokohama, Y., & Tanaka, J. (2001). The cellular ability of desiccation tolerance in Japanese intertidal algae. *Botanica Marina, 44*, 125–131.

Apel, K., & Hirt, H. (2004). Reactive oxygen species: metabolism, oxidative stress and signal transduction. *Annual Review of Plant Biology, 55*, 373–399.

Bhattacharjee, S. (2005). Reactive oxygen species and oxidative burst: roles in stress, senescence and signal transduction in plant. *Current Science, 89*, 1113–1121.

Bhattacharya, D., Price, D. C., Chan, C. X., Qiu, H., Rose, N., Ball, S., et al (2013) Genome of the red alga Porphyridium purpureum. Nature Communications, 4, 1941.

Bischof, K., & Rautenberger, R. (2012). Seaweed responses to environmental stress: reactive oxygen and antioxidative strategies. In C. Wiencke, & K. Bischof (Eds.), *Seaweed biology. Novel insights into ecophysiology, ecology and utilization Ecological studies: Vol. 219*. (pp. 109–132). Heidelberg: Springer.

Blouin, N. A., Brodie, J. A., Grossman, A. C., Xu, P., & Brawley, S. H. (2011). Porphyra: a marine crop shaped by stress. *Trends in Plant Science, 16*, 29–37.

Burritt, D. J., Larkindale, J., & Hurd, K. (2002). Antioxidant metabolism in the intertidal red seaweed *Stictosiphonia arbuscula* following desiccation. *Planta, 215*, 829–838.

Choi, T. S., Kang, E. J., Kim, J. H., & Kim, K.Y. (2010). Effect of salinity on growth and nutrient uptake of *Ulva pertusa* (Chlorophyta) from an eelgrass bed. *Algae, 25*, 17–26.

Cock, J. M., Sterck, L., Rouzé, P., Scornet, D., Allen, A. E., Amoutzias, G., et al. (2010). The *Ectocarpus* genome and the independent evolution of multicellularity in brown algae. *Nature, 465*, 617–621.

Collen, J., & Davison, I. R. (1999a). Reactive oxygen production and damage in intertidal *Fucus* spp. (Phaeophyceae). *Journal of Phycology, 35*, 54–61.

Collen, J., & Davison, I. R. (1999b). Stress tolerance and reactive oxygen metabolism in the intertidal red seaweeds *Mastocarpus stellatus* and *Chondrus crispus*. *Plant Cell and Environment, 22*, 1143–1151.

Collen, J., Guisle-Marsollier, I., Leger, J. J., & Boyen, C. (2007). Response of the transcriptome of the intertidal red seaweed *Chondrus crispus* to controlled and natural stresses. *New Phytologist, 176*, 45–55.

Collen, J., Porcel, B., Carre, W., Ball, S., Chaparro, C., Tonon, T., et al. (2013). Genome structure and metabolic features in the red seaweed *Chondrus crispus* shed light on evolution of the Archaeplastida. *Proceedings of the National Academy of Sciences*. http://dx.doi.org/10.1073/pnas.1221259110.

Contreras-Porcia, L., López-Cristoffanini, C., Lovazzano, C., Flores-Molina, M. R., Thomas, D., Núñez, A., et al. (2013). Differential gene expression in *Pyropia columbina* (Bangiales, Rhodophyta) under natural hydration and desiccation conditions. *Latin American Journal of Aquatic Research, 41*, 933–958.

Contreras, L., Mella, D., Moenne, A., & Correa, J. A. (2009). Differential responses to copper induced oxidative stress in the marine macroalgae *Lessonia nigrescens* and *Scytosiphon lomentaria* (Phaeophyceae). *Aquatic Toxicology, 94*, 94–102.

Contreras-Porcia, L., Thomas, D., Flores, V., & Correa, J. A. (2011). Tolerance to oxidative stress induced by desiccation in *Porphyra columbina* (Bangiales, Rhodophyta). *Journal of Experimental Botany, 62*, 1815–1829.

Davison, I. R., & Pearson, G. A. (1996). Stress tolerance in intertidal seaweeds. *Journal of Phycology, 32*, 197–211.

De Casabianca, M. L. (1989). Degradation of *Ulva* (*Ulva rotundata*, Prévost Lagoon, France). *Comptes Rendus Academie des Sciences, Paris, 308*, 155–160.

Dittami, S. M., Gravot, A., Renault, D., Goulitquers, S., Eggert, A., Bouchereau, A., et al. (2011). Integrative analysis of metabolite and transcript abundance during the short-term response to saline and oxidative stress in the brown alga *Ectocarpus siliculosus*. *Plant Cell & Environment, 34*, 629–642.

Dittami, S. M., Scornet, D., Petit, J. L., Segurens, B., Silva, C. D., Corre, E., et al. (2009). Global expression analysis of the brown alga *Ectocarpus siliculosus* (Phaeophyceae) reveals large-scale reprogramming of the transcriptome in response to abiotic stress. *Genome Biology, 10*, R66.

Dring, M., & Brown, F. (1982). Photosynthesis of intertidal brown algae during and after periods of emersion: a renewed search for physiological causes of zonation. *Marine Ecology Progress Series, 8*, 301–308.

Dromgoole, F. I. (1980). Desiccation resistance of intertidal and subtidal algae. *Botanica Marina, 23*, 149–159.

Eklund, B. T., & Kautsky, L. (2003). Review on toxicity testing with marine macroalgae and the need for method standardization – exemplified with copper and phenol. *Marine Pollution Bulletin, 46*, 171–181.

Fillit, M. (1995). Seasonal changes in the photosynthetic capacities and pigment content of *Ulva rigida* in a Mediterranean coastal lagoon. *Botanica Marina, 38*, 271–280.

Gao, S., Niu, J., Chen, W., Wang, G., Xie, X., Pan, G., et al. (2013). The physiological links of the increased photosystem II activity in moderately desiccated *Porphyra haitanensis* (Bangiales, Rhodophyta) to the cyclic electron flow during desiccation and re-hydration. *Photosynthesis Research. 116*, 45–54. http://dx.doi.org/10.1007/s11120-013-9892-4.

Gao, S., Shen, S., Wang, G., Niu, J., Lin, A., & Pan, G. (2011). PSI-driven cyclic electron flow allows intertidal macro-algae *Ulva* sp. (Chlorophyta) to survive in desiccated conditions. *Plant and Cell Physiology, 52*, 885–893.

Garcia-Jimenez, P., Just, M. P., Delgado, M. A., & Robaina, R. R. (2007). Transglutaminase activity decrease during acclimation to hyposaline conditions in marine seaweed *Grateloupia doryphora* (Rhodphyta, Halymeniaceae). *Journal of Plant Physiology, 364*, 367–370.

Gill, S. S., & Tuteja, N. (2010). Reactive oxygen species and antioxidant machinery in abiotic stress tolerance in crop plants. *Plant Physiology and Biochemistry, 48*, 909–930.

Gylle, M.A., Nygard, C.A., Svan, C. I., Pocock, T., & Ekelund, N. G.A. (2013). Photosynthesis in relation to D1, PsaA and Rubisco in marine and brackish water ecotypes of *Fucus vesiculosus* and *Fucus radicans* (Phaeophyceae). *Hydrobiologia, 700*, 109–119.

Hayat, S., Hayat, Q., Alyemeni, M. N., Wani, A. S., Pichtel, J., & Ahmad, A. (2012). Role of proline under changing environments: a review. *Plant Signaling and Behavior, 7*, 1456–1466.

Holzinger, A., & Karsten, U. (2013). Desiccation stress and tolerance in green algae: consequences for ultrastructure, physiological, and molecular mechanisms. *Frontiers in Plant Science, 4*, 327.

Hsu, Y. T., & Lee, T. M. (2012). Nitric oxide up-regulates the expression of methionine sulfoxide reductase genes in the intertidal macroalga *Ulva fasciata* for high light acclimation. *Plant and Cell Physiology, 53*, 445–456.

Hurd, C. L., & Dring, M. J. (1991). Desiccation and phosphate uptake by intertidal fucoid algae in relation to zonation and season. *British Journal of Phycology, 26*, 327–333.

Ichihara, K., Miyaji, K., & Shimada, S. (2012). Comparing the low-salinity tolerance of *Ulva* species distributed in different environments. *Phycological Research, 61*, 52–57.

Ji, Y., Gao, K., & Tanaka, J. (2005). Photosynthetic recovery of desiccated intertidal seaweeds after dehydration. *Progress in Natural Science, 15*, 689–693.

Ji, Y., & Tanaka, J. (2002). Effect of desiccation on the photosynthesis of seaweeds from the intertidal zone of Honshu, Japan. *Phycological Research, 50*, 145–153.

Jiang, H., Gao, K., & Helbling, E.W. (2008). UV-absorbing compounds in *Porphyra haitanensis* (Rhodophyta) with special reference to effects of desiccation. *Journal of Applied Phycology, 20*, 387–395.

Karsten, U. (2012). Seaweed acclimation to salinity and desiccation stress. In C. Wiencke & K. Bischof (Vol. Eds.), *Ecological studies. Seaweed ecophysiology and ecology* (Vol. 219) (pp. 87–107). Berlin: Springer.

Karsten, U., Wiencke, C., & Kirst, G. O. (1991). The effect of salinity changes upon the physiology of eulittoral green macroalgae from Antarctica and Southern Chile. I. Cell viability, growth, photosynthesis and dark respiration. *Journal of Plant Physiology, 138*, 667–673.

Kim, J. K., Kraemer, G. P., & Yarish, C. (2009). Comparison of growth and nitrate uptake by New England *Porphyra* species from different tidal elevations in relation to desiccation. *Phycological Research, 57*, 152–157.

Kim, J. K., Kraemer, G. P., & Yarish, C. (2013). Emersion induces nitrogen release and alteration of nitrogen metabolism in the intertidal genus *Porphyra*. *PLoS One, 8*, e69961.

Kim, K. Y., & Garbary, D. J. (2007). Photosynthesis in *Codium fragile* (Chlorophyta) from a Nova Scotia estuary: responses to desiccation and hyposalinity. *Marine Biology, 151*, 99–107.

Kirst, G. O. (1990). Salinity tolerance of eukaryotic marine algae. *Annual Review of Plant Physiology and Plant Molecular Biology, 41*, 21–53.

Kostamo, K., Olsson, S., & Korpelainen, H. (2011). Search for stress-responsive genes in the red alga *Furcellaria lumbricalis* (Rhodophyta) by expressed sequence tag analysis. *Journal of Experimental Marine Biology and Ecology, 404*, 21–25.

Kumar, M., Bijo, A. J., Baghel, R. S., Reddy, C. R. K., & Jha, B. (2012). Selenium and spermine alleviate cadmium induced toxicity in the red seaweed *Gracilaria dura* by regulating antioxidants and DNA methylation. *Plant Physiology and Biochemistry, 51*, 129–138.

Kumar, M., Gupta, V., Trivedi, N., Kumari, P., Bijo, A. J., Reddy, C. R. K., et al. (2011). Desiccation induced oxidative stress and its biochemical responses in intertidal red alga *Gracilaria corticata* (Gracilariales, Rhodophyta). *Environmental and Experimental Botany, 72*, 194–201.

Kumar, M., Kumari, P., Gupta, V., Reddy, C. R. K., & Jha, B. (2010). Biochemical responses of red algae *Gracilaria corticata* (Gracilariales, Rhodophyta) to salinity induced oxidative stress. *Journal of Experimental Marine Biology and Ecology, 391*, 27–34.

Kumar, M., & Reddy, C. R. K. (2012). *Oxidative stress tolerance mechanisms in marine macroalgae (seaweeds): Oxidative stress in seaweeds*. Germany: LAP LAMBERT Academic Publisher. ISBN-13: 978-3-8473-7432-9.

Kupper, F. C., Carpenter, L. J., McFiggans, G. B., Palmer, C. J., Waite, T. J., Boneberg, E. M., et al. (2008). Iodide accumulation provides kelp with an inorganic antioxidant impacting atmospheric chemistry. *Proceedings of the National Academy of Sciences, 105*, 6954–6958.

Lee, T. M. (1998). Investigations of some intertidal green macroalgae to hyposaline stress: detrimental role of putrescine under extreme hyposaline conditions. *Plant Science, 138*, 1–8.

Lee, T. M., & Liu, C. H. (1999). Correlation of decreased calcium contents with proline accumulation in the marine green macroalga *Ulva fasciata* exposed to elevated NaCl contents in seawater. *Journal of Experimental Botany, 50*, 1855–1862.

Li, L. C., Hsu, Y. T., Chang, H. L., Wu, T. M., Sung, M. S., Cho, C. L., et al. (2013). Polyamine effects on protein disulfide isomerase expression and implications for hypersalinity stress in the marine alga *Ulva lactuca* Linnaeus. *Journal of Phycology. 49*, 1181–1191. http://dx.doi.org/10.1111/jpy.12129.

Li, X. C., Xing, Y. Z., Jiang, X., Qiao, J., Tan, H. L., Tian, Y., et al. (2012). Identification and characterization of the catalase gene PyCAT from the red alga *Pyropia yezoensis* (Bangiales, Rhodophyta). *Journal of Phycology, 48*, 664–669.

Liu, F., & Pang, S. J. (2010). Stress tolerance and antioxidant enzymatic activities in the metabolisms of the reactive oxygen species in two intertidal red algae *Grateloupia turuturu* and *Palmaria palmata*. *Journal of Experimental Marine Biology and Ecology, 382*, 82–87.

Liu, F., Sun, X., Wang, W., Liang, Z., & Wang, F. (2013). De novo transcriptome analysis-gained insights into physiological and metabolic characteristics of *Sargassum thunbergii* (Fucales, Phaeophyceae). *Journal of Applied Phycology*. http://dx.doi.org/10.1007/s10811-013-0140-2.

Lu, I. F., Sung, M. S., & Lee, T. M. (2006). Salinity stress and hydrogen peroxide regulation of antioxidant defense system in *Ulva fasciata*. *Marine Biology, 150*, 1–15.

Luo, M. B., & Liu, F. (2011). Salinity-induced oxidative stress and regulation of antioxidant defense system in the marine macroalga *Ulva prolifera*. *Journal of Experimental Marine Biology and Ecology, 409*, 223–228.

Mittler, R. (2002). Oxidative stress, antioxidants and stress tolerance. *Trends in Plant Science, 7*, 405–410.

Murthy, M. S., Rao, A. S., & Reddy, E. R. (1986). Dynamics of nitrate reductase activity in two intertidal algae under desiccation. *Botanica Marina, 29*, 471–474.

Murthy, M. S., & Sharma, C. L. N.S. (1989). Peroxidase activity in *Ulva lactuca* under desiccation. *Botanica Marina, 32*, 511–513.

Murthy, M. S., Sharma, C. L. N.S., & Rao, Y. N. (1988). Salinity induced changes in peroxidase activity in the green seaweed *Ulva lactuca*. *Botanica Marina, 31*, 307–310.

Myndee, M. N. (2010). Vertical zonation: studying ecological patterns in the rocky intertidal zone. *Science Activities, 47*, 814.

Nelson, T. A., Olson, J., Imhoff, L., & Nelson, A. V. (2010). Aerial exposure and desiccation tolerances are correlated to species composition in "green tides" of the Salish Sea (northeastern Pacific). *Botanica Marina, 53*, 103–111.

Parages, M. L., Heinrich, S., Wiencke, C., & Jimnez, C. (2013). Rapid phosphorylation of MAP kinase-like proteins in two species of Arctic kelps in response to temperature and UV radiation stress. *Environmental and Experimental Botany, 91*, 30–37.

Pearson, G., Hoarau, G., Lago-Leston, A., Coyer, J. A., Kube, M., Reinhardt, R., et al. (2010). An expressed sequence tag analysis of the intertidal brown seaweed *Fucus serratus* (L.) and *F. vesiculosus* (L.) (Heterokontophyra, Phaeophyceae) in response to abiotic stressors. *Marine Biotechnology, 12*, 195–213.

Proctor, M. C. F., & Pence, V. C. (2002). Vegetative tissues: Bryophytes, vascular resurrection plants, and vegetative propogules. In M. Black, & H. W. Pritchard (Eds.), *Desiccation and survival in plants: Drying without dying* (pp. 207–237). Wallingford, Oxon: CABI Publishing.

Qiao, K., Li, S. G., Li, H.Y., Tong, S. M., & Hou, H. S. (2013). Molecular cloning and sequence analysis of methionine adenosyltransferase from the economic seaweed *Undaria pinnatifida*. *Journal of Applied Phycology*, *25*, 81–87.

Romero-Puertas, M. C., Corpas, F. J., Sandalio, L. M., Leterrier, M., Rodriguez-Serrano, M., del Rio, L. A., et al. (2006). Glutathione reductase from pea leaves: response to abiotic stress and characterization of the peroxisomal isozyme. *New Phytologist*, *170*, 43–52.

Ross, C., & Van Alstyne, K. L. (2007). Intraspecific variation in stress-induced hydrogen peroxide scavenging by ulvoid macro alga *Ulva lactuca*. *Journal of Phycology*, *43*, 466–474.

Rousvoal, S., Groisillier, A., Dittami, S. M., Michel, G., Boyen, C., & Tonon, T. (2011). Mannitol-1-phosphate dehydrogenase activity in *Ectocarpus siliculosus*, a key role for mannitol synthesis in brown algae. *Planta*, *233*, 261–273.

Russell, G. (1987). Salinity and seaweed vegetation. In R. M. M. Crawford (Ed.), *The physiological ecology of amphibious and intertidal plants* (pp. 35–52). Oxford: Blackwell.

Sampath-Wiley, P., Neefus, C. D., & Jahnke, L. S. (2008). Seasonal effects of sun exposure and emersion on intertidal seaweed physiology: fluctuation in antioxidant contents, photosynthetic pigments and photosynthetic efficiency in the red algae *Porphyra umbilicalis* Kutzing (Rhodophyta, Bangiales). *Journal of Experimental Marine Biology and Ecology*, *361*, 83–91.

Shi, H., & Chan, Z. (2013). Improvement of plant abiotic stress tolerance through modulation of the polyamine pathway. *Journal of Integrative Plant Biology*. http://dx.doi.org/10.1111/jipb.12128.

Shu, S., Guo, S. R., & Yuan, L. Y. (2012). A review: polyamines and photosynthesis. In M. Najafpour (Ed.), *Advances in photosynthesis – Fundamental aspects*. InTech http://dx.doi.org/10.5772/26875. (ISBN: 978-953-307-928-8).

Sung, M. S., Hsu, Y. T., Wu, T. M., & Lee, T. M. (2009). Hypersalinity and hydrogen peroxide upregulation of gene expression of antioxidant enzymes in *Ulva fasciata* against oxidative stress. *Marine Biotechnology*, *11*, 199–209.

Tanou, G., Ziogas, V., Belghazi, M., Christou, A., Filippou, P., Job, D., et al. (2013). Polyamines reprogram oxidative and nitrosative status and the proteome of citrus plants exposed to salinity stress. *Plant Cell & Environment*. *37*, 864–885. http://dx.doi.org/10.1111/pce.12204.

Teo, S. S., Ho, C. L., Teoh, S., Rahim, R. A., & Phang, S. M. (2009). Transcriptomic analysis of *Gracilaria changii* (Rhodophyta) in response to hyper-and hypo-osmotic stresses. *Journal of Phycology*, *45*, 1093–1099.

Thomas, T. E., Turpin, D. H., & Harrison, P. J. (1987). Desiccation enhanced nitrogen uptake rates in intertidal seaweeds. *Marine Biology*, *94*, 293–298.

Uji, T., Hirata, R., Mikami, K., Mizuta, H., & Saga, N. (2012). Molecular characterization and expression analysis of sodium pump genes in the marine red alga *Porphyra yezoensis*. *Molecular Biology Report*, *39*, 7973–7980.

Uji, T., Monma, R., Mizuta, H., & Saga, N. (2012). Molecular characterization and expression analysis of two Na^+/H^+ antiporter genes in the marine red alga *Porphyra yezoensis*. *Fisheries Science*, *78*, 985–991.

Verret, F., Wheeler, G., Taylor, A., Farnham, G., & Brownlee, C. (2010). Calcium channels in photosynthetic eukaryotes: implications for evolution of calcium-based signalling. *New Phytologist*, *187*, 23–43.

Weinberger, F., Lion, U., Delage, L., Kloareg, B., Potin, P., Beltrán, J., et al. (2011). Up-regulation of lipoxygenase, phospholipase, and oxylipin production in the induced chemical defense of the red alga *Gracilaria chilensis* against epiphytes. *Journal of Chemical Ecology*, *37*, 677–686.

Wong, J. W. (2009). *Dehydrin-like proteins in desiccation tolerance in intertidal seaweeds* (Biology Master's theses). Paper 3, http://hdl.handle.net/2047/d1001911x.

Wright, D. G., Pawlowicz, R., McDougall, T. J., Feistel, R., & Marion, G. M. (2010). Absolute salinity, "density salinity" and the reference-composition salinity scale: present and future use in the seawater standard TEOS-10. *Ocean Science Discuss*, 7, 1559–1625.

Xia, J., Li, Y., & Zou, D. (2004). Effects of salinity stress on PSII in *Ulva lactuca* observed by chlorophyll fluorescence measurements. *Aquatic Botany*, *80*, 129–137.

Xie, X., Gao, S., Gu, W., Pan, G., & Wang, G. (2013). Desiccation induces accumulations of antheraxanthin and zeaxanthin in intertidal macro-alga *Ulva pertusa* (Chlorophyta). *PLoS One*, *8*, e72929.

Xu, J., Fan, X., Zhang, X., Xu, D., Mou, S., Cao, S., et al. (2012). Evidence of coexistence of C_3 and C_4 photosynthetic pathways in a green-tide-forming alga, *Ulva prolifera*. *PLoS One*, 7, e37438.

Yu, C. H., Lim, P. E., & Phang, S. M. (2013). Effects of irradiance and salinity on the growth of carpospore-derived tetrasporophytes of *Gracilaria edulis* and *Gracilaria tenuistipitata* var liui (Rhodophyta). *Journal of Applied Phycology*, *25*, 787–794.

Zhang, L. P., Mehta, S. K., Liu, Z. P., & Yang, Z. M. (2008). Copper induced proline synthesis is associated with nitric oxide generation in *Chlamydomonas reinhardtii*. *Plant Cell and Physiology*, *49*, 411–419.

Zhang, X., Cao, S., Li, Y., Mou, S., Xu, D., Fan, X., et al. (2012). Expression of three putative early light-induced genes under different stress conditions in the green alga *Ulva linza*. *Plant Molecular Biology Report*, *30*, 940–948.

Zou, D., & Gao, K. (2002). Effects of desiccation and CO_2 concentration on emersed photosynthesis in *Porphyra haitanensis* (Bangiales, Rhodophyta), a species farmed in China. *European Journal of Phycology*, *37*, 587–592.

CHAPTER FIVE

Plant Growth Regulators in Seaweeds: Occurrence, Regulation and Functions

Wendy A. Stirk[1] and Johannes Van Staden
Research Centre for Plant Growth and Development, School of Life Sciences, University of KwaZulu-Natal Pietermaritzburg Campus, Scottsville, South Africa
[1]Corresponding author: e-mail address: stirk@ukzn.ac.za

Contents

Advances in Botanical Research, Volume 71
ISSN 0065-2296
http://dx.doi.org/10.1016/B978-0-12-408062-1.00005-6

Abstract

Plant growth regulators (PGRs) play an important role in vascular plants, coordinating growth and stress responses and regulating most physiological processes. Recent advances have identified genes and enzymes involved in their biosynthesis and signalling pathways. In contrast, little is known about PGRs in seaweeds and there is still debate as to whether seaweeds are able to synthesise their own PGRs. In this chapter, the five classic plant hormones, namely auxin, cytokinins, gibberellins, abscisic acid and ethylene as well as more recently discovered brassinosteroids, jasmonates, salicylic acid and strigolactones are considered in terms of their occurrence in seaweeds, physiological responses and their biosynthesis and regulation. These are compared to PGRs in nonvascular and vascular plants. In addition, potential practical uses of PGRs in seaweed biotechnology are highlighted.

5.1 INTRODUCTION

Although seaweeds have a simpler morphology, they have many similarities with vascular plants such as analogous reproductive events and apical dominance (Bradley, 1991; Evans & Trewavas, 1991) that require a certain level of organisation. Endogenous plant growth regulators (PGRs) control most growth and developmental processes in vascular plants and it is possible that they have a similar function in seaweeds. However, research on PGRs in seaweeds is not nearly as advanced as in vascular plants, especially at the molecular level and there is still debate as to whether seaweeds are able to synthesise PGRs and about their mechanisms of action.

While knowledge of PGRs in vascular plants provides a useful framework from which to investigate the role of PGRs in seaweeds, their divergent phylogeny (Section 5.1.1) needs to be borne in mind when comparing PGRs in seaweeds and vascular plants. PGRs have a dual function in plants – regulation of growth processes such as cell division (Section 5.1.2) and in the perception of environmental changes (Section 5.1.3). Both these functions of PGRs need to be considered when investigating the role of PGRs in seaweeds.

5.1.1 Phylogeny

It is estimated that the monophyletic Archaeplastida arose 1–1.5 billion years ago due to endosymbiotic events between eukaryotic cells and cyanobacteria to form the first photosynthetic eukaryotic cells (Collén et al., 2013; Leliaert et al., 2012). Shortly thereafter, divergence gave rise to the green lineage (Viridiplantae) and red seaweed ancestors with the fossil record for the red seaweed ancestor dating back approximately 1.2 billion years. This red seaweed ancestor gave rise to several other lineages including the Phaeophyceae via secondary endosymbiosis (Collén et al., 2013; Popper et al., 2011). Rhodophyceae are the most species-rich group of marine seaweeds and have unique features such as unstacked thylakoids and no flagella (Collén et al., 2013).

Early divergence (approximately 725 million years ago) of the green Viridiplantae lineage resulted in two discreet clades – the Chlorophyta and Streptophyta. The diverse Chlorophyta include most of the green algae such as the primitive unicellular Prasinophyceae, the Trebouxiophyceae, Chlorophyceae and Ulvophyceae, which encompasses both unicellular species and multicellular green seaweeds. The Streptophyta clade consists of the Charophyceae, the freshwater *Klebsormidium* algae, nonvascular plants which separated from seed plants more than 400 million years ago and vascular plants (Leliaert et al., 2012; Yevdakova et al., 2008). Genomic studies show that fundamental traits of vascular plants were inherited from the green algal ancestor (Leliaert et al., 2012).

5.1.2 PGRs and the Cell Cycle

In plant cells, coordination of cell division and differentiation is influenced by PGRs (De Veylder, Beeckman, & Inzé, 2007). The cell cycle includes re-entry of the cell into the cell cycle, four phases – growth phase 1 (G_1), DNA synthesis (S), growth phase 2 (G_2) and mitosis (M) and exit from the cell cycle (De Veylder et al., 2007). Cell cycle regulators are highly conserved in all eukaryotic cells and transitions are tightly regulated by cyclin-dependent kinases (CDKs; del Pozo, Lopez-Matas, Ramirez-Parra, & Gutierrez, 2005). Auxins and cytokinins control the transcriptional expression of some cell cycle genes and the balance between cytokinins and auxins concentrations is important for the induction of cell division (Perrot-Rechenmann, 2010; del Pozo et al., 2005). In addition, PGRs influence the progression of the cell cycle. For example, auxins act on multiple targets during the cell cycle, influencing both transcriptional and post-transcriptional regulation and protein turnover (Perrot-Rechenmann, 2010); cytokinins are involved in the G_1/S and G_2/M transitions; ethylene and abscisic acid (ABA) inhibit DNA replication; the G_1/S transition is inhibited by ABA and jasmonates; jasmonates also have a negative effect on the G_2/M transition (del Pozo et al., 2005). The role of PGRs in coordinating the cell cycle in seaweeds needs to be considered.

5.1.3 PGRs and Environmental Signals

PGRs also coordinate responses of vascular plants to environmental changes. Plants have a multidimensional, integrated network of hormonal and non-hormonal signals in order to quickly respond to continually changing environmental conditions (Chico, Chini, Fonseca, & Solano, 2008). These responses are tightly regulated, either by auto-regulatory negative feed-back loops or by interactions (either additive, synergistic or antagonistic) with other PGRs in order to achieve the correct balance for optimum growth, specific developmental processes and/or stimulation of defense processes to improve their chances of survival (Chico et al., 2008; Lyons, Manners, & Kazan, 2013). PGR pathways overlap with specific PGR combinations required for different responses (Weiss & Ori, 2007) with antagonistic responses allowing for the allocation of resources into either growth or defence as required (Lyons et al., 2013). There are many examples of 'hormonal crosstalk' in vascular plants where the biosynthetic and/or signalling pathways are influenced by other PGRs. For example, auxins and cytokinins are inversely correlated with auxins inhibiting cytokinin biosynthesis (Woodward & Bartel, 2005); auxin induces gibberellin (GA) biosynthesis (Yamaguchi, 2008); benzyladenine (BA) an aromatic cytokinin, upregulates proteins involved in ABA and ethylene biosynthesis (Žd'árská et al., 2013); GA and ABA act antagonistically to control seed development and germination (Yamaguchi, 2008); cytokinins inhibit GA biosynthesis and promote GA-deactivation (Weiss & Ori, 2007); ethylene and GA may either interact positively or negatively depending on environmental cues (Weiss & Ori, 2007).

The shift towards moist terrestrial habitats and ultimately the colonisation of dry land approximately 450 million years ago (Chico et al., 2008) exposed plants to a wide range of harsh environmental conditions such as a larger temperature range, desiccation, osmotic stress and high UV radiation (Leliaert et al., 2012; Popper et al., 2011). This gave rise to many physiological and morphological adaptations in order to survive and was associated with a large expansion in the genome and gene expression (Leliaert et al., 2012). Many authors propose that one such physiological adaptation was the biosynthesis and accumulation of PGRs (reviewed in Leliaert et al., 2012; Lyons et al., 2013). However, comparative plant genomic studies between Charophyceae and terrestrial plants indicate that certain physiological changes such as regulation of photosynthesis and photorespiration, cellulose synthesis and asymmetrical cell division occurred prior to the shift to the terrestrial habitat. There is also some evidence that PGRs evolved from pre-existing elements in the ancestral alga (Leliaert et al., 2012; Pils & Heyl, 2009).

Seaweeds also respond to environmental changes and have seasonal growth and reproductive cycles analogous to vascular plants (Bradley, 1991;

Evans & Trewavas, 1991). Seaweeds, especially intertidal species, are exposed daily to highly variable and rapid environmental fluctuations such as desiccation, high UV radiation and extreme temperatures, salinity and pH. Seaweeds are also exposed to biotic stresses from marine herbivores and pathogens. Thus, the involvement of PGRs in seaweed responses to environmental changes needs to be considered.

In this chapter, the five classic plant hormones, namely auxins, cytokinins, GAs, ABA and ethylene as well as more recently discovered brassinosteroids, jasmonates, salicylic acid (SA) and strigolactones are discussed in terms of their occurrence in seaweeds, their physiological effects and their biosynthesis and regulation. Polyamines are not included in this review as they have been extensively reviewed elsewhere (Fuell, Elliot, Hanfrey, Franceschetti, & Michael, 2010). Examples of PGRs in microalgae are only included when information on their occurrence in seaweeds is limited. Comparisons to nonvascular and vascular plants are also discussed.

5.2 AUXINS

Auxins are key regulators for almost every aspect of plant growth and development, being involved in cell division, elongation and differentiation as well as in tropic responses, promoting root growth, apical dominance, vascular development, flowering and senescence (Mano & Nemoto, 2012; Murphy, 2002; Woodward & Bartel, 2005).

5.2.1 Endogenous Auxins in Seaweeds

Endogenous indole-3-acetic acid (IAA) and various conjugates are present in a wide range of seaweeds. For example, IAA was detected in *Kappaphycus alvarezii* (Rhodophyceae) and *Sargassum tenerrimum* (Phaeophyceae) and indole-3-pyruvic acid (IPA) in *Gracilaria edulis* (Rhodophyceae; Prasad et al., 2010). IAA and indole-3-butyric acid (IBA) were present in four wild harvested *Ulva* species and in laboratory-cultured *Monostroma oxyspermum* (Chlorophyceae; Gupta et al., 2011). IAA and indole-3-acetamide (IAM) were the main indoles present in 11 Rhodophyceae species from the Bangiales, Gelidiales, Gracilariales and Gigartinales with low concentrations of other conjugates including indole-3-ethanol (IEt) in eight species, indole-3-acetyl glutamic acid (IAGlu) in three species and indole-3-acetyl leucine (IALeu) in *Hypnea musciformis* (Yokoya et al., 2010). Analysis of *Ulva fasciata* (Chlorophyceae) and *Dictyota humifusa* (Phaeophyceae) from six bimonthly collections detected IAA and lower concentrations of IAM. Auxin concentrations were lower during the cooler months and increased during the warmer months in *U. fasciata* but did not follow a seasonal trend in *D. humifusa* (Stirk et al., 2009).

5.2.2 Physiological Responses to Auxins

Further evidences that auxins are involved in seaweed physiology are reports of exogenous auxin application eliciting physiological responses in seaweeds. For example, decapitation experiments with *Chara globularis* rhizoids demonstrated the physiological requirement for auxin in rhizoid growth with exogenous IAA stimulating rhizoid growth and naphthylphthalamic acid (NPA), an auxin efflux transport inhibitor, inhibiting growth (Klämbt, Knauth, & Dittmann, 1992). In contrast, IAA application to sporophyte discs of *Laminaria japonica, Undaria pinnatifida* and *Alaria crassifolia* (Phaeophyceae) had no effect on the growth rate apart from being inhibitory at higher concentrations and delaying sorus formation (Kai, Nimura, Yasui, & Mizuta, 2006).

There are also examples of auxin influencing zygote development in Phaeophyceae. Endogenous IAA was detected in the mature thallus and in synchronised zygote cultures of *Fucus distichus*. The first cell division of the zygote established the apical basal pattern that formed the thallus and rhizoid tissue, respectively. Exogenous IAA altered the zygote orientation and morphology with zygotes developing multiple, randomly distributed rhizoids that lead to early branching. These effects were partially mitigated by light. Radiolabelled experiments demonstrated that IAA was transported by passive polar diffusion across the zygote membrane with NPA blocking IAA efflux from the zygote into the media (Basu, Sun, Brian, Quatrano, & Muday, 2002). While IAA stimulated embryo growth and apical hair formation in developing *Fucus vesiculosus* embryos, it had an inhibitory effect on photosynthesis – decreasing the maximum quantum yield (F_v/F_m), the rate of electron transfer under saturation irradiances ($rETR_{max}$) and photosynthetic efficiency under nonsaturating irradiances (αETR; Tarakhovskaya, Kang, Kim, & Garbary, 2013).

5.2.3 Auxin Biosynthesis and Metabolism

There are multiple pathways for the biosynthesis of active IAA in vascular plants with many of the genes and enzymes involved in these pathways identified. Tryptophan (Trp)-dependent pathways include:

1. the IPA pathway from Trp → IPA → indole-3-acetaldehyde (IAAld) → IAA is found in microorganisms as well as being the most common pathway in vascular plants;

2. the IAM pathway from Trp → IAM → IAA was first described in microorganisms but has since be found in many diverse vascular plant species;
3. the tryptamine (TAM) pathway from Trp → TAM → *N*-hydroxyl-TAM → indole-3-acetaldoxime (IAOx) → IAAld → IAA is present in some vascular plants;
4. the IAOx pathway from Trp → IAOx → indole-3-acetonitrile (IAN) or IAAld → IAA is restricted to Brassicaceae species and some grasses.

A Trp-independent pathway that converts the precursor indole-3-glycerol phosphate to IAA has also been identified in vascular plants (Mano & Nemoto, 2012; Murphy, 2002; Woodward & Bartel, 2005).

In vascular plants, the IAM pathway involves two enzymes – Trp-2-monooxygenase and IAM hydrolase. While IAM is found in relatively high concentrations in seaweeds (Stirk et al., 2009; Yokoya et al., 2010) and microalgae (Jirástová et al., 2009; Stirk, Ördög, et al., 2013), analysis of several Chlorophyceae microalgae with complete genomic sequences (*Chlorella vulgaris*, *Micromonas pusilla*, *Chlamydomonas reinhardtii* and *Ostreococcus lucimarinus*) found no orthologs for genes that encode Trp-2-monooxygenase and IAM hydrolase (De Smet et al., 2011). However, there was some evidence that alternative mechanisms for auxin biosynthesis and mediating auxin responses were present in these microalgae (Lau, Shao, Bock, Jürgens, & De Smet, 2009). In agreement with this was the analysis of IAA biosynthesis in the seaweed *Ectocarpus siliculosus* (Phaeophyceae) where the complete genome, which is relatively small (200 Mb), has been sequenced. IAA, indole-3-carboxylic acid (ICA) and IPA were present but no IAM or conjugates were detected. It was likely that ICA and IPA were degradation products of IAA. Homologues of genes encoding for enzymes involved in the Trp-dependent TAM and IAOx biosynthetic pathways were identified with no molecular evidence for the IPA and IAM pathways (Le Bail et al., 2010). While there is some evidence that genes involved in auxin signalling are present in members of the Chlorophyta and Streptophyta (De Smet et al., 2011) and analysis of the genome of the moss *Physcomitrella patens* found that all the main components of auxin signalling in vascular plants are present in the moss (Lau et al., 2009), more genome sequences of a wider range of seaweeds is required before the origins of auxin signalling in seaweed can be elucidated.

As high concentrations of IAA can be inhibitory, endogenous levels are tightly controlled by biosynthesis, conjugation and degradation in vascular plants (Ludwig-Müller, 2011). IAA can be converted to an ester-linked, amide-linked or high molecular weight peptide or protein conjugate, which is readily converted to IAA by hydrolysis (Woodward & Bartel, 2005). IAA is also permanently inactivated via ring oxidation to oxindole-3-acetic acid (oxIAA) or by conjugation to indole-3-acetylaspartate (IAA-Asp) and indole-3-acetylglutamate (IAA-Glu; Murphy, 2002; Woodward & Bartel, 2005). Auxin conjugate profiles are species-specific (Woodward & Bartel, 2005) with ester, amide, protein and peptide conjugates present in a wide diversity of vascular plants (Ludwig-Müller, 2011), fewer auxin conjugates as well as IAA-Asp and IAA-Glu present in ferns, mosses and hornworts in similar concentrations as in vascular plants while conjugate levels were much lower in liverworts (Sztein, Cohen, de la Fuente, & Cooke, 1999; Sztein, Cohen, Slovin, & Cooke, 1995). Auxins are present in seaweeds but only a few conjugates have been reported (Yokoya et al., 2010) with no molecular evidence to support conjugation in *E. siliculosus* although there was some evidence for the conversion of IAA to IBA (Le Bail et al., 2010).

Thus, auxins are present in seaweeds and elicit specific growth responses but to date, there are no reports of changes in auxin composition in response to environmental signals. In addition, biosynthetic and metabolic pathways appear more rudimentary than in vascular plants. *De novo* biosynthesis in seaweeds appears to favour the TAM pathway. IAA homeostasis in seaweeds appears to be regulated by biosynthesis and degradation rather than conjugation. Taken together, this suggests that adaptation to the terrestrial environment was associated with a diversification of auxin biosynthesis and signalling pathways. However, this conclusion is based on the genomic sequence of only a few microalgal and one seaweed species. More seaweed species from diverse taxonomic classes need to be analysed before such generalisations can be accepted. In addition, conflicting reports on the widespread occurrence of IAM in seaweeds and microalgae, but with no molecular evidence for the IAM biosynthetic pathway needs to be addressed.

5.3 CYTOKININS

Cytokinins are essential for cell division as well as influencing many developmental processes in vascular plants such as controlling the root:shoot ratio, delaying senescence, breaking bud dormancy, influencing nutrient mobilisation, apical dominance, floral development and seed germination (Kieber, 2002; Sakakibara, 2006). Most naturally occurring cytokinins are adenine derivatives with either an isoprenoid side chain, i.e. isopentenyladenine (iP), *trans*-zeatin (*t*Z), *cis*-zeatin (*c*Z) and dihydrozetin (DHZ), or an aromatic side chain, i.e. BA and topolins (Sakakibara, 2006; Yevakova et al., 2008). There is also evidence that cytokinin biosynthesis occurs in the slime mold *Dictyostelium discoideum* (Sakakibara, 2006).

5.3.1 Endogenous Cytokinins in Seaweeds

Cytokinins are present in many diverse seaweeds. For example, *t*Z was present in *K. alvarezii*, *G. edulis* and *S. tenerrimum* (Prasad et al., 2010); iPR, *t*Z and *t*ZR were identified in *L. japonica* (Duan et al., 1995); iPR was detected in *C. globularis* (Zhang, Yamane, Takahashi, Chapman, & Phinney, 1989) and *Porphyra perforata* while iPA and *c*ZR were present in *Sargassum muticum* (Zhang, Chapman, Phinney, & Spray, 1991). Endogenous cytokinin profiles of 31 seaweeds (5 Chlorophyceae; 7 Phaeophyceae and 19 Rhodophyceae) collected from warmer KwaZulu-Natal waters and cooler Cape waters in South Africa were reported. These seaweeds contained 13 isoprenoid cytokinin conjugates with higher concentrations of iP and *c*Z conjugates, moderate concentrations of *t*Z conjugates and very low amounts of DHZ conjugates. In addition, aromatic cytokinins (BA and 5 topolins) were also present (Stirk, Novák, Strnad, & van Staden, 2003). Analysis of 11 Rhodophyceae species identified 11 isoprenoid cytokinin conjugates – five *c*Z conjugates with highest concentrations of *cis*-zeatin riboside-5′-monophosphate (*c*ZRMP), three *t*Z conjugates with highest concentrations of *t*Z and three iP conjugates with highest concentrations of iP. In addition, three DHZ conjugates were present in nine of these species (Yokoya et al., 2010; further discussed in Section 5.3.3).

5.3.2 Physiological Responses to Cytokinins

There are a few examples in the literature on the physiological effect of exogenous cytokinin application to seaweeds. For example, exogenous application of iPR and kinetin increased growth of *Ectocarpus confervoides* and *Pylaiella litoralis* (Phaeophyceae) while higher concentrations were inhibitory (Pedersén, 1973). Exogenous application of Z and iP to excised discs of *Macrocystis pyrifera* (Phaeophyceae) stimulated growth (de Nys, Jameson, & Brown, 1991). Application of kinetin offset the degradation of phycoerythrin in excised apical segments of *H. musciformis* as well as increasing the protein and chlorophyll content (Jennings, Broughton, & McComb, 1972). In the microalga *C. vulgaris*, exogenous application of BA, kinetin and Z had a positive effect, increasing cell number, chlorophyll and carotenoid content as well as being involved in carbon metabolism (Piotrowska & Czerpak, 2009).

The cytokinin-like activity of kelps (Phaeophyceae) has been extensively studied using bioassays such as the *Amaranthus* betacyanin and soybean callus bioassays. In *M. pyrifera*, the basal portion of young blades had the fastest growth with the highest rate of cell division and little growth in the old blades. Cytokinin-like activity was highest in the old blades and co-chromatographed with O-glucosides. There was also a gradient of cytokinin-like activity with the basal portion of the young blades having the highest activity which co-chromatographed with free bases and ribosides and the tip having the lowest activity. These activities suggested that as in vascular plants, O-glucosides may act as storage forms and free bases are the active cytokinins in *M. pyrifera* (de Nys *et al.*, 1991). There were seasonal changes in cytokinin-like activity in *Ecklonia maxima* stripes with higher activity in summer and an increase in activity that co-chromatographed with O-glucosides in winter (Featonby-Smith & van Staden, 1984). Similarly, highest cytokinin-like activity occurred in *M. pyrifera* collected in autumn, which coincided with maximum growth and lowest activity in summer and winter, which coincided with minimum growth (de Nys, Jameson, Chin, Brown, & Sanderson, 1990). Seasonal changes in the cytokinin-like activity of *Sargassum heterophyllum* showed a peak of activity in spring, which coincided with regenerative growth and receptacle initiation and a second peak in autumn, which coincided with gamete release. There was a low activity in summer during periods of active growth and in winter during senescence (Mooney & van Staden, 1984b). Cytokinin-like activity in *E. maxima* also showed a lunar rhythm with maximum activity during the last quarter and full moon. Gamete initiation and release are often linked to tidal rhythms and these results suggest that cytokinins may be involved in the regulation of reproductive events in *E. maxima* (Hofman, Featonby-Smith, & van Staden, 1986). There was also a lunar rhythm in *S. heterophyllum* with highest cytokinin-like activity in the reproductive laterals at full and new moon, which coincided with gamete release (Mooney & van Staden, 1984a). Although these studies used bioassays rather than positive identification of cytokinins, the reported changes in cytokinin-like activity do suggest that cytokinin levels and profiles are linked to growth phases and environmental conditions.

In a more recent study, 11 isoprenoid and lower concentrations of four aromatic cytokinins were identified in *U. fasciata* and *D. humifusa* from six bimonthly collections. The predominant cytokinins were *c*Z forms (mainly *c*ZRMP) and iP forms (mainly isopentenylsdenosine-5′-monophosphate (iPRMP) and iP) with lower concentrations of cytokinin O-glucosides and no N-glucosides.

Cytokinin concentrations followed a seasonal trend in *U. fasciata* with lower concentrations during the cooler months while there was less variation in cytokinin concentrations in *D. humifusa* (Stirk et al., 2009).

5.3.3 Cytokinin Biosynthesis and Metabolism

In vascular plants, the initial step of isoprenoid cytokinin biosynthesis is the prenylation of nucleotides (AMP, ADP and ATP) to cytokinin ribotides by adenylate-isopentenyl transferases (IPT; Sakakibara, 2006). The ribotides are further modified to interconvertible cytokinin-free bases and ribosides (Kieber, 2002; Sakakibara, 2006; Yevdakova et al., 2008). Cytokinin homeostasis is maintained by the rate of biosynthesis, conjugation, inactivation and degradation with structural modification to the side chain-altering physiological activity (Sakakibara, 2006). iP- and *t*Z-free bases are the most biological activities with *c*Z having little biological activity (Kieber, 2002; Sakakibara, 2006). Glycosylation may occur on either the side chain (O-glucosides) or at the N^3, N^7 or N^9 positions on the purine moiety (N-glucosides). O-glucosylation is reversible and forms inactive storage cytokinins while N-glucosylation is irreversible and is considered as a deactivation step (Sakakibara, 2006). CK oxidase/dehydrogenase (CKX) catalyses the irreversible degradation by the cleavage of the side chain with iP and *t*Z readily catalysed while DHZ and *c*Z forms are less susceptible (Sakakibara, 2006). In addition to the isoprenoid biosynthetic pathway, *c*Z-type cytokinins may be derived from tRNA degradation although based on tRNA turnover rates, this is not a major source of cytokinins in vascular plants (Kieber, 2002; Sakakibara, 2006). The biosynthetic pathway for aromatic cytokinins has still to be elucidated (Sakakibara, 2006).

Isoprenoid cytokinins are generally more abundant than aromatic cytokinins in vascular plants although the cytokinin profile depends on the plant species, tissue type and developmental stage (Sakakibara, 2006). A review by Auer (1997) highlighted the trend of vascular plants (both gymnosperms and angiosperms) having mainly Z- and DHZ-type cytokinins with a complex set of conjugates while nonvascular plants and green algae having mainly iP- and Z-type cytokinins with fewer conjugates. The present review of the literature (Section 5.3.1) confirms that iP- and *c*Z-type cytokinins are the most commonly detected cytokinins in seaweeds with moderate levels of *t*Z-type cytokinins and only a few reports of low concentrations of DHZ-type cytokinins. There were generally higher concentrations of ribotide conjugates, moderate amounts of free bases, ribosides and O-glucosides and no N-glucosides (Stirk et al., 2003).

iPR and *c*ZR were isolated from tRNA of *Porphyra perforata* indicating that cytokinins may be formed via tRNA degradation (Zhang, Yamane, Spray, Chapman, & Phinney, 1992). Twenty isoprenoid and aromatic cytokinins were identified in *Physcomitrella patens* although no homologues of adenylate-IPTs were found. Use of a cytokinin-overproducing mutant suggested that tRNA degradation may be an important biosynthetic pathway in moss (Yevdakova et al., 2008).

The cytokinin-signalling pathway in vascular plants is controlled by a negative feedback mechanism and involves more than 30 proteins with hormone-binding receptors and negative regulators (Pils & Heyl, 2009; Žd'árská et al., 2013). Cytokinin receptors are histidine kinases with signal transduction involving two types of proteins – histidine phosphotransfer proteins (HPt) and type-B response regulators (RR; Steklov, Lomin, Osolodkin, & Romanov, 2013). In a comparative study, the various elements in the cytokinin-signalling pathway were compared in *C. reinhardtii* and *Volvox carteri* and five land plants, namely the moss *P. patens*, lycophyte *Selaginella moellendorfii* and monocotyledonous rice and dicotyledonous *Arabidopsis thaliana* and *Populus trichocarpa*. Analysis showed that different components of the signalling pathway evolved at different times with *P. patens* being the most basal plant species that contained all the components required for cytokinin signalling. However, the two algal species did contain at least one gene that coded for HPt with the number of HPTs increasing from two in *Selaginella* up to 10 in *Populus*. Similarly, the algae contained two to four RR genes with *Populus* having 32 RR genes (Pils & Heyl, 2009). A later study also found histidine kinases in the filamentous *Ectocarpus siliculosis* (Steklov et al., 2013). This genomic data indicates that cytokinin receptors probably have a monophyletic origin (Steklov et al., 2013) although more extensive research is required to elucidate all the possible components of the cytokinin signalling pathway in seaweeds.

Thus cytokinins are present in a diverse range of seaweeds and are able to influence both growth responses and other physiological processes such as pigment and protein metabolism. Cytokinin profiles change in response to environmental conditions suggesting a similar growth regulating function in seaweeds as in vascular plants. As with auxins, cytokinin biosynthesis and metabolism appears less complex in seaweeds compared to vascular plants. The biosynthetic pathways are unknown but could be via tRNA degradation. Fewer conjugate types suggest that homeostasis is by biosynthesis and interconversion between ribotides, free bases and ribosides with lower rates of inactivation to O-glucosides and no deactivation to N-glucosides.

Molecular evidence indicates that a simplified signalling pathway is present in seaweeds with both HPt and RR genes detected with the shift to terrestrial habitats resulting in an expansion of the biosynthetic and signalling pathways.

5.4 GIBBERELLINS

GAs are tetracyclic diterpenoid carboxylic acids involved in diverse physiological processes in vascular plants such as seed germination, stem elongation, leaf expansion and flower and seed development (Urbanová, Tarkowská, Novák, Strnad, & Hedden, 2013; Yamaguchi, 2008). GAs are key mediators in the perception of environmental signals with many examples of interactions with light, e.g. GAs are implicated in the phytochrome system in seed germination, photomorphogenesis of etiolated seedlings and in photoperiod control of stem elongation and flowering. GAs are also involved in temperature responses in seed germination and in certain abiotic stress responses (Yamaguchi, 2008).

5.4.1 Endogenous GAs in Seaweeds

There are a few reports of endogenous GAs in seaweeds. GA_3 was positively identified in wild harvested *Ulva* species (*U. fasciata*, *U. lactuca*, *U. taeniata*, *U. reticulata* and *U. linza*) but was below the limit of detection in cultured *M. oxyspermum* (Gupta et al., 2011); GA_3 was identified in *K. alvarezii* and *G. edulis* and GA_9 was identified in *S. tenerrimum* (Prasad et al., 2010). Recently, 18 GAs were identified in the kelp *E. maxima* (Stirk, Tarkowská, Turečová, Strnad, & van Staden, 2014).

5.4.2 Physiological Responses to GAs

There are limited and conflicting reports on the physiological effects elicited with GA application in seaweeds and microalgae. Exogenously applied GA_3 stimulated cell division in *Enteromorpha prolifera* and *Ecklonia radiata* gametophytes (Jennings, 1968) but had no effect on the morphology and photosynthetic physiology of developing *Fucus vesiculosus* embryos (Tarakhovskaya et al., 2013). GA_3 and to a lesser extent GA_9 promoted cell division and dry weight accumulation in the microalgae *C. vulgaris*, *Scenedesmus quadricauda* and *Dictyosphaerium pulchellum* (Burkiewicz, 1987). Total endogenous GA content showed a negative correlation to growth rate in 24 axenic microalgal strains with slower growing strains having a higher GA content (Stirk, Bálint, et al., 2013).

5.4.3 GA Biosynthesis and Metabolism

In vascular plants, GA biosynthesis can be separated into three stages. Firstly, geranylgeranyl diphosphate, a C_{20} precursor for diterpenoids is converted in a multistep pathway to *ent*-kaurene by a class of terpene cyclases which are located in the plastids. Secondly, oxidation steps involving Cyt P450 mono-oxygenases that are associated with the endoplasmic reticulum convert *ent*-kaurene to GA_{12}-aldehyde. These two stages are common to all vascular plants (Weiss & Ori, 2007). The final step in the biosynthetic pathway is the conversion of GA_{12}-aldehyde via GA_{12} to bioactive GAs by 2-oxoglutarate dependent dioxygenases located in the cytoplasm (Yamaguchi & Kamiya, 2000). GA_{12} is converted via two main oxidation pathways to the bioactive GA_4 and GA_1. These are:

1. $GA_{12} \rightarrow GA_{15} \rightarrow GA_{24} \rightarrow GA_9 \rightarrow GA_4$
2. $GA_{12} \rightarrow GA_{53} \rightarrow GA_{44} \rightarrow GA_{19} \rightarrow GA_{20} \rightarrow GA_1$

In addition, two minor pathways synthesise GA_3 and GA_6 from GA_{20} via GA_5 (Yamaguchi, 2008). This stage varies with species favouring different pathways (Weiss & Ori, 2007). Over 100 GAs have been identified but only a few are biologically active, i.e. GA_1, GA_3, GA_4, GA_5, GA_6 and GA_7 with GA_1 and GA_4 being the most widespread in vascular plants (Yamaguchi, 2008). Deactivation by methylation and glucosylation to form inactive GAs such as GA_{51}, GA_{34}, GA_{29} and GA_8 occur at specific steps along these pathways in order to regulate the levels of active GAs (Urbanová et al., 2013; Yamaguchi, 2008).

Unlike vascular plants where GA_4 and GA_1 are the main biologically active GAs, GA_4 and GA_6 were the predominant active GAs identified in the kelp *E. maxima*. Although no GA_{12}-ald and GA_{12} were detected, other components of both major biosynthetic pathways such as $GA_{15} > GA_9 > GA_{24}$ and $GA_{53} > GA_{44} > GA_{20} = GA_{19}$ were detected. In addition, GA_{51}, a catabolite in vascular plants was present in high concentrations in *E. maxima* (Stirk et al., in 2014). Similar GA profiles comprising of 20–22 GAs (including GA_{12}-ald and GA_{12}) were detected in 24 axenic microalgae strains (Stirk, Bálint, et al., 2013).

In *P. patens*, the gene encoding *ent*-kaurene synthase is present, indicating that early GA intermediates are present in moss. However, DELLA and GID1 proteins involved in GA signalling pathways in vascular plants (Weiss & Ori, 2007) were not present (Yamaguchi, 2008). GA_1 and GA_4 were identified in the sporophyte of some tree ferns while functional GID1 and DELLA homologues were present in some *Selaginella* species. In combination, these findings suggest that the GA signalling pathways via GID1/DELLA interactions arose in land plants after bryophyte divergence (Yamaguchi, 2008). The presence or absence of GID1 and DELLA homologues in seaweeds has not been reported to date.

Although not nearly as extensively investigated as auxins and cytokinins, these few reports do indicate that GAs are present in seaweeds and that they are likely to have a physiological function. Based on the GA profiles, it would appear that GA biosynthesis favours the GA_6 pathway which is a minor pathway in vascular plants. Future experiments with exogenous GAs should include GA_6 and GA_4 treatments to investigate if these GAs are more biologically active than GA_3 in seaweeds.

5.5 ABSCISIC ACID

ABA is a C_{15} sesquiterpene plant hormone with a duel function in vascular plants. It is involved in seed development, promoting early embryo growth but inducing and maintaining seed dormancy during late embryogenesis. ABA is also associated with senescence, increasing protein and chlorophyll degradation and inhibiting photosynthesis as well as acting as a stress hormone with up regulation of ABA synthesis genes in response to environmental stress, particularly water stress (Nambara & Marion-Poll, 2005).

5.5.1 Endogenous ABA in Seaweeds and Their Physiological Responses

ABA is found in all classes of seaweeds. For example, ABA and lower concentrations of 2-*trans*-methyl ABA were identified in *Ascophyllum nodosum* (Phaeophyceae; Boyer & Dougherty, 1988), ABA was identified in *U. fasciata* and *D. humifusa* (Stirk et al., 2009), in 10 Rhodophyceae species (Yokoya et al., 2010) and in five *Ulva* species and in cultured *Monostroma oxyspermum* (Gupta et al., 2011). There are numerous examples of ABA in microalgae (reviewed in Takezawa, Komatsu, & Sakata, 2011).

Examples of physiological responses to ABA are limited mainly to Phaeophyceae and microalgae. For example, ABA was present in the cultured sporophytes of *Laminaria hyperborea*, *Laminaria digitata* and *Laminaria saccharina* which showed seasonal growth patterns with a negative correlation between endogenous ABA and growth rate. ABA concentrations increased at the beginning of the resting phase and decreased prior to new blade growth. Responses to exogenous ABA application to *L. hyperborea* sporophytes was influenced by daylength with ABA inhibiting growth in short day treatments and having no effect on long day treatments (Schaffelke,

1995). Growth of *L. japonica* sporophyte discs was suppressed with exogenous ABA treatments while sorus formation was accelerated (Nimura & Mizuta, 2002). Fucoid eggs develop externally in seawater and therefore need to be photosynthetically competent to provide energy for the developing embryo. Exogenous application of ABA to eggs of *F. vesiculosus* had an inhibitory effect on chl *a* fluorescence as well as inhibiting apical hair number and length (Tarakhovskaya et al., 2013). There are also examples of ABA levels in microalgae increasing under salt stress conditions and inducing oxidative stress mechanisms via reactive oxygen species (ROS; Yoshida, Igarashi, Wakatsuki, Miyamoto, & Hirata, 2004) and increasing carotene accumulation (Cowan & Rose, 1991).

5.5.2 ABA Biosynthesis and Metabolism

The ABA biosynthetic pathway is well established in vascular plants. The synthesis of ABA from the C_{15} precursor xanthoxin is a two-step process via the intermediate abscisic aldehyde (ABAld). These two steps are catalysed by AtABA2 (isolated from *Arabidopsis*) and abscisic aldehyde oxidases respectively (Nambara & Marion-Poll, 2005). ABA catabolism is via three hydroxylation pathways to form dihydrophaseic acid (DPA), phaseic acid (PA) and neoPA. In addition, conjugation with glucose to form inactive ABA-glucose esters is catalysed by ABA glucosyltransferases (Nambara & Marion-Poll, 2005). Most of the enzymes involved in these pathways have been identified and characterised (Nambara & Marion-Poll, 2005).

Exogenous ABA elicits stress-related physiological responses in mosses and liverworts (reviewed in Takezawa et al., 2011). Molecular analysis showed that ABA biosynthesis via the carotenoid pathway and ABA signalling components are present in nonvascular plants, i.e. *P. patens* and *S. moellendorfii* while algae do not synthesise ABA via the carotenoid pathway. The authors speculated that the development of the carotenoid biosynthetic pathway for ABA in nonvascular plants enabled ABA to evolve as a signalling substance which was an important stress adaptation to the terrestrial environment (Takezawa et al., 2011).

In a study to determine the functional origin of ABA-related genes, *A. thaliana*, *Arabidopsis lyrata*, *P. trichocarpa*, *Oryza sativa* and nonvascular plants *S. moellendorffi* and *P. patens* were compared. The authors concluded that the ABA-related genes responsible for 11 key protein functions were derived from a common ancestor and that the key functions were mostly conserved in all the species. However, duplicate genes, especially those related to the ABA-signalling pathway, were greatly expanded following morphological divergence, suggesting that the ABA-signalling pathway contributed to the sophisticated mechanisms of stress tolerance in plant evolution (Hanada, Hase, Toyoda, Shinozaki, & Okamoto, 2011). Taken together, these results suggest that the ABA metabolic pathway is similar in nonvascular and vascular plants (Hanada et al., 2011). However, ABA-related genes in algae were not investigated and these need to be compared in order to gain more conclusive proof of the evolutionary origin of ABA-related genes (Hanada et al., 2011).

While endogenous ABA is present in seaweeds from diverse taxonomic classes and is able to elicit physiological responses related to both growth and photosynthesis, the biosynthetic and metabolic pathways still need to be determined.

5.6 ETHYLENE

Ethylene is a simple, unsaturated hydrocarbon gas that elicits physiological responses in plants such as fruit ripening, senescence of flowers and leaves and inhibition of shoot and root elongation. It is a 'stress hormone', being generated in response to various biotic and abiotic stresses (Bleecker & Kende, 2000). There appears to be no transport pathways between plant tissues but rather ethylene readily diffuses across membranes from cell to cell (Bleecker & Kende, 2000). In vascular plants, ethylene is derived from methionine in a two-step pathway via *S*-adenosyl-L-methionine (SAM) and 1-aminocyclopropane-1-carboxylic acid (ACC) with the steps controlled by ACC synthase and ACC oxidase, respectively. Expression of genes involved in the ACC pathway is regulated by both environmental and developmental signals, including PGR signals (Bleecker & Kende, 2000). In a study involving many genera of nonvascular land plants (two liverworts, two mosses and four ferns) and a range of vascular plants including the more primitive Gnetaceae, Ephedraceae as well as Gingkoaceae, cycads, gymnosperms and angiosperms, all plants produced ethylene at comparable rates. While exogenous ACC enhanced ethylene production in the vascular plants, it had no effect on ethylene production in the nonvascular plants. This suggested that the ethylene biosynthetic pathway via SAM and ACC is highly conserved in vascular plants but followed a different pathway in nonvascular land plants. However, the precursor for this pathway was not identified (Osborne, Walters, Milborrow, Norville, & Stange, 1996).

Seaweeds produce many volatile compounds, including ethylene (García-Jiménez, Brito-Romano, & Robaina, 2013). There are two proposed ethylene biosynthetic pathways in seaweeds – the ACC pathway as in vascular plants and the acrylate decarboxylase pathway. This alternative two-step pathway converts methionine to dimethylsulphoniopropionate (DMSP) and then the enzyme DMSP lyase converts DMSP to acrylate and the volatile dimethyl sulphide (DMS). Acrylate is further converted by decarboxylation to ethylene (Plettner, Steinke, & Malin, 2005). There are only a few studies on ethylene production in seaweeds and four such studies are discussed below:

1. Ethylene concentrations increased throughout the day in rock pools containing various seaweeds with lower rates of ethylene production during the night. In addition to being light-dependent, ethylene production was also temperature-dependent. Further incubation experiments in closed vessels showed that all 10 seaweed species investigated (7 Phaeophyceae, 2 Rhodophyceae and 1 Chlorophyceae) produced ethylene with *Asparagopsis armata* (Rhodophyceae) and *Ulva intestinalis* having the highest production rates and the Phaeophyceae lower production rates (Broadgate, Malin, Küpper, Thompson, & Liss, 2004).
2. *U. intestinalis* produced ethylene under normal growth conditions and this increased along with DMS production, when cultures were stressed by transfer to high light conditions. Exogenous ACC increased ethylene production while L-alpha-(2-(2-aminoethoxy) vinyl) glycine hydrochloride (AVG), an inhibitor of ACC synthase activity decreased ethylene production. Application of various inducers and inhibitors of the acrylate decarboxylase pathway had little effect on ethylene production, suggesting that the ACC pathway is the major ethylene-producing pathway in *U. intestinalis*. Ethephon (an ethylene releasing agent) increased chl *a* degradation in *U. intestinalis* (Plettner et al., 2005).
3. *Gelidium arbuscula* (Rhodophyceae) released a number of volatile compounds including ethylene and DMS when exposed to different light and salinity treatments. Exposure to exogenous ethylene increased the rate of ethylene production but did not alter DMS production, indicating that the acrylate decarboxylation pathway was not a major ethylene biosynthesis pathway in *G. arbuscula* (García-Jiménez et al., 2013).
4. Ethylene affected tetrasporogenesis in *Pterocladiella capillaceae* (Rhodophyceae) where the ethylene production rate was correlated to ACC synthase and ACC oxidase activity. Exposure to ethylene for various lengths of time (from 30 s to 3 h) increased the number of mature tetrasporangial branches by up to 200-fold in a time-dependent manner with an optimum exposure time of 15–30 min. Ethylene and

polyamines also compete for the same intermediate (SAM) and optimal ethylene treatments increased putrescine concentrations in *P. capillaceae* (García-Jiménez & Robaina, 2012).

These studies provide evidence that seaweeds do synthesise ethylene and that production rates are influenced by environmental factors such as light, temperature and salinity. Ethylene appears to have similar physiological functions in seaweeds as in vascular plants, increasing chlorophyll degradation (Plettner et al., 2005) and influencing maturation of reproductive structures (García-Jiménez & Robaina, 2012). It would also appear that as in vascular plants, the ACC pathway is the favoured ethylene biosynthesis route in seaweeds rather than the acrylate decarboxylase pathway. However, ethylene production and the physiological responses evoked by ethylene needs to be investigated in a greater diversity of seaweed species before such generalised statements can be confirmed. In addition, the presence (or absence) of enzymes such as ACC synthase and ACC oxidase in seaweeds needs to be established.

5.7 BRASSINOSTEROIDS

Brassinosteroids are steroidal compounds consisting of a C_{27}, C_{28} or C_{29} side-chain and two ring structures – A and B rings (Bagjuz & Tretyn, 2003). They have a duel function in plants, promoting growth and providing protection against environmental stresses. Brassinosteroids positively regulate cell division and expansion (Gudesblat & Russinova, 2011) and are involved in a number of developmental processes such as stem and root growth, floral initiation and flower and fruit development (Bajguz & Hayat, 2009). They are found in all plant organs but generally occur in higher concentrations in actively growing tissues such as pollen and immature seeds compared to mature vegetative tissues (Bajguz & Tretyn, 2003). Brassinosteroids also play a role in protection against biotic and abiotic stresses where they stimulated the production of low molecular-weight metabolites such as ROS and proline (Bajguz & Hayat, 2009).

There are two parallel biosynthetic pathways for C_{28} brassinosteroids in vascular plants – the early and late C_6-oxidation pathways where campestanol is converted to castasterone (CS) and finally to brassinolide (BL). There is a crosstalk between these two pathways with both pathways being common in vascular plants. Brassinosteroid metabolism to regulate the bioactive pool includes hydroxylation, dimethylation, side chain cleavage and conjugation while oxidation and epimerisation, often accompanied by conjugation with fatty acids and glucose, are important deactivation steps (Fujioka & Yokota, 2003). A number of the biosynthetic enzymes and genes have been identified in angiosperms (Fujioka & Yokota, 2003).

Brassinosteroids are widely distributed, occurring in angiosperms, gymnosperms (6 conifer species) and in the nonvascular *Equisetum arvense* (Pteridophyte) and *Marchantia polymorpha* (Bryophyte) with over 70 brassinosteroids being identified to date (Bajguz & Hayat, 2009). CS is the most widely distributed brassinosteroid, followed by BL, typhasterol (TY), 6-deoxoCS, teasterone (TE) and 28-norCS (Bajguz & Tretyn, 2003). To date, there are only two reports of endogenous brassinosteroids in algae – two C_{28} brassinosteroids (24-*epi*CS and 28-HomoCS) were isolated from the freshwater green alga *Hydrodictyon reticulatum* (Yokota et al., 1987) and BL and CS were detected in similar quantities in the stipe and fronds of the kelp *E. maxima* (Stirk et al., 2014). Brassinosteroids also occur in microalgae with BL and CS detected in 24 microalgae strains (Stirk, Bálint, et al., 2013) and seven C_{28} brassinosteroids (TE, TY, 6-deoxoTE, 6-deoxoTY, 6-deoxoCS, CS and BL) isolated from *C. vulgaris* (Bajguz, 2009). These compounds are intermediates in the early and late C_6-oxidation pathways.

To date, there are no reports on the effects of exogenous application of BRs in seaweeds with only reports on microalgae. Exogenous application of BL, CS and homoCS-stimulated growth (cell number) in *C. vulgaris* as well as stimulating DNA, RNA and protein content (Bajguz, 2000) and overcame heavy metal stress by inhibiting heavy metal accumulation and reducing chlorophyll, protein and monosaccharide loss (Bajguz, 2011).

While brassinosteroids are present in seaweeds, their possible physiological functions have still to be investigated along with the biosynthetic and signalling pathways.

5.8 JASMONATES

The free acid jasmonic acid (JA) and its derivatives including the volatile methyl jasmonate (MeJA) are widely distributed in vascular plants. Jasmonates are involved in developmental processes such as seed germination, root growth, tendril coiling, flower development, fruit ripening and senescence (Cheong & Choi, 2003; Kombrink, 2012). In addition, they are also involved in defence mechanisms of plants, increasing in concentration in response to biotic wounding by pathogens and herbivore as well as to abiotic stresses such as drought, salinity and UV radiation (Cheong & Choi, 2003; Kombrink, 2012).

One of the primary functions of jasmonates in vascular plants is inducing chemical defence mechanisms against grazers and activating defence-related genes. There are similar responses in seaweeds. Seaweeds increase their secondary metabolites such as terpenoids and polyphenols as a chemical defence mechanism against herbivores such as littorine snails and various pathogens (Arnold, Targett, Tanner, Hatch, & Ferrari, 2001; Küpper et al., 2009). Experiments with *in situ F. vesiculosus* exposed to air-borne MeJA during low tide resulted in an increase in the polyphenolic content in the apical tips after 12 days to levels similar to those induced by stimulated grazing, providing evidence that jasmonates may be involved in the chemical defence mechanism of *F. vesiculosus* (Arnold et al., 2001). In vascular plants, stress induces an oxidative cascade of free fatty acids, which leads to the synthesis of stress hormones. This oxidative burst also occurs in seaweeds, mostly in response to pathogen attack and wounding. Exogenous linolenic acid and MeJA stimulated an oxidative burst in *L. japonica* sporophytes (Küpper et al., 2009) and MeJA activated oxidative metabolism of fatty acids in *Chondrus crispus* (Rhodophyceae), stimulating enzymes involved in defence reactions and transcription of defence-related genes (Collén, Hervé, Guisle-Marsollier, Léger, & Boyen, 2006). Following incubation of *C. crispus* gametophytes with exogenous MeJa, microarray and real-time PCR analysis showed that 6% of the genes responded to MeJA with stress-related genes being up regulated after 2 h incubation while after 6 h incubation, metabolism-related genes were up regulated and stress-related genes were down regulated. In addition, expression data from the antioxidative genes indicated that the MeJA treatment induced reactive oxygen metabolism in *C. crispus* in a similar manner as in vascular plants (Collén et al., 2006). JA and MeJA have also been identified in microalgae, e.g. two *Dunaliella* species (Fujii, Yamamoto, Miyamoto, & Ueda, 1997) and *Chlorella* species (Ueda et al., 1991). Jasomonate treatments increased the cell growth and metabolite content in *C. vulgaris* in a dose-dependent manner (Czerpak et al., 2006) and reduced bacterial infection and improved temperature and salinity tolerance in *Scenedesmus incrassulatus* (Christov et al., 2001; Fedina & Benderliev, 2000).

However, there are also conflicting reports where JA and the JA precursor oxo-phytodienoic acid (OPDA) were not detected in *Dictyota dichotoma*, which produces terpenoids as a chemical defence mechanism (Wiesemeier, Jahn, & Pohnert, 2008). In addition, JA and MeJA treatments did not change the profile of metabolites such as terpenes, phenolics, lipids and fatty acids in seven Phaeophyceae (*D. dichotoma*, *Colpomenia peregrine*, *Ectocarpus fasciculatus*,

F. vesiculosus, Himanthalia elongate, Saccharina latissima and *S. muticum*) suggesting that jasmonates do not play a role in inducing chemical defence responses in these species (Wiesemeier et al., 2008).

Biosynthesis of jasmonates is via the oxylipin octadecanoid pathway where initially α-linolenic acid is released from membrane lipids by phospholipases and then catalysed by lipoxygenases in a multistep process involving reduction and oxidation to JA. The enzyme JA carboxyl methyltransferase further metabolises JA to the volatile MeJA (Cheong & Choi, 2003; Kombrink, 2012). Apart from MeJA, there are many other jasmonate metabolites that are formed by conjugation with amino acids and glucose esters and hydroxylation, O-glycosylation and decarboxylation products (Cheong & Choi, 2003; Kombrink, 2012). JA-amino acid conjugates are bioactive (Chico et al., 2008) while other conjugates may function as inactivated storage forms (Kombrink, 2012). Two allene oxide cyclases that are key enzymes in the synthesis of OPDA were isolated from the moss *P. patens* as well as OPDA being detected. However, no JA or JA-amino acids were detected. Knockout mutants show that OPDA is involved in reproductive events in *P. patens*, suggesting that the role of oxylipins in reproductive events is evolutionary conserved (Stumpe et al., 2010). Seaweeds also have a primitive oxylipin-signalling component. Oxylipins were identified in *C. crispus* as well as two genes encoding for lipoxygenase. However, while MeJA was detected when *C. crispus* was incubated with linolenic acid, no candidates for the associated enzymes (allene oxide synthase, allene oxide cyclase and jasmonic acid methyltransferase) were found (Collén et al., 2013). In contrast, two enzymes, a 12-lipoxygenase and a hydroperoxide isomerase, involved in the biosynthetic pathway of JA from linolenic acid were identified in *Gracilariopsis lemaneiformis* (Rhodophyceae; Hamberg & Gerwick, 1993).

Although reports on the occurrence of jasmonates in seaweeds is limited, physiological responses backed by molecular data support the concept that jasmonates do play a similar stress-relieving role in seaweeds as in vascular plants. Further investigations are required.

5.9 SALICYLIC ACID

SA is a phenolic compound derived from the shikimate-phenylpropanoid pathway. It acts as a signal molecule for the induction of defence mechanisms against biotic and abiotic stresses in vascular plants and is involved in the protection against oxidative damage (Horváth, Szalai, & Janda, 2007).

ROS generated by plants in response to stress can cause cellular damage. SA is able to regulate the concentrations of ROS by controlling antioxidant enzyme activity (Zhou, Tang, & Wang, 2010) as well as crosstalk with other PGRs such as auxins and jasmonates (Staswick, Tiryaki, & Rowe, 2002).

Endogenous SA has been reported in a few seaweed species such as *Ulva* (*U. fasciata*, *U. lactuca*, *U. taeniata* and *U. linza*) and laboratory cultivated *M. oxyspermum* (Gupta et al., 2011) and was a major component of the total phenolic acid complement in the subtidal *Halimeda incrassate* and *Halimeda monile* (Halimedaceae; Mancini-Filho et al., 2009; Novoa et al., 2011). There is also evidence that SA is involved in the oxidative defence mechanisms for protection against environmental stresses in seaweeds in a similar manner as in vascular plants. For example, SA-pretreatment of healthy *L. japonica* juvenile sporophytes prior to heat stress improved thermotolerance by altering antioxidant enzymatic activity with increased superoxidase dismutase (SOD), peroxidase (POD) and catalase (CAT) activity (Zhou et al., 2010).

These results indicate a possible role of SA in improving the tolerance of seaweeds to various environmental stresses but this needs further investigation.

5.10 STRIGOLACTONES

A group of tricyclic lactones known as strigolactones were initially isolated from the root exudates of parasitic weeds. This group of PGRs consists of strigol and to date, 19 other strigol-related compounds (de Saint Germain et al., 2013). They occur in a wide variety of monocotyledonous and dicotyledonous species (Yoneyama, Xie, Yoneyama, & Takeuchi, 2009). Strigolactones function as germination stimulants of root parasitic weeds and act as host recognition signals for arbuscular mycorrhizal fungi (Kapulnik et al., 2011) with plants exuding strigolactones into the rhizosphere to facilitate this symbiotic relationship (Ruyter-Spira & Bouwmeester, 2012). Strigolactones also influence root and shoot growth, being negative regulators of lateral root initiation, positive regulators of root-hair elongation (Kapulnik et al., 2011) and inhibiting shoot branching (Yoneyama et al., 2009). Crosstalk between strigolactones and auxins has been demonstrated (Kapulnik et al., 2011).

Strigolactones are derived from the carotenoid pathway from the putative precursor all-*trans*-β-carotene. Following isomerisation to 9-*cis*-β-carotene by the iron-containing protein D27, two enzymatic cleavage steps catalysed by carotenoid cleavage dioxygenases CCD7 and CCD8 yield the intermediate carlactone. Several orthologs of genes encoding these two proteins have been characterised in a number of different species, e.g. *MAX3* and *MAX4* in *Arabidopsis*, *RMS5* and *RMS1* in pea and *D17* and *D10* in rice. Further conversion mediated by a cytochrome P450 enzyme MAX1 results in 5-deoxystrigol and other strigolactones. Perception of strigolactones and downstream signalling is controlled by an F-box leucine rich protein (MAX2/RMS4/D3 in *Arabidopsis*, pea and rice respectively) and α/β-fold hydrolases (D14 and D14-like; Delaux et al., 2012; Ruyter-Spira & Bouwmeester, 2012; de Saint Germain et al., 2013).

Early basal land plants from the Streptophyta produce strigolactones. Endogenous strigolactones are produced by the moss *P. patens* and extruded into the medium where they act as signalling molecules to control protonema growth and branching (Proust et al., 2011). Genome analysis found unique *CCD7*- and *CCD8*-like sequences and several homologues for *D27* although no *MAX1* homologues were detected (Proust et al., 2011). Strigolactones are also present in the liverworts *Marchantia* spp. and *Lunularia cruciate* and the Charales *Nitella hyalina and Nitella pseudoflabellata* with the purified extracts being active in the germ tube-branching assay. However, no strigolactones were identified in the Chlorophyta *C. reinhardtii* although there was the possibility of unknown strigolactones being overlooked. In addition, the extracts were not active in the germ tube-branching assay (Delaux et al., 2012). Analysis for strigolactone-related genes from available protein sequences, and genome and transcript assemblies of various lower order plants showed that *D27* is highly conserved throughout the green lineage including Chlorophyceae, Prasinophyceae and Trebouxiophyceae while *CCD7* and *CCD8* are less so. Homologues for downstream strigolactone signalling also appear to be absent in Chlorophyta but present in the Charophyta (Delaux et al., 2012).

Taken together, these results suggest that strigolactones are an ancient group of molecules that arose early on in the land plant lineage, appearing first in the Charophyta. They initially functioned as PGRs, stimulating rhizoid growth (Delaux et al., 2012) and later played a role in promoting the symbiotic relationship with arbuscular mycorrhizal fungi that stimulated the adaption of plants for the colonisation of land (Proust et al., 2011). Fossil evidence for symbiotic relationships with arbuscular mycorrhizal fungi date back 430–480 million years showing this relationship predated the appearance of roots but coincided with the first bryophyte-like land plants (Proust et al., 2011).

Based on the current literature, it would appear that strigolactones are not present in seaweeds. However, the lack of complete genome sequences for lower order plants including seaweeds, make such conclusions difficult.

5.11 POTENTIAL USES OF PGRs IN SEAWEED BIOTECHNOLOGY

Seaweeds have many industrial applications with phycocolloids (alginates, carrageenan and agar) used as structural, coagulation and gelling agents. They are also used extensively in agriculture as plant biostimulants due to the presence of PGRs (Stirk et al., 2014). There is also a large global industry for seaweeds used as food, for human consumption as well as animal and fish feed. More recently, there has been a growing demand for seaweeds to be used as functional foods and beverages, food supplements and nutraceuticals as seaweeds are rich sources of bioactive secondary metabolites such as phenolics, polysaccharides and organic acids (Gupta & Abu-Ghannam, 2011). This section briefly highlights the potential role of PGRs in seaweed biotechnology.

5.11.1 Mariculture and Aquaculture

Increased demand for seaweeds combined with declining natural seaweed populations has led to the development of ocean-based mariculture and more recently, land-based aquaculture. Ocean-based seaweed mariculture is a large industry where anchored nets and lines are seeded with the seaweed species being cultivated (Zhou et al., 2010). Major problems associated with mariculture include the production of robust plants for seeding the nets and lines, epiphyte growth (Hafting, Critchley, Cornish, Hubley, & Archibald, 2012) and biotic diseases, which are often triggered by abiotic factors. These factors all lead to poor quality plants and ways to improve crop yields are being investigated (Kai et al., 2006; Zhou et al., 2010). PGRs are able to promote reproduction and growth in seaweeds. For example, mariculture of *Porphyra* (in demand for 'nori') relies on the production of high-quality conchospores, which are seeded onto nets. Growth of the conchocelis phase of three *Porphyra* species was promoted by the exogenous application of IAA, kinetin and to a lesser extent, by GA_3. These effects were further modified by environmental factors of temperature and photoperiod (Lin & Stekoll, 2007). ABA treatments accelerated sorus formation in *L. japonica* and thus may be a useful technology to produce 'seedlings' for mariculture (Nimura & Mizuta, 2002).

Land-based aquaculture involves the pumping of water from the ocean or secondary source in integrated aquaculture with fish and abalone, into a holding tank where the seaweed is grown (Hafting et al., 2012). This practise has potential to treat the water with PGRs to improve seaweed growth and prevent diseases. Evidence suggests that PGRs such as ABA, brassinosteroids, jasmonates and SA could increase the tolerance of seaweeds to biotic and abiotic stresses as in vascular plants but this needs further investigation. However, toxicity studies will have to be conducted to confirm that seaweeds treated with exogenous PGRs are safe for both human and animal consumption and use.

Thus PGRs could potentially be used to alleviate some of the major problems associated with cultivation of seaweeds.

5.11.2 Tissue Culture

Tissue culture techniques are successfully used in commercial propagation of vascular plants. This technology uses PGRs, especially auxins and cytokinins with different PGR-containing media used to induce either rooting or shooting. One advantage of tissue culture is the faster, large-scale production of clonal material. Applications of similar techniques for commercial seaweed species would have many advantages including allowing for the selection of desirable seaweed characteristics such as high yielding and fast growing strains as well as reducing the pressure on natural seaweed populations.

Callus induction and thallus regeneration have been reported for a number of commercially utilised seaweeds, mainly Rhodophyceae species (reviewed in Reddy, Jha, Fujita, & Ohno, 2008) and a few examples are listed below. Positive results were reported for *Gracilaria tenuistipitata* and *Gracilaria perplexa* where auxin (IAA and 2,4-D) and kinetin stimulated the induction of both apical and intermediate callus formation. Following transfer to liquid culture, the regenerated plants developed tetrasporangia and released tetraspores that gave rise to adult gametophytes (Yokoya, West, & Luchi, 2004). Auxins and cytokinins also stimulated callus formation in *Gracilaria vermiculophylla*, eliciting a larger response in the intercalary segments compared to the apical segments (Yokoya, Kakita, Obika, & Kitamura, 1999). Explants of *K. alvarezii* formed callus from intact apical tips. When the callus was excised, it formed clumps of filaments that lead to the formation of micropropagules that developed into young plantlets that were successfully cultivated *in situ*. Treatments with naphthaleneacetic acid (NAA) and BA had no effect on callus induction and growth but did increase rates of cell division in the microproagules and young plantlets (Reddy, Kumar, & Tewari, 2003). Auxins and cytokinins were stimulatory in *Grateloupia dichotoma* when applied in combination with a high light intensity (Yokoya & Handro, 1997). In *Gracilaria dichotoma*, only IAA, isopentenyladenosine (iPR) and iP added in combination had a slight stimulatory effect on callus growth while other auxins (NAA and phenylacetic acid (PAA)), cytokinins (BA, kinetin and kinetin riboside) and GA_3 and ABA had no effect (Kaczyna & Megnet, 1993).

Thus, there is a potential to utilise tissue culture for large-scale production of 'seedling' stock to be used in mariculture and aquaculture. However, results are variable from species to species, the source of the explant tissue type (apical tip vs intercalary segments), specific PGRs and combinations thereof, as well as environmental conditions such as light, temperature, nutrient medium and pH. All these parameters could potentially influence endogenous PGR biosynthesis and metabolism. A more comprehensive understanding of the role of PGRs in seaweed physiology would enable the manipulation of tissue culture systems to improve the success rate and quality of 'plantlet' production for use in mariculture and aquaculture.

5.11.3 Edible Seaweeds and Functional Foods

There is a large demand for fresh, edible seaweeds but a short shelf-life limits their marketing and shipping to larger markets. Post-harvest monitoring of *Gracilaria* species showed that ethylene production occurred at similar rates whether stored in light or dark conditions while increased temperature accelerated ethylene production. The most successful post-harvest treatment was a 5 min dip into heated seawater (38–42 °C; Paull & Chen, 2008). A better understanding of the physiological role of ethylene in seaweed senescence and elucidating its biosynthetic and inactivation pathways may provide a practical approach to prolonging the shelf-life of edible seaweeds.

As many seaweeds are rich in phenolic compounds including salicylic acid which have antioxidant activity, they have potential to be used in the food industry to protect foods from oxidative processes. There is also a growing demand for seaweeds as functional foods due to the presence of other secondary metabolites (Novoa et al., 2011). Understanding the interactions between growth, PGRs and environmental cues could potentially enable the secondary metabolite content to be manipulated for maximum yield. However, there are stringent quality and safety issues with 'Good Manufacturing Practices' required especially in the growing of seaweeds for human use (Gellenbeck, 2012). PGRs are extensively used in agriculture crops and thus should not be a problem if used in seaweed aquaculture. However, cytotoxicity studies will have to be conducted to confirm this.

5.12 CONCLUSIONS

There are examples of all groups of PGRs being present in a diverse range of seaweeds from the Chlorophyceae, Rhodophyceae and Phaeophyceae. However, in the case of some PGR groups such as auxins and cytokinins, there were fewer conjugate types than found in nonvascular and vascular land plants, indicating more rudimentary biosynthetic and metabolic pathways.

In vascular plants, PGRs control essential growth processes. Auxins and cytokinins are essential for cell viability (Murphy, 2002), regulating cell division by transcriptional expression of some cell cycle genes and by post-transcriptional regulation and protein turnover (Perrot-Rechenmann, 2010). Many of the physiological responses elicited in seaweeds with PGR application either stimulated (auxins, cytokinins, GAs) or inhibited (GAs, ABA) growth and reproductive events where high rates of cell division can be expected. Thus, the presence of PGRs in seaweeds may be associated with essential physiological processes such as cell division and photosynthetic metabolism necessary for growth.

There was substantial diversification and specialisation with plants becoming larger and more complex following colonisation of the terrestrial environment. Most components involved in biosynthetic and signalling pathways are expanded in vascular plants, suggesting that adaptation to drier conditions drove the evolution of an integrated network of PGR pathways for the perception of and rapid response to environmental changes. However, it is still unclear if PGR signalling pathways were prerequisites for movement onto land or if they arose after colonisation of the terrestrial environment (De Smet et al., 2011; Lau et al., 2009). Seaweeds are also exposed to highly variable environmental conditions and stress-related responses to ABA, ethylene, brassinosteroids, jasmonates and SA have been reported in seaweeds and microalgae. Components of the cytokinin signalling pathway (HPt and RR proteins) and the jasmonate pathway are present in microalgae and seaweeds while the presence (or absence) of DELLA and GID1 proteins involved in GA signalling, ACC synthase and ACC oxidase enzymes involved in ethylene production and ABA- and brassinosteroid-related genes have not been confirmed in seaweeds. More genome sequences combined with additional PGR profiles and physiological studies are necessary in a wider range of seaweed and nonvascular plant groups before the evolutionary origins of the various PGR signals in response to environmental fluctuations can be fully elucidated.

Although arising from the same ancient green algal ancestor, early divergence of the various seaweed classes means that pre-existing genes in the green algal ancestor may have evolved to recognise different hormonal conjugates in seaweeds (Chico et al., 2008). This may explain the differences in the various biosynthetic pathways in seaweeds compared to vascular plants. For example, seaweeds appear to favour the TAM and GA_6 pathways, which are minor auxin and GA biosynthetic pathways in vascular plants. The genome of *E. siliculosus* was recently sequenced with 214 million base pairs that encoded for 16,256 proteins. The genome reflected both its ancient evolutionary past as well as the emergence of newer gene families associated with novel proteins (Cock et al., 2010). Given that many seaweeds inhabit rocky coastal zones and have adapted to these harsh and highly variable tidal environments, these novel proteins could potentially be associated with as yet unknown PGR-type compounds involved in adaptive responses to unique environmental conditions not encountered by terrestrial plants. Current analytical techniques such as GC–MS and UPLC-MS are limited to detecting known PGRs. Axenic seaweed-based bioassays need to be developed in order to detect novel seaweed-specific PGRs. There are a few early reports of PGR-type compounds specific to seaweeds, e.g.'rhizin' that was involved in the orientation of polarity of *Fucus* zygotes (in Bradley, 1991) and such avenues of research need to be explored further.

A better understanding of biosynthesis, homeostasis and physiological functions of PGRs in seaweeds would greatly benefit seaweed biotechnology, allowing for the manipulation of seaweed growth in commercial mariculture and aquaculture practices and so improve both the yield and quality of the seaweeds harvested.

ACKNOWLEDGEMENTS

The University of KwaZulu-Natal is thanked for financial support.

REFERENCES

Arnold, T. M., Targett, N. M., Tanner, C. E., Hatch, W. I., & Ferrari, K. E. (2001). Evidence for methyl jasmonate-induced phlorotannin production in *Fucus vesiculosus* (Phaeophyceae). *Journal of Phycology*, *37*, 1026–1029.

Auer, C. A. (1997). Cytokinin conjugation: recent advances and patterns in plant evolution. *Plant Growth Regulation*, *23*, 17–31.

Bajguz, A. (2000). Effect of brassinosteroids on nucleic acids and protein content in cultured cells of *Chlorella vulgaris*. *Plant Physiology and Biochemistry*, *38*, 209–215.

Bajguz, A. (2009). Isolation and characterization of brassinosteroids from algal cultures of *Chlorella vulgaris* Beijerinck (Trebouxiophyceae). *Journal of Plant Physiology*, *166*, 1946–1949.

Bajguz, A. (2011). Suppression of *Chlorella vulgaris* growth by cadmium, lead and copper stress and its restoration by endogenous brassinolide. *Archives of Environmental Contamination and Toxicology*, *60*, 406–416.

Bajguz, A., & Hayat, S. (2009). Effects of brassinosteroids on the plant responses to environmental stresses. *Plant Physiology and Biochemistry*, *47*, 1–8.

Bajguz, A., & Tretyn, A. (2003). The chemical characteristics and distribution of brassinosteroids in plants. *Phytochemistry*, *62*, 1027–1046.

Basu, S., Sun, H., Brian, L., Quatrano, R. L., & Muday, G. K. (2002). Early embryo development in *Fucus distichus* is auxin sensitive. *Plant Physiology*, *130*, 292–302.

Bleecker, A. B., & Kende, H. (2000). Ethylene: a gaseous signal molecule in plants. *Annual Review of Cell and Developmental Biology*, *16*, 1–18.

Boyer, G. L., & Dougherty, S. S. (1988). Identification of abscisic acid in the seaweed *Ascophyllum nodosum*. *Phytochemistry*, *27*, 1521–1522.

Bradley, P. M. (1991). Plant hormones do have a role in controlling growth and development of algae. *Journal of Phycology*, *27*, 317–321.

Broadgate, W. J., Malin, G., Küpper, F. C., Thompson, A., & Liss, P. S. (2004). Isoprene and other non-methane hydrocarbons from seaweeds: a source of reactive hydrocarbons to the atmosphere. *Marine Chemistry*, *88*, 61–73.

Burkiewicz, K. (1987). The influence of gibberellins and cytokinins on the growth of some unicellular Baltic algae. *Botanica Marina*, *30*, 63–69.

Cheong, J. J., & Choi, Y. D. (2003). Methyl jasmonate as a vital substance in plants. *Trends in Genetics*, *19*, 409–413.

Chico, J. M., Chini, A., Fonseca, S., & Solano, R. (2008). JAZ repressors set the rhythm in jasmonate signaling. *Current Opinion in Plant Biology*, *11*, 486–494.

Christov, C., Pouneva, I., Bozhkova, M., Toncheva, T., Fournadziev, S., & Zafirova, T. (2001). Influence of temperature and methyl jasmonate on *Scenedesmus incrassulatus*. *Biologia Plantarum*, *44*, 367–371.

Cock, J. M., Sterck, L., Rouzé, P., Scornet, D., Allen, A. E., Amoutzias, G., et al. (2010). The *Ectocarpus* genome and the independent evolution of multicellularity in brown algae. *Nature*, *465*, 617–621.

Collén, J., Hervé, C., Guisle-Marsollier, I., Léger, J. J., & Boyen, C. (2006). Expression profiling of *Chondrus crispus* (Rhodophyta) after exposure to methyl jasmonate. *Journal of Experimental Botany*, *57*, 3869–3881.

Collén, J., Porcel, B., Carré, W., Ball, S. G., Chaparro, C., Tonon, T., et al. (2013). Genome structure and metabolic features in the red seaweed *Chondrus crispus* shed light on evolution of the Archaeplastida. *Proceedings of the National Academy of Sciences*, *11*, 5247–5252.

Cowan, A. K., & Rose, P. D. (1991). Abscisic acid metabolism in salt-stressed cells of *Dunaliella salina*. *Plant Physiology*, *97*, 798–803.

Czerpak, R., Piotrowska, A., & Szulecka, K. (2006). Jasmonic acid affects changes in the growth and some components content in alga *Chlorella vulgaris*. *Acta Physiologiae Plantarum*, *28*, 195–203.

Delaux, P.-M., Xie, X., Timme, R. E., Puech-Pages, V., Dunand, C., Lecompte, E., et al. (2012). Origins of strigolactones in the green lineage. *New Phytologist*, *195*, 857–871.

De Smet, I., Voβ, U., Lau, S., Wilson, M., Shao, N., Timme, R. E., et al. (2011). Unraveling the evolution of auxin signaling. *Plant Physiology*, *155*, 209–221.

De Veylder, L., Beeckman, T., & Inzé, D. (2007). The ins and outs of the plant cell cycle. *Nature Reviews*, *8*, 655–665.

Duan, D., Liu, X., Pan, F., Liu, H., Chen, N., & Fei, X. (1995). Extraction and identification of cytokinin from *Laminaria japonica* Aresch. *Botanica Marina, 38*, 409–412.

Evans, L.V., & Trewavas, A. J. (1991). Is algal development controlled by plant growth substances? *Journal of Phycology, 27*, 322–326.

Featonby-Smith, F. C., & van Staden, J. (1984). Identification and seasonal variation of endogenous cytokinins in *Ecklonia maxima* (Osbeck) Papenf. *Botanica Marina, 27*, 527–531.

Fedina, I. S., & Benderliev, K. M. (2000). Response of *Scenedesmus incrassatulus* to salt stress as affected by methyl jasmonate. *Biologia Plantarum, 43*, 625–627.

Fuell, C., Elliot, K. A., Hanfrey, C. C., Franceschetti, M., & Michael, A. J. (2010). Polyamine biosynthetic diversity in plants and algae. *Plant Physiology and Biochemistry, 48*, 513–520.

Fujii, S., Yamamoto, R., Miyamoto, K., & Ueda, J. (1997). Occurrence of jasmonic acid in *Dunaliella* (Dunaliellales, Chlorophyta). *Phycological Research, 45*, 223–226.

Fujioka, S., & Yokota, T. (2003). Biosynthesis and metabolism of brassinosteroids. *Annual Review of Plant Biology, 54*, 137–164.

García-Jiménez, P., Brito-Romano, O., & Robaina, R. R. (2013). Production of volatiles by the red seaweed *Gelidium arbuscula* (Rhodophyta): emission of ethylene and dimethyl sulphide. *Journal of Phycology, 49*, 661–669.

García-Jiménez, P., & Robaina, R. R. (2012). Effects of ethylene on tetrasporogenesis in *Pterocladiella capillacea* (Rhodophyta). *Journal of Phycology, 48*, 710–715.

Gellenbeck, K. W. (2012). Utilization of algal materials for nutraceutical and cosmeceutical applications – what do manufacturers need to know? *Journal of Applied Phycology, 24*, 309–313.

Gudesblat, G. E., & Russinova, E. (2011). Plants grow on brassinosteroids. *Current Opinion in Plant Biology, 14*, 530–537.

Gupta, S., & Abu-Ghannam, N. (2011). Recent developments in the application of seaweeds or seaweed extracts as a means for enhancing the safety and quality attributes of food. *Innovative Food Science and Emerging Technologies, 12*, 600–609.

Gupta, V., Kumar, M., Brahmbhatt, H., Reddy, C. K. R., Seth, A., & Jha, B. (2011). Simultaneous determination of different endogenetic plant growth regulators in common green seaweeds using dispersive liquid-liquid microextraction method. *Plant Physiology and Biochemistry, 49*, 1259–1263.

Hafting, J. T., Critchley, A. T., Cornish, M. L., Hubley, S. A., & Archibald, A. F. (2012). On-land cultivation of functional seaweed products for human usage. *Journal of Applied Phycology, 24*, 385–392.

Hamberg, M., & Gerwick, W. H. (1993). Biosynthesis of vicinal dihydroxy fatty acids in the red alga *Gracilariopsis lemaneiformis*: identification of a sodium-dependent 12-lipoxygenase and a hydroperoxide isomerase. *Archives of Biochemistry and Biophysics, 305*, 115–122.

Hanada, K., Hase, T., Toyoda, T., Shinozaki, K., & Okamoto, M. (2011). Origin and evolution of genes related to ABA metabolism and its signaling pathways. *Journal of Plant Research, 124*, 455–465.

Hofman, P. J., Featonby-Smith, B. C., & van Staden, J. (1986). The development of ELISA and IRA for cytokinin estimation and their application to a study of lunar periodicity in *Ecklonia maxima* (Osbeck) Papenf. *Journal of Plant Physiology, 122*, 455–466.

Horváth, E., Szalai, G., & Janda, T. (2007). Induction of abiotic stress tolerance by salicylic acid signaling. *Journal of Plant Growth Regulation, 26*, 290–300.

Jennings, R. C. (1968). Gibberellins as endogenous growth regulators in green and brown algae. *Planta, 80*, 34–42.

Jennings, R. C., Broughton, W. J., & McComb, A. J. (1972). Effect of kinetin on the phycoerythrin and chlorophyll content of a red alga. *Phytochemistry, 11*, 1937–1943.

Jirásková, D., Poulíčková, A., Novák, O., Sedláková, K., Hradecká, V., & Strnad, M. (2009). High-through put screening technology for monitoring phytohormones production in microalgae. *Journal of Phycology, 45*, 108–118.

Kaczyna, F., & Megnet, R. (1993). The effects of glycerol and plant growth regulators on *Gracilaria verrucosa* (Gigartinales, Rhodophyceae). *Hydrobiologia, 268*, 57–68.

Kai, T., Nimura, K., Yasui, H., & Mizuta, H. (2006). Regulation of sorus formation by auxin in Laminariales sporophyte. *Journal of Applied Phycology, 18*, 95–101.

Kapulnik, Y., Delaix, P. M., Resnick, N., Mayzlish-Gati, E., Wininger, S., Bhattacharya, C., et al. (2011). Strigolactones affect lateral root formation and root-hair elongation in *Arabidopsis*. *Planta, 233*, 209–216.

Kieber, J. (2002). Cytokinins: regulators of cell division. In L. Taiz, & E. Zeiger (Eds.), *Plant physiology* (3rd ed.) (pp. 493–517). Sunderland Massachusetts: Sinauer Associates Inc.

Klämbt, D., Knauth, B., & Dittmann, I. (1992). Auxin dependent growth of rhizoids of *Chara globularis*. *Physiologia Plantarum, 85*, 537–540.

Kombrink, E. (2012). Chemical and genetic exploration of jasmonate biosynthesis and signaling paths. *Planta, 236*, 1351–1366.

Küpper, F. C., Gaqueral, E., Cosse, A., Adas, F., Peters, A. F., Müller, D. G., et al. (2009). Free fatty acids and methyl jasmonate trigger defence reactions in *Laminaria digitata*. *Plant and Cell Physiology, 50*, 789–800.

Lau, S., Shao, N., Bock, R., Jürgens, G., & De Smet, I. (2009). Auxin signaling in algal lineages: fact or myth? *Trends in Plant Science, 14*, 182–188.

Le Bail, A., Billoud, B., Kowalczyk, N., Kowalczyk, M., Gicquel, M., Le Panse, S., et al. (2010). Auxin metabolism and function in the multicellular brown alga *Ectocarpus siliculosus*. *Plant Physiology, 153*, 128–144.

Leliaert, F., Smith, D. R., Moreau, H., Herron, M. D., Verbruggen, H., Delwiche, C. F., et al. (2012). Phylogeny and molecular evolution of the green algae. *Critical Reviews in Plant Sciences, 31*, 1–46.

Lin, R., & Stekoll, M. S. (2007). Effects of plant growth substances on the conchocelis phase of Alaskan *Porphyra* (Bangiales, Rhodophyta) species in conjunction with environmental variables. *Journal of Phycology, 43*, 1094–1103.

Ludwig-Müller, J. (2011). Auxin conjugates: their role for plant development and in the evolution of land plants. *Journal of Experimental Botany, 62*, 1757–1773.

Lyons, R., Manners, J. M., & Kazan, K. (2013). Jasmonate biosynthesis and signaling in monocots: a comparative overview. *Plant Cell Reports, 32*, 815–827.

Mancini-Filho, J., Novoa, A. V., González, A. E. B., Andrade-Wartha, E. R. S., Silva, A. M. O., Pinto, J. R., et al. (2009). Free phenolic acids from the seaweed *Halimeda monile* with antioxidant effect protecting against liver injury. *Zeitschrift Fuer Naturforschung, 64*, 657–663.

Mano, Y., & Nemoto, K. (2012). The pathway of auxin biosynthesis in plants. *Journal of Experimental Botany, 63*, 2853–2872.

Mooney, P. A., & van Staden, J. (1984a). Lunar periodicity of the levels of endogenous cytokinins in *Sargassum heterophyllum* (Phaeophyceae). *Botanica Marina, 27*, 467–472.

Mooney, P. A., & van Staden, J. (1984b). Seasonal changes in the levels of endogenous cytokinins in *Sargassum heterophyllum* (Phaeophyceae). *Botanica Marina, 27*, 437–442.

Murphy, A. (2002). Auxin: the growth hormone. In L. Taiz, & E. Zeiger (Eds.), *Plant physiology* (3rd ed.) (pp. 423–460). Sunderland Massachusetts: Sinauer Associates Inc.

Nambara, E., & Marion-Poll, A. (2005). Abscisic acid biosynthesis and catabolism. *Annual Review of Plant Biology, 56*, 165–185.

Nimura, K., & Mizuta, H. (2002). Inducible effects of abscisic acid on sporophyte discs from *Laminaria japonica* Areschoug (Laminariales, Phaeophyceae). *Journal of Applied Phycology, 14*, 159–163.

Novoa, A. V., Andrade-Wartha, E. R. S., Linares, A. F., Silva, A. M. O., Genovese, M. I., González, A. E. B., et al. (2011). Antioxidant activity and possible bioactive components in hydrophilic and lipophilic fractions from the seaweed *Halimeda incrassate*. *Brazilian Journal of Pharmacognosy, 21*, 53–57.

de Nys, R., Jameson, P. E., & Brown, M.T. (1991). The influence of cytokinins on the growth of *Macrocystis pyrifera*. *Botanica Marina, 34*, 465–467.

de Nys, R., Jameson, P. E., Chin, N., Brown, M. T., & Sanderson, K. J. (1990). The cytokinins as endogenous growth regulators in *Macrocystis pyrifera* (L.) C. Ag. (Phaeophyceae). *Botanica Marina, 33*, 467–475.

Osborne, D. J., Walters, J., Milborrow, B.V., Norville, A., & Stange, L. M. (1996). Evidence for a non-ACC ethylene biosynthesis pathway in lower plants. *Phytochemistry, 42*, 51–60.

Paull, R. E., & Chen, N. J. (2008). Postharvest handling and storage of the edible red seaweed *Gracilaria*. *Postharvest Biology and Technology, 48*, 302–308.

Pedersén, M. (1973). Identification of a cytokinin, 6-(3 methyl-2-butenylamino) purine, in seawater and the effect of cytokinins on brown algae. *Physiologia Plantarum, 28*, 101–105.

Perrot-Rechenmann, C. (2010). Cellular responses to auxin: division versus expansion. *Cold Spring Harbour Perspectives in Biology, 2*, a001446.

Pils, B., & Heyl, A. (2009). Unravelling the evolution of cytokinin signaling. *Plant Physiology, 151*, 782–791.

Piotrowska, A., & Czerpak, R. (2009). Cellular response of light/dark-grown green alga *Chlorella vulgaris* Beijerinck (Chlorophyceae) to exogenous adenine- and phenylurea-type cytokinins. *Acta Physiologiae Plantarum, 31*, 573–585.

Plettner, I., Steinke, M., & Malin, G. (2005). Ethene (ethylene) production in the marine macroalgae *Ulva (Enteromorpha) intestinalis* L. (Chlorophyta, Ulvophyceae): effect of light-stress and co-production with dimethyl sulphide. *Plant, Cell and Environment, 28*, 1136–1145.

Popper, Z. A., Michel, G., Hervé, C., Domozych, D. S., Willats, W. G. T., Tuohy, M. G., et al. (2011). Evolution and diversity of plant cell walls: from algae to flowering plants. *Annual Review of Plant Biology, 62*, 567–590.

del Pozo, J. C., Lopez-Matas, M. A., Ramirez-Parra, E., & Gutierrez, C. (2005). Hormonal control of the plant cell cycle. *Physiologia Plantarum, 123*, 173–183.

Prasad, K., Das, A. K., Oza, M. D., Brahmbhatt, H., Siddhanta, A. P., Meena, R., et al. (2010). Detection and quantification of some plant growth regulators in a seaweed-based foliar spray employing a mass spectrometric technique sans chromatographic separation. *Journal of Agricultural and Food Chemistry, 58*, 4594–4601.

Proust, H., Hoffmann, B., Xie, X., Yoneyama, K., Schaefer, D. G., Yoneyama, K., et al. (2011). Strigolactones regulate protonema branching and act as a quorum sensing-like signal in the moss *Physcomitrella patens*. *Development, 138*, 1531–1539.

Reddy, C. R. K., Jha, B., Fujita, Y., & Ohno, M. (2008). Seaweed micropropagation techniques and their potentials: an overview. *Journal of Applied Phycology, 20*, 609–617.

Reddy, C. R. K., Kumar, G. R. K., & Tewari, A. K. S.A. (2003). *In vitro* somatic embryogenesis and regeneration of somatic embryos from pigmented callus of *Kappaphycus alvarezii* (Doty) Doty (Rhodophyta, Gigartinales). *Journal of Phycology, 39*, 610–616.

Ruyter-Spira, C., & Bouwmeester, H. (2012). Strigolactones affect development in primitive plants. The missing link between plants and arbuscular mycorrhizal fungi? *New Phytologist, 195*, 730–733.

de Saint Germain, A., Bonhomme, S., Boyer, F.-D., & Rameau, C. (2013). Novel insights into strigolactone distribution and signaling. *Current Opinion in Plant Biology, 16*, 583–589.

Sakakibara, H. (2006). Cytokinins: activity, biosynthesis, and translocation. *Annual Review of Plant Biology, 57*, 431–449.

Schaffelke, B. (1995). Abscisic acid in sporophytes of three *Laminaria* species (Phaeophyta). *Journal of Plant Physiology, 146*, 453–458.

Staswick, P. E., Tiryaki, I., & Rowe, M. L. (2002). Jasmonate response locus *JAR1* and several related Arabidopsis genes encode enzymes of the firefly luciferase superfamily that show activity on jasmonic, salicyclic, and indole-3-acetic acids in an assay for adenylation. *The Plant Cell, 14*, 1405–1415.

Steklov, M.Y., Lomin, S. N., Osolodkin, D. I., & Romanov, G. A. (2013). Structural basis for cytokinin receptor signaling: an evolutionary approach. *Plant Cell Reports*, *32*, 781–793.

Stirk, W. A., Bálint, P., Tarkowská, D., Novák, O., Strnad, M., Ördög, V., et al. (2013). Hormone profiles in microalgae: gibberellins and brassinosteroids. *Plant Physiology and Biochemistry*, *70*, 348–353.

Stirk, W. A., Novák, O., Hradecká, V., Pěnčík, A., Rolčík, J., Strnad, M., et al. (2009). Endogenous cytokinins, auxins and abscisic acid in *Ulva fasciata* (Chlorophyta) and *Dictyota humifusa* (Phaeophyta): towards understanding their biosynthesis and homeostasis. *European Journal of Phycology*, *44*, 231–240.

Stirk, W. A., Novák, O., Strnad, M., & van Staden, J. (2003). Cytokinins in macroalgae. *Plant Growth Regulation*, *41*, 13–24.

Stirk, W. A., Ördög, V., Novák, O., Rolčík, J., Strnad, M., Bálint, P., et al. (2013). Auxin and cytokinin relationships in 24 microalgal strains. *Journal of Phycology*, *49*, 459–467.

Stirk, W. A., Tarkowská, D., Turečová, V., Strnad, M., van Staden, J. (2014). Abscisic acid, gibberellins and brassinosteroids in Kelpak®, a commercial seaweed extract made from *Ecklonia maxima*. *Journal of Applied Phycology*, *26*, 561–567. doi:10.1007/s10811-013-0062-z.

Stumpe, M., Göbel, C., Faltin, B., Beike, A. K., Hause, B., Himmelsbach, K., et al. (2010). The moss *Physcomitrella patens* contains cyclopentenones but no jasmonates: mutations in allene oxide cyclase lead to reduced fertility and altered sporophyte morphology. *New Phytologist*, *188*, 740–749.

Sztein, A. E., Cohen, J. D., de la Fuente, I. G., & Cooke, T. J. (1999). Auxin metabolism in mosses and liverworts. *American Journal of Botany*, *86*, 1544–1555.

Sztein, A. E., Cohen, J. D., Slovin, J. P., & Cooke, T. J. (1995). Auxin metabolism in representative land plants. *American Journal of Botany*, *82*, 1514–1521.

Takezawa, D., Komatsu, K., & Sakata, Y. (2011). ABA in bryophytes: how a universal growth regulator in life became a plant hormone? *Journal of Plant Research*, *124*, 437–453.

Tarakhovskaya, E. R., Kang, E. J., Kim, K.Y., & Garbary, D. J. (2013). Influence of phytohormones on morphology and chlorophyll *a* fluorescence parameters in embryos of *Fucus vesiculosus* L. (Phaeophyceae). *Russian Journal of Plant Physiology*, *60*, 176–183.

Ueda, J., Miyamoto, K., Aoki, M., Hirata, T., Sato, T., & Momotani, Y. (1991). Identification of jasmonic acid in *Chlorella* and *Spirulina*. *Bulletin of the University of Osaka Prefecture Series B, Agriculture and Biology*, *43*, 103–108.

Urbanová, T., Tarkowská, D., Novák, O., Strnad, M., & Hedden, P. (2013). Analysis of gibberellins as free acids by ultra performance liquid chromatography-tandem mass spectrometry. *Talanta*, *112*, 85–94.

Weiss, D., & Ori, N. (2007). Mechanisms of cross-talk between gibberellin and other hormones. *Plant Physiology*, *144*, 1240–1246.

Wiesemeier, T., Jahn, K., & Pohnert, G. (2008). No evidence for the induction of brown algal chemical defence by the phytohormones jasmonic acid and methyl jasmonate. *Journal of Chemical Ecology*, *34*, 1523–1531.

Woodward, A. W., & Bartel, B. (2005). Auxin: regulation, action, and interaction. *Annals of Botany*, *95*, 707–735.

Yamaguchi, S. (2008). Gibberellin metabolism and its regulation. *Annual Review of Plant Biology*, *59*, 225–251.

Yamaguchi, S., & Kamiya, Y. (2000). Gibberellin biosynthesis: its regulation by endogenous and environmental signals. *Plant Cell Physiology*, *41*, 251–257.

Yevdakova, N. A., Motyka, V., Malbeck, J., Trávníčková, A., Novák, O., Strnad, M., et al. (2008). Evidence for importance of tRNA-dependent cytokinin biosynthetic pathway in the moss *Physcomitrella patens*. *Journal of Plant Growth Regulation*, *27*, 271–281.

Yokota, T., Kim, S. K., Fukui, Y., Takahashi, N., Takeuchi, Y., & Takematsu, T. (1987). Brassinosteroids and sterols from a green alga, *Hydrodictyon reticulatum*: configuration at C-24. *Phytochemistry, 26*, 503–506.

Yokoya, N. S., & Handro, W. (1997). Thallus regeneration and growth induced by plant growth regulators and light intensity in *Grateloupia dichotoma* (Rhodophyta). In T. Kitamura (Ed.), *Proceedings of I.T.I.T. international symposium on new technologies from marine sphere* (pp. 83–86).

Yokoya, N. S., Kakita, H., Obika, H., & Kitamura, T. (1999). Effects of environmental factors and plant growth regulators on growth of the red alga *Gracilaria vermiculophylla* from Shikoku Island, Japan. *Hydrobiologia, 398/399*, 339–347.

Yokoya, N. S., Stirk, W. A., van Staden, J., Novák, O., Turečková, V., Pěnčík, A., et al. (2010). Endogenous cytokinins, auxins, and abscisic acid in red algae from Brazil. *Journal of Phycology, 46*, 1198–1205.

Yokoya, N. S., West, J. A., & Luchi, A. E. (2004). Effects of plant growth regulators on callus formation, growth and regeneration in axenic tissue cultures of *Gracilaria tenuistipitata* and *Gracilaria perplexa* (Gracilariales, Rhodophyta). *Phycological Research, 52*, 244–254.

Yoneyama, K., Xie, X., Yoneyama, K., & Takeuchi, Y. (2009). Strigolactones: structures and biological activities. *Pest Management Science, 65*, 467–470.

Yoshida, K., Igarashi, E., Wakatsuki, E., Miyamoto, K., & Hirata, K. (2004). Mitigation of osmotic and salt stresses by abscisic acid through reduction of stress-derived oxidative damage in *Chlamydomonas reinhardii*. *Plant Science, 167*, 1335–1341.

Žd'árská, M., Zatloukalová, P., Benítez, M., Šedo, O., Potěšil, D., Novák, O., et al. (2013). Proteome analysis of Arabidopsis reveals shoot- and root-specific targets of cytokinin action and differential regulation of hormonal homeostasis. *Plant Physiology, 161*, 918–930.

Zhang, W., Chapman, D. J., Phinney, B. O., & Spray, C. R. (1991). Identification of cytokinins in *Sargassum muticum* (Phaeophyta) and *Porphyra perforata* (Rhodophyta). *Journal of Phycology, 27*, 87–91.

Zhang, W., Yamane, H., Spray, C. R., Chapman, D. J., & Phinney, B. O. (1992). Cytokinins from the tRNA of the red alga *Porphyra perforata* J. Ag. *Plant Cell Physiology, 33*, 197–199.

Zhang, W., Yamane, H., Takahashi, N., Chapman, D. J., & Phinney, B. O. (1989). Identification of a cytokinin in the green alga *Chara globularis*. *Phytochemistry, 28*, 337–338.

Zhou, B., Tang, X., & Wang, Y. (2010). Salicylic acid and heat acclimation pretreatment protects *Laminaria japonica* sporophyte (Phaeophyceae) from heat stress. *Chinese Journal of Oceanology and Limnology, 28*, 924–932.

CHAPTER SIX

The Status of Kelp Exploitation and Marine Agronomy, with Emphasis on *Macrocystis pyrifera*, in Chile

Alejandro H. Buschmann*[,1], Steven Prescott[†], Philippe Potin[‡], Sylvain Faugeron[§], Julio A. Vásquez[‖], Carolina Camus[#], Javier Infante[#], María Carmen Hernández-González*, Alfonso Gutíerrez* and Daniel A. Varela*

*Centro i-mar, Universidad de Los Lagos, Puerto Montt, Chile
[†]Institute of Aquaculture, University of Stirling, Stirling, UK
[‡]Centre National de la Recherche Scientifique, Sorbonne Universités UPMC Univ. Paris 06, Roscoff cedex, Brittany, France
[§]Centro de Conservación Marina, Facultad de Ciencias Biológicas, Pontificia Universidad Católica de Chile, Santiago, Chile[¶]
[‖]Facultad de Ciencias del Mar, Universidad Católica del Norte, Center for Advanced Studies in Arid Zones CEAZA, Coquimbo, Chile
[#]Consorcio BalBiofuel, Camino Chinquihue km 6, Puerto Montt, Chile
[1]Corresponding author: e-mail address: abuschma@ulagos.cl

Contents

Abstract

Kelp cultivation started in Japan, China and Korea, mainly for human consumption; new applications are still expanding. In Chile, three 'wild' *Lessonia* species and *Macrocystis pyrifera* are under a strong and increasing pressure of exploitation mainly for alginate production and as a source of feed for abalone. Regulatory restrictions for kelp

Advances in Botanical Research, Volume 71
ISSN 0065-2296
http://dx.doi.org/10.1016/B978-0-12-408062-1.00006-8

exploitation and the increased demand for biomass provided a positive environment for the installation of a kelp farming industry. Pilot-production studies demonstrated that 200 tonnes (fresh)/ha/year can be achieved and genetic diversity and breeding studies suggested that this volume could be increased. Kelp disease research is a necessary condition for securing the future development of this industry, as are environmental studies on the impacts of large-scale aquaculture. Beyond the positive bioremediation, ecosystem service effects that kelp farming can provide, especially in a region such as in southern Chile, where intensive salmon and mussel cultivation occurs. Life Cycle Assessment suggests that the energy returns on investment in kelp farming are positive, but more detailed data are still required.

6.1 INTRODUCTION

The cultivation of brown seaweeds started about 300 years ago in Japan, China and Korea, mainly for human consumption (e.g. Hanisak, 1998; Kaneko, 1999; Tseng, 1987). More recently, the use of natural stocks of kelp expanded to the majority of temperate regions, where such species occur naturally. This geographic expansion was driven by an increased demand for brown seaweed biomass for the extraction of alginate (Bixler & Porse, 2011), and potentially by interests in the development of new uses (e.g. alginate to increase lithium batteries' performance: Kovalenko et al., 2011; as 'health' ingredient of food products: Prabhasankar et al., 2009; or extracts for agronomic applications: Craigie, 2011). Traditional and new applications are increasing the pressures on natural stocks worldwide, including on the Chilean coast (e.g. Ugarte & Sharp, 2012; Vásquez, 2008; Vega, Broitman, & Vásquez, 2014). During the last decade, abalone farming in Chile grew significantly, from 73 tonnes in 2001, to 840 tonnes in 2011 (Sernapesca, 2013). The expanding abalone industry alone demanded high volumes seaweed; 3.6–8.9 g of kelp per day are required to produce a 1 g increase in abalone weight (Greenier & Takekawa, 1992). Harvesting to meet this demand has already caused some deterioration of natural kelp populations (Vásquez, Piaget, & Vega, 2012) and considering the increased biomass demand associated to new applications that has actually taken place, the farming of *Macrocystis pyrifera* (L.) C. Agardh is being developed in Chile (Buschmann, Varela, & Hernández-González, 2008).

In addition, Chile is one of the largest mariculture producers of the western world, producing 361,000 tonnes of products in 1998 and over 870,000 tonnes in 2010 (Buschmann et al., 2013). Salmonid culture accounts for the majority of Chilean aquaculture production, and it grew rapidly since the 1980s, with Chile becoming the worlds' second largest producer of farmed salmon, after Norway (Buschmann et al., 2013). In Chile, rapid growth and concentration of aquaculture activities within enclosed areas have led to concerns regarding potential environmental impacts (Buschmann et al., 2009). To mitigate the effects associated with dissolved inorganic and suspended organic emissions of the high-trophic level organisms, Integrated Multi-Trophic Aquaculture (IMTA) models were developed (see reviews by Chopin et al., 2001; Neori et al., 2004; Troell et al., 2003). The initiation of IMTA research in Chile began in the late 1980s using land-based systems (Buschmann et al., 1994; Buschmann, Troell, Kautsky, & Kautsky, 1996). The first published research of IMTA within an open-water context was by Troell et al. (1997). The red alga *Gracilaria chilensis* Bird, McLachlan, & Oliveira was found to have increased growth rates (20–40%) when cultured in proximity to trout and *Atlantic salmon* grow-out sites. Buschmann, Varela, Hernández-González, and Huovinen (2008) demonstrated that both *G. chilensis* and *M. pyrifera* were suitable candidates for bioremediation within Chilean waters, and the co-culture of these two different species has been proposed as a means of increasing biomass productivity, therefore maximising the potential for bioremediation. However, due to inadequate regulations and the lack of financial incentives for seaweed aquaculture (Buschmann et al., 2013), industrial cultivation of kelp has only been carried out at experimental and pilot-scale (Correa, et al., in press; Gutiérrez et al., 2006; Macchiavello, Araya, & Bulboa, 2010; Westermeier, Patiño, Piel, Maier, & Müller, 2006), and official data from Chilean aquaculture indicated a total production of only 1 tonne (wet weight) in 2010 (Sernapesca, 2013).

Despite a research demonstrating the operational feasibility and advantages of integration, purposely constructed IMTA systems are as yet to be developed within Chile. Economic considerations remain a challenge to the adoption of IMTA technology by the salmon industry. To attract investors, pilot demonstrations are required to show how profitable IMTA approaches are, as has been shown in Canada (Ridler et al., 2007). In the case of *M. pyrifera,* this encouraged efforts to develop novel products in order to add value and stimulate demand (Buschmann, Varela, Hernández-González, & Huovinen, 2008; Gutiérrez et al., 2006). Furthermore, current research is investigating the efficacy of bioethanol production from cultured *M. pyrifera*. It has been recently demonstrated that the carbohydrates of giant kelp can be fermented and transformed into bioethanol (Wargacki et al., 2012). The variety of possible end-uses of *M. pyrifera* provides interesting opportunities for research on different aspects of kelp biology for improving sustainability beyond pilot-biomass production studies.

A considerable amount of information on *M. pyrifera* cultivation already exists (see as an example North, Gerard, & Kuwabara, 1982). However, additional biological data are required to make successful commercial cultivation a viable option. It is especially important to acquire knowledge of the differences in environmental requirements and the complex morphological and reproductive variability between differing populations of the giant kelp (see Graham, Vásquez, & Buschmann, 2007). The anticipated market driving forces mentioned above, and the potential impacts of cultivation upon natural populations and the wider environment, and the limited biological knowledge necessary to produce a high quality product, all present areas of important consideration. Although there have been recent reviews on seaweed aquaculture in Chile (e.g. Buschmann, Varela, & Hernández-González, 2008), there is a need for an in-depth analysis of new perspectives, challenges and considerations, in light of the additional kelp research and development that has taken place in recent years. This paper reviews recent literature on kelp fisheries and aquaculture production in Chile. It also presents new information gathered during the development of a 21-ha pilot cultivation site for *M. pyrifera*, as well as from associated research activities regarding kelp genetics and diseases, and environmental impact analysis using Life Cycle Assessment (LCA). In addition, an economic analysis is performed, which takes into account different production scales and algal price scenarios.

6.2 KELP FISHERIES IN CHILE

Harvesting and collection from natural populations of Chilean brown seaweeds, mainly for alginate extraction, represents 10% of the world supply; with landings of up to 300,000 dry tonnes per year, the industry was valued at US$ 60 million (Vásquez et al., 2012). Until 2000, the Chilean brown seaweed fishery was mainly sustained by natural mortality, whereby artisanal fishermen collected all individuals, which had been cast on the shore. Since then, driven mainly by international demands for raw material, three kelp species of economic importance (*Lessonia nigrescens*,[1] *Lessonia trabeculata* Villouta & Santelices and *M. pyrifera*) have been intensively harvested, from coastal areas between 18° and 42° S (Vásquez et al., 2012; Vega et al., 2014) (Figure 6.1).

Despite the fact that *Lessonia* and *Macrocystis* are distributed along the entire Chilean continental coast (Hoffmann & Santelices, 1997),

[1] *Lessonia nigrescens* was recently split in two cyptic species: *Lessonia berteroana* Montagne and L. spicata (Suhr) Santelices (González et al., 2012).

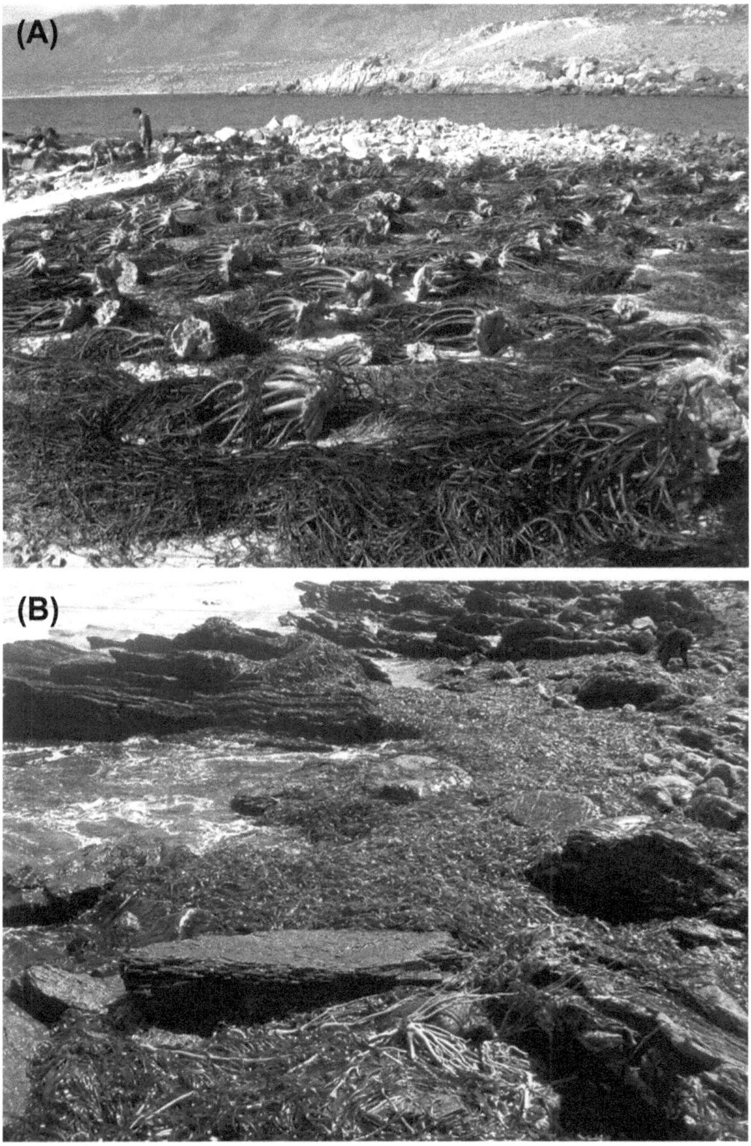

Figure 6.1 Kelp landings in the north of Chile: *Lessonia trabeculata* (A) and *Macrocystis pyrifera* (B). (See the colour plate.)

their collection and extraction is concentrated between 18° and 32° S (Vásquez, 2008). The geographical location and the desert climatic conditions of northern Chile allow for a significant cost reduction in the drying of raw material and, consequently, in the total processing cost. In the last decade, exploitation of brown seaweeds of economic importance in northern Chile was carried out using best practices, agreed upon by consensus between fishermen, industry, government and scientists (Vásquez, 2008; Vásquez et al., 2012). These practices focused on the selective harvest of sporophytes to allow for the maintenance of a reproductive stock, facilitating recruitment and minimising the impacts of grazing, benthic herbivores. Recently, Vega et al. (2014) introduced several ecological indicators to monitor the sustainability of kelp species. These indicators are related to density, biomass, recruitment and the population size structure, all used for comparisons of coastal areas with different regimens of harvesting administration (e.g. areas with co-management practices, versus open access areas, or marine protected areas without human intervention). Harvesting strategies for *Lessonia* (e.g. Santelices, 1982; Vega et al., 2014) and *Macrocystis* (e.g. Westermeier et al., 2012) were developed to ensure the sustainability of harvesting practices. Ecological indicators, including demographic variables, should reinforce the co-management administration of kelp resources (see Castilla, Campo, & Bustamante, 2007; Gelcich, Godoy, & Castilla, 2009). This would facilitate a viable harvesting administration system along the entire Chilean coast, allowing for the sustainability of a brown seaweed fishery. Moreover, ecological indicators, together with other harvesting parameters, could be useful for justification of the application of other administration strategies, such as quotas, bans or access restrictions for fishermen.

Various *Lessonia* species represents more than 70% of the total brown seaweed landings, which go as raw material to the alginate industry. On the other hand, at present, the availability of fresh *M. pyrifera*, as a feed input, is one of the most important constraints in the emerging abalone aquaculture industry. Between 18° and 32° S, *M. pyrifera* has low abundance and a fragmented distribution (Vásquez, 2008). The demands for kelp for abalone cultivation in Chile greatly exceeds the production potential of the natural kelp populations. At high latitudes, south of 40° S, a second epicenter for abalone cultivation is located; here *M. pyrifera* forms continuous belts that comprise the most important reserve of brown algae in the world (Graham et al., 2007). However, in contrast to the perennial populations in northern Chile, *M. pyrifera* in southern locations is annual and therefore minimally available in winter (Buschmann, Moreno, Vásquez, & Hernández-González, 2006). Considering this natural constraint, the feasibility of abalone cultivation depends on the future culture of this kelp species and should not be centred exclusively within Northern Chile.

Kelp beds along the Chilean coast present a relevant ecological, economical, and social value. Vásquez et al. (2014) estimated that kelp beds along the northern coast of Chile, alone, had a value of US$ 540 million. The study indicated that kelp fisheries accounted for 75% of this value, and that the environmental services kelps provided for CO_2 capture represented only 9% of that total value. For these reasons, to secure the sustainability of Chilean kelp fisheries, it is highly relevant to introduce conservation targets and a higher demand for the utilisation of biomass should be entertained using re-population techniques or by the development of large-scale, intensive kelp aquaculture.

6.3 THE DEVELOPMENT OF KELP AQUACULTURE

Considering the large-scale kelp fisheries, which exploit the natural beds in Chile, and also the susceptibility of brown algae to oceanographic events such as El Niño (Martínez, Cárdenas, & Pinto, 2003; Vásquez, Vega, & Buschmann, 2006; Vega, Vázquez, & Buschmann, 2005), kelp farming appears as to be a logical developmental initiative. Kelp sporophyte production, under laboratory conditions, as well as in the natural environment, can help the natural recovery of coastal rocky areas after mass mortality events. More importantly, it could allow for the development of a sustainable biomass resource for a variety of applications. However, some unknown biological features of kelps, such as its life history strategies and responses under culture conditions, are still constraints for large-scale cultivation. The technological development of large scale kelp faming in oriental countries has demonstrated that this activity is technically and economically feasible. However, due to the different market conditions and cost structures of mariculture sectors in the Western countries, kelp farming is still underdeveloped. Presently, there are several kelp research and pilot farming projects, worldwide (e.g. Buck & Buchholz, 2004; Kraan, 2013; Peteiro, Salinas, Freire, & Fuertes, 2006). However, various characteristics of *M. pyrifera* (i.e. its unique morphology, genetic and reproductive factors, and regrowth capacity) necessitate the development of new culture technologies and farming strategies, within the limits set by environmental regulations, if the farming of this species is to be successful.

In general, two different aquaculture production approaches have been tested in Chile. In one method, free-floating cultivation of the sexual phase, producing unattached, floating sporophytes, which are grown in tanks for several months, after which individual juvenile kelps can be attached manually to floating long-line ropes in the sea (Westermeier et al., 2006). The second method is closer to those applied in East Asian regions (see Sahoo & Yarish, 2005), which allow the gametophytes to settle on strings installed in tanks. The gametophyte and juvenile sporophytes develop attached to artificial substrata (e.g. ropes), and after 30–40 days are installed in sea cultivation sites (Gutiérrez et al., 2006; Macchiavello et al., 2010).

By using the second approach, a 4 ha, suspended pilot-farm of *M. pyrifera* was established in southern Chile. The biomass was cultured on 100-m long-lines, and a yield of 41 kg wet weight/m of culture line per year was achieved (Correa et al., in press). As this pilot trial allowed for the manipulation of some cultivation variables, such as kelp density, we can predict that a final production of up to 200 fresh tonnes/ha/year could be obtained in a suspended culture system. Higher production values were obtained by Westermeier, Patiño, Müller, and Müller (2010), Westermeier, Murúa, Patiño, and Müller (2011) on an experimental scale, using isolated, individual plants, produced through controlled sexual crosses. These results suggested that through genetic selection and breeding programs, the production values indicated previously can still be improved upon.

Under farmed conditions, the protein content of cultivated *M. pyrifera* reached a value of 9% dry weight, which was significantly higher than that of tested, wild populations (Gutiérrez et al., in press). If the kelp farm was situated near salmon cages, with higher nitrogen availability, the protein content could increase to more than 13% of dry biomass (Buschmann, Varela, Hernández-González, & Huovinen, 2008). This is highly relevant to the value of the kelp biomass as abalone feed. At this stage of development, we can conclude that farmed production can be increased further if a breeding program is developed and if the abalone food quality can be improved by placing suspended *M. pyrifera* cultivation systems in proximity to salmonid cages sites, in southern Chile. By doing so, production could be increased, while also performing bioremediation services through the removal of fish nutrient emissions (Buschmann, Varela, Hernández-González, & Huovinen, 2008).

Using the above results and the 4-ha pilot-scale cost information, it was possible to calculate the economic profitability of a *M. pyrifera* farm, located in Chile, when the crop was used as a feed input, for abalone cultivation (Correa et al., in press). The main costs for a 10 ha farm are given in Figure 6.2. The culture installation represented the highest cost, at 44% of the total. Other significant costs were personnel, purchase of the aquaculture

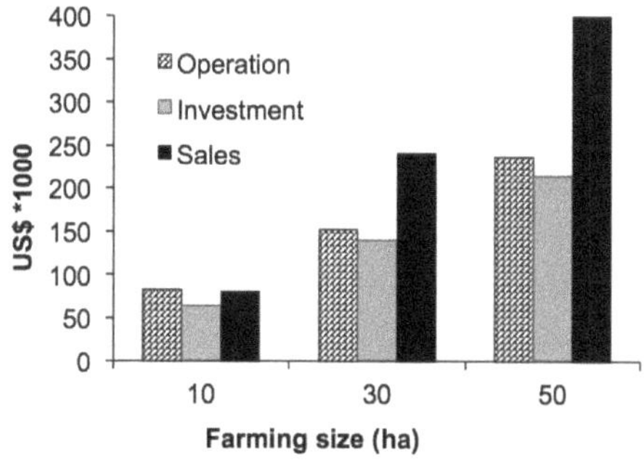

Figure 6.2 Capital and operational costs and revenues values in US$ estimated for a culture system estimated from a 4 ha kelp farm established in southern Chile.

license and equipment required (e.g. boats, harvesting platforms; Correa et al., in press). The economic analysis shows that the farm size and crop price were important aspects to be considered in order to achieve economic profitability (Correa et al., in press). The results indicated that only a culture system greater than 30 ha was economically viable, based on the prevailing price of US$ 78/tonne of *M. pyrifera*. These results demonstrated that *M. pyrifera* farming could be commercially viable in southern Chile. If goods, with added value were to be developed (e.g. human food; Gutiérrez et al., 2006), the commercial interest in this economic activity would increase.

6.4 KELP GENETICS AND STRAIN SELECTION

There are relatively few published studies regarding kelp breeding and genetic improvement, most of which are focused on *Saccharina japonica* (for a review see Robinson, Winberg, & Kirkendale, 2013). In China and neighbouring countries, where the species is extensively cultivated, the main strategy has been to perform repetitive breeding generations of a few inbred lines (Robinson et al., 2013). This has led either to improved biomass production, or the increase of specific traits/molecules. To date, knowledge of kelp genetics in Chile is focused on evolutionary and ecological topics: (1) global phylogeography and parapatric speciation; (2) regional phylogeography and post-glacial recolonisation; and (3) local genetic diversity and structure following natural and human perturbations. The first published studies were aimed at understanding the effects of natural (i.e. ENSO, Martínez et al., 2003) and anthropogenic disruptions (i.e. copper mining wastes, Faugeron, Martínez, Correa, & Billot, 2005) on the spatial and temporal dynamics of the kelp *L. nigrescens* in northern Chile. Both studies characterised population genetic structure, within and between continuous stands, separated by the impacted areas. They showed that the disruption of an otherwise continuous distribution of this kelp coincided with increased genetic differentiation between stands occurring northward and southwards from the impacted area. These authors demonstrated that gene flow occurred mainly between populations separated by short distances (i.e. a few km at most) as a result of their relatively short dispersal capacity. Indeed, after the 1982/83 ENSO event, even after 20 years *L. nigrescens* had not recolonised more than 50 km of coastline. This is a slow rate considering that the event killed extensive kelps beds along 600 km of coastline, leaving only sparse relic populations in upwelling centres (Martínez et al., 2003). These results were further confirmed by phylogeographic studies embracing most of the species distribution (Tellier, Meynard, Correa, Faugeron, & Valero, 2009). Based upon nuclear, mitochondrial and plastidial markers, a low genetic diversity was observed in each sampled locality, and shared only with neighbouring populations. In addition, strong genetic divergence was observed in the 30° S biogeographic transition zone described by Camus (2001) and Thiel et al. (2007). This genetic divergence corresponded to a speciation process (Tellier, Tapia, Faugeron, Destombe, & Valero, 2011), as a result of both the reduced dispersal and the ecological divergence (López-Cristoffanini, Tellier, Otaiza, Correa, & Contreras-Porcia, 2013; Oppliger et al., 2012, 2011). Overall, these results showed that specific traits (in this case, traits related to temperature and desiccation tolerance) could evolve in response to pressures of natural selection in wild populations.

Genetic studies in *M. pyrifera*, along the coast of Chile, focused on global phylogeography. As for *L. nigrescens,* Macaya and Zuccarello (2010) showed poor local (i.e. intrapopulation) genetic diversity and a strong genetic differentiation between the populations of *M. pyrifera*, and/or between regions along the Chilean coast north of the 42° S. This was surprising for a species capable of long-distance dispersal through rafting (Macaya et al., 2005; Thiel & Haye, 2006), and again suggested that local, rather than global, processes were shaping the genetic diversity predominantly because of restricted dispersal. On a global scale, the reduced genetic diversity was emphasised by the relatively recent arrival of the genus to the southern hemisphere, as a product of *trans*-tropical dispersal, from the northern hemisphere, during late Pleistocene (see Coyer, Smith, & Andersen, 2001; Macaya & Zuccarello, 2010). The situation is different south of 42° S where post-glacial colonisation led to the presence of a single haplotype in the region of the Magellan Strait (Macaya & Zuccarello, 2010). There, the importance of long-distance dispersal has been a determining factor in shaping the genetic diversity of such a large geographic region. Indeed, this single haplotype is shared with other sub-Antarctic regions, including south New Zealand, a pattern very similar to the one reported for the bull kelp, *Durvillaea antarctica* (Fraser, Nikula, Spencer, & Waters, 2009). However, in persistent populations, for which equilibrium is reached between the respective influence of gene flow, genetic drift and mutation, it seemed that restricted dispersal of spores was the dominant process shaping local and regional genetic diversity (see Alberto et al., 2010, 2011). This was a seemingly common pattern in most kelp species (see Valero et al., 2011).

One interesting feature of the phylogeographic studies was that they reported sharing haplotypes between all known morphotypes, previously considered as different species (e.g. *Macrocystis integrifolia*, *Macrocystis angustifolia* and *Macrocystis laevis sensu* Demes, Graham, & Suskiewicz, 2009), which suggested that they shared a very recent ancestor. However, this result is still under discussion, as suggested by Astorga, Hernández, Valenzuela, Avaria-Llautureo, and Westermeier (2012) who used nuclear markers (ITS-2) and representatives of *Macrocystis* from Canada, USA, South Africa, Australia and Chile. The authors observed a strong phylogenetic signal of morphological differences, which indicates that historical processes associated with diversification may have determined the current differentiation of the morphotype. Recent evidence of heterosis in the giant kelp (Westermeier et al., 2010, 2011) was an additional indication that phenotypic differentiation existed between genetically differentiated populations in Chile, and that local diversity could be re-arranged by controlled crosses into new genotypes, producing valuable phenotypes (Westermeier et al., 2011). This contrasted strongly with the paradigm of a high phenotypic plasticity underlying *Macrocystis* diversity, as suggested by Graham et al. (2007) and Demes et al. (2009). In addition to morphological plasticity, plastic phenotypic responses in terms of physiological and life history strategies were observed in different coastal habitats (Buschmann et al., 2006, 2014, 2004). Therefore, the relative roles of plasticity and selection in shaping kelp phenotypes are still a debate that requires dedicated scientific attention.

Future prospects for the establishment of a breeding program in *Macrocystis* should involve a detailed analysis of the phenotypic (morphology, chemical composition and other relevant production traits) differentiation between natural populations, together with an estimation of the genetic contribution (i.e. heritability) to trait values. It is important to pursue research on natural populations because the domestication process of this species is incipient, and up to now, most algal production has been based on spores obtained from natural stands. On the other hand, genetic interaction of cultivated and natural stocks will require attention in order to avoid potential genetic modifications of natural stocks through gene flow from modified cultivars, when cultivated outdoors, which has also been suggested for algae such as *G. chilensis* (Guillemin et al., 2008). Finally, kelp life cycle management is an issue that requires special attention for its potential to use parthenogenesis to produce pure genetic lines (i.e. fully homozygous genotypes) in a single generation, as suggested by recent studies showing that parthenogenetic sporophytes can produce single-sexed (e.g. 100% females) gametophytic progenies (Shan, Pang, & Gao, 2013). Pure genetic lines are essential for the quantitative genetics of specific traits (e.g. QTL analysis) and the definition of genetic markers to assist breeding strategies.

6.5 DISEASES

Information on the pathogens and diseases of brown algal kelps, of the order of Laminariales, is limited (Andrews, 1976; Correa, 1996; Gachon, Sime-Ngando, Strittmatter, Chambouvet, & Kim, 2010; Ishikawa & Saga, 1989; Potin, Bouarab, Salaun, Pohnert, & Kloareg, 2002; Wang et al., 2008). In natural populations, kelps are plagued by prokaryotes (Goecke, Labes, Wiese, & Imhoff, 2010; Michelou, Caporaso, Knight, & Palumbi, 2013), viruses (Easton, Lewis, & Pearson, 1997) and eukaryotic colonisers such as fungi and oomycetes (Li, Zhang, Tang, & Wang, 2010; Schatz, 1984). Other microscopic filamentous algae (Eggert, Peters, & Küpper, 2010; Potin, 2012) and invertebrates (Wahl & Mark, 1999) may also affect kelps. In the last 10 years, the effects of disease on natural marine communities have become increasingly apparent, mainly due to human impacts, via pollution or climatic change. For example, recent studies demonstrate how increased temperatures can negatively affect the chemical defences that ultimately result in diseases of natural algal populations (Harder, Campbell, Egan, & Steinberg, 2012).

Reports of major pathogen outbreaks in kelp natural populations are rare. However, *Ecklonia radiata,* the dominant habitat-forming seaweed in temperate Australian and New Zealand waters, was reported to experience high rates of mortality (Lachnit et al., 2013), which was attributed to a combination of environmental stress and viral infections, leading to discolouration and bleaching. Laboratory inoculation experiments with the extracted virii of 'sick', infected tissue resulted in cross-infection of healthy tissue of *E. radiata*. The frequency of putative disease morphotypes, and the relative abundance of virii in *E. radiata* around Sydney, appeared to be higher at sites that were impacted by untreated, wastewater discharge (Lachnit et al., 2013). This was the second reported case of virii found in *Ecklonia* (Easton et al., 1997). The absence of reports of virii in other kelp species is, most likely, due to lack of research, rather than to the absence of viral infections (Dodds, 1979). Previous virus and virus-like particle discoveries in algae have occurred during microscopic investigations of the ultrastructural aspects of algal life histories, rather than an investigation of a disease (van Etten, Lane, & Meints, 1991).

During the last 50 years in Asia, the rapid development of intensive mariculture has led to an increase in the number of diseases that attack commercially cultivated kombu (*Saccharina (Laminaria) japonica*). Disease symptoms were described affecting both sporelings during cultivation in nurseries and maturing sporophytes in the grow-out stages on long-line cultures (Wang et al., 2008). Among the pathologies recorded in kombu, several studies showed marine bacteria as causative agents. In China, South Korea and Japan, various bacterial pathogens (e.g. *Pseudoalteromonas*, *Alteromonas*) caused rot/hole disease of the kelp blades and significant loss of biomass (Ezura, Yamamoto, & Kimura, 1988; Yumoto, Ezura, & Kimura, 1989). Recently, by using phenotypic features and molecular identification, *Pseudoalteromonas bacteriolytica* was verified to be correlated with red-spot disease of *S. japonica* (Sawabe et al., 1998). In addition, another marine bacterium, *Pseudoalteromonas elyakovii* was considered to be associated with the spot-wounded fronds of *S. japonica* (Sawabe et al., 2000). In the Eastern Pacific, kelp-associated bacteria were recently reinvestigated using molecular ecology methods (Hengst, Andrade, González, & Correa, 2010), including *M. pyrifera* beds in California (Michelou et al., 2013). However, no bacterial disease has yet been described in Chilean kelps.

Although infections by oomycetes do occur in kelps, these eukaryotic pathogens have not yet been described as pathogens of maricultured kelps (Strittmatter, Gachon, & Küpper, 2009). However, they were shown in natural populations to invade the kelp stipe and frond and severely perturb its growth. Kelp endophytes were recently reviewed by Eggert et al. (2010). Some of these endophytes caused only minor changes in their hosts, such as tissue darkening, whereas others are known to produce either degradative losses or tumor-like lesions, accompanied by severe deformation of the thalli (Apt, 1988; Peters & Schaffelke, 1996). In Europe, epidemiological studies of kelp populations demonstrated that endophytic infections may affect 100% of the host population (Andrews, 1976; Ellertsdottir & Peters, 1997; Lein, Sjoetun, & Wakili, 1991; Peters & Schaffelke, 1996). Such endophytes have been found in the South-East Pacific Ocean and Antarctic regions. Along the Pacific coasts of South America, *M. pyrifera* is typically infected by *Laminariocolax macrocystis* (Burkhardt & Peters, 1998), initially described as *Streblonema macrocystis* (Peters, 1991). Recent results showed that galls, affecting natural populations of *L. nigrescens* were associated with infection by a filamentous brown algal endophyte, of the genus *Laminariocolax* (Thomas et al., 2009). Similar patterns of a few filamentous algae growing as endophytes in Antarctic subtidal habitats has been found, except within the thalli of the larger chemically defended macroalgae (Peters, 2003).

During a 21-ha pilot suspended farming experience of *M. pyrifera,* installed in the Chiloé region of southern Chile (42° 29′27″S; 73° 18′28″W), no deformative or tissue destructive diseases were found to affect any of the biomass production, over a 3-year period. Nevertheless, some preliminary symptoms were found (see Figure 6.3). These symptoms never attained a prevalence above 10%, of the culture population and no potential associated pathogens were found. Compared to the progress in genetics and breeding and ecology of farmed kelps in Chile, the study of pathogens and diseases is still in its infancy. Considering the potential negative impacts and economic losses of diseases in farmed Asian kelps, there is an urgent need to study the causative agents/virulence factors, modes of transmission, genetic basis of virulence, and resistance and environmental factors that lead to symptoms in cultivated kelps. These investigations will help to avoid major mistakes and will lead to the application of biosecurity and epidemiological programs designed to mitigate the risk factors for disease emergence in kelps. Other outputs may also lead to the development of guidelines for the selection of disease-resistant kelp varieties

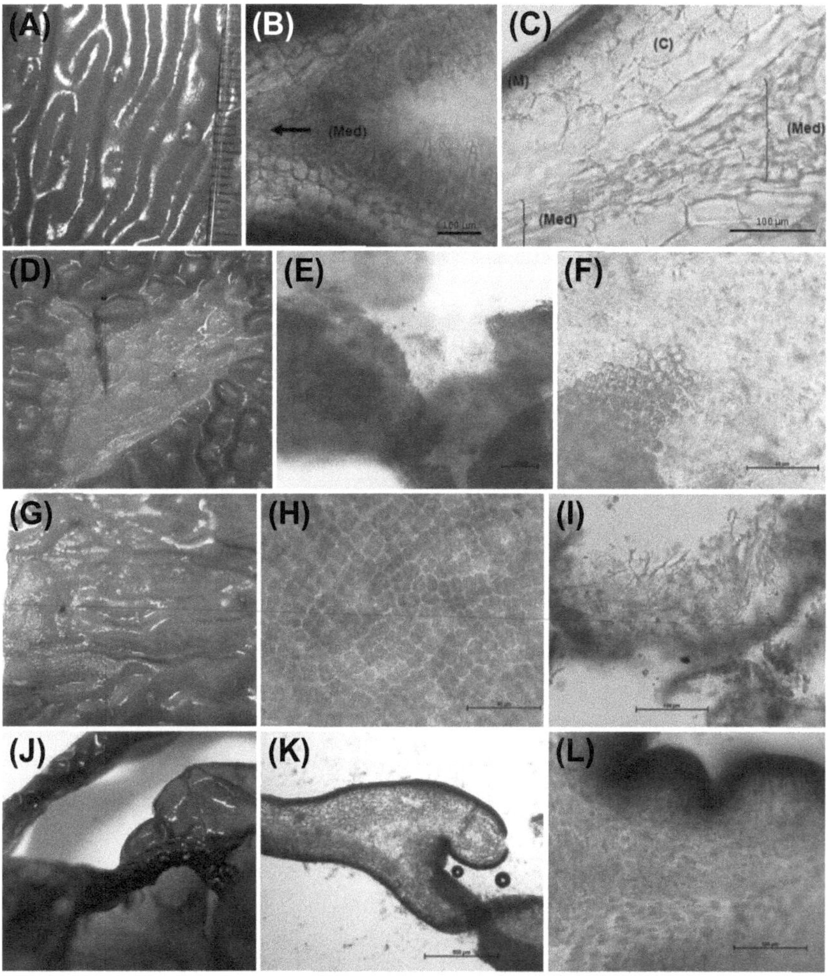

Figure 6.3 Kelp disease symptoms found in kelp cultivated in a 21 ha farm located in southern Chile. Protuberance of the lamina: (A) Superficial view of protuberance on the lamina; (B) Cross-section of the protuberance; (C) Opening of the medulla due to protuberance ((Med) = medulla and (C)= cortex). White spot on the lamina: (D) Superficial view of white spot; (E) and (F) Cross-section showing presence of bacteria. Green spot on the lamina: (G) and (H) Superficial view of green spot; (I) Cross-section of lamina showing decompression of the cells of the meristoderm, breakdown and loss of cortical cells. Plication of the lamina: (J) Superficial view of the frond; (K) and (L) Cross-section of twisted showing overgrowth of the meristodermatic area. (See the colour plate.)

and promote the exploration and interest of host immune-stimulation in land-based aquaculture facilities. It must be pointed out that as the scale of commercial cultivation increases, disease risks will most likely become more relevant and therefore kelp disease studies will require more urgent attention in the near future.

6.6 ENVIRONMENTAL IMPLICATIONS OF LARGE-SCALE AQUACULTURE

Seaweed farming is often considered as the least environmentally damaging form of aquaculture (Folke, Kautsky, Berg, Jansson, & Troell, 1998): it requires little or no input of fertilizers, fresh water resources, or medicines, and does not cause any major physical alterations of the environment (Bryceson, 2002; Johnstone & Ólafsson, 1995). However, awareness is developing regarding the environmental side effects, which may occur as a consequence of large-scale farming, occupying extensive marine areas (e.g. Zemke-White & Smith, 2006). By altering the habitat and the involuntary spread of cultivars of algae to adjacent areas, seaweed farming appears to affect many components of natural communities, e.g. bacteria (Johnstone & Ólafsson, 1995), meiofauna (Ólafsson, Johnstone, & Ndaro, 1995), benthic macrofauna (Eklöf, de la Torre, Adelsköld, Jiddawi, & Kautsky, 2005), fish (Bergman, Svensson, & Öhman, 2001) and corals (Russell, 1983). The installation of large beds of seaweed can contribute to complex modifications of local species interactions, which are not yet well understood (Buschmann, Correa, Westermeier, Hernández-González, & Norambuena, 2001). For example, *Eucheuma denticulatum* (Burman) Collins & Harvey and *Kappaphycus alvarezii* (Doty) have been farmed in shallow lagoons for the polysaccharide, carrageenan, using the 'off-bottom' method where algal thalli, tied to strings stretched between wooden sticks, driven into the sea bottom, are harvested every 2–3 months. In the mid-1990s, farms covered a total area of approximately 1000 ha (Ólafsson et al., 1995). If left unmanaged, such a development could, in the worst case, contribute to large-scale ecosystem changes (e.g. extensive seagrass loss, or spread of farmed algae to adjacent coral reefs), with implications for the production of several ecological goods and services (Eklöf, Henriksson, & Kautsky, 2006; Eklöf et al., 2005). Another study showed that kelp farming in Sungou Bay in China has low benthic environment impacts, but the results also showed that the impacts in summer and autumn were likely to be greater than in the winter and spring (Zhang, Hansen, Fang, Wang, & Jiang, 2009). As environmental impacts of open ocean seaweed cultivation are still disputed to some degree (see e.g. Ask, 2001), there remains a clear need for experimental studies in order to validate environmental effects of large-scale seaweed aquaculture systems (Trono, 1990; Zemke-White & Smith, 2006).

The environmental benefits of recycling inorganic nutrients and preventing eutrophication conditions through seaweed farming have been reviewed by various authors (e.g. Buschmann, Troell, & Kautsky, 2001; Chopin et al., 2001; Troell, Rönnbäck, Halling, Kautsky, & Buschmann, 1999). In the case of kelp farming, several studies point out that these large brown algae can be efficient removers of nitrogen inputs from fish aquaculture, in different regions of the world (e.g. Ahn, Petrell, & Harrison, 1998; Petrell & Alie, 1996), as also with the case of Chile (Buschmann, Varela, Hernández-González, & Huovinen, 2008). However, large-scale kelp cultivation could have unknown impacts, as a consequence of biomass losses, dissolved polysaccharide losses or an undesirably high nutrient uptake with consequences to local primary productivity. As these aspects are not well understood, they require attention. Preliminary environmental studies at the 21 ha pilot-farm of *M. pyrifera*, installed in southern Chile (42° 29′27″S; 73° 18′28″W) indicated that no benthic modifications were found, even after 3 years of cultivation (Figure 6.3). Few algal fronds have been found on the bottom under the kelp farm. The organic matter under the culture site also did not show a significant trend of increase over time, always under 2% (Figure 6.4). In

Figure 6.4 Underwater bottom images under the 21 ha kelp farm (42° 29′27″S; 73° 18′28″W). Images correspond to transects made at: (A) summer 2012, (B) winter 2012, (C) spring 2012, (D) summer 2013. (See the colour plate.)

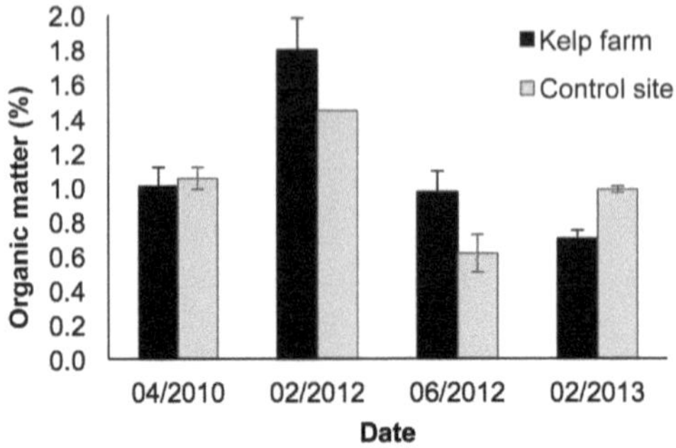

Figure 6.5 Mean (±1SD) total organic matter (%) in sediment under a 21 ha kelp farm and a control site 650 m from the kelp farming area in southern Chile (42° 29′27″S; 73° 18′28″W).

addition seasonally repeated measurements over a 2 year period have shown that nitrogen were not affected and slight increases of oxygen levels could be measured, at the outflow of the farm during summer. As this pilot-farm for kelp was installed at a depth of 30–40 m, light attenuation was considered not to be an issue. However, if the scale of cultivation were to be increased further studies should be taken. It may also be necessary to include a study of the genetic impacts of farmed beds of kelp may have on natural stocks, especially if a breeding and genetic selection program is considered. In the case of the red seaweed *G. chilensis*, losses in genetic diversity, in natural population processes, have been observed. This could be avoided in kelps through actions, such as maintaining the diversity of the species in a germplasm (Guillemin et al., 2008) (Figure 6.5).

6.7 LIFE-CYCLE ASSESSMENT (LCA) OF KELP AQUACULTURE

Environmental impacts are not only confined to local areas, aquaculture production systems also contribute to regional and global scale impacts (e.g. climate change or regional coastal eutrophication) (Liu et al., 2013; Pelletier et al., 2007; Pelletier & Tyedmers, 2008; Pelletier et al., 2009). Life Cycle Assessment (LCA) is a tool used to investigate and analyse the impacts of at these scales. When performing LCA, the material and energy inputs and outputs associated with the studied product system, are quantified. The contributions of these material and energy flows to a variety of environmental impacts are then modelled. In 2010, the global production of macroalgae was approximately 18.1 million tonnes, more than 98.9% of which was seaweed (FAO, 2012). Despite this level of production, among the published LCAs of aquaculture, macroalgal systems have been the focus of relatively few (e.g. Aresta, Dibenedetto, & Baeberio, 2005; Fry, Joyce, & Aumônier, 2012; Langlois et al., 2012). This information, combined with a renewed interest in product development and integrated fish/seaweed systems, presents a context in which LCA can provide some useful insights.

Some authors link a reduction of surface water carbon levels, at large-scale seaweed cultivation sites, to the sequestration of atmospheric CO_2 (Hughes et al., 2012; Tang, Zhang, & Fang, 2011). Even if this is realistic, expanding the focus of assessment beyond the cultivation phase will likely show such a perspective to be limited, as the crops' end use will involve emissions. For example, carbon emissions will be released during the production and subsequent combustion of kelp-derived biofuel. Similarly, when considering the potential of kelp cultivation to remove nutrients released into coastal waters by anthropogenic sources, life-cycle thinking should be applied. In this case, nutrients extracted by kelp cultivation may be transferred to other ecosystem compartments during its utilisation (e.g. agronomic nutrient products entering into freshwater compartments after application; feeding of cultured abalone resulting in some nutrients being released back into the ocean). Thus, it can be seen that LCA is useful for highlighting potential cases of shifting environmental problems. LCA methodology can also be used to compare the environmental performance of products (Gorrée et al., 2002; ISO, 2006a, 2006b). In the case of biofuels, it has been suggested that macroalgae cultivation is a means of producing feedstock that avoids the fertiliser inputs and competition for land-use with food-crop systems, that is associated with terrestrial biofuel feedstock production (Wargacki et al., 2012; Wei, Quarterman, & Jin, 2013). LCA provides a framework, with which such considerations, as well as others, can be explored; a low contribution to one environmental impact category may be offset by contributions to others. In order to provide meaningful comparisons, it is important that equivalent units of comparison are selected for each product (Gorrée et al., 2002; ISO, 2006a, 2006b). For example, in the case of biofuel, this unit may be 'the provision of one MJ of energy'. However, various, complex challenges will be faced when making these comparisons, which requires expert application of LCA methodology (see Wardenaar et al., 2012).

Partnership trials using the cultivated kelp crop for abalone feed and as a feedstock for the production of bioethanol in a purpose built commercial scale refinery are among the advances being made by the Centro i-mar of the Universidad de Los Lagos towards the development of *M. pyrifera* cultivation. These activities, including other advances described in this paper, present various stages of an experimental product value chain for which analyses using LCA are highly appropriate. The first stage of the assessment showed that the infrastructure is an important contributor to the life cycle impacts attributed to the seed production and cultivation stages. Thus, an increase in crop yield would reduce these impacts, and infrastructure alternatives should be explored. These outcomes also agree with other publications based on hypothetical and small-scale cultivation (Fry et al., 2012; Langlois et al., 2012) and are also in accordance with kelp farming and a hypothetical biofuel production in Denmark, which shows that the kelp production phase for biogas and bioethanol was the most energy intensive processing step (50–57%; Alvarado-Morales et al., 2013). The next stage of the LCA will compare the life cycle impacts of the different kelp products. The bioethanol refinery is an opportunity of particular interest. LCA will be used to calculate the cumulative energy demand of all the stages involved in the production of the crop and bioethanol, including the production and transport of energy and infrastructure inputs. This should provide a good indication of the potential for this bioethanol product to contribute positively to energy supplies within Chile. In a previous study, estimations of energy return on investment (EROI) for the hypothetical production of bioethanol using *Saccharina latissima* as a feedstock in Canada, produced an EROI of 1.78, but this value varied greatly as a factor of methods and assumptions (Philippsen, 2010).

6.8 CONCLUSIONS

The exploitation of Chilean natural kelp populations in central-northern Chile has reached maximum levels and an increase in biomass production seems achievable only if farming activities are implemented. Pilot-scale cultivation studies of *M. pyrifera* demonstrated that maximum fresh biomass production yields of 200 tonnes/ha/year could be achieved and might be improved incrementally through a breeding programme. However, kelp disease research is a necessity in order to secure the development of kelp aquaculture. Also relevant are some environmental impacts of large-scale aquaculture, which may arise beyond the positive bioremediation aspects that kelp farming can provide in a region such as in southern Chile, already experiencing the effects of intensive salmon and mussels production. Initial Life Cycle Assessment suggested that the energy returns on investment from kelp farming are positive, and continued research is investigating the impacts of the full macroalgae farming value chain.

ACKNOWLEDGEMENTS

This study presents results from grants supported by FONDEF D04I1067 and FONDEF D03I1152 and AHB acknowledge also the support of FONDECYT 1050550 & 1110845 and Corfo-Innova. The help of Luis González during the preparation of this paper is also appreciated. This review is a contribution of the international scientific cooperation network GDRI (GDRI N° 0803) Diversity, Evolution and Biotechnology of Marine Algae (DEBMA) supported by Chile, Brazil and France.

REFERENCES

Ahn, O., Petrell, R. J., & Harrison, P. J. (1998). Ammonium and nitrate uptake by *Laminaria saccharina* and *Nereocystis luetkeana* originating from a salmon sea cage farm. *Journal of Applied Phycology, 10*, 333–340.

Alberto, F., Raimondi, P. T., Reed, D. C., Coelho, N. C., Leblois, R., Whitmer, A., et al. (2010). Habitat continuity and geographic distance predict population genetic differentiation in giant kelp. *Ecology, 91*, 49–56.

Alberto, F., Raimondi, P. T., Reed, D. C., Watson, J. R., Siegel, D. A., Mitarai, S., et al. (2011). Isolation by oceanographic distance explains genetic structure for *Macrocystis pyrifera* in the Santa Barbara Channel. *Molecular Ecology, 20*, 2543–2554.

Alvarado-Morales, M., Boldrin, A., Karakashev, D. B., Holdt, S. L., Angelidaki, I., & Astrup, T. (2013). Life cycle assessment of biofuel production from brown seaweed in Nordic conditions. *Bioresource Technology, 129*, 92–99.

Andrews, J. H. (1976). The pathology of marine algae. *Biology Review, 51*, 211–253.

Apt, K. E. (1988). Galls and tumor-like growths on marine macroalgae. *Diseases of Aquatic Organisms, 4*, 211–217.

Aresta, M., Dibenedetto, A., & Baeberio, G. (2005). Utilization of macro-algae for enhanced CO_2 fixation and biofules production: development of a computing software for an LCA study. *Fuel Processing Technology, 86*, 1679–1693.

Ask, E. I. (2001). Creating a sustainable commercial *Eucheuma* cultivation industry: The importance and necessity of the human factor. In A. R. O. Chapman, R. J. Anderson, V. J. Vreeland, & I. R. Davison (Vol. Eds.), *Proceedings of the 17th international seaweed symposium* (pp. 13–18). Oxford: University Press.

Astorga, M. P., Hernández, C. E., Valenzuela, C. P., Avaria-Llautureo, J., & Westermeier, R. (2012). Origin, diversification, and historical biogeography of the giant kelps genus *Macrocystis*: evidences from Bayesian phylogenetic analysis. *Revista de Biología Marina y Oceanografía, 47*, 573–579.

Bergman, K. C., Svensson, S., & Öhman, M. C. (2001). Influence of algal farming on fish assemblages. *Marine Pollution Bulletin, 42*, 1379–1389.

Bixler, H. J., & Porse, H. (2011). A decade of change in the seaweed hydrocolloids industry. *Journal of Applied Phycology, 23*, 321–335.

Bryceson, I. (2002). Coastal aquaculture developments in Tanzania: sustainable and non-sustainable experiences. *Western Indian Ocean Journal of Marine Science, 1*, 1–10.

Buck, B. H., & Buchholz, C. M. (2004). The offshore-ring: a new system design for the open ocean aquaculture of macroalgae. *Journal of Applied Phycology, 16*, 355–368.

Burkhardt, E., & Peters, A. F. (1998). Molecular evidence from nrDNA sequences that *Laminariocolax* (Phaeophyceae, Ectocarpales *sensu lato*) is a worldwide clade of closely related kelp endophytes. *Journal of Phycology, 34*, 682–691.

Buschmann, A. H., Cabello, F., Young, K., Carvajal, J., Varela, D. A., & Henríquez, L. (2009). Salmon aquaculture and coastal ecosystem health in Chile: analysis of regulations, environmental impacts and bioremediation systems. *Coastal and Ocean Management, 52*, 243–249.

Buschmann, A. H., Correa, J., Westermeier, R., Hernández-González, M. C., & Norambuena, R. (2001). Cultivation of red algae in Chile: a review. *Aquaculture, 194*, 203–220.

Buschmann, A. H., Mora, O., Gómez, P., Böttger, M., Buitano, S., Retamales, C. A., et al. (1994). *Gracilaria chilensis* outdoor tank cultivation in Chile: use of land-based salmon culture effluents. *Aquacultural Engineering, 13*, 283–300.

Buschmann, A. H., Moreno, C., Vásquez, J. A., & Hernández-González, M. C. (2006). Reproduction strategies of *Macrocystis pyrifera* (Phaeophyta) in southern Chile: the importance of population dynamics. *Journal of Applied Phycology, 18*, 575–582.

Buschmann, A. H., Pereda, S. V., Varela, D. A., Rodríguez-Maulén, J., López, A., González-Carvajal, L., et al. (2014). Ecophysiological plasticity of anual populations of giant kelp (*Macrocystis pyrifera*) in a seasonally variable coastal environment in the Northern Patagonian Inner Sears of southern Chile. *Journal Applied Phycology, 26*, http://dx.doi.org/10.1007/s10811-013-0070-z.

Buschmann, A. H., Stead, R. A., Hernández-González, M. C., Pereda, S. V., Paredes, J. E., & Maldonado, M. A. (2013). Un análisis crítico sobre el uso de macroalgas como base para una acuicultura sustentable. *Revista Chilena de Historia Natural, 86*, 251–264.

Buschmann, A. H., Troell, M., & Kautsky, N. (2001). Integrated algal farming: a review. *Cahiers de Biologie Marine, 42*, 83–90.

Buschmann, A. H., Troell, M., Kautsky, N., & Kautsky, L. (1996). Integrated tank cultivation of salmonids and *Gracilaria chilensis* (Rhodophyta). *Hydrobiologia, 326/327*, 75–82.

Buschmann, A. H., Varela, D. A., & Hernández-González, M. C. (2008). Seaweed future cultivation in Chile: perspectives and challenges. *International Journal of Environment and Pollution, 33*, 432–456.

Buschmann, A. H., Varela, D. A., Hernández-González, M. C., & Huovinen, P. (2008). Opportunities and challenges for the development of an integrated seaweed-based aquaculture activity in Chile: determining the physiological capabilities of *Macrocystis* and *Gracilaria* as biofilters. *Journal of Applied Phycology, 20*, 571–577.

Buschmann, A. H., Vásquez, J. A., Osorio, P., Reyes, E., Filún, L., Hernández-González, M. C., et al. (2004). The effect of water movement, temperature and salinity on abundance and reproductive patterns of *Macrocystis* spp (Phaeophyta) at different latitudes in Chile. *Marine Biology, 145*, 849–862.

Camus, P. (2001). Marine biogeography of continental Chile. *Revista Chilena de Historia Natural, 74*, 587–617.

Castilla, J. C., Campo, M. A., & Bustamante, R. H. (2007). Recovery of *Durvillaea antarctica* (Durvilleales) inside and outside Las Cruces marine reserve, Chile. *Ecological Application, 17*, 1511–1522.

Chopin, T., Buschmann, A. H., Halling, C., Troell, M., Kautsky, N., Neori, A., et al. (2001). Integrating seaweeds into marine aquaculture systems: a key toward sustainability. *Journal of Phycology, 37*, 975–986.

Correa, J. A. (1996). Infectious diseases of marine algae: Current knowledge and approaches. In F. E. Round & D. J. Chapman (Series Eds.), *Progress in phycological research* (pp. 149–180). Bristol: Biopress Ltd.

Correa, T., Gutiérrez, A., Flores, R., Buschmann, A. H., Cornejo, P., & Bucarey, C. *Production and economic assessment of giant kelp Macrocystis pyrifera cultivation for abalone feed in the south of Chile. Aquaculture Research* (in press).

Coyer, J. A., Smith, G. J., & Andersen, R. A. (2001). Evolution of *Macrocystis* spp. (Phaeophyceae) as determined by ITS1 and ITS2 sequences. *Journal of Phycology, 37*, 574–585.

Craigie, J. S. (2011). Seaweed extract stimuli in plant science and agriculture. *Journal of Applied Phycology, 23*, 371–393.

Demes, K. W., Graham, M. H., & Suskiewicz, T. S. (2009). Phenotypic plasticity reconciles incongruous molecular and morphological taxonomies: the giant kelp, *Macrocystis* (Laminariales, Phaeophyceae), is a monospecific genus. *Journal of Phycology, 45*, 1266–1269.

Dodds, J. A. (1979). Viruses of marine algae. *Experientia, 35*, 440–442.
Easton, L. M., Lewis, G. D., & Pearson, M. N. (1997). Virus-like particles associated with dieback symptoms in the brown alga *Ecklonia radiata*. *Diseases of Aquatic Organisms, 30*, 217–222.
Eggert, A., Peters, A. F., & Küpper, F. C. (2010). The potential impact of climate change on endophyte infections in kelp sporophytes. In J. Seckbach (Vol. Ed.), *Seaweeds and their role in globally changing environments* (pp. 141–154). Berlin: Springer Verlag.
Eklöf, J. S., de la Torre, M., Adelsköld, L., Jiddawi, N. S., & Kautsky, N. (2005). Differences in macrofaunal and seagrass assemblages in seagrass beds with and without seaweed farms. *Estuarine, Coastal and Shelf Science, 63*, 385–396.
Eklöf, J. S., Henriksson, R., & Kautsky, N. (2006). Effects of tropical open-water seaweed farming on seagrass ecosystem structure and function. *Marine Ecology Progress Series, 325*, 73–84.
Ellertsdottir, E., & Peters, A. F. (1997). High prevalence of infection by endophytic brown algae in populations of *Laminaria* spp. (Phaeophyceae). *Marine Ecology Progress Series, 146*, 135–143.
Ezura, Y., Yamamoto, H., & Kimura, T. (1988). Isolation of a marine bacterium that produces red-spots on the culture bed of makonbu *Laminaria japonica* cultivation. *Nippon Suisan Gakkaishi, 54*, 665–672.
FAO. (2012). *The state of world fisheries and agriculture*. Rome.
Faugeron, S., Martínez, E. A., Correa, J. A., & Billot, C. (2005). Long-term copper mine waste disposal in northern Chile associated with gene flow disruption of the intertidal kelp *Lessonia nigrescens*. *Marine Ecology Progress Series, 288*, 129–140.
Folke, C., Kautsky, N., Berg, H., Jansson, A., & Troell, M. (1998). The ecological footprint concept for sustainable seafood production: a review. *Ecological Application, 8*, S63–S71.
Fraser, C. I., Nikula, R., Spencer, H. G., & Waters, J. M. (2009). Kelp genes reveal effects of sub-Antarctic sea ice during the Last Glacial Maximum. *Proceedings of the National Academy of Sciences of the United States of America, 106*, 3249–3253.
Fry, J. M., Joyce, P. J., & Aumônier, S. (2012). *Carbon footprint of seaweed as a biofuel*. London: Environmental Resources Management Limited (ERM).
Gachon, C. M., Sime-Ngando, T., Strittmatter, M., Chambouvet, A., & Kim, G. H. (2010). Algal diseases: spotlight on a black box. *Trends in Plant Science, 15*, 633–640.
Gelcich, S., Godoy, N., & Castilla, J. C. (2009). Artisanal fishers' perceptions regarding coastal co-management policies in Chile and their potentials to scale-up marine biodiversity conservation. *Ocean and Coastal Management, 52*, 424–432.
Goecke, F., Labes, A., Wiese, J., & Imhoff, J. F. (2010). Chemical interactions between marine macroalgae and bacteria. *Marine Ecology Progress Series, 409*, 267–299.
González, A. V., Beltrán, J., Hiriart-Bertrand, L., Flores, V., de Reviers, B., Correa, J. A., et al. (2012). Identification of cryptic species in the *Lessonia nigrescens* complex (Phaeophyceae, Laminariales). *Journal of Phycology, 48*, 1153–1165.
Gorrée, M., Heijungs, R., Huppes, G., Kleijn, R., de Koning, A., van Oers, L., et al. (2002). In J. B. Guinée (Ed.), *Handbook on life cycle assessment: Operational guide to the ISO standards*. Dordrecht: Kluwer Academic Publishers.
Graham, M. H., Vásquez, J. A., & Buschmann, A. H. (2007). Global ecology of the giant kelp *Macrocystis*: from ecotypes to ecosystems. *Oceanography and Marine Biology: An Annual Review, 45*, 39–88.
Greenier, J. J., & Takekawa, J. Y. (1992). Growth models and food conversion of cultured juvenile red abalone (*Haliotis rufescens*). In S. A. Shepard, M. J. Tegner, & S. Guzman del Proo (Vol. Eds.), *Abalone of the world: Biology, fisheries and culture* (pp. 527–537). Oxford: Blackwell.
Guillemin, M. L., Faugeron, S., Destombe, C., Viard, F., Correa, J. A., & Valero, M. (2008). Genetic variation in wild and cultivated populations of the haploid-diploid red alga *Gracilaria chilensis*: how farming practices favor asexual reproduction and heterozygosity. *Evolution, 62*, 1500–1519.

Gutiérrez, A., Correa, T., Muñoz, V., Santibañez, A., Marcos, R., Cáceres, C., et al. (2006). Farming of the giant kelp *Macrocystis pyrifera* in southern Chile for development of novel food products. *Journal of Applied Phycology, 18*, 259–267.

Hanisak, M. D. (1998). Seaweed cultivation: global trends. *World Aquaculture, 29*, 18–21.

Harder, T., Campbell, A. H., Egan, S., & Steinberg, P. (2012). Chemical mediation of ternary interactions between marine holobionts and their environment as exemplified by the red alga *Delisea pulchra*. *Journal of Chemical Ecology, 38*, 442–450.

Hengst, M. B., Andrade, S., González, B., & Correa, J. A. (2010). Changes in epiphytic bacterial communities of intertidal seaweeds modulated by host, temporality, and copper enrichment. *Microbial Ecology, 60*, 282–290.

Hoffmann, A. J., & Santelices, B. (1997). *Flora Marina de Chile Central*. Santiago: Ediciones Universidad Católica de Chile.

Hughes, A. D., Black, K. D., Campbell, I., Davidson, K., Kelly, M. S., & Stanly, M. S. (2012). Does seaweed offer a solution for bioenergy with biological carbon capture and storage? *Greenhouse Gases-Science and Technology, 2*, 402–407.

Ishikawa, Y., & Saga, N. (1989). The diseases of economically valuable seaweeds and pathology in Japan. In S. Miyachi, I. Karube, & Y. Ishida (Vol. Eds.), *Current topics in marine biotechnology* (pp. 215–218). Tokyo: Fuji Technology Press Ltd.

ISO. (2006a). *Environmental management—Life cycle assessment—Principles and framework (ISO 14040:2006)*. Geneva: International Organization for Standardization.

ISO. (2006b). *Environmental management—Life cycle assessment—Requirements and guidelines (ISO 14044:2006)*. Geneva: International Organization for Standardization.

Johnstone, R., & Ólafsson, E. (1995). Some environmental aspects of open water algal cultivation: Zanzibar, Tanzania. *Ambio, 24*, 465–469.

Kaneko, T. (1999). Kelp cultivation in Japan. *World Aquaculture, 30*. 62–65±68.

Kovalenko, I., Zdyrko, B., Magasinski, A., Hertzberg, B., Milicev, Z., Burtovyy, R., et al. (2011). A major constituent of brown algae for use in high-capacity Li-ion batteries. *Science, 334*, 75–79.

Kraan, S. (2013). Mass-cultivation of carbohydrate rich macroalgae, a possible solution for sustainable biofuel production. *Mitigation and Adaptation Strategies for Global Change, 18*, 27–46.

Lachnit, T., Hill, R., Marzinelli, E. M., Campbell, A. H., Harder, T., Thomas, T., et al. (2013). ss DNA viruses the causative agents of bleaching and reduced photosynthesis in *Ecklonia radiata*. In *21st International seaweed symposium Bali, abstract book*. www.xxiseaweedsymposium.org/.../download.php?fil...21stISS.pdf.

Langlois, L., Sassi, J.-F., Jard, G., Steyer, J.-P., Delgenes, J.-P., & Hélias, A. (2012). Life cycle assessment of biomethane from offshore-cultivated seaweed. *Biofules Bioproducts and Biorefining-Biofpr, 6*, 387–404.

Lein, T. E., Sjoetun, K., & Wakili, S. (1991). Mass-occurrence of a brown filamentous endophyte in the lamina of the kelp *Laminaria hyperborea* (Gunnerus) Foslie along the southwestern coast of Norway. *Sarsia, 76*, 187–193.

Liu, F., Pang, S., Chopin, T., Gao, S., Shan, T., Zhao, X., et al. (2013). Understanding the recurrent large-scale green tide in the Yellow Sea: temporal and spatial correlations between multiple geographical, aquacultural and biological factors. *Marine Environmental Research, 83*, 38–47.

Li, W., Zhang, T., Tang, X., & Wang, B. (2010). Oomycetes and fungi: important parasites on marine algae. *Acta Oceanologica Sinica, 29*, 74–81.

López-Cristoffanini, C., Tellier, F., Otaiza, R., Correa, J. A., & Contreras-Porcia, L. (2013). Tolerance to air exposure: a feature driving the latitudinal distribution of two sibling kelp species. *Botanica Marina, 56*, 431–440.

Macaya, E. C., Boltaña, S., Hinojosa, I. A., Macchiavello, J. E., Valdivia, N. A., Vásquez, N. R., et al. (2005). Presence of sporophylls in floating kelp rafts of *Macrocystis* spp. (Phaeophyceae) along the Chilean Pacific coast. *Journal of Phycology, 41*, 913–922.

Macaya, E. C., & Zuccarello, G. C. (2010). DNA barcoding and genetic divergence in the giant kelp *Macrocystis* (Laminariales). *Journal of Phycology, 420*, 103–112.

Macchiavello, J., Araya, E., & Bulboa, C. (2010). Production of *Macrocystis pyrifera* (Laminariales; Phaeophyceae) in northern Chile on spore based culture. *Journal of Applied Phycology, 22*, 691–697.
Martínez, E. A., Cárdenas, L., & Pinto, R. (2003). Recovery and genetic diversity of the intertidal kelp *Lessonia nigrescens* (Phaeophyceae) 20 years after El Niño 1982/83. *Journal of Phycology, 39*, 504–508.
Michelou, V. K., Caporaso, J. G., Knight, R., & Palumbi, S. R. (2013). The ecology of microbial communities associated with *Macrocystis pyrifera*. *PLoS One, 19*, e67480.
Neori, A., Chopin, T., Troell, M., Buschmann, A. H., Kraemer, G. P., Halling, C., et al. (2004). Integrated aquaculture: rationale, evolution and state of the art emphasizing seaweed biofiltration in modern aquaculture. *Aquaculture, 231*, 361–391.
North, W. J., Gerard, V. A., & Kuwabara, J. (1982). Farming of *Macrocystis* at coastal and oceanic sites. In L. M. Srivastava (Vol. Ed.), *Synthetic and degradative processes in marine macrophytes* (pp. 247–262). Berlin: Walter de Gruyter.
Ólafsson, E., Johnstone, R. W., & Ndaro, S. G. M. (1995). Effects of intensive seaweed farming on the meiobenthos in a tropical lagoon. *Journal of Experimental Marine Biology and Ecology, 191*, 101–117.
Oppliger, L. V., Correa, J. A., Engelen, A. H., Tellier, F., Vieira, V., Faugeron, S., et al. (2012). Temperature effects on life-history traits and geographic distribution of the kelp species complex *Lessonia nigrescens*. *PLoS One*, 7, e39289.
Oppliger, L. V., Correa, J. A., Valero, M., Faugeron, S., Tellier, F., Beltrán, J., et al. (2011). Sex ratio variation in the *Lessonia nigrescens* complex (Laminariales, Phaeophyceae): effects of latitude, temperature and marginality. *Journal of Phycology, 47*, 5–12.
Pelletier, N. L., Ayer, N. W., Tyedmers, P. H., Kruse, S. A., Flysjo, A., Robillard, G., et al. (2007). Impact categories for life cycle assessment research of seafood production systems: review and prospectus. *International Journal of Life Cycle Assessment, 12*, 414–421.
Pelletier, N., & Tyedmers, P. (2008). Life cycle considerations for improving sustainability assessments in seafood awareness campaigns. *Environmental Management, 42*, 918–931.
Pelletier, N., Tyedmers, P., Sonesson, U., Scholz, A., Ziegler, F., Flysjo, A., et al. (2009). Not all salmon are created equal: Life Cycle Assessment (LCA) of global salmon farming systems. *Environmental Science & Technology, 43*, 8730–8736.
Peteiro, C., Salinas, J. M., Freire, Ó., & Fuertes, C. (2006). Cultivation of the autoctonous seaweed *Laminaria saccharina* off the Galician coast (NW Spain): production and features of the sporophytes for an annual and biennial harvest. *Thalassas, 220*, 45–53.
Peters, A. F. (1991). Field and culture studies of *Streblonema macrocystis* sp. nov. (Ectocarpales, Phaeophyceae) from Chile, a sexual endophyte of giant kelp. *Phycologia, 30*, 365–377.
Peters, A. F. (2003). Molecular identification, taxonomy and distribution of brown algal endophytes with emphasis on species from Antarctica. In A. R. O. Chapman, R. J. Anderson, V. Vreeland, & I. R. Davison (Vol. Eds.), *Proceedings of the 17th international seaweed symposium* (pp. 293–302). New York: Oxford University Press.
Peters, A. F., & Schaffelke, B. (1996). *Streblonema* (Ectocarpales, Phaeophyceae) infection in the kelp *Laminaria saccharina* (Laminariales, Phaeophyceae) in the western Baltic. *Hydrobiologia, 326/327*, 111–116.
Petrell, R. J., & Alie, S. Y. (1996). Integrated cultivation of salmonids and seaweeds in open systems. *Hydrobiologia, 326/327*, 67–73.
Philippsen, A. (2010). *Energy input, carbon intensity, and cost for ethanol produced from brown seaweed* (Master thesis). Victoria: University of Victoria.
Potin, P. (2012). Intimate associations between epiphytes, endophytes, and parasites of seaweeds. In C. Wiencke & K. Bischof (Vol. Eds.), *Seaweed biology, ecological studies* (Vol. 219, pp. 202–233). Berlin: Springer.
Potin, P., Bouarab, K., Salaun, J. P., Pohnert, G., & Kloareg, B. (2002). Biotic interactions of marine algae. *Current Opinion in Plant Biology, 5*, 308–317.

Prabhasankar, P., Ganesan, P., Bhaskar, N., Hirose, A., Stephen, N., Gowda, L. R., et al. (2009). Edible Japanese seaweed, wakame (*Undaria pinnatifida*) as an ingredient in pasta: chemical, functional and structural evaluation. *Food Chemistry, 115*, 501–508.

Ridler, N., Wowchuk, M., Robinson, B., Barrington, K., Chopin, T., Robinson, S., et al. (2007). Integrated multi-trophic aquaculture (IMTA): a potential strategic choice for farmers. *Aquaculture Economics and Management, 11*, 99–110.

Robinson, N., Winberg, P., & Kirkendale, L. (2013). Genetic improvement of macroalgae: status to date and needs for the future. *Journal of Applied Phycology, 25*, 703–716.

Russell, D. J. (1983). Ecology of the imported red seaweed *Eucheuma striatum* Schmitz on Coconut Island, Oahu, Hawaii. *Pacific Science, 37*, 87–107.

Sahoo, D., & Yarish, C. (2005). Mariculture of seaweeds. In R. A. Anderson (Vol. Ed.), *Algal culturing techniques* (pp. 219–237). Oxford: Elsevier.

Santelices, B. (1982) Biological bases for the management of *Lessonia nigrescens* (Phaeophyta, Laminariales) in central Chile. In J. C. Castilla (Vol. Ed.), *Monografías Biológicas. Bases biológicas para el uso y manejo de recursos biológicos renovables marinos* (pp. 135–150). Santiago: Pontificia Universidad Católica de Chile.

Sawabe, T., Makino, H., Tatsumi, M., Nakano, K., Tajima, K., Iqbal, M. M., et al. (1998). *Pseudoalteromonas bacteriolytica* sp. nov. a marine bacterium that is the causative agent of red spot disease of *Laminaria japonica*. *International Journal of Systematic Bacteriology, 48*, 747–755.

Sawabe, T., Tanaka, R., Iqbal, M. M., Tajima, K., Ezura, Y., Ivanova, E. P., et al. (2000). Assignment of *Alteromonas elyakovii* KMM 162T and five strains isolated from spot-wounded fronds of *Laminaria japonica* to *Pseudoalteromonas elyakovii* comb. nov. and the extended description of the species. *International Journal of Systematic and Evolution Microbiology, 50*, 264–271.

Schatz, S. (1984). Degradation of *Laminaria saccharina* by saprobic fungi. *Mycologia, 76*, 426–432.

Sernapesca. (2013). *Anuario Estadístico de Pesca*. Ministerio de Economía, Fomento y Reconstrucción. http://www.sernapesca.cl/index.php?option=com_remository&Itemid=54&func=select&id=2.

Shan, T. F., Pang, S. J., & Gao, S. Q. (2013). Novel means for variety breeding and sporeling production in the brown seaweed *Undaria pinnatifida* (Phaeophyceae): crossing female gametophytes from parthenosporophytes with male gametophytes clones. *Phycological Research, 61*, 154–161.

Strittmatter, M., Gachon, C. M. M., & Küpper, F. C. (2009). Ecology of lower oomycetes. In S. Kamoun & K. Lamour (Vol. Eds.), *Oomycete genetics and genomics: Diversity, plant and animal interactions, and toolbox*. Hoboken: John Wiley & Sons.

Tang, Q., Zhang, J., & Fang, J. (2011). Shellfish and seaweed mariculture increase atmospheric CO_2 absorption by coastal ecosystems. *Marine Ecology Progress Series, 424*, 97–104.

Tellier, F., Meynard, A. P., Correa, J. A., Faugeron, S., & Valero, M. (2009). Phylogeographic analyses of the 30°S south-east Pacific biogeographic transition zone establish the occurrence of a sharp genetic discontinuity in the kelp *Lessonia nigrescens*: vicariance or parapatry? *Molecular Phylogenetic Evolution, 53*, 679–693.

Tellier, F., Tapia, J., Faugeron, S., Destombe, C., & Valero, M. (2011). The *Lessonia nigrescens* species complex (Laminariales, Phaeophyceae) shows strict parapatry and complete reproductive isolation in a secondary contact zone. *Journal of Phycology, 47*, 893–907.

Thiel, M., & Haye, P. A. (2006). The ecology of rafting in the marine environment. III. Biogeographical and evolutionary consequences. *Oceanography and Marine Biology: An Annual Review, 44*, 323–429.

Thiel, M., Macaya, E. C., Acuña, E., Arntz, W. E., Bastias, H., Brokordt, K., et al. (2007). The Humboldt current system of northern and central Chile. *Oceanography and Marine Biology: An Annual Review, 45*, 195–344.

Thomas, D., Beltrán, J., Flores, V., Contreras, L., Bollmann, E., & Correa, J. A. (2009). *Laminariocolax* sp. (Phaeophyceae) associated with gall developments in *Lessonia nigrescens* (Phaeophyceae). *Journal of Phycology, 45*, 1252–1258.

Troell, M., Halling, C., Neori, A., Buschmann, A. H., Chopin, T., Yarish, C., et al. (2003). Integrated mariculture: asking the right questions. *Aquaculture, 226*, 69–80.

Troell, M., Halling, C., Nilsson, A., Buschmann, A. H., Kautsky, N., & Kautsky, L. (1997). Integrated open sea cultivation of *Gracilaria chilensis* (Gracilariales, Rhodophyta) and salmons for reduced environmental impact and increased economic output. *Aquaculture, 156*, 45–62.

Troell, M., Rönnbäck, P., Halling, C., Kautsky, N., & Buschmann, A. H. (1999). Ecological engineering in aquaculture: use of seaweeds for removing nutrients from intensive mariculture. *Journal of Applied Phycology, 11*, 89–97.

Trono, G. C. (1990). Environmental impact of seaweed farming. *Journal of Phillipine Geographical Journal, 34*, 23–26.

Tseng, C. K. (1987). Some remarks on the kelp cultivation industry of China. In K. T. Bird & P. H. Benson (Vol. Eds.), *Seaweed cultivation for renewable resources* (pp. 147–153). Amsterdam: Elsevier.

Ugarte, R., & Sharp, G. (2012). Management and production of the brown algae *Ascophyllum nodosum* in the Canadian maritimes. *Journal of Applied Phycology, 24*, 409–416.

Valero, M., Destombe, C., Mauger, S., Ribout, C., Engel, C. R., Daguin-Thiebaut, C., et al. (2011). Using genetic tools for sustainable management of kelps: a literature review and the example of *Laminaria digitata*. *Cahiers Biologie Marine, 52*, 467–484.

van Etten, J. L., Lane, L. C., & Meints, R. H. (1991). Viruses and virus-like particles of eukaryotic algae. *Microbiological Reviews, 55*, 586–620.

Vásquez, J. A. (2008). Production, use and fate of Chilean brown seaweeds: re-sources for a sustainable fishery. *Journal of Applied Phycology, 20*, 457–467.

Vásquez, J. A., Piaget, N., & Vega, J. M. A. (2012). Chilean *Lessonia nigrescens* fishery in northern Chile: how do you harvest is more important than how much do you harvest. *Journal of Applied Phycology, 24*, 417–426.

Vásquez, J. A., Vega, J. M. A., & Buschmann, A. H. (2006). Long term studies on El Niño-La Niña in northern Chile: effects on the structure and organization of subtidal kelp assemblages. *Journal of Applied Phycology, 18*, 505–519.

Vásquez, J. A., Zuñiga, S., Tala, F., Piaget, N., Rodriguez, D. C., & Vega, J. M. A. (2014). Economic evaluation of kelp forest in northern Chile: values of good and service of the ecosystem. *Journal of Applied Phycology*. http://dx.doi.org/10.1007/s10811-013-0173-6.

Vega, J. M. A., Broitman, B., & Vásquez, J. A. (2014). Monitoring the sustainability of *Lessonia nigrescens* complex (Laminariales, Phaeophyta) in northern Chile under strong harvest pressure. *Journal Applied Phycology*. http://dx.doi.org/10.1007/s10811-013-0167-4.

Vega, J. M. A., Vázquez, J. A., & Buschmann, A. H. (2005). Population biology of the subtidal kelps *Macrocystis integrifolia* and *Lessonia trabeculata* (Laminariales, Phaeophyceae) in an upwelling ecosystem of northern Chile: interannual variability and El Niño 1997-98. *Revista Chilena de Historia Natural, 78*, 33–50.

Wahl, M., & Mark, O. (1999). The predominantly facultative nature of epibiosis: experimental and observational evidence. *Marine Ecology Progress Series, 187*, 59–66.

Wang, G. G., Shuai, L., Li, Y., Lin, W., Zhao, X. W., & Duan, D. L. (2008). Phylogenetic analysis of epiphytic marine bacteria on the hole-rotten diseased sporophytes of *Laminaria japonica*. *Journal of Applied Phycology, 20*, 403–409.

Wardenaar, T., van Ruijven, T., Beltran, A. M., Vad, K., Guinée, J., & Heijungs, R. (2012). Differences between LCA for analysis and LCA for policy: a case study on the consequences of allocation choices in bio-energy policies. *International Journal of Life Cycle Assessment, 17*, 1059–1067.

Wargacki, A. J., Leonard, E., Win, M. N., Regitsky, D. D., Santos, C. N. S., Kim, P. B., et al. (2012). An engineered microbial platform for direct biofuel production from brown macroalgae. *Science, 335*, 308–313.

Wei, N., Quarterman, J., & Jin, Y. S. (2013). Marine macroalgae: an untapped resource for producing fuels and chemicals. *Trends in Biotechnology*, *31*, 70–77.
Westermeier, R., Murúa, P., Patiño, D. J., & Müller, D. G. (2011). *Macrocystis* mariculture in Chile: growth performance of heterosis genotype constructs under field conditions. *Journal of Applied Phycology*, *23*, 819–825.
Westermeier, R., Patiño, D. J., Müller, H., & Müller, D. G. (2010). Towards domestication of giant kelp (*Macrocystis pyrifera*) in Chile: selection of haploid parent genotypes, outbreeding, and heterosis. *Journal of Applied Phycology*, *22*, 357–361.
Westermeier, R., Patiño, D. J., Murúa, P., Muñoz, L., Ruiz, A., Atero, C., et al. (2012). Utilization of holdfast fragments for vegetative propagation of *Macrocystis integrifolia* in Atacama, Northern Chile. *Journal of Applied Phycology*, *25*, 639–642.
Westermeier, R., Patiño, D., Piel, M., Maier, I., & Müller, D. G. (2006). A new approach to kelp mariculture in Chile: production of free-floating sporophyte seedlings from gametophyte cultures of *Lessonia trabeculata* and *Macrocystis pyrifera*. *Aquaculture Research*, *37*, 164–171.
Yumoto, I., Ezura, Y., & Kimura, T. (1989). Distribution of the *Alteromonas* sp., the causative agent of red-spots on the culture bed of makonbu *Laminaria japonica*, in the coastal area of Funka Bay. *Nippon Suisan Gakkaishi*, *55*, 453–462.
Zemke-White, W. L., & Smith, J. E. (2006). Environmental impacts of seaweed farming in the tropics. In A. T. Critchley, M. Ohno, & D. B. Largo (Vol. Eds.), *World seaweed resources—An authorative reference system*. Wokingham: ETI BioInformatics, DVD-Rom.
Zhang, J., Hansen, P. K., Fang, J., Wang, W., & Jiang, Z. (2009). Assessment of the local environmental impact of intensive marine shellfish and seaweed farming—Application of the MOM system in the Sungo Bay, China. *Aquaculture*, *287*, 304–310.

CHAPTER SEVEN

Seaweeds (Macroalgae) and Their Extracts as Contributors of Plant Productivity and Quality: The Current Status of Our Understanding

Jatinder Singh Sangha*, Stephen Kelloway*, Alan T. Critchley† and Balakrishnan Prithiviraj*,1

*Department of Environmental Sciences, Faculty of Agriculture, Dalhousie University, Truro, NS, Canada
†Acadian Seaplants Limited, Dartmouth, NS, Canada
[1]Corresponding author: e-mail address: bprithiviraj@dal.ca

Contents

Abstract

The beneficial effects of seaweeds (macroalgae) and their extracts on plant systems has been used for many centuries to improve soil properties and to enhance the productivity and quality of agricultural crops and ornamental plants and turf grass. However, it

Advances in Botanical Research, Volume 71
ISSN 0065-2296
http://dx.doi.org/10.1016/B978-0-12-408062-1.00007-X

has only been recently that advances in scientific research have uncovered the potential applications for commercially produced macroalgal extracts as plant biostimulants. Extensive literature on seaweed-derived extracts as biostimulants to improve plant tolerance to abiotic and/or biotic stress, plant growth promotion and their effects on root/microbe interactions, have demonstrated the varied roles in plant health, quality/biofortification and soil improvement. These effects have been supported by physiological, biochemical and molecular mechanisms, as evident through investigations using model plants. This review discusses selected aspects of the role of seaweeds and their extracts on plant growth; it also provides an overview of mechanisms of activity and potential bioactive components. The review is presented to stimulate further studies on the use of macroalgal-derived extracts for applications in agriculture and horticulture.

7.1 INTRODUCTION

Seaweeds are macroscopic, multicellular marine macroalgae that form an integral part of coastal ecosystems of the world, providing essential ecosystem services to the in-shore marine environment. There are about 10,000 species of macroalgae. Based on their pigmentation, they are classified into the phyla: Chlorophyta (the 'green' algae), Ochrophyta (the 'brown' algae) and Rhodophyta (the 'red' algae) (see: Guiry & Guiry, 2013: www.algaebase.org). Most macroalgae grow naturally attached to hard substrata, however, a limited number of selected (semi-domesticated) species are cultivated in open water or land-based facilities for commercial applications (food and extracts).

Plant growth and development relies on the availability of a favourable growing environment, which includes healthy soils, availability of nutrients, also protection from pests and other stresses. These conditions can be met naturally or provided artificially. Amongst various methods used to improve crop growth and development the use of plant biostimulants have gained importance in the recent years of crops. Biostimulants are materials, other than fertilisers or plant nutrients, which are able to stimulate plant growth and development when applied in small quantities. They are also referred to as 'metabolic enhancers' (Zhang & Schmidt, 1997). Amongst a wide variety of benefits, biostimulants may enhance fertiliser use efficiency, enhance tolerance to nutrient, water and salinity stress. (Russo & Berlyn, 1991). Macroalgae are rich sources of active compounds and minor nutrients, with the ability to produce a wide variety of secondary metabolites, which are implicated in a broad spectrum of biological activities (Kulik, 1995; Lordan et al., 2011). Seaweeds and their (commercial) extracts have considerable future potential for agricultural applications.

Biostimulants may consist of various organic components; seaweeds and their commercial extracts are such a category of product commonly used in various plant production industries. In fact, whole seaweeds (as mulch) and their components (as a compost) have been used for centuries as organic inputs to improve soil health and so to enhance the growth of agricultural crops. More recently, various industrial-scale and commercial processes have been developed and refined to manufacture extracts (liquid and powder forms) from a wide range of seaweed. Brown macroalgae such as *Ascophyllum nodosum*, *Fucus* spp., *Laminaria* spp., *Sargassum* spp., *Ecklonia maxima*, and *Durvillaea* spp. etc. in particular have been used as biofertilisers in agriculture, largely based on natural, extensive coastal populations (Hong, Hien, & Son, 2007).

Commercial extracts of a wide variety of seaweeds are increasingly popular, not only for use as (bio)stimulants of plant growth, but also to impart tolerance to several abiotic, environmental stresses. Several commercial seaweed extracts are reported to enhance multiple agronomically beneficial parameters such as root growth, plant health (immune status), biomass and quality (value) particularly under stressed conditions/regimes, e.g. saline soils and periodic exposure to low/high temperatures. The effect of defined applications (rates and timings are critical for particular responses to be elicited) of seaweed extract is further reported to be related to the activation of several molecular and biochemical changes in the treated plants, thereby indicating the likely impact of seaweed extracts on genetic pathways.

The biostimulatory effect of seaweeds and their extracts on various crops, and the protection imparted on exposure to a variety of abiotic stress conditions, suggest a review is appropriate to update on the use of commercial seaweed extracts in plant production. This article provides highlights of selected research performed to date, to try to stimulate discussions and future research on seaweeds and their (commercial) extracts.

7.2 THE BENEFICIAL EFFECTS OF SEAWEED EXTRACTS ON GERMINATION AND EARLY PLANT GROWTH AND ESTABLISHMENT

Seed germination, root and shoot growth, biomass and yield are all dependent on multiple factors such as the genetic potential of the plant, the immediate growth environment, and agricultural inputs and practices. Whole seaweed mulch or extracts have recently been included as 'plant biostimulants', with wide applications in agricultural systems (Craigie, 2011; Khan et al., 2009; Spann & Little, 2011; Zhang & Schmidt, 1997).

Several reports demonstrated that application of varied extracts enhanced root vigour, plant growth and development, plus potentiated plant hardiness and increased shoot/root biomass and leaf numbers, resulting in higher yields of vegetables, fruit and field crops under normal and a variety of stress conditions (Arthur, Stirk, & Van Staden, 2003; Crouch & Van Staden, 1992, 1993; Khan, Hiltz, Critchley, & Prithiviraj, 2011; Khan et al., 2009; Khan, Zhai, Souleimanov, Critchley, Smith, & Prithiviraj, 2012; Kumari, Kaur, & Bhatnagar, 2011; Kumar & Sahoo, 2011; Rayorath, Jithesh, et al., 2008; Rayorath, Khan, et al., 2008; Stirk, Novák, Strnad, & Van Staden, 2003).

Rayorath, Jithesh, et al. (2008) and Rayorath, Khan, et al. (2008) evaluated the plant growth promoting activity of commercial extracts of the brown alga *A. nodosum* (labelled *Ascophyllum nodosum extract-1* (ANE1) and ANE2 in the experiment), using the model plant *Arabidopsis thaliana*, in both laboratory and greenhouse bioassays. For the laboratory experiments, liquid (half-strength Murashige and Skoog (MS) basal medium containing 1% sucrose) or solid (with the addition of 0.8% agar in half-strength MS basal medium) medium amended with 0.01 and 0.1 g/l aqueous extract (0.01 and 0.1 g/l of ANE1, or ANE2) or 2 g/l of methanolic extract was used. *Arabidopsis thaliana* (Col-0) as grown on 1/2MS medium and then transferred, after 5 days, to new plates containing treatments of *Ascophyllum* extract. Subsequently, observations on root elongation were taken daily up to 7 days. The results showed that the two treatments (ANE1 and ANE2), stimulated both root and shoot growth, as compared to the controls. After 1 week of *A. nodosum* extract (0.01 g/l) treatment, ANE1 and ANE2 showed 15% and 38% increases in root length, respectively. A similar trend was observed with 0.1 g/l of both ANE1 and ANE2, which showed root length increases of 16.9% and 34%, respectively. In the greenhouse, the *Arabidopsis* plants were planted in to peat pellets, maintained at 22 ± 2 °C, with a photoperiod of 16:8 h. Plants were irrigated with 10 ml solution of *A. nodosum*, commercial extract dissolved in water (1 g/l), 14 days after germination. Three additional treatments followed at weekly intervals. In the greenhouse, plants treated with ANE1 showed an increase in height of 57% over untreated controls. On the other hand, ANE2-treated plants showed a 16% increase over the untreated controls. ANE1 treatment also increased the average number of leaves over the control plants by 20%; however, this effect was not evident with ANE2. Further analysis with a transgenic *Arabidopsis* line, carrying DR5: GUS reporter gene construct, indicated the role of auxin synthesis in enhanced plant growth, which was attributed to endogenous plant synthesis in response to the *A. nodosum* extract treatment.

Contradictory reports also appeared, showing the negative effect of seaweed extract on plant root growth and development (Wally et al., 2012). It was found that the addition of aqueous ANE (0.1 g/l) to plant growth media (Murashige and Skoog), decreased root length by over 20%, 3–5 days after treatment; the total number of secondary roots was reduced by 35%. Observations on root morphology and architecture showed that the secondary roots grew more vertically. To further explain the mechanisms of *Ascophyllum* extract on *Arabidopsis*, mutants and transgenic plants were used. Observations obtained from abi4-1 mutants showed an increased lateral root ratio in ANE-treated abi4-1 plants, e.g. a 50%, as compared to a 35%, increase in the control. On the other hand, Col-0 plants showed a reduction in the length of their primary roots with ANE treatment. This effect was not evident with the mutant; this suggested the influence of specific genes during this interaction. Additionally, *Arabidopsis* plants treated with the synthetic auxin-responsive promoter DR5, and the cytokinin-sensitive promoter ARR5, driving the GUS reporter gene in *Arabidopsis*, were used to determine the role of these hormones. Treatment with 0.1 g/l ANE reduced the GUS activity in the DR5:GUS seedlings (33–42%), whereas ARR5 expression was increased. It was hypothesized that the phenotype observed was not a sole result of the extract, but possibly also due to increased biosynthesis of endogenous phyto-hormones such as cytokinins, and abscisic acid (ABA) within the plant (this was further investigated by Wally et al., 2012).

Rayorath, Jithesh, et al. (2008) and Rayorath, Khan, et al. (2008) investigated commercial *A. nodosum* extract as a biostimulant for the promotion of growth and productivity in the agricultural crop, barley (cv. AC Sterling). Seeds were planted in sterile vermiculite and the pots then irrigated with 50 ml aqueous *A. nodosum* solution or an aqueous solution of varied organic sub-fractions, at 0.1, 0.5, and 1 g/l. Seedling emergence and shoot and root length were quantified. The barley seeds, when treated with an aqueous solution of ANE and its organic fractions, improved emergence, i.e. germination, shoot length and dry weight. ANE-treated seeds showed 87% emergence, as compared to 69% in the water control. Seeds treated with organic fractions of ANE also showed an increase in shoot length of 13–20%, as compared to the control. It was found that the *A. nodosum* extract-induced α-amylase activity in barley seeds. A concentration of 0.5 g/l ANE gave a 13-fold increase in α-amylase activity, whereas the organic fractions, at the equivalent of 1 g/l of ANE, induced a four to sevenfold increase. Rayorath, Jithesh, et al. (2008) and Rayorath, Khan, et al. (2008), suggested that the organic components of the *A. nodosum* extracts induced α-amylase activity, independent of gibberellic acid-3 (GA3), that might act, in concert with GA-dependent α-amylase production, resulting in enhanced germination and seedling vigour in barley.

The above findings corroborated a number of other publications, which demonstrated the effects of various types of seaweed commercial extracts on seedling development and establishment, including some on germination, plant growth and yield. Unfortunately, the mechanisms (or 'mode of action') of such activities have only been determined in a few investigations. One direct effect was the 'plant growth regulator-like activity' of some seaweed extracts (Khan et al., 2009). Khan et al. (2011), using *Arabidopsis*, showed a higher level of 'cytokinin-like activity' in roots and shoots, following treatment with a commercial, *A. nodosum*-derived liquid concentrate (i.e. Stimplex®). The results were supported by a biochemical analysis of plant GUS activity, and a higher level of 'cytokinin-like activity' was detected with an application of a 3 ml/l concentration of Stimplex®. A similar trend was found when plants were given foliar spray treatments demonstrating a similar 'cytokinin-like activity' at a concentration of 5 ml/l. It is important to note that we clearly differentiated a 'cytokinin-like response' in the plant to the application of extract (i.e. the extract contains elicitors, which stimulate endogenous production of phyto-hormones), from the situation where the commercial extract is thought to actually contain a cytokinin.

Not all commercial extracts come from brown seaweeds. In the following example, a foliar spray (5% v/v) of the red alga *Kappaphycus alvarezii* sap, applied to tomato plants resulted in improved plant growth and yield of the fruit (60–89%), as compared to control plants sprayed with water (Zodape et al., 2011). Bi, Iqbal, Arman, Ali, and Hassan (2011) used kappa-carrageenan, an extracted red algal polysaccharide, (from *Kappaphycus* a seaweed that is commercially, extensively cultivated in open water on industrial scales), to induce secondary metabolites in chickpea and maize plants, and observed a significant change in plant height, stem diameter and number of leaves per plant, although the number of cobs per plant and flowering time were not changed by the treatments. Vinoth, Gurusaravanan, and Jayabalan (2012) used extracts of two seaweeds (red alga – *Gracilaria edulis* and brown alga – *Sargassum wightii*) in an *in vitro* propagation system, for tomato, with a high success rate for the survival of transgenic plants. Sharma et al. (2012) investigated extracts of five brown macroalgae (i.e. *A. nodosum*, *Fucus serratus*, *Fucus vesiculosus*, *Laminaria hyperborea* and *Sargassum muticum*) in mung bean and pak choi bioassays. The alkaline extracts from *F. vesiculosus* and *A. nodosum* stimulated higher dry matter accumulation in the mung bean plants. The majority of their acidic extracts enhanced root formation in the mung bean stem cuttings, as compared to the alkaline or neutral extracts. Using the pak choi bioassay, as non-hydroponic system, acidic extracts were used as foliar spray that affected the vegetative growth of the treated plants. Among the five extracts tested, the *A. nodosum* and *F. vesiculosus* ones provided the highest dry matter yields for the treated pak choi. The effect of an *A. nodosum* extract was studied on rapeseed and revealed a significant effect on root (102%) and shoot (23%) growth; Jannin et al. (2012) suggested that one of their products (AZAL5) promoted plant growth and nutrient uptake.

Commercially produced extracts of *A. nodosum* are effective on fruit trees. For example, Spinelli, FiorI, Noferini, Sprocatti, and Costa (2009) demonstrated positive effects of Actiwave®, on alternate-bearing apple trees. Trees treated with Actiwave® decreased oscillations in yield between their 'on' and 'off' years, and increased the average fruit weight. In addition, the leaf chlorophyll content increased by 12%, which led to increased rates of photosynthesis and respiration. Interestingly, under standard conditions, it was shown that the extract did not have a significant effect on well-fertilised unstressed plants. MacDonald, Hacking, Weng, & Norrie, 2012 investigated the effects of *A. nodosum* extract applied to lodge-pole pine seedlings, grown under nursery conditions. It was reported that pretreatment with a commercial extract, in the finisher fertiliser, at rates of 1:750, 1:500 and 1:250, after which the seedlings were freezer stored, significantly reduced the total length of the root system. However, post-overwintering and under favourable environmental conditions for 21 d, after being removed from freezer storage, for the 1:500 rate, the total number of white roots, as well as the number of both short and long white roots, were significantly increased. This suggested that the use of the *A. nodosum* treatment improved root, growth after planting, thereby contributing to enhanced establishment of the seedlings and further enhanced drought resistance mediated by an enhanced root growth.

The mechanism (i.e. mode of action), how seaweed extracts improve the growth and vigour of plants has been difficult to elucidate as extracts contain numerous bioactive compounds, the complexity of which is not fully researched nor understood, although it is believed that macro- and micro-element nutrients, amino acids, vitamins, and hormone-like growth substances and elicitors present in seaweed (whole seaweeds and their extracts) may lead to the multiplicity of biostimulatory activities experienced by plants when treated with commercial extracts (see Crouch & Van Staden, 1993; Ördög, Stirk, Van Staden, Novák, & Strnad, 2004; Stirk et al.,

2003; Wally et al., 2012). For example, higher chlorophyll levels in seaweed extract-treated plants were reported due to the presence of betaines in some seaweed extracts (Blunden, Jenkins, & Liu, 1996). The presence of phyto-hormones in seaweed extracts have been used to explain the biostimulatory effects observed. However this argument has recently been questioned (Wally et al. 2012). This conundrum needs to be resolved by further collaborative and multidisciplinary research.

Wally et al. (2012) quantified the phyto-hormone levels present in the rosette leaves of *Arabidopsis* at 24, 96, and 144 h following treatment with a commercial extract of *A. nodosum*. In this work, the total concentration of *in planta* cytokinins, particular of the *trans*-zeatin- and *cis*-zeatin types were increased; this was further correlated with an increased transcript of cytokinin biosynthesis genes. The study also showed that Abscisic Acid (ABA) and ABA-catabolite levels were increased, however auxin levels were reduced. Interestingly, the authors showed that higher plant, identified and catalogued phyto-hormone levels in different commercial seaweed extracts were relatively low (<10 ng/g Dry weight (DW) of sample), and the concentrations of ABA, ABA catabolites and GAs were less than 5 ng/g DW, whereas the levels of indole acetic acid (IAA) ranged from 20 to 25 ng/g DW, depending on the source of seaweed raw material used to manufacture the particular extract. Wally et al. (2012) concluded that these phyto-hormones, present in such low amounts within the commercial extracts cannot/influence plant growth and development and that chemical components present in the extracts stimulated the biosynthesis of endogenous phyto-hormones in *Arabidopsis*. The concentration of growth-stimulating compounds in different seaweed extracts was variable, which could also have had variable effects on the plant responses. The 'cytokinin-like activity' of a commercial extract manufactured from *A. nodosum* was detected in tobacco callus, this was calculated to be equivalent to 5.4 μg kinetin equivalent activity per litre of the extract (Sanderson & Jameson, 1985). Featonby-Smith and Van Staden (1983) estimated a total cytokinin-activity equivalent of 516 ng kinetin in 20 g of Kelpak™ (manufactured from *E. maxima* by a cell burst process). Stirk, Arthur, Lourens, Novák, Strnad, & Van Staden, 2004 used seaweed concentrates made from the kelps *E. maxima* and *Macrocystis pyrifera* and reported several different cytokinins in both the concentrates, of which *trans*-zeatin-O-glucoside was the predominant form. The data on the beneficial effects of seaweed extracts is a significant body of research and supports their use for plant growth stimulation and mitigation of the impacts of abiotic stresses.

In addition to the presence of elicitors or primers for *in planta* phyto-hormone synthesis, other factors for such effects could be due to the effect of seaweed extracts on the interactions of roots and their microbial communities, i.e. improved activity of symbionts (Khan et al., 2012). In addition to the microbes, the soil rhizosphere is also very rich in fauna, which contribute to a healthy soil ecosystem. The role of beneficial organisms in such interactions is well known. The nod-forming bacteria such as *Sinorhizobium meliloti* are well known to fix nitrogen (N). Recent reports suggested that a commercial extract from *A. nodosum* stimulated both root nodulation and growth of the alfalfa plants (Khan et al., 2012).

7.3 SEAWEEDS AND THEIR EXTRACTS IMPROVE PLANT NUTRIENT CONTENT/BIOFORTIFICATION

One of the major targets in agricultural production is the introduction of value-added traits, especially nutritional factors (McGloughlin, 2010; Newell-McGloughlin, 2008). Improved crop nutrition may have a direct effect on consumer's health. Biofortification, a process of nutrient enrichment and bioavailability of health promoting minerals in food crops, has been commonly targeted through fertilisation, plant breeding or biotechnological approaches (Holm, Kristiansen, & Pedersen, 2002; White & Broadley, 2009). Therefore, the use of seaweed extracts, on a variety of crops, has enormous potential. Fan et al. (2011) demonstrated that the nutrient composition of treated plants with *Ascophyllum* extract was increased, and was also shown to have health benefits when consumed using a model animal system to consumers. Dobromilska, Mikiciuk, & Gubarewicz, 2008 reported increased mineral components (i.e. N, P, K, Ca, Zn and Fe) in tomatoes after multiple treatments of Bio-algeen (a product made of seaweed extract containing brown pigment). The nutritional quality of okra was improved by the application of liquid seaweed (Zodape, Kawarkhe, Patolia, & Warade, 2008). Foliar applications of *K. alvarezii* sap, at 7.5%, increased the nutrient contents in grains from 7.91% (K) to 31.82% (S) whereas the sap increased the nutrient content from 5.72% (N) to 37.54% (Mg; Shah, Zodape, Chaudhary, Eswaran, & Chikara, 2013). In another study on wheat, the yield and nutritional quality, e.g. carbohydrates, protein and minerals of the grains were improved after spray treatment of the plants with 0.25%, 0.50% and 1.0% of seaweed *K. alvarezii* (Zodape, Mukherjee, Reddy, & Chaudhary, 2009). It was suspected that the increased yield was due to the presence of some growth promoting substances (the authors speculated IAA and IBA, gibberellins, cytokinins, micronutrients, vitamins and amino acids). Taken together, these studies present strong evidence on the biostimulatory effect of various seaweed extracts and positive opportunities in agriculture for the improvement of both crop yield and the quality of the produce for consumption.

The application of seaweed extracts, or their components, to food crops also provide has health benefits to consumers. Studies on the effect of *A. nodosum* extract treatment to spinach (*Spinacia oleracea* L.) showed that the treatment improved postharvest storage quality and at the same time enhanced flavonoid content in spinach leaves (Fan et al., 2011). Spinach plants were treated with 50 ml of ANE solutions (i.e. 0, 0.1, 1.0, or 5.0 g/l) at 14 and 7 days prior to harvest. Applications of 1.0 g/l extract increased the total phenolic and flavonoid content of the treated spinach leaves. The biological effect of ANE-enhanced polyphenols was tested using the *Caenorhabditis elegans* nematode model, i.e. extract-treated spinach material was incorporated into the diet of the model invertebrate. The increased flavonoid content was responsible for better performance of *C. elegans* under oxidative (50%) and thermal (61%) stress.

The mechanisms underlining seaweed extract-induced plant nutrient composition or 'biofortification' are not well understood. However, several factors can be attributed to such types of activity in general. As mentioned in previous sections, seaweeds themselves are rich sources of macro- and micro-elemental nutrients, amino acids, vitamins and compounds which may have effects on enhancing the nutritional value of the treated plants.

The majority of the world's population is deprived of essential nutrients in their diet (White & Broadley, 2009). The poor nutritional content of food crops has contributed to this situation (Graham et al., 2007; Welch & Graham, 2005). One of many ways to alleviate this issue would be to increase the concentrations of mineral elements in common food crops through the management of crop inputs, i.e. applications of biostimulants, such as seaweed extracts, into varied cultivation programmes at different stages of the plant growth cycle. The literature suggests that application of seaweed extracts could increase the mineral contents, crop quality and also shelf life of agricultural produce. Additional experiments are needed to determine if seaweed extracts can be used as a biofortification agent.

7.4 SEAWEED EXTRACTS AS BIOSTIMULANTS OF STRESS TOLERANCE IN PLANTS

7.4.1 Abiotic Stress

The adverse impact of abiotic stress on plant health and production is a great concern to the advancement of agriculture, which needs immediate attention and suitable solutions. It is estimated that soil salinity alone could limit the yield of major crops for human consumption (Zhu et al., 2000). Frost damage is another concern to food production, particularly in temperate regions. Similarly, drought and heat stresses are major threats to agricultural production in many parts of the world. In the absence of alternative, effective technologies to mitigate such abiotic stresses, applications of seaweed extracts to reduce the impact of plant stress.

7.4.2 Seaweed Extracts Help Mitigate Freezing Stress

Cold stress alter fluidity of membrane lipids and the structural properties of cell membranes (Marschner, 1995). Bioactive substances present in seaweed extracts may enhance the performance of plants under abiotic stress. Spray applications of extracts have been shown to improve plant tolerance to freezing temperature stress (Mancuso et al., 2006). Freezing tolerance in grapes was improved with application of *A. nodosum* extract formulation, which resulted in reduced osmotic potential of the leaves, a key indicator of osmotic tolerance (Wilson, 2001). Burchett, Fuller, and Jellings (1998) studied winter hardiness and frost tolerance of winter barley (*Hordeum vulgare* cv. Igri) treated with *A. nodosum* extract and concluded that *A. nodosum* extract improved tolerance to cold stress.

In a study using the model plant, *Arabidopsis thaliana* Rayorath et al. (2009) reported that lipophilic components of a commercial *A. nodosum* extract improved the freezing tolerance. The ANE-treated plants recovered from a freezing stress of −7.5 °C *in vitro* and −5.5 °C *in vivo*. In this study, the extract of *Ascophyllum* appeared to protect the membrane integrity of the treated plants, as cell damage was minimized under freezing stress (Rayorath et al. 2009). Chlorophyll damage was reduced by 70% during freezing recovery, as compared to the controls, which correlated with a reduced expression of the chlorophyllase genes *AtCHL*1 and *AtCHL2*, and an upregulation of the cold response genes *RD29A, CBF3* and *COR15A* (Rayorath et al., 2009).

The mechanisms behind seaweed extract-induced freezing tolerance have been elucidated in some of the recent reports. One of which suggested the role of proline in plant tolerance to low temperature stress (Nair et al., 2012). These authors concluded that freezing tolerance in *Arabidopsis*, after application with lipophilic components of a commercial *A. nodosum* extract, was through a combination of increased accumulation of osmo-protectants (such as proline) and altered cellular fatty acid composition. The expression of proline synthesis genes, notably P5CS1 and P5CS2, was higher in treated plants, whereas expression of the proline dehydrogenase (ProDH) gene was reduced. In addition, the total soluble sugars were increased in treated plants during freezing stress. The Proton Nuclear Magnetic resonance (1H NMR) metabolite profile of ANE-treated *Arabidopsis* plants revealed a spectrum dominated by chemical shifts (delta), representing soluble sugars, sugar alcohols, organic acids and lipophilic components, such as fatty acids in LPC, significantly differed from the control plants subjected to freezing stress. Analysis of 2D NMR spectra suggested an increase in the degree of unsaturation of fatty acids in the treated plants, under freezing stress. Based on the whole genome transcriptome analysis, the expression of several genes was found to be altered in extract treated plants. These data provide insights into multiple modes of action of *A. nodosum* extract-induced freeze tolerance in *Arabidopsis*.

7.4.3 Seaweed Extracts Mitigate Salinity Stress in Plants

Soil salinity is one of the most serious environmental problems influencing agriculture on a global scale (Zhu et al., 2000). Relatively little is known to establish if seaweed extracts, and their chemical components, could protect plants experiencing salinity stresses. Yan (1993) demonstrated that the uptake of Na ions was reduced in grass treated with seaweed. In addition, seaweed extract treatments have been reported to increase the tolerance of turfgrass to salinity (Nabati, Schmidt, & Parrish, 1994). In both, field and greenhouse experiments, the effect of seaweed extracts on the survival of Kentucky bluegrass (*Poa pratensis* L. cv. Plush), grown under increasing levels of salinity, were investigated. Kentucky bluegrass treated with *A. nodosum* extract, at 38 L/ha, enhanced growth and stimulated rooting under 0.15 S/m salinity conditions (Nabati et al., 1994). This indicated that the extract could potentially be used in Kentucky bluegrass production under saline soil conditions or where groundwater used for irrigation was saline. Recent studies showed that applications of a commercial *A. nodosum* extract could reduce NaCl stress in the *Arabidopsis* model. It was found that commercial extracts, or partially purified chemical components of the extract, reduced the accumulation of Na^+ in plant leaves and reduced the net uptake of Na^+ ions by roots (Jithesh et al., 2012). To further understand the molecular mechanisms underlying such a response, the whole genome transcriptome of *Arabidopsis*, undergoing salt stress, was analysed by microarray, after treatment with the chemical components of commercial extract of *A. nodosum* (Jithesh et al., 2012). The study elegantly revealed an upregulation of a class of genes known as dehydrins (known to be involved in alleviating salt, drought and cold stresses, Close, 2006), while at the same time negative regulators of salt stress were downregulated. Furthermore, the net K^+ loss from the roots was reduced, thereby enabling the plants to maintain ionic homeostasis. All these factors, taken together, assisted extract-treated plants to better survive NaCl stress (Jithesh et al., 2012). Using T-DNA insertion mutants of Arabidopsis these authors examined the role of down regulated genes in salinity stress tolerance. Several mutations were identified with enhanced salt-tolerant phenotypes. The results revealed a role for a pectin methylesterase inhibitor (PMEI) as a negative regulator of NaCl resistance. This finding stressed the need to understand the effect of *A. nodosum* extract on plant stress-induced transcriptome analysis, for the identification of novel regulators of seaweed extract mediated abiotic stress tolerance. This information can be useful to further elucidate, or use, further biotechnological approaches in general to develop salt hardiness in plants.

7.4.4 Seaweed Extracts Help to Mitigate Water Stress in Plants

Plant growth is dependent on the availability of water. Water stress hampers plant performance through disruption of metabolic pathways. The loss of integrity of biological membranes, principally due to oxidative damage, is another negative impact of drought stress in plants (Liu et al., 1998; Feng, Li, Xue, An, & Wang, 2007). Any technology that mitigate water stress, could contribute substantially to sustainable crop production.

In greenhouse studies, the treatment of vegetables, bedding plants and turf crops with a commercial extract of *A. nodosum* significantly delayed wilting, decreased water use (i.e. better water use efficiency), increased leaf water content and improved the recovery of drought-wilted plants, as compared to the control (Little & Neily, 2010; Neily, Shishkov, Tse, & Titus, 2008). Root applications of a commercial extract of *Ascophyllum nodosum* to almonds, at two week intervals, enhanced the negative midday stem water potential (Little & Neily, 2010). In another example, under rain-fed conditions, foliar applications with extracts of the red seaweed *K. alvarezii,* on soybean, improved nutrient uptake, growth and yield (Rathore et al., 2009).

Li and Van Staden (1998) suggested that improved water stress tolerance of commercial *Ecklonia* extract-treated plants was likely to be primarily due to increased antioxidant activity, leading to increased stress tolerance, in the treated plants. Likewise, an *A. nodosum* preparation, increased the activity of glutathione reductase and ascorbate peroxidase in tall fescue (Ayad, 1998). Tall fescue, with a higher concentration of superoxide dismutase, was tolerant to drought.

Two seaweed suspensions prepared from *A. nodosum* and from *L. hyperborea* were evaluated for their effects on the water sensitivity of barley

(*H. vulgare* L.) (Moller & Smith, 1999). Priming with *A. nodosum* extract was beneficial and improved seed germination, under elevated water stress. The suspension prepared from *A. nodosum* reduced the water sensitivity of barley seeds more than either water or polyethylene glycol treatments. The *A. nodosum* extract also reduced the microbial population on the seeds by 86%. It was suggested that the hygroscopic properties and the antimicrobial effect of the suspension of *A. nodosum* provided greater oxygen availability to the seed embryos, enabling more seeds to germinate under oxygen deficient conditions, while submerged (Moller & Smith, 1999).

7.4.5 Seaweed Extracts Help Mitigate Thermal Stress in Plants

Global climate change is a major challenge to increasing agricultural production and quality of crops. It is known that thermal stress limits the performance of a number of important crops during summer, in many temperate and subtropical regions (Kauffman, Kneivel, & Watschke, 2007; Zhang & Ervin, 2008). Seaweed extracts have been reported to improve thermal stress tolerance in several crop plants. Applications of these extracts induced heat tolerance in creeping bentgrass and these benefits have been attributed to 'cytokinin-like' substances present in the extract (see Ervin, Zhang, & Fike, 2004; Zhang & Ervin, 2008). However, the synthetic cytokinin benzylaminopurine, when applied at similar concentrations to those present in a commercial extract of *E. maxima,* showed no effect on yield, irrespective of the K^+ supply. It is possible that the beneficial effects of various seaweed extracts, against thermal stress in plants, are partially elicited by bioactive components other than the plant hormone cytokinin (Beckett & Van Staden, 1989). The beneficial effects of seaweed extract with regards to increased heat tolerance in plants, appeared to be associated with the organic and stimulatory, or elicitating components, which may not be the mineral fraction of the seaweed extract (Zhang & Erwin, 2008). Application of *A. nodosum* extract also increased superoxide dismutase activity and alleviated the decline in turfgrass quality while also increasing photochemical efficiency and root viability (Li & Van Staden, 1998; Zhang & Schmidt, 1999).

Zhang, Ervin, and Schmidt (2003a) investigated the effects of a commercial seaweed extract, prepared from *A. nodosum,* along with plant growth regulators (PGRs) viz. propiconazole (PPC) and humic acid (HA), on the tolerance of Kentucky bluegrass (*P. pratensis* L.) sod to thermal stress during storage. The PGR-treated sod was exposed to 37 °C for 72 or 96 h and then replanted in the field. It was reported foliar applications of *A. nodosum* extract at 0.5 kg/ha, plus humic acid at 1.50 kg/ha, propiconazole at 0.44 kg/ha alone, or a combination of *A. nodosum* extract + HA with 0.22 kg/ha propiconazole, enhanced the photochemical efficiency of the pre-harvest sod, reduced the visual injury of the turf and enhanced transplant rooting by 21.8 (SWE + HA), 34.7 (PPC) and 44.2% (SWE + HA + PPC), respectively. In an additional study, Zhang, Ervin, and Schmidt (2003b) studied the effect of seaweed extract on heat tolerance and post-transplant quality of tall fescue. Their data suggested that foliar applications of an *A. nodosum* extract alone, or in combination with humic acid, and propiconazole enhanced photochemical efficiency of pre-harvest sod. These treatments reduced heat injury resulting in improved, post-transplant rooting and overall quality of the tall fescue sod. The commercial *A. nodosum* extract was also reported to promote the performance of lettuce seeds under high temperature stress. In addition, seed germination of lettuce was influenced by priming with *A. nodosum* extract in that germination improved under high temperature conditions (Moeller & Smith, 1998).

The growth promoting, biostimulatory, activity of seaweeds could be attributed to the presence of bioactive components present in the seaweed extracts (Arthur et al., 2003; Crouch & Van Staden, 1992, 1993; Khan et al., 2009; Kumari et al., 2011; Kumar & Sahoo, 2011; Stirk et al., 2003; Rayorath, Jithesh, et al., 2008; Rayorath, Khan, et al., 2008). Several phytohormone-related plant responses are activated by seaweed extracts. Wally et al. (2012) quantified the concentration of ABA and its catabolites in the rosette leaves of *Arabidopsis* after seaweed extract treatment. It was shown that the ABA levels increased, with time, in both the control and the seaweed extract-treated plants. However, an increase in the ABA levels was observed in seaweed extract-treated plants, that was 40% at 24, 23% at 96% and 25% increase at 144 h. The effects of *A. nodosum* extract on the metabolites of ABA (i.e. phaseic acid, dihydrophaseic acid and ABA glucose-ester (ABA-GE)) were also examined by Wally et al. (2012), which also showed an increasing trend. These increases were related to *A. nodosum* extract-regulated conversion of the enhanced ABA produced to less active forms, which was required to alleviate the growth-reducing effects of ABA. Other plant hormones, viz. gibberellins following treatment with seaweed extract were also determined in those plants treated with *A. nodosum* extract (Wally et al., 2012). Data suggested that the *A. nodosum* extract treatment slightly increased the overall levels of GAs, in the treated plant tissue.

Study of the abiotic stress tolerance in plants is a challenging task. Around the world, a major portion of research activities is focused to mitigate abiotic stress, particularly drought and salinity, in food crops. There has been limited success with the biotechnological approach (i.e. Genetically modified plants). Biostimulants, such as seaweed extracts show evidence of alleviating some of these stresses in plants. The transcriptome analysis of seaweed extract-treated plants could also help to isolate novel genes that mediate abiotic stresses. Seaweed extracts, from a variety of sources, integrated with other approaches of mitigating abiotic stress in plants, will help contribute to improve crop survival and performance under abiotic stresses.

7.5 BIOTIC STRESS

Plant growth and development is often threatened by microbial pathogens (Sangha, Ravichandran, Prithiviraj, Critchley, & Prithiviraj, 2010). However, plants have developed inducible defence mechanisms, which are capable of being activated by various elicitors and in-turn trigger a defence response against pathogenic organisms (Dixon, Harrison, &Lamb, 1994; Nuernberger & Lipka, 2005). Plant-induced defences can be described as a coordination of molecular (Pieterse et al., 1998), biochemical (Dixon et al., 1994; Lorenzo & Solano, 2005) and physiological responses (Sangha et al., 2010; Subramanian et al., 2011).

7.5.1 Seaweed Extracts Induce Plant Defences against Bacteria and Fungi

In a growth chamber experiment, Subramanian et al., (2011) demonstrated the potential for application of a seaweed extract to manage plant pathogens. An extract of *A. nodosum* was used to suppress infection by *Pseudomonas syringae* pv. *tomato* DC3000 (Pst DC300) and a necrotroph *Sclerotinia sclerotiorum* (Lib.) de Bary, both pathogens of agricultural importance (Subramanian et al., 2011). In both cases the infection was suppressed by seaweed extract treatment. The disease severity of *Pst* DC3000 on *Arabidopsis* plants was reduced with a root treatment of the *A. nodosum* extract. On day four, post-inoculation, the disease severity was only 35–43%, as compared to 57% in the control. The *A. nodosum* extract did not inhibit *Pst* DC3000 growth *in vitro*, which suggested that reduction in the severity of the disease was primarily mediated by *A. nodosum* extract-elicited physiological changes in the plant, rather than via a direct antimicrobial effect on *Pst*. Similar results were observed against *S. sclerotiorum* infection. The *A. nodosum* extract treatment delayed the development of *Sclerotinia* lesions on the leaf. The average lesion size on *A. nodosum* extract-treated plants, 3 days after inoculation, was significantly less than the control plants.

In addition, various algal polysaccharides, including laminarin (from brown seaweeds) and carrageenans (from red seaweeds) have been shown to have the potential to induce disease resistance in plants and animals. Sangha et al. (2010) examined the effect of carrageenans on *Arabidopsis* resistance to the necrotropic fungal pathogen *S. sclerotiorum*. Carrageenans are a collective family of linear, sulphated galactans found in a number of commercially important species of marine red algae. The authors used two types of carrageenans viz. (iota-(i-), and lambda (λ)-carrageenan), which differ in their levels of sulphation. Three-week-old *Arabidopsis* plants, with fully expanded leaves, were sprayed twice with the test solutions, at 5-day intervals, and inoculated with *S. sclerotiorum*, 48 h after the second spray. Interestingly, the *Arabidopsis* plants treated with the i- and λ-carrageenans exhibited differential response to *S. sclerotiorum* infections. Plants sprayed with λ-carrageenan (1 g/l) showed enhanced resistance against *S. sclerotiorum*, delayed lesion development and a characteristic yellow halo around the lesions, with similarity to the suppressive effects shown with the *A. nodosum* extract treatment. It was also suggested that the necrosis in the infected area was possibly due to the accumulation of phenolic compounds. In contrast, the i-carrageenan-treated plants exhibited enhanced susceptibility to *S. sclerotiorum* resulting in a rapid spread of the infection. The mycelial growth was not inhibited *in vitro* suggesting that the polysaccharide was not an antimicrobial agent against *S. sclerotiorum*, but rather the plant's own defence mechanism was the main factor in suppression of the pathogen infection. These findings were further supported by an increased expression of the defence-related genes; allene oxide synthase (AOS), pathogenesis-related 3 (PR3) and plant defensin 1.2 (PDF1.2), in λ-carrageenan-treated plants, at 24 h after *S. sclerotiorum* infection. In addition, λ-carrageenan treatment to ics1 (a mutant defective in salicylic acid (SA)-dependent resistance), exhibited resistance against *S. sclerotiorum* and the lesion size was significantly less than the control, 48 h after inoculation. However, λ-carrageenan treatment did not elicit resistance in ics1. These data suggested that the highly sulphated λ-carrageenan-induced-resistance in *Arabidopsis*, via multiple biochemical and molecular defence responses, was largely affecting the jasmonic acid (JA)/ethylene mediated defence pathways.

Jaulneau et al. (2011) studied the effect of an aqueous extract of the green alga, *Ulva armoricana*, rich in ulvan (a sulphated polysaccharide), against three powdery mildew pathogens, e.g. *Erysiphe polygoni*, *Erysiphe necator* and *Sphareotheca fuliginea*, on the common bean, grapevine and cucumber; these authors reported a noticeable difference in efficacy based on the concentrations used. Weekly spraying at 3 g/l dry matter, provided 50% protection, whereas the severity of the symptoms was reduced by up to 90% at double the concentration (i.e. 6 g/l dry matter). A reporter gene, tagged to a defence-gene promoter, in a transgenic tobacco line, was elicited by treatment with extracts of *U. armoricana,* which probably contributed to induced plant defence, against the powdery mildew pathogen. *U. armoricana* is a reproducible source of active compounds, which can be used to efficiently protect crop plants against powdery mildew diseases. Similar results were observed with ethanolic fractions of *Ulva* spp., which also activated plant defence enzymes in *Arabidopsis* (Jaulneau et al., 2010). Paulert, Ebbinghaus, Urlass, and Moerschbacher (2010) showed that ulvan could prime the chitin- and chitosan-elicited oxidative burst in wheat and rice cells. The pretreatment of wheat cells with ulvan increased the chitin-elicited oxidative burst by five- to six-fold, as compared to chitosan alone (a two-fold increase). Similarly, in rice cells, the elicitation of H_2O_2 production by chitin or chitosan was increased by 150 and 80 times, respectively, after an ulvan pretreatment. Furthermore, ulvan-treated plants showed reduced symptoms of *Blumeria graminis* infection, by 45% in wheat and by 80% in barley.

Different mechanisms have been reported to be behind disease resistance associated with elicitation by various seaweed extracts. Using the model plant *Arabidopsis*, it was shown that the JA pathway played a critical role in plant defence responses elicited by a commercial extract of *A. nodosum* (Subramanian et al., 2011). Inoculation of *P. syringae* pv. tomato DC3000 on *A. nodosum* extract-treated plants (by root irrigation), showed a JA-dependent, induced systemic resistance. This was evident when the genes associated in JA-signalling, such as AOS and PDF1.2 were strongly induced in the *A. nodosum* extract-treated plants. Pathogens that are inhibited by the JA pathway are most often necrotrophs. Many devastating diseases in agricultural crops are caused by necrotrophic fungal and oomycete pathogens such as: *Alternaria*, *Fusarium*, *Botrytis*, *Verticillium*; *Phytophthora* and *Pythium* and bacterial pathogens such as: *Xanthomonas*, *Pseudomonas* and *Erwinia*. Thus, seaweed extracts such as *A. nodosum,* which elicit JA responses in plants may offer protection to many of these phytopathogens. These findings were further corroborated with observations that jar1, a mutant compromised in the JA-dependent pathogen resistance pathway, was susceptible to *Pst* DC3000, even after treated with *A. nodosum* extracts. These results indicated the potential application of seaweed extract to manage other pathogens in crops, particularly members of the mustard family (of which there are many, e.g. cauliflower, broccoli, etc.). It is also possible that sterols and fatty acids in the seaweed extracts trigger non-specific lipid transfer protein (nsLTPs), in the plasma membrane, which might potentiate disease resistance.

In a similar manner, seaweed polysaccharides, such as the ulvans from green algae, have been shown to activate plant defences through the JA-signalling pathway (Cluzet et al., 2004; Jaulneau et al., 2010) and JA-dependent pathways shown in carrageenan-induced plant defence responses to pathogens (Sangha et al., 2010) and insects (Sangha et al., 2011). These observations seem to suggest that stimulation of the JA-pathway could be a general feature of sulphated polysaccharides in plants. Other mechanisms also operate in seaweed extract-induced plant responses, such as oxidative bursts in tobacco cell suspensions and induced the SA-signalling pathway in infiltrated tobacco and *A. thaliana* leaf tissues (Menard et al., 2005). Induction of H_2O_2 production, callus and phenol depositions coupled to the hyper-sensitive response at the infection sites were also reported when sulphated laminarin (PS3)-treated grapevine (*Vitis vinifera*) against downy mildew (*Plasmopara viticola*; Trouvelot et al., 2008).

The antioxidant properties of the polyphenols contained in some seaweed extracts may have acted against certain pathogens (Zhang et al., 2006). Phytoalexins produced in grapevines, with seaweed extract treatment were associated with reduced severity of grey mold (Jeandet, Adrian, Joubert, Hubert, & Bessis, 1996). Laminarin, from *A. nodosum* was also shown to elicit a plant defence response by increasing antimicrobial phytoalexins (Patier, Yvin, Kloareg, Liénart, & Rochas, 1993). Aziz et al (2003) reported laminarin significantly reduced the severity of grey mold in grapes.

7.5.2 Seaweed Extracts Induce Plant Defence against Viruses and Viroids

Polysaccharides present in seaweeds are potent elicitors of plant defences against viruses and viroids. Vera, Castro, Gonzalez, & Moenne, 2011 studied the antiviral effect of the Poly-Ga in tobacco leaves infected with Tobacco mosaic virus (TMV). The number of necrotic lesions was lower in the treatment than in the control. The levels of TMV-capsid protein (CP) transcripts decreased in the distal leaves, indicating that Poly-Ga induced systemic protection against TMV. It was shown that increased activity of the defence enzymes correlated with decreased numbers of necrotic lesions and TMV-CP transcript levels. We demonstrated that pretreatment of tomato plants with λ-carrageenan significantly reduced the symptoms of the Tomato chlorotic dwarf viroid, eight weeks after inoculation. Taken together, the above studies suggested that seaweed extracts have considerable potential to protect plants against viruses and viroids.

Vera et al. (2011) showed that applications of sulphated galactans to tobacco and subsequent infection with TMV, resulted in the induction of the defence enzymes, phenylalanine ammonia lyase and lipoxygenase, with a decrease in the TMV-CP transcript level. An increased activity of defence-related enzymes, with seaweed extract treatment is another mechanism, which could operate to enhance plant resistance to pathogens. Disease resistance is also reported due to higher levels of total phenols and peroxidase activity in the plant, after the application of seaweed extract (Goicoechea, Aguirreolea, & García-Mina, 2004; Raghavendra, Lokesh, & Prakash, 2007).

7.5.3 Seaweed Extracts Induce Plant Defences against Insects

Seaweed extracts have been shown to have a suppressive effect against red spider mites (Stephenson, 1966; Hankins & Hockey, 1990). Spray applications of hydrolysed seaweed extract on apple trees reduced the red spider mite population. This effect was observed for a period of two to three years and was comparable to some acaricides tested in the experiment (Stephenson, 1966). Organic extracts of a green marine alga (i.e. *Codium* sp.) were used on lima bean leaf discs, which were then infested with adult females of *Tetranychus urticae*. The extracts, at 4 g/ml, were highly toxic to the female adults after 96 h, whereas a concentration of 1 g/ml resulted in only 48% mortality. The extracts prepared with chloroform and ethyl alcohol were least effective against the females. However, the chloroform extract was found to be more effective against the egg stage suggesting specificity of the extracts on various stages of the pest. In another study, the efficiency of a commercial extract from *A. nodosum* was studied against the two-spotted red spider mite *T. urticae* (Hankins & Hockey, 1990). It was shown that seven applications of the treatment distributed over a 20-day period considerably reduced the two-spotted red spider mites in strawberry grown under high polythene tunnels.

Reitz and Trumble (1996) reported that lima bean (*Phaseolus lunatus* L.) and tomato plants treated with *A. nodosum* extract were consumed more by a generalist insect herbivore, *Spodoptera exigua* (Hübner). Furthermore, it was shown that treatment with this seaweed extract altered the 'attractiveness' of the tomato foliage to the *S. exigua* larvae. It was also shown that root applications of *A. nodosum* led to increased consumption and preference by the larvae, whereas foliar applications did not. Moreover, the larvae gained more weight on *A. nodosum* extract-treated plants compared with controls. In a recent study, Weeraddana, (2012) investigated the effect of a commercial *A. nodosum* extract on plant responses to a phloem-feeding, green peach aphid (*Myzus persicae*) using the *Arabidopsis* model. It was shown that the ANE did not confer protection against *M. persicae*. However, plants did not alter the insect biology, delayed senescence in *Arabidopsis*, during *M. persicae* infestation, could have played role in plant tolerance to this phloem-feeding insect. Applications of seaweed extracts to plants are known to improve quality, succulency and chlorophyll content, all of which could promote insect feeding. However, a clear understanding of the mechanisms of seaweed extract-induced plant defence responses to insects could generate information on the role of seaweed extracts in insect management.

The extracts of a variety of seaweeds do not seem to exert any direct insecticidal activity. As in the case of other known elicitors such as JA, SA, and 2,6-dichloroisonicotinic acid that can also induce insect defence responses in plants, seaweed extracts, or their active components, have been shown to have such activity against insect pests. Sangha et al. (2011) investigated the effect of foliar applications of *i*-, *k*- and λ-carrageenans in *Arabidopsis* against a generalist insect, *Trichoplusia ni* (cabbage looper), which is known to cause economic losses in crop plants. In a no-choice experiment, plants treated with *i*- and *k*-carrageenan incurred reduced feeding damage by *T. ni* larvae. In contrast, the damage recorded to those plants treated with λ- carrageenan, was similar to that of the control. Larval weight was also reduced by more than 20% in *i*- and *k*-carrageenan treatments. Further studies revealed that defence genes; *PR1*, *PDF1.2*, and trypsin inhibitor (*TI*) were induced with *i*-carrageenan but not by *k*- and λ-carrageenan treatment. Biochemical analyses revealed increased concentrations of both isothiocyanates and nitriles with *i*-carrageenan treatment, at 48 h post infestation, which also matched with the expression of indole glucosinolate biosynthesis genes CYP79B2, CYP83B1 and the glucosinolate hydrolysing QTL, ESM1. Taken together, the results suggested that carrageenans have differential effects on *Arabidopsis* resistance to *T. ni* and that the degree of sulphation of the polysaccharide chain may well mediate this effect. This is an interesting area for future studies.

Plant defence mechanisms can be modulated with applications of seaweed extracts, some of these responses may act against insect pests. Additional work is required to elucidate the mechanisms by which seaweed extracts alter the insect infestation.

7.5.4 Seaweed Extracts Affect Rhizosphere Microbes

One of the beneficial effects of whole seaweed and seaweed extract application to the soil is the modulation of the activities and composition of the rhizosphere. Applications of seaweed extracts to soil have resulted in a better performance of plants, which could be attributed to the effect on the soil environment, including the interactive biotic communities. Soil applications of liquid seaweed extract stimulated microbes that were antagonistic to *Pythium ultimum*, resulting in the reduced incidence of damping-off disease in cabbage seedlings (Dixon & Walsh, 2002). Plant-microbe interactions were influenced by seaweed extract treatment to the soil. Khan et al. (2012) showed that alfalfa roots inoculated with *S. meliloti* had more bacterial colonies, with *A. nodosum* extract treatments. A translational fusion of *S. meliloti* NodC:LacZ (strain JM57) was used to observe the effects of *A. nodosum* extract on bacterial gene expression and compounds reflected the bioactivity of *A. nodosum* extracts, similar to the plant-signalling molecule luteolin for eliciting Nod factors.

Zhai, (2013) also observed that soil treatment with an *A. nodosum* extract increase the activity of nod-forming bacteria, which improved plant growth, probably due to increased N fixation. Plants treated with ANE at 1 g/l demonstrated a higher number (74) of nodules, over the control (42) plants. Plants treated with 2 g/l ANE exhibited increased leaf area (87 cm^2) over the controls (71 cm^2). These plants also showed trends in increased root length in the *A. nodosum* extract-treated plants, although the difference was not significant from the control plants. Analysis of the β-galactosidase activity showed that the NodC genes were not expressed in untreated bacterial cells whereas the *A. nodosum* extract and luteolin (positive control) treatment resulted in a blue colour suggesting the expression of the NodC genes. Quantification of the β-galactosidase activity revealed that the Miller units were higher in the luteolin treatment than in the *A. nodosum* extract treatments. The *A. nodosum* extract treatment, at 0.5 g/l, significantly increased (almost two-fold), as compared to the controls (Khan et al. 2012).

These reports suggested that ANE may contain compound(s) that promote the legume–rhizobia symbiotic signalling.

Antagonism is another factor that could alter the soil rhizosphere activity. Soil amendments using dry powder from brown, green and red seaweeds were shown to have broader activities against root pathogens. Soil treatment with 1% extracts of the seaweeds *Stokeyia indica* (*Cystoseira indica*), *Padina pavonia* (brown seaweeds), and *Solieria robusta* (red alga), protected plants from infection by *Macrophomina phaseolina*, *Rhizoctonia solani* and *Fusarium solani*. In contrast, an extract of *Codium iyengarii* (a green seaweed) was effective at the lower dose of 0.5% w/w against *F. solani* (Sultana, Ehteshamul-Haque, Ara, & Athar, 2005). Similarly, soil amendment with the red seaweed *S. indica* and *S. robusta,* suppressed root-infecting fungi *M. phaseolina*, *R. solani* and *F. solani* in chilli (Sultana, Ara, & Ehteshamul–Haque, 2008). Similar results were observed in tomatoes using powdered brown seaweeds: *Spatoglossum asperum* and *Sargassum swartzii* as soil amendments, against these root rotting fungi (Sultana, Ehteshamul-Haque, Ara, & Athar, 2009).

7.6 CHEMICAL COMPONENTS OF SEAWEEDS THAT MITIGATE PLANT STRESSES

Seaweeds contain a wide range of natural bioactive compounds including plant growth promoting compounds with cytokinin-like, auxin-like and gibberellin-like activities (Crouch & Van Staden, 1993; Khan et al., 2009). Additionally, biological stimulants, which elicit metabolic signalling pathways in plants have been identified in seaweeds including polysaccharides, fatty acids, betaines, sterols and polyamines (Khan et al., 2009). Whole seaweeds, in themselves, are also rich in macro- and micro-elemental nutrients, such as amino acids and vitamins and some of these may be included in their commercial extracts, which can be directly beneficial to plants or help support beneficial soil microbes (Liu et al., 1998; Zhang & Schmidt, 1999; Khan et al., 2009). Macroalgal extracts are also known to enhance antioxidant activities in plants (Zhang & Schmidt, 2000; Zhang & Ervin, 2004; Nakayasu, Fukushima, Sasaki, Tanaka, & Nakamura, 2009). The stimulatory effect of commercial seaweed extracts may be attributed to single, additive, synergistic or even antagonistic actions of these components (Norrie & Keathley 2006; Craigie, MacKinnon, & Walter, 2008; Craige, 2011). In fact, there is evidence that seaweed extract-induced, abiotic stress tolerance is due in part, to the cumulative effect of PGR-like activities, biostimulants (or biological elicitors), osmo-protectants and other compounds that can elicit antioxidant activity (Allen et al., 2001; Zhang & Ervin, 2004). The overall bioactivity attributed to seaweed extracts varies widely between the types of seaweed, particularly due differences in their chemical components, the methods of commercial extract production, the modes of action of various seaweed compounds plants and the perception/receptor mechanisms of individual plant species (Durand, Briand, & Meyer, 2003; Stirk et al., 2003; Craigie et al., 2008; Khan et al., 2009).

7.7 THE FUTURE OF SEAWEEDS AND SEAWEED EXTRACTS IN AGRICULTURE AND HORTICULTURE

Advances in algal biotechnology have revealed tremendous potential for the use of seaweed extracts and their components in agriculture and Horticulture. From being collected from the shores and applied to the field as a mulch/compost, to a well-developed, high technology seaweed extract product, the metamorphosis of the seaweed industry has changed the perception about seaweed uses and their benefits as crop biostimulants. The applications of seaweeds, or their extract products, have shown specific effects against various plant stresses caused by abiotic and biotic factors. The breeding methods for crop improvement are, although useful, however, time consuming and laborious. Seaweeds and their commercial extracts generally contain numerous chemical components, which have already been demonstrated to provide plant growth effects and alleviate certain abiotic and biotic stresses. Detailed information on the mechanisms of activity (modes of action) of seaweed extracts is emerging and new genetic pathways activated in plants are being discovered, during plant-seaweed and environmental stress interactions (Figure 7.1). The science of the biostimulatory effects of seaweed extract will have multiple, beneficial impacts on agricultural production. The broad spectrum activity of seaweeds and their extracts as plant biostimulants, as evidenced from the above-ground (aerial) and below-ground effects, have clearly illustrated the need to focus on further detailed studies on their use in agriculture and horticultre.

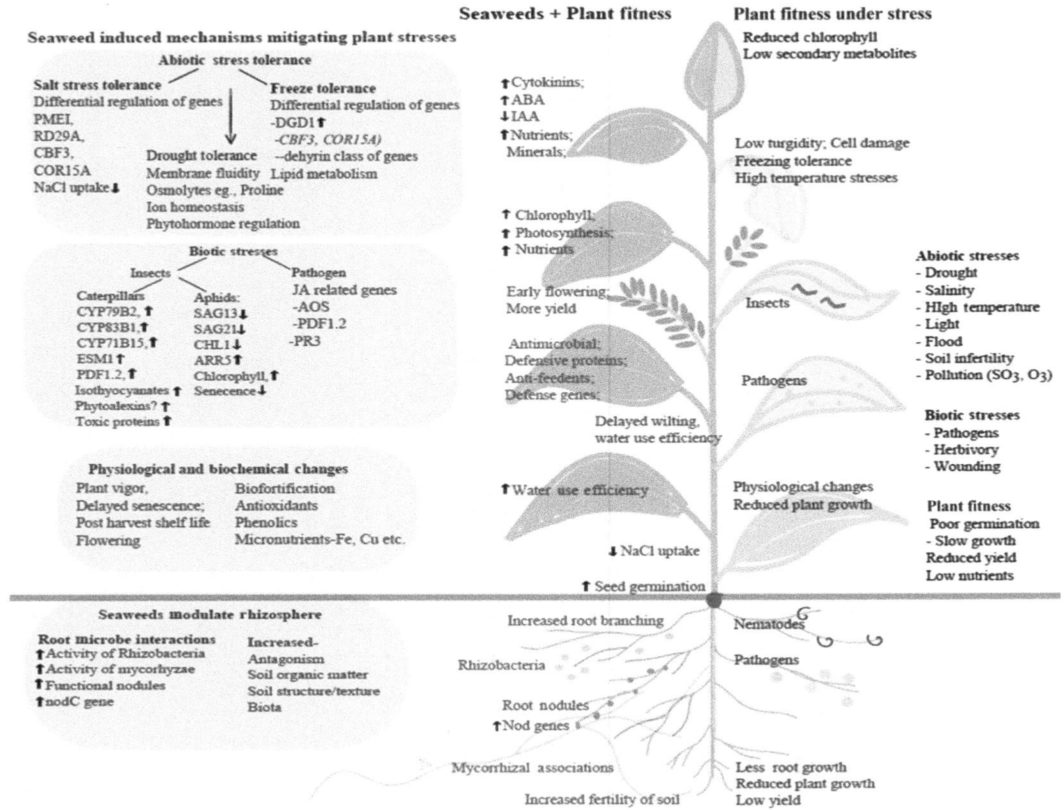

Figure 7.1 A schematic representation of effect and mechanism(s) of activity of seaweed extracts. (See the colour plate.)

ACKNOWLEDGEMENTS

BP's lab is supported by grants from the Natural Sciences and Engineering Research Council of Canada, Nova Scotia Department of Agriculture and Marketing and Acadian Seaplants Limited. Authors are grateful to Abhinandan Kumar for his assistance with illustrations.

REFERENCES

Allen, V., Pond, K., Saker, K., Fontenot, J., Bagley, C., Ivy, R., et al. (2001). Tasco: Influence of a brown seaweed on antioxidants in forages and livestock—A review. *Journal of Animal Sciences*, *79*, 21–31.

Arthur, G., Stirk, W., & Van Staden, J. (2003). Effect of a seaweed concentrate on the growth and yield of three varieties of *Capsicum annuum*. *South African Journal of Botany*, *69*, 207–211.

Ayad J. (1998). *The effect of seaweed extract (Ascophyllum nodosum) extract on antioxidant activities and drought tolerance of tall fescue (Festuca arundinacea Schreb.)* (Ph.D. dissertation). Lubbock: Texas Tech University.

Aziz, A., Poinssot, B., Daire, X., Adrian, M., Be zier, A., Lambert, B., et al. (2003). Laminarin elicits defence responses in grapevine and induces protection against *Botrytis cinerea* and *Plasmopara viticola*. *Molecular Plant–microbe Interactions*, *16*, 1118–1128.

Beckett, R., & Van Staden, J. (1989). The effect of seaweed concentrate on the growth and yield of potassium stressed wheat. *Plant Soil*, *116*, 29–36.

Bi, F., Iqbal, S., Arman, M., Ali, A., & Hassan, M. (2011). Carrageenan as an elicitor of induced secondary metabolites and its effects on various growth characters of chickpea and maize plants. *Journal of Saudi Chemical Society*, *15*, 269–273.

Blunden, G., Jenkins, T., & Liu, Y.W. (1996). Enhanced leaf chlorophyll levels in plants treated with seaweed extract. *Journal of Applied Phycology*, *8*, 535–543.

Burchett, S., Fuller, M., & Jellings, A. (1998). *Application of seaweed extract improves winter hardiness of winter barley cv Igri*. The Society for Experimental Biology, Annual Meeting, York University.

Close, T. J. (2006). Dehydrins: emergence of a biochemical role of a family of plant dehydration proteins. *Physiology Plantarum*, *97*, 795–803.

Cluzet, S., Torregrosa, C., Jacquet, C., Lafitte, C., Fournier, J., Mercier, L., et al. (2004). Gene expression profiling and protection of *Medicago truncatula* against a fungal infection in response to an elicitor from green algae *Ulva spp*. *Plant Cell and Environment*, *27*, 917–928.

Craigie, J. S. (2011). Seaweed extracts stimuli in plant science and agriculture. *Journal of Applied Phycology*, *23*, 371–393.

Craigie, J. S., MacKinnon, S. L., & Walter, J. A. (2008). Liquid seaweed extracts identified using 1 H NMR profiles. *Journal of Applied Phycology*, *20*, 665–671.

Crouch, I. J., & Van Staden, J. (1992). Effect of seaweed concentrate on the establishment and yield of greenhouse tomato plants. *Journal of Applied Phycology*, *4*, 291–296.

Crouch, I. J., & Van Staden, J. (1993). Evidence for the presence of plant growth regulators in commercial seaweed products. *Plant Growth Regulation*, *13*, 21–29.

Dixon, R., Harrison, M., & Lamb, C. (1994). Early events in the activation of plant defense responses. *Annual Review of Phytopathology*, *32*, 479–501.

Dixon, G. R., & Walsh, U. F. (2002). Suppressing *Pythium ultimum* induced damping-off in cabbage seedlings by biostimulation with proprietary liquid seaweed extracts managing soil-born pathogens: a sound rhizosphere to improve productivity in intensive horticultural systems. In *Proceedings of the XXVI International Horticultural Congress, Toronto, Canada 11–17 August. 103-106.*

Dobromilska, R., Mikiciuk, M., & Gubarewicz, K. (2008). Evaluation of cherry tomato yielding and fruit mineral composition after using of Bio-algeen S-90 preparation. *Journal of Elementology*, *13*, 491–499.

Durand, N., Briand, X., & Meyer, C. (2003). The effect of marine bioactive substances (N PRO) and exogenous cytokinins on nitrate reductase activity in *Arabidopsis thaliana*. *Physiologia Plantarum*, *119*, 489–493.

Ervin, E. H., Zhang, X., & Fike, J. H. (2004). Ultraviolet-B radiation damage on Kentucky Bluegrass II: hormone supplement effects. *HortScience*, *39*, 1471–1474.

Fan, D., Hodges, D. M., Zhang, J., Kirby, C. W., Ji, X., Locke, S. J., et al. (2011). Commercial extract of the brown seaweed *Ascophyllum nodosum* enhances phenolic antioxidant content of spinach (*Spinacia oleracea* L.) which protects *Caenorhabditis elegans* against oxidative and thermal stress. *Food Chemistry*, *124*, 195–202.

Featonby-Smith, B., & Van Staden, J. (1983). The effect of seaweed concentrate and fertilizer on the growth of *Beta vulgaris*. *Z. Pflanzenphysiol*, *112*, 155–162.

Feng, H., Li, S., Xue, L., An, L., & Wang, X. (2007). The interactive effects of enhanced UV-B radiation and soil drought on spring wheat. *South African Journal of Botany*, *73*, 429–434.

Goicoechea, N., Aguirreolea, J., & García-Mina, J. (2004). Alleviation of *Verticillium* wilt in pepper (*Capsicum annuum* L.) by using the organic amendment COA H of natural origin. *Scientia Horticulturae*, *101*, 23–37.

Graham, R. D., Welch, R. M., Saunders, D. A., Ortiz-Monasterio, I., Bouis, H. E., Bonierbale, M., et al. (2007). Nutritious subsistence food systems. *Advances in Agronomy*, *92*, 1–74.

Guiry, M. D., & Guiry, G. (2013). *AlgaeBase*. . www.algaeBase.org.

Hankins, S., & Hockey, H. (1990). The effect of a liquid seaweed extract from *Ascophyllum nodosum* (Fucales, Phaeophyta) on the two-spotted red spider mite *Tetranychus urticae*. *Hydrobiologia*, *204*, 555–559.

Holm, P. B., Kristiansen, K. N., & Pedersen, H. B. (2002). Transgenic approaches in commonly consumed cereals to improve iron and zinc content and bioavailability. *Journal of Nutrition*, *132*, 514S–516S.

Hong, D., Hien, H., & Son, P. (2007). Seaweeds from Vietnam used for functional food, medicine and biofertilizer. *Journal of Applied Phycology*, *19*, 817–826.

Jannin, L., Arkoun, M., Etienne, P., Laîné, P., Goux, D., Garnica, M., et al. (2012). *Brassica napus* growth is promoted by *Ascophyllum nodosum* (L.) Le Jol. seaweed extract: microarray analysis and physiological characterization of N, C, and S metabolisms. *Journal of Plant Growth Regulation*, *32*, 31–52.

Jaulneau, V., Lafitte, C., Corio-Costet, M. F., Stadnik, M. J., Salamagne, S., Briand, X., et al. (2011). An *Ulva armoricana* extract protects plants against three powdery mildew pathogens. *European Journal of Plant Pathology*, *131*, 393–401.

Jaulneau, V., Lafitte, C., Jacquet, C., Fournier, S., Salamagne, S., Briand, X., et al. (2010). Ulvan, a sulfated polysaccharide from green algae, activates plant immunity through the jasmonic acid signaling pathway. *Journal of Biomedicine and Biotechnology*, *525291*, 1–11.

Jeandet, P., Adrian, M., Joubert, J., Hubert, F., & Bessis, R. (1996). Induction of phytoalexin (resveratrol) synthesis in grapevines treated with Synermix [iprodione]. *Phytoma La Defense des Vegetaux*, *48*, 21–25.

Jithesh, M. N., Wally, O. S. D., Manfield, I., Critchley, A. T., Hiltz, D., & Prithiviraj, B. (2012). Analysis of seaweed extract-induced transcriptome leads to identification of a negative regulator of salt tolerance in *Arabidopsis*. *HortScience*, *47*, 704–709.

Kauffman, G. L., Kneivel, D. P., & Watschke, T. L. (2007). Effects of a biostimulant on the heat tolerance associated with photosynthetic capacity, membrane thermostability, and polyphenol production of perennial ryegrass. *Crop Science*, *47*, 261–267.

Khan, W., Hiltz, D., Critchley, A. T., & Prithiviraj, B. (2011). Bioassay to detect *Ascophyllum nodosum* extract-induced cytokinin-like activity in *Arabidopsis thaliana*. *Journal of Applied Phycology*, *23*, 409–414.

Khan, W., Rayirath, U. P., Subramanian, S., Jithesh, M. N., Rayorath, P., Hodges, D. M., et al. (2009). Seaweed extracts as biostimulants of plant growth and development. *Journal of Plant Growth Regulation*, *28*, 386–399.

Khan, W., Zhai, R., Souleimanov, A., Critchley, A. T., Smith, D. L., & Prithiviraj, B. (2012). Commercial extract of *Ascophyllum nodosum* improves root colonization of alfalfa by its bacterial symbiont *Sinorhizobium meliloti. Communications in Soil and Plant Analysis, 43,* 2425–2436.

Kulik, M. M. (1995). The potential for using cyanobacteria (blue-green algae) and algae in the biological control of plant pathogenic bacteria and fungi. *European Journal of Plant Pathology, 101,* 585–599.

Kumari, R., Kaur, I., & Bhatnagar, A. (2011). Effect of aqueous extract of *Sargassum johnstonii* Setchell & Gardner on growth, yield and quality of *Lycopersicon esculentum* Mill. *Journal of Applied Phycology, 23,* 623–633.

Kumar, G., & Sahoo, D. (2011). Effect of seaweed liquid extract on growth and yield of *Triticum aestivum* var. Pusa Gold. *Journal of Applied Phycology, 23,* 251–255.

Little, H., & Neily, W. (2010). *Commercial extracts of the brown seaweed Ascophyllum nodosum improve plant water use and drought stress resistance in the greenhouse and filed.* Western Plant Growth Society (Annual Meetings, Davis, California).

Liu, Q., Kasuga, M., Sakuma, Y., Abe, H., Miura, S., Yamaguchi-Shinozaki, K., et al. (1998). Two transcription factors, DREB1 and DREB2, with an EREBP/AP2 DNA binding domain separate two cellular signal transduction pathways in drought-and low-temperature-responsive gene expression, respectively, in *Arabidopsis. The Plant Cell Online, 10,* 1391–1406.

Li, L., & Van Staden, J. (1998). Effects of plant growth regulators on drought resistance of two maize cultivars. *South African Journal of Botany, 64,* 116–120.

Lordan, S., Ross, R. P., & Stanton, C. (2011). Marine bioactives as functional food ingredients: potential to reduce the incidence of chronic diseases. *Marine drugs, 9,* 1056–1100.

Lorenzo, O., & Solano, R. (2005). Molecular players regulating the jasmonate signalling network. *Current Opinion in Plant Biology, 8,* 532–540.

MacDonald, J. E., Hacking, J., Weng, Y., & Norrie, J. (2012). Root growth of containerized lodgepole pine seedlings in response to *Ascophyllum nodosum* extract application during nursery culture. *Canadian Journal of Plant Science, 92,* 1–6.

Mancuso, S., Azzarello, E., Mugnai, S., & Briand, X. (2006). Marine bioactive substances (IPA extract) improve foliar ion uptake and water stress tolerance in potted *Vitis vinifera* plants. *Advances in Horticultural Science, 2,* 156–161.

Marschner, H. (1995). *Mineral nutrition of higher plants* (2nd ed.). San Diego, CA (USA): Academic Press.

McGloughlin, M. N. (2010). Modifying agricultural crops for improved nutrition. *Nature Biotechnology, 27,* 494.

Ménard, R., De Ruffray, P., Fritig, B., Yvin, J. C., & Kauffmann, S. (2005). Defense and resistance-inducing activities in tobacco of the sulfated β-1, 3 glucan PS3 and its synergistic activities with the unsulfated molecule. *Plant Cell Physiology, 46,* 1964–1972.

Moeller, M., & Smith, M. (1998). The significance of the mineral component of seaweed suspensions on lettuce (*Lactuca sativa* L.) seedling growth. *Journal of Plant Physiology, 153,* 658–663.

Möller, M., & Smith, M. (1999). The effects of priming treatments using seaweed suspensions on the water sensitivity of barley (*Hordeum vulgare* L.) caryopses. *Annals of Applied Biology, 135,* 515–521.

Nabati, D., Schmidt, R., & Parrish, D. (1994). Alleviation of salinity stress in Kentucky Bluegrass by plant growth regulators and iron. *Crop Science, 34,* 198–202.

Nair, P., Kandasamy, S., Zhang, J., Ji, X., Kirby, C., Benkel, B., et al. (2012). Transcriptional and metabolomic analysis of *Ascophyllum nodosum* mediated freezing tolerance in *Arabidopsis thaliana. BMC Genomics, 13,* 643.

Nakayasu, K., Fukushima, M., Sasaki, K., Tanaka, S., & Nakamura, H. (2009). Comparative studies of the reduction behavior of chromium (VI) by humic substances and their precursors. *Environmental Toxicology and Chemistry, 18*, 1085–1090.

Neily, W., Shishkov, L., Tse, T., & Titus, D. (2008). Acadian LSC helps reduce salinity stress in pepper seedlings cv. California Wonder. *PGRSA News, 1*, 1.

Newell-McGloughlin, M. (2008). Nutritionally improved agricultural crops. *Plant Physiology, 147*, 939–953.

Norrie, J., & Keathley, J. (2006). Benefits of *Ascophyllum nodosum* marine-plant extract applications to 'Thompson seedless' grape production. *Acta Horticulturae, 727*, 243–245.

Nuernberger, T., & Lipka, V. (2005). Non–host resistance in plants: new insights into an old phenomenon. *Molecular Plant Pathology, 6*, 335–345.

Ördög, V., Stirk, W. A., Van Staden, J., Novák, O., & Strnad, M. (2004). Endogenous cytokinins in three genera of microalgae from the Chlorophyta. *Journal of Phycology, 40*, 88–95.

Patier, P., Yvin, J. C., Kloareg, B., Liénart, Y., & Rochas, C. (1993). Seaweed liquid fertilizer from *Ascophyllum nodosum* contains elicitors of plant d-glycanases. *Journal of Applied Phycology, 5*, 343–349.

Paulert, R., Ebbinghaus, D., Urlass, C., & Moerschbacher, B. (2010). Priming of the oxidative burst in rice and wheat cell cultures by ulvan, a polysaccharide from green macroalgae, and enhanced resistance against powdery mildew in wheat and barley plants. *Plant Pathology, 59*, 634–642.

Pieterse, C. M. J., Van Wees, S. C. M., Van Pelt, J. A., Knoester, M., Laan, R., Gerrits, H., et al. (1998). A novel signaling pathway controlling induced systemic resistance in *Arabidopsis*. *The Plant Cell Online, 10*, 1571–1580.

Raghavendra, V. B., Lokesh, S., & Prakash, H. (2007). Dravya, a product of seaweed extract (*Sargassum wightii*), induces resistance in cotton against *Xanthomonas campestris* pv. *malsvacearum*. *Phytoparasitica, 35*, 442–449.

Rathore, S. S., Chaudhary, D. R., Boricha, G. N., Ghosh, A., Bhatt, B. P., Zodape, S. T., et al. (2009). Effect of seaweed extract on the growth, yield and nutrient uptake of soybean (*Glycine max*) under rainfed conditions. *South African Journal of Botany, 75*, 351–355.

Rayorath, P., Benkel, B., Mark Hodges, D., Allan-Wojtas, P., MacKinnon, S., Critchley, A. T., et al. (2009). Lipophilic components of the brown seaweed, *Ascophyllum nodosum*, enhance freezing tolerance in *Arabidopsis thaliana*. *Planta, 230*, 135–147.

Rayorath, P., Jithesh, M. N., Farid, A., Khan, W., Palanisamy, R., Hankins, S. D., et al. (2008). Rapid bioassays to evaluate the plant growth promoting activity of *Ascophyllum nodosum* (L.) Le Jol. using a model plant, *Arabidopsis thaliana* (L.) Heynh. *Journal of Applied Phycology, 20*, 423–429.

Rayorath, P., Khan, W., Palanisamy, R., MacKinnon, S. L., Stefanova, R., Hankins, S. D., et al. (2008). Extracts of the brown seaweed *Ascophyllum nodosum* induce gibberellic acid (GA 3)-independent amylase activity in barley. *Journal of Plant Growth Regulation, 27*, 370–379.

Reitz, S. R., & Trumble, J. T. (1996). Tritrophic interactions among linear furanocoumarins, the herbivore *Trichoplusia ni* (Lepidoptera: Noctuidae), and the polyembryonic parasitoid *Copidosoma floridanum* (Hymenoptera: Encyrtidae). *Environmental Entomology, 25*, 1391–1397.

Russo, R. O., & Berlyn, G. P. (1991). The use of organic biostimulants to help low input sustainable agriculture. *Journal of Sustainable Agriculture, 1*, 19–42.

Sanderson, K., & Jameson, P. (1985). The cytokinins in a liquid seaweed extract: could they be the active ingredients? *Vth international symposium on growth regulators in fruit production, 179*, 113–116.

Sangha, J. S., Khan, W., Ji, X., Zhang, J., Mills, A. A. S., Critchley, A. T., et al. (2011). Carrageenans, sulphated polysaccharides of red seaweeds, differentially affect *Arabidopsis thaliana* resistance to *Trichoplusia ni* (Cabbage Looper). *PloS One, 6*, e26834.

Sangha, J. S., Ravichandran, S., Prithiviraj, K., Critchley, A. T., & Prithiviraj, B. (2010). Sulfated macroalgal polysaccharide λ-carrageenan induces resistance against *Sclerotinia sclerotiorum* in *Arabidopsis thaliana* whereas i-carrageenan enhances susceptibility. *Physiological and Molecular Plant Pathology*, *75*, 38–45.

Shah, M. T., Zodape, S. T., Chaudhary, D. R., Eswaran, K., & Chikara, J. (2013). Seaweed sap as an alternative liquid fertilizer for yield and quality improvement of wheat. *Journal of Plant Nutrition*, *36*, 192–200.

Sharma, S. H. S., Lyons, G., McRoberts, C., McCall, D., Carmichael, E., Andrews, F., et al. (2012). Biostimulant activity of brown seaweed species from Strangford Lough: compositional analyses of polysaccharides and bioassay of extracts using mung bean (*Vigno mungo* L.) and pak choi (*Brassica rapa chinensis* L.). *Journal of Applied Phycology*, *24*, 1081–1091.

Spann, T. M., & Little, H. A. (2011). Applications of a commercial extract of the brown seaweed *Ascophyllum nodosum* increases drought tolerance in container-grown 'Hamlin' sweet orange nursery trees. *HortScience*, *46*, 577–582.

Spinelli, F., FiorI, G., Noferini, M., Sprocatti, M., & Costa, G. (2009). Perspectives on the use of a seaweed extract to moderate the negative effects of alternate bearing in apple trees. *Journal of Horticultural Science and Biotechnology*, *84*, 131–137.

Stephenson, W. (1966). The effect of hydrolysed seaweed on certain plant pests and diseases. *Proceedings of the International Seaweed Symposium*, *5*, 405–415.

Stirk, W., Arthur, G., Lourens, A., Novák, O., Strnad, M., & Van Staden, J. (2004). Changes in cytokinin and auxin concentrations in seaweed concentrates when stored at an elevated temperature. *Journal of Applied Phycology*, *16*, 31–39.

Stirk, W., Novák, O., Strnad, M., & Van Staden, J. (2003). Cytokinins in macroalgae. *Plant Growth Regulation*, *41*, 13–24.

Subramanian, S., Sangha, J. S., Gray, B. A., Singh, R. P., Hiltz, D., Critchley, A. T., et al. (2011). Extracts of the marine brown macroalga, *Ascophyllum nodosum*, induce jasmonic acid dependent systemic resistance in *Arabidopsis thaliana* against *Pseudomonas syringae* pv. tomato DC3000 and *Sclerotinia sclerotiorum*. *European Journal of Plant Pathology*, *131*, 237–248.

Sultana, V., Ara, J., & Ehteshamul–Haque, S. (2008). Suppression of root rotting fungi and root knot nematode of chili by seaweed and *Pseudomonas aeruginosa*. *Journal of Phytopathology*, *156*, 390–395.

Sultana, V., Ehteshamul-Haque, S., Ara, J., & Athar, M. (2005). Comparative efficacy of brown, green and red seaweeds in the control of root infecting fungi and okra. *International Journal of Environmental Science and Technology*, *2*, 129–132.

Sultana, V., Ehteshamul-Haque, S., Ara, J., & Athar, M. (2009). Effect of brown seaweeds and pesticides on root rotting fungi and root-knot nematode infecting tomato roots. *Journal of Applied Botany and Food Quality*, *83*, 50–53.

Trouvelot, S., Varnier, A. L., Allegre, M., Mercier, L., Baillieul, F., Arnould, C., et al. (2008). A β-1, 3 glucan sulfate induces resistance in grapevine against *Plasmopara viticola* through priming of defense responses, including HR-like cell death. *Molecular Plant–Microbe Interactions*, *21*, 232–243.

Vera, J., Castro, J., Gonzalez, A., & Moenne, A. (2011). Seaweed polysaccharides and derived oligosaccharides stimulate defense responses and protection against pathogens in plants. *Marine drugs*, *9*, 2514–2525.

Vinoth, S., Gurusaravanan, P., & Jayabalan, N. (2012). Effect of seaweed extracts and plant growth regulators on high-frequency in vitro mass propagation of *Lycopersicon esculentum* L. (tomato) through double cotyledonary nodal explant. *Journal of Applied Phycology*, *24*, 1329–1334.

Wally, O. S. D., Critchley, A. T., Hiltz, D., Craigie, J. S., Han, X., Zaharia, L. I., et al. (2012). Regulation of phytohormone biosynthesis and accumulation in *Arabidopsis* following treatment with commercial extract from the marine macroalga *Ascophyllum nodosum*. *Journal of Plant Growth Regulation*, 1–16.

Weeraddana CDS. (2012). *Extracts of the brown seaweed, Ascophyllum nodosum, effect Arabidopsis thaliana–Myzus persicae interaction* (MSc. Thesis). Canada: Dalhousie University.

Welch, R. M., & Graham, R. D. (2005). Agriculture: the real nexus for enhancing bioavailable micronutrients in food crops. *Journal of Trace Elements in Medicine and Biology, 18*, 299–307.

White, P. J., & Broadley, M. R. (2009). Biofortification of crops with seven mineral elements often lacking in human diets–iron, zinc, copper, calcium, magnesium, selenium and iodine. *New Phytologist, 182*, 49–84.

Wilson, S. (2001). Frost management in cool climate vineyards. In *University of Tasmania Research Reports- UT 99/1*. Australia: Grape and Wine Research & Development Corporation.

Yan JY. (1993). *Influence of plant growth regulators on turfgrass polar lipid composition, tolerance to drought and salinity stresses, and nutrient efficiency* (Ph.D. dissertation). Blacksburg: Virginia Polytechnic Institute and State University.

Zhai R. (2013). *Effects of the brown seaweed, Ascophyllum nodosum, on the nodulation and growth of alfalfa* (MSc. Thesis). Canada: Dalhousie University.

Zhang, X., & Ervin, E. (2004). Cytokinin-containing seaweed and humic acid extracts associated with creeping bentgrass leaf cytokinins and drought resistance. *Crop Science, 44*, 1737–1745.

Zhang, X., & Ervin, E. (2008). Impact of seaweed extract-based cytokinins and zeatin riboside on creeping bentgrass heat tolerance. *Crop Science, 48*, 364–370.

Zhang, X., Ervin, E., & Schmidt, R. (2003a). Plant growth regulators can enhance the recovery of Kentucky Bluegrass sod from heat injury. *Crop Science, 43*, 952–956.

Zhang, X., Ervin, E., & Schmidt, R. (2003b). Physiological effects of liquid applications of a seaweed extract and a humic acid on creeping bentgrass. *Journal of American Society of Horticultural Science, 128*, 492–496.

Zhang, X., & Schmidt, R. (1997). The impact of growth regulators on the α-tocopherol status in water-stressed *Poa pratensis*. *International Turfgrass Society Research Journal, 8*, 1364–1371.

Zhang, X., & Schmidt, R. (1999). Antioxidant response to hormone-containing product in Kentucky Bluegrass subjected to drought. *Crop Science, 39*, 545–551.

Zhang, X., & Schmidt, R. (2000). Hormone-containing products' impact on antioxidant status of tall fescue and creeping bentgrass subjected to drought. *Crop Science, 40*, 1344–1349.

Zhang, Q., Zhang, J., Shen, J., Silva, A., Dennis, D.A., & Barrow, C. J. (2006). A simple 96-well microplate method for estimation of total polyphenol content in seaweeds. *Journal of Applied Phycology, 18*, 445–450.

Zhu, Y., Chen, H., Fan, J., Wang, Y., Li, Y., Chen, J., et al. (2000). Genetic diversity and disease control in rice. *Nature, 406*, 718–722.

Zodape, S., Gupta, A., Bhandari, S., Rawat, U., Chaudhary, D., Eswaran, K., et al. (2011). Foliar application of seaweed sap as biostimulant for enhancement of yield and quality of tomato (*Lycopersicon esculentum* Mill.). *Journal of Science and Industrial Research, 70*, 215–219.

Zodape, S., Kawarkhe, V., Patolia, J., & Warade, A. (2008). Effect of liquid seaweed fertilizer on yield and quality of okra (*Abelmoschus esculentus* L.). *Journal of Science and Industrial Research, 67*, 1115–1117.

Zodape, S., Mukherjee, S., Reddy, M., & Chaudhary, D. (2009). Effect of *Kappaphycus alvarezii* (Doty) Doty ex Silva. extract on grain quality, yield and some yield components of wheat (*Triticum aestivum* L.). *International Journal of Plant Production, 3*, 97–101.

CHAPTER EIGHT

Controlling Seaweed Biology, Physiology and Metabolic Traits in Production for Commercially Relevant Bioactives in Glycobiology

Pia C. Winberg[*,†,1], **Helen J. Fitton**[‡], **Damien Stringer**[‡], **Sam S. Karpiniec**[‡] **and Vicki-Anne Gardiner**[‡]

[*]School of Medicine, University of Wollongong, Wollongong, NSW, Australia
[†]Venus Shell Systems Pty. Ltd, Narrawallee, NSW, Australia
[‡]Marinova Pty. Ltd Cambridge, Tasmania, Australia
[1]Corresponding author: e-mail address: pia@uow.edu.au

Contents

Abstract

This review considers a plethora of multidisciplinary literature relating to seaweeds and their bioactive metabolites. Rather than providing another stocktake of the diverse metabolites and applications, we undertake a meta-analysis of the status and direction of research and development towards biological applications. Of significance is that for health and medical applications specifically, the dominant focus remains on

Advances in Botanical Research, Volume 71
ISSN 0065-2296
http://dx.doi.org/10.1016/B978-0-12-408062-1.00008-1

the unique and abundant sulfated polysaccharides. However, there is a shift from the historical publications that emerged during the development of the hydrocolloid industry in the last century. Recent development focuses on the huge diversity of these metabolites across more species and at finer structural scale. This is concurrent with the frontier of research in glycobiology where sulfated polysaccharide macromolecules are shown to have specific and important biological activities related to key molecular traits. Phycologists, biochemists, chemical engineers and health and medical researchers will build the teams that can deliver production and consistency of these complex molecules.

8.1 INTRODUCTION

The historical applications of macroalgae (seaweeds) for humans are indeed diverse and date back to some of the earliest records in human history (Table 8.1). However, as an industrial scale crop, seaweed cultivation is still a recent development following its rapid expansion of production and technological developments during the past half century. Currently, the scale of seaweed cultivation is considerable, contributing about $4.2B of trade accounting for approximately 46% of the biomass (18M tonnes) produced in marine aquaculture globally; the bulk of which is for food (FAO, 2010; FAO, 2011). This rapid development has occurred in highly industrialised nations such as Japan; from large to small nations such as China and Fiji; and has been vital to the employment of 100,000s of families, socio-economic progress and the sustainable industry development of developing nations, particularly in South East Asia. The industry cultivates diverse species, approximately 20 (FAO, 2011), from arctic water kelps such as *Laminaria* spp., to tropical red species such as *Kappaphycus* spp. and produces mostly foods including the most valued nori (*Porphyra* or *Pyropia* spp.), as well as commodity ingredients from the refinement of abundant hydrocolloids (again *Laminaria* and *Kappaphycus* spp.).

Despite this considerable scale and progress, there are good reasons to take stock of the current status and recent developments in this industry. Seaweed as a crop has an established track record as one of the globally transitional industries with rapid development potential (Table 8.1) that addresses the long-term issues of nutritional and food security as well as environmental sustainability (assuming best practice management). This includes addressing the issues of current global malnutrition (esp. vitamins, minerals and other trace elements) (Traka & Mithen, 2011), the future challenge of feeding nine billion people (Godfray et al., 2010), as well as the emerging evidence for macroalgal chemical traits that can contribute to

Table 8.1 The Use of Seaweeds by Humans Over Time with a Focus on the Modern History Trajectory of Applied Seaweed Research and Industry

Period	Status or Stage of Development in Applied Seaweed Research and Industry
Up to 35,000BC	Tasmanian aboriginal cultures utilise the hygroscopic properties of kelp to make water carriers (Dillehay et al., 2008)
13,000BC	Seaweed used and traded in nutrition and health of ancient civilisations in Chile (Alexander, 2006)
0–300	Seaweeds used as medicines in Greece (Rindi et al., 2012)
	Roman Columella and Paldeus recommend the application of marine algae to the roots of agricultural crops (Henderson, 2004)
	Documented nutritional use of seaweeds in China for iodine supplmentation (Volpe, 2011)
	Seaweed ranching for food production in Japan (Elisabeth Mann, 1980)
1700s	Seaweed hygroscopic properties utilised in medical dilating devices (Johnson, 1990)
1940s	Industrial development of hydrocolloids from seaweeds; seaweed biology; research on seaweed as food (Craigie, 2010; Tseng, 1944)
	Reproduction breakthrough on *Porphyra* life history and intensification of *Porphyra* industry in Japan (Harris, Matsuda, & Sattelle, 2013)
	Technology to liquefy brown seaweeds for agricultural crop application developed (Chan et al., 2006)
1950s	First international seaweed symposium in Scotland (Chan et al., 2006; Chengkui, 1984)
	Peoples republic of China establishes large-scale, *Laminaria* raft cultivation (Chengkui, 1984)
1960s	Taxonomic classification; biogeography and ecophysiology of seaweeds continues and industry grows (Chan et al., 2006)
	Early publications on bioactive fucoidan polysaccharides emerge
1970s	Tank-based cultivation developed in Canada to secure source of carageenan from *Chondrus crispus* (Moseley, 1990)
	Micropropagation techniques developed (Reddy, Jha, Fujita, & Ohno, 2008)
	Establishment of commercial phycocultivation of carageen seaweeds in the Philippines (Hurtado, Montaño, & Martinez-Goss, 2013)

Continued

Table 8.1 The Use of Seaweeds by Humans Over Time with a Focus on the Modern History Trajectory of Applied Seaweed Research and Industry—cont'd

Period	Status or Stage of Development in Applied Seaweed Research and Industry
1980s	Mutant studies; seaweed cultivation and biotechnology growth; properties of hydrocolloids in industries; drug discovery and bioactive compounds; classical genetics (Chan et al., 2006)
	Phycoculture in China reaches over 1 million tonnes per annum (Chengkui, 1984)
	Seaweed carageen cultivation introduced to Indonesia and Africa (Hurtado et al., 2013; Reddy et al., 2008)
1990	Genetic transformation and tissue culture; molecular phylogenetics; small-scale gene cloning and characterisation (Chan et al., 2006)
2000	Introduction of expressed sequence tags approach in large-scale study on molecular genetics (Halling et al., 2012)
	Barcode of life established and first seaweed barcodes published
	Micropropagation technology expands (Reddy et al., 2008)
	Evidence for the bioactivity of seaweed-sulfated polysaccharides (Chan et al., 2006)
	Search of a model plant for large-scale genomic study; shift to multidisciplinary research (Chan et al., 2006)
	Increased interest in biorefinery and chemical engineering for high-value products or biofuels
	Porphyra as proposed candidate for genome sequence (Chan et al., 2006)
2010	First model seaweed genome sequenced–brown, *Ectocarpus* (Cock et al., 2010)
	Genetic tools applied to reclassify taxonomy of some of the biggest commercial species (Sutherland et al., 2011)
	Seaweed, molecular techniques and glycobiology meet
2013	*Chondrus crispus* (red) genome sequenced (Collén et al., 2013)

the prevention of epidemic disease (e.g. obesity, diabetes and inflammatory diseases), as well as animal nutrition and health and large-scale agricultural applications (see reviews by Craigie (2010) and Holdt & Kraan (2011)). At the same time, global scale production contributes to considerable environmental benefits including nutrient remediation, local carbon balances and buffering of acidification effects (Jiang, Fang, Mao, Han, & Wang, 2012).

The publication rate is currently 50% higher (0.25% growth in MEDLINE) for research in the health and medical fields using seaweeds compared to the publication rate in the total seaweed-related research (0.17% growth through Google Scholar). However, if the production of macroalgae for health and medical applications is to progress into real world outcomes, the production industry must become more sophisticated. Specifically, the biomass requires increased consistency, quality control and evidence of biological functionality alongside claims, similar to other emerging phytochemical sectors (Traka & Mithen, 2011). Understanding and controlling the genetic and environmental influences on biological traits, and further development of cultivation systems that can control these traits (Hafting, Critchley, Cornish, Hubley, & Archibald, 2011), are essential for progress.

Over the last century, most research has focussed on the extraction of the sulfated polysaccharide (SPS) components of seaweeds for industrial hydrocolloid applications. This work has delivered over 337,000 publications that refer to the polysaccharide category alginate, 95,300 to carrageenan, and more recently there is increasing effort on other categories of polysaccharides such as 15,300 publications that refer to fucoidan and 968 on ulvan. Indeed the review of literature reveals that the core interest of medical and bioactive applications remains on SPS metabolites in seaweeds, compared to a more evenly distributed effort across other metabolites in the broader literature on seaweeds (Figure 8.1).

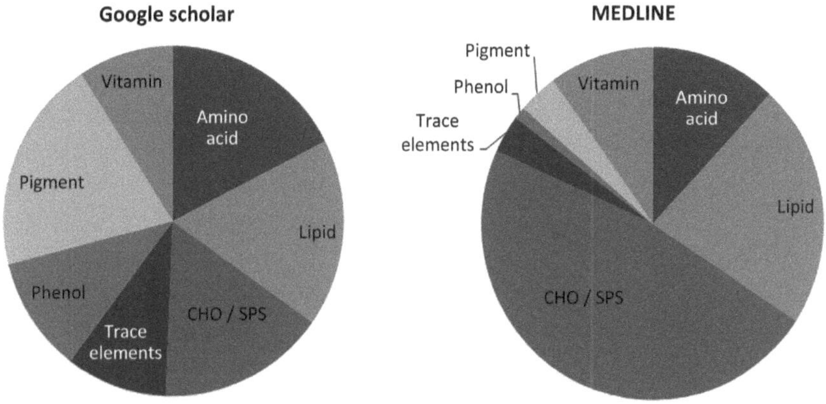

Figure 8.1 The relative focus on diverse metabolite categories of seaweeds in the broad peer-reviewed scientific literature (Google Scholar $n = 23{,}800$) vs the nutritionally and medically focused literature (MEDLINE $n = 2370$). Review based on search words for each of the metabolite categories within literature related to "seaweed"; CHO/SPS is abbreviated for the search terms carbohydrates or sulfated polysaccharides.

This expanding interest in the SPSs of seaweeds in the medical literature is concurrent with the growing field of research in glycobiology. The term glycobiology was coined in 1988 and has until recently lived on the fringe of medical and physiological research. Medical science has tended to focus on disease and the biochemistry of proteins and lipids more than polysaccharides, which have been regarded as an inactive bulk compared to proteins and lipids. Glycobiology is a more complicated science than protein biology as glycometabolites are quantitative traits, a step downstream from protein synthesis from DNA. Glycometabolites are thus controlled by multiple genes, and are often large and complicated molecules to work with. Therefore progress in this field of research is considered very recent and was previously ignored by metabolic chemists due to the large molecular weight and complex chemistry of carbohydrate molecules (Alper, 2001a,b).

Now, recent reviews and scientific publications are highlighting the progression of glycobiology as the current frontier in medical research, following in the footsteps of and interacting with nucleic acids, proteins and functional lipids (Alper, 2001a,b). We know that proteins and lipids do not operate in isolation and that the SPSs or glycol components are essential. Multiple areas of expertise are now attempting to put the pieces of that biochemical puzzle together and to emphasise the need to consider the biological function of SPSs in "macromolecular crowds" (Foffi, Pastore, Piazza, & Temussi, 2013). It is not difficult to understand why the current approach to medical research has not paid much attention to the importance of polysaccharides in health, as their function is often not a direct one. For example, the glycosaminoglycan hyaluronan in human milk, a structure similar to many of the SPSs in seaweeds, has been shown to be an important trigger for antipathogenic status of the infant gut and the expression of antimicrobial peptides (Hill et al., 2013). Thus, it appears that the action is not direct. Adding this glycosaminoglycan *in vitro,* to a petri dish of bacteria, would not have produced the important peptide response that would have existed *in vivo.* Similar indirect, but essential action, has been found from seaweed metabolites in animal guts and in the digestive world of plants, the soil (Khan et al., 2012). Here seaweed polysaccharides have worked as prebiotics or elicitors of metabolic pathways that protect both plants and animals from disease (for examples or reviews see Craigie, 2010; Jaulneau et al., 2010; O'Sullivan et al., 2010).

The roles of SPSs across all lifeforms, including seaweeds, are diverse. They append and stabilize most of the other complex biological molecules such as proteins (glycoproteins) and lipids (glycolipids), functioning as the pilots or the keys in molecular connections. Such functioning is important to protect against viral and bacterial enzymes such as proteases, cellulases and glycosidases from degrading proteins or other carbohydrates. Most immune system-related molecules are glycosylated, have SPS attached, and glycans are an essential part of physiological systems. Hence the development of recent vaccines that function by exploiting glycosignalling pathways against viruses such as the flu (Alper, 2001a,b) and against dengue fever (Hidari, Abe, & Suzuki, 2013). Further, genetic mutations in humans where normal carbohydrate synthesis pathways are lacking have been identified as the cause behind a number of serious disorders (Alper, 2001a,b). Further research addressed the role of carbohydrate polymers in autoimmune diseases, Rudd, Elliott, Cresswell, Wilson, and Dwek (2001). Microorganisms, normal cell communication, cancer cells, pathogens and symbionts all evolved in a world of diverse carbohydrate signalling. However, a challenge has been to understand the complexity of size and structure in these molecules. Access to large enough quantities of such polysaccharides to drive this field of medical research has been limiting (Alper, 2001a,b). This is where the controlled production of seaweeds, and a quality controlled and traceable vertical chain of development to molecular structures, can contribute to this important field of research.

8.2 SULFATED POLYSACCHARIDES

8.2.1 Function in Seaweeds

The roles of polysaccharides in seaweeds are diverse and complex. These large molecules, at up to 1000s kD (Kraan, 2012) are particularly abundant and diverse in marine seaweeds (Lobban & Harrison, 2000). Energy storage in carbohydrate form is of course one strategy common to most lifeforms and intracellular storage is observed in seaweeds (Gomez & Huovinen, 2012). Such storage molecules can be similar to those in land plants such as starch, especially for green seaweeds, but also unique molecules such as laminarin exist in some brown seaweeds. Cell energy storage forms reflect the ambient growing and nutritional conditions for seaweeds, which can transfer carbohydrate storage to lipid forms when nutrition becomes limited (Kumari, Kumar, Reddy, & Jha, 2013). Therefore the relative proportions of storage carbohydrates are generally inverse to that of other storage such as lipids. These storage polysaccharides will also vary in molecular weight for osmotic function; as for small molecules there is a higher osmotic potential than for larger molecules (Lobban & Harrison, 2000). These components of seaweeds can fluctuate widely according to season or environmental conditions. A key example of such fluctuation is that of cold water *Laminaria* spp., which builds up the brown storage forms of carbohydrates, laminarin, in the summer, and survives through darker winter to a large degree by utilizing this stored energy (Gomez & Huovinen, 2012).

Structure is of course a core requirement for the strength and morphology of the holdfasts and thalli of seaweeds; thus cellulose is a component of seaweed molecular structure; however, it is a much lower percentage compared to terrestrial plants where resistance to gravity requires a larger proportion of structural carbohydrates. Seaweeds instead require flexibility, and in this way alginates are important in the kelps for rigidity that is flexible depending on the ionic composition of the water (Smidsrød & Grasdalen, 1984). Similarly, the degree of sulfation of polysaccharide molecules delivers variable flexibility and gel strength for red seaweeds (Villanueva, Montaño, Romero, Aliganga, & Enriquez, 2005). In addition, uronic acid components are known for their adhesive properties in different parts of seaweed extracellular matrix regions and provide cohesion (Lobban & Harrison, 2000).

Further and promising roles of seaweed sulfated polysaccharides are functional ones, beyond structure and energy storage. As for the structural and storage polysaccharides, there are higher taxonomic specific polysaccharides (Table 8.2) that have other functions, and broadly these are components of the extracellular matrix of seaweeds and exist side by side with purely structural, fibrillar ones (Lobban & Harrison, 2000). It is thought that this charged network of polysaccharides, in some ways, is the functional equivalent to the vascular network of higher order plants in keeping the cells hydrated and neutralizing the chemical potential (charge) of water (Smidsrød & Grasdalen, 1984). For example, specific positions of sulfation and iduronic acid containing sequences of SPS control the anticoagulant properties in animals, and there would be similar function/structure requirements in seaweeds (Chen & Liu, 2005). Some of these functional polysaccharide groups are well known, have substantial work completed on molecular structure and are produced at a scale of commodity such as alginate, agar and carrageenan.

More recent SPS categories of interest include fucoidan and ulvan (Jiao, Yu, Zang, & Ewart, 2011; Wijesekara, Pangestuti, & Kim, 2011); where there is increasing evidence for a diverse range of health benefits linked to a diversity of detailed differences in the molecular weights, sugar composition and order, and degree of sulfation; and how these slight differences are triggers for very specific metabolic and cellular function.

Table 8.2 Diverse Forms of Other Functional Carbohydrates Across the Different Groups (Red, Green and Brown) of Seaweeds, Illustrating Higher Taxonomic Categories of Carbohydrates, but also Lower Taxonomic Level Variation Within These Categories

	Polysaccharide categories	Repeating units
Intracellular storage	floridean starch	glucopyranose, amylose
	laminaran	glucopyranose
	starch	glucose
Extracellular matrix	carageenans (galactans) cellulose mannans xylans	glucose, mannose, xylose, galactopyranose, glucuronic acid,
	alginic acid fucoidan	fucose, uronic acids, xylose, glucose, galactose, guluronic acids
	ulvans, cellulose, xyloglucan	rhamnose, glucose, galactose, mannose, xylan, uronic acids (glucuronic and iduronic)

Adapted from the literature including Usov and Zelinsky (2013).

However, as indicated by the publication rates and focus in the medical literature vs phycological literature, it is sometimes less well understood how these molecules function in the seaweeds (Lobban & Harrison, 2000) compared to how they function out of seaweeds in other biological systems. The development of extra-seaweed applications (for example patents for triggering plant defences (Briand, Cluzet, Dumas, Esquerre-Tugaye, & Salamagne, 2010)) might indicate the type of roles that these SPS have in seaweeds. Specifically, these seaweed molecules are often considered to function as Microbial Associated Molecular Patterns (MAMP) (Jaulneau et al., 2010) that can elicit specific function in relation to a range of microbial interactions at the molecular level. This is an important function throughout the evolution of seaweeds, which have to constantly evolve to protect themselves against biological and chemical attacks from the water (Rindi, Sloer-Vila, & Guiry, 2012). In addition to single-celled organism interactions, such molecules are also important in cellular and molecular communication in the multicelled organisms.

8.2.2 Function in Anthropogenic Applications

The primary and well-established industry of seaweed polysaccharides as a functional ingredient in foods is astonishingly poorly appreciated in the west, despite being one of the most commonly consumed emulsifying and textural additives. Such applications in food have been well reviewed elsewhere (Venugopal, 2011), and applications also extend to a broader hydrocolloid industry where seaweed sulfated polysaccharides contribute to the spray and fluidity of products such as fire retardants, paints and printer ink. This sector is undergoing some decline in growth due to a range of factors as reviewed by Bixler and Porse (2011); however, these challenges need to be addressed on a range of fronts to deliver continuation of socioeconomic progress of the south east Asian nations that rely on this industry. This includes increasing the genetic diversity of existing seaweed crop stocks, diversification of species and product development (Robinson, Kirkendale, & Winberg, 2012).

Recent developments in chemical engineering and molecular technology are elucidating quite specific and high-end applications of seaweed sulfated polysaccharides. The breadth of cause-and-effect relationships being established between specific polysaccharides structures and other biological systems is expanding rapidly. The challenge is to embrace these biotechnological tools, but also to focus on commercially relevant applications and to deliver evidence of efficacy (Chan, Ho, & Phang, 2006). For example, animal and agricultural applications are well established (Craigie, 2010; Khan et al., 2009) and the specificity of seaweed polysaccharide molecular structure for applications in crop diseases continues to grow (see Paulert, Ebbinghaus, Urlass, and Moerschbacher (2010); Subramanian et al. (2011); Wally et al. (2012) for examples). Animal health studies are now abundant and seaweed provides a regular component of animal feeds (Allen et al., 2001), especially with a focus on gut health and consequent general wellbeing (O'Sullivan et al., 2010) as well as the reduction in use of subtherapeutic antibiotics. Holdt & Kraan provide a substantial review of the nutritional value and metabolic profiles of diverse species. Kraan (2012) overviewed diverse applications but particularly medical and health research. Of note is that Kraan's review is one of the two, addressing the role of algal carbohydrates in a book reviewing the broad and growing field of glycobiology and glycotechnology (Chang, 2012). Thus seaweeds and algae are an important part of this growing field of research. Further to the glycobiology and the crowd-molecules concept, algal lectins (a glycoprotein) in glycobiological research are targeting diseases based on antiviral, antinociceptive, anti-inflammatory, antitumor activities, and for fabrication of protein expression systems and nutraceuticals (Singh, Thakur, & Bansal, 2013).

The sulfated polysaccharide functional molecular traits across a range of medical applications are diverse, and sometimes one particular functional application can have the inverse molecular trait requirements of another. For example, the polysaccharides of many species of brown seaweeds have been identified to have blood anticoagulant properties; a property that is positively correlated, with, and can be enhanced by, the degree of molecular sulfation (Shanmugam & Mody, 2000). This makes sense as a more charged molecule will increase fluidity, which has been well established for the different forms of carrageenan (Silva, Basson, & Moe, 1996). Lower degrees of sulfation increase the viscosity and thus functional role in the gut; however, it might also reduce some of the chemical signalling function that the degree of sulfation imparts (Leiro, Castro, Arranz, & Lamas, 2007). However, this effect can also be achieved with large molecular weight as reviewed by Kraan (2012), which generally increases the viscosity of soluble dietary fibres. Higher molecular weight and viscous dietary fibres have a role in the regulation of metabolic disorders involving glucose and cholesterol imbalances (Chen, Li, & Yu, 2008; Liu, Wu, Mao, Wu, & Ouyang, 2010).

SPSs have also been shown to block the progression of viral infections (Witvrouw & De Clerck, 1997) by binding to the positively charged glycoprotein of the viral envelope, thus interfering with viral attachment to heparin-type mammalian host cell-wall antigens. Such pathogen (viral) resistance linked to seaweed SPSs has been demonstrated in numerous studies, and considering the long evolutionary timeline of hosts and pathogens in the marine environment, we may assume SPS evolution and pathogen response has evolved multiple times, in multiple forms with diverse specificity across many species of seaweeds.

It has been suggested that the composition of the sugars in SPSs might only function as a structural backbone to design the pattern of active sulfation, and antiviral effects have been shown to increase with increase molecular weight and the degree of sulfation (Witvrouw & De Clerck, 1997). However, specific components such as glucuronic acid have been shown to be essential to other functions (Hallak, Collins, Knudson, & Peeples, 2000) and reduction in lower molecular weight fractions have been shown to have increased\anti-oxidant activity (Qi et al., 2005a, 2005b).

More recently and under controlled conditions, seaweed SPS are being used to look for not only cell compatible components of drug delivery systems and biological implants (Silva et al., 2012), but also cell recognition, developmental triggers and protective properties (Hanbo, Xuan, Bin, & Chunlan, 2014). There are these plus many more applications that are seeking biologically and environmentally friendly, functional molecular building blocks, provided by a long trajectory of evolution, where seaweed SPS have an important role in a matrix blend or as a dopant. For these types of applications, there are multiple aspects of the polysaccharides that are of interest. First the actual gelling consistency and containment of bioactive ingredients for cells is important. The slow release or regulation of compounds from gels can be controlled by a number of factors such as molecular weights, cations, temperature and pH (Hanbo et al., 2014). Further, cell communication within a cellular matrix is important and it is known that mammalian glycosaminoglycans modulate the adhesion of stem cells, their subsequent differentiation and gene expression. Thus sulfated polysaccharides can be used as doping agents in a blend of other matrix polysaccharides and mimic the heterogeneous polymer environment that naturally occurs. Recent work demonstrates that cell and biological protein recognition and cell differentiation are all important activities that seaweed polysaccharides can impart. In particular to cell communication systems and behaviour are molecular traits such as the degree of sulfation, molecular weight and structural composition. These traits have very specific cell recognition function in biological systems, with changes being observed in different physiological but also pathological processes.

Thus, across the applications of textural and flexible materials, animal, agricultural and human biological systems, there are a number of key molecular traits of seaweed SPS that need to be considered. This is broadly published and has been well reviewed by Jiao et al. (2011). However, the complexity of the topic is evident and there is a huge amount of work to be done before the opportunities to apply this emerging field of phycoglycobiology can be realised. Attributing to this complexity are several dominant key factors of glycomolecules that affect biological applications including:

8.2.2.1 Molecular Weight

The molecular weight of matrix polysaccharides can be very large, and they are very polydisperse. For example the molecular weight range of ulvan has been reported to range from the 100s kDa to over 8000 kDa (Lahaye, 1998), although for some published ranges it is thought that these might be confounded by agglomeration properties for some of the matrix polysaccharides and the molecular weights are in the 100s kDa (Paradossi, Cavalieri, & Chiessi, 2002). The gelling capacity and broad range of strengths are known to increase with molecular weight from a certain point (e.g. few gels form with molecular weights smaller than 70 kDa) and to a maximum weight (e.g. larger than 500 kDa) (Smidsrød & Grasdalen, 1984). Then, added to this, temperature, ionic environment, presence of metals, pH and the diverse branching or charged structures further affect gelling capacity and strength. Thus, there is an incredible breadth of diverse mechanical properties of matrix polysaccharides that has not yet been fully appreciated, and is contributing to the growing sector of intelligent polymer and biomaterials research.

Aside from the physical gelling properties, the biological function is of increasing interest and value. Low molecular weight molecules have shown higher anti-oxidant activities (Qi et al., 2005a, 2005b), while in contrast high molecular weights have been associated with higher antiviral activities (Holtkamp, Kelly, Ulber, & Lang, 2009). Recent work has shown that the higher molecular weights are essential in the cell protein recognition ability of gels for medical culture (Hanbo et al., 2014).

Although we have highlighted the consistency of sulfated polysaccharides within species, that structural consistency is stratified into a range of molecular weight fractions with their own composition. So what the actual large macromolecules are made up of in terms of sugar units differs across species (Ale, Mikkelsen, & Meyer, 2011), but also fractions within species (for example fucoidan diversity within species fractions (Chizhov et al., 1999)). This has substantial consequences in biological applications as reviewed by Fitton (2011). For example, an arthritis study by Park et al. (2010) showed opposing effects of high molecular weight and low molecular weight fractions of the same fucoidan extract. The high molecular weight fraction activated the inflammatory process, whereas the low molecular weight fraction inhibited the disease.

8.2.2.2 Monomer Composition

It is apparent that seaweeds have specific polysaccharide composition that triggers unique biofunctionality. For example, uronic acids are important factors in the binding of certain viruses. Glucuronic acid was demonstrated to be a very specific cell membrane molecule used by respiratory viruses for cell invasion; and consequently, incubation of cells with iduronic acid could prevent this invasion by blocking the viral glucuronic acid receptors (Hallak et al., 2000). Ulvan for example is unique among algae in its glucuronic and iduronuic acid composition. Further to these acidic dimer carbohydrates, seaweeds species feature unique diversity of neutral carbohydrate monomers such as fucose, xylose, galactose, rhamnose and glucose (Table 8.2). This diversity offers a wide range of molecular building blocks for targeted applications. These structures are also similar to human and animal glycosaminoglycans, such as heparin, heparin sulfate and chondroitin sulfate (Hallak et al., 2000).

8.2.2.3 Molecular Branching

In addition to monomer composition, the polysaccharides can be branched or unbranched providing increased solubility and altered bioactivity with branching (Rupérez, Ahrazem, & Leal, 2002).

8.2.2.4 Degree of Sulfation or Alternative Substituting Branch

It is known that sulfation profile of extracellular matrix carrageenan can shift across seasons (Villanueva et al., 2005); however, whether this is due to an active sulfation and sulfation replacement activity or whether it is simply the characteristic of seasonal biomass growth that shifts this profile is not fully understood yet. The degree of sulfation is deemed to be very important to many biological activities (Zhao, Li, Xue, & Sun, 2011). In fact, the degree of sulfation seems to play such important role that recent efforts have been focusing on the preparation and further use of highly sulfated glycosaminoglycans derivatives, in particular by introducing sulfate features in hyaluronic acid of increasing sulfate degree of chondroitin sulfate. This work has been extended to seaweed molecules (Arlov, 2012).

In summary, specific activities can be controlled by a range of molecular traits, and inversely, specific molecular traits contribute to a range of biological activities. For example, low molecular weights, increased degree of sulfation as well as the uronic acid content all appear to increase antioxidant activity in the case of Ulvan (Qi et al., 2005a, 2005b; Redouan E., Emmanuel P., Christine B., et al., 2010; Redouan E., Emmanuel P., Michelle P., et al., 2010). However, targeting bioactivity requires that these traits are increasingly consistent and controlled to maintain consistency in the activity. In the absence of consistent and controlled production of many of the heteropolysaccharides from seaweeds, effort is being made to alter and optimise the chemical structures of commercially scaled and more consistent polysaccharides such as the alginates, to deliver heparin analogues for nonanimal medical applications. However, understanding the factors that influence the variability of seaweed sulfated polysaccharides will contribute to the development of integrated, traceable and quality-controlled production systems that can deliver more consistent and active products.

8.3 CONTROL IN PRODUCTION

8.3.1 Low-Level Taxonomy

Generally, the selection of green, brown or red seaweed species provides a simple and clear categorisation of the monomer composition of polysaccharides (Table 8.2). For example, brown seaweeds rich in alginate and fucoidan uniquely contain fucose, red seaweeds and carrageenan contain galactose, and green seaweeds contain rhamnose as well as some other common and specific oligosaccharides and uronic acids, hence the sourcing of carrageen from red seaweeds and alginate from brown seaweeds. Thus, the consistency composition exists to a degree at this taxonomic level and has been exploited commercially very successfully in the hydrocolloid industry. However, industry research has established clearly that the diversity of the relative monomer abundances, molecular organisation and additional components, such as sulfation, will vary within genera and species (Ale et al., 2011).

8.3.2 Species Level Taxonomy

There is a perception that seaweed polysaccharides are extremely heterogeneous in size and structure, particularly for the green seaweeds (Jiao et al., 2011). However, some of this heterogeneity arises where chemists have not been resourced with phycological taxonomists to qualify species. This has been a major impediment to the further development of polysaccharide molecules from seaweeds (Alves, A., Sousa, R., & Reis, 2012). For example, Kirkendale, Saunders, and Winberg (2013) showed that for publications relating to the genus *Ulva* in Australia, 30% identified a common species. However, in surveys using molecular barcode markers to determine species, this apparently common species was not identified from any samples and it can be assumed that many previous identifications that did not apply molecular markers were incorrect. However, the taxonomy of seaweeds has taken great leaps forward since the development of these molecular tools which should now be an essential component of any seaweed bioactive's development programme (Rindi et al., 2012).

The importance of accurate species and even strain identification has been shown in a number of studies dating back to the beginnings of the hydrocolloid industry (Chopin, Hanisak, Koehn, & Mollion, 1990). Of importance to the polysaccharides of seaweeds is that the polysaccharides are broadly consistent within, and specific to species (Ale et al., 2011). Although the polysaccharide categories such as ulvan or fucoidan are often discussed as if they were a consistent composition, the polysaccharides of ulvan and fucoidan are unique to and within each species (Ale et al., 2011; Paradossi, Cavalieri, Pizzoferrato, & Liquori, 1999; Robic, Sassi, Dion, Lerat, & Lahaye, 2009). We have tested this finding on an increased diversity of *Ulva* species, sampling times and conditions; qualified by molecular barcoding taxonomy (Kirkendale, Saunders, & Winberg, 2013). The results lend strong support to the context that despite various environmental conditions, location and time, key molecular traits of seaweed sulfated polysaccharides remain consistent to and within species (Figure 8.2). Thus compared to other metabolites that can be quite variable and respond rapidly to environmental shifts, the extracellular matrix polysaccharides are relatively consistent.

8.3.3 Life History

The control of life history stages to deliver increased consistency of the sulfated polysaccharide composition and functional properties was well established and commercialised during the development of carrageen from

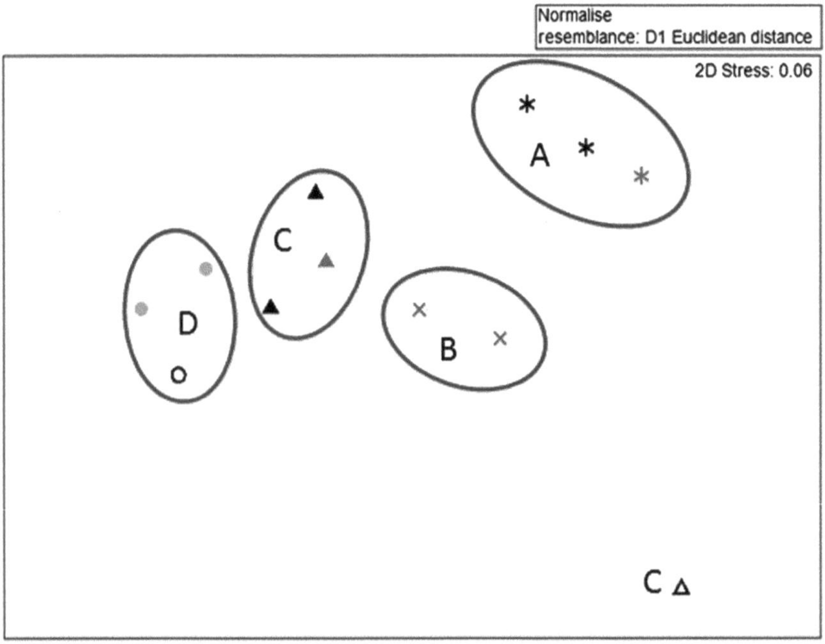

Figure 8.2 Concurrent and consistent extraction methods of four species of ulvan biomass, where each species was represented by samples from different locations or wild vs cultivated biomass. Species consistent traits of molecular weight range, monomer ratios and uronic acid composition and yields were consistent to species despite the diverse sources (unpublished data). Ulvan A was extracted from three samples of wild harvest ($n=2$) and cultivated biomass ($n=1$), ulvan B was extracted from biomass of species B from two locations within 10 km, ulvan C was extracted from three samples of species C within 10 km and across 1000 km distances, while the empty triangle represented an aged sample of species C with reduced molecular weight after two years of storage, and ulvan D represented two samples of wild biomass from the same location at different times as well as one biomass that had been spored in culture.

Chondrus crispus in the last century (Chen et al., 1975; Moseley, 1990). It became evident that the male and female biomass as well as the diploid sporophytes contain different ratios of the various categories of carrageen with different degrees of sulfation and structure (Figure 8.3). Not all seaweeds have the same types of distinct life history stages or differences in composition across them. However, green seaweeds with distinct life histories, like the reds, have also been shown to have distinct isomer forms of the structural types of monomers across the life stages (Figure 8.3) (Gordon-Mills & McCandless, 1975).

For seaweed species that have evolved a higher degree of tissue specialisation, for example in many of the brown species with distinct holdfasts, stipes, thalli and reproductive components, the content or composition of the sulfated polysaccharides can also differ across the specialised morphologies. For example, the reproductive structure, sporophyll, of *Undaria* species is known and marketed separately, as mekabu, to the rest of the seaweed biomass. This extends to the extraction and marketing for sulfated polysaccharides from this seaweed (Kim et al., 2010). Similar to this, the age of the seaweed influences the types of extracellular matrix that is developed (Skriptsova, Shevchenko, Zvyagintseva, & Imbs, 2010). Thus, there can be a sort of ripening or optimal time for harvest based on the life history and age of the seaweed, as well as the harvested component of the seaweed.

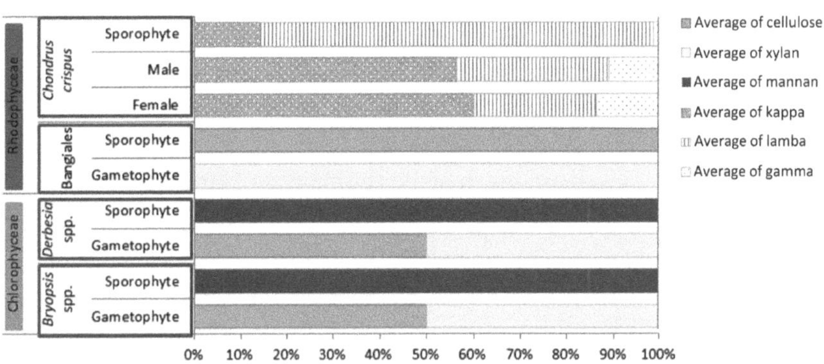

Figure 8.3 This relative and qualitative composition of extracellular matrix seaweed polysaccharides as found in the peer-reviewed literature does not provide a definitive generalisation of class or lower taxonomic order of polysaccharide composition. However, it does serve to illustrate that despite species consistency of polysaccharide composition as indicated above, the different life stages contribute an added dimension of diversity that needs to be considered for certain species. *Data edited from Lobban and Harrison (2000) and Chen et al. (1975).*

However, this is only relevant for species that have developed such specialisation, and many species can be cultivated where there is little differentiation of morphology across reproductive stages or structures.

Adding to this complexity and warranty, further consideration is a recent evidence that some seaweed-associated metabolites are not necessarily a product of the seaweed themselves, but rather from symbionts of seaweeds (Miranda, Hutchison, Grossman, & Brawley, 2013; Yung, Burke, Lewis, Kjelleberg, & Thomas, 2011). Increasingly, we are being made aware that biodiversity operates at a much smaller scale than at the organismal level, and that microbial ecology or even chemical ecology is the driving force behind biological systems. For example, it is known that some seaweeds species require specific microorganisms on their surface to develop typical morphologies (as reviewed by Miranda, Hutchison, Grossman, & Brawley, 2013), and also to function as a defence against pathogenic attack. This symbiotic relationship has demonstrated that effectively, species of seaweeds are hosts to a species specific and unique set of microflora that are often the source of bioactivity that functions both in the metabolism of the seaweeds and potentially in a clinical or other applied anthropogenic application (Yung, Burke, Lewis, Kjelleberg, & Thomas, 2011).

8.3.4 Environment/Cultivation

It is well established that the storage polysaccharides fluctuate widely in terms of yields depending on the season and environment as discussed above, as well as some composition based on age and development stage of the seaweeds. However, for extracellular-matrix polysaccharides, there has been little evidence that much can be changed about these once they have been placed in the matrix (Lobban & Harrison, 2000). It is evident that the yield can shift, usually increasing per gram of tissue due to seasons and/or nutritional starvation (Chopin, Hanisak, Koehn, & Mollion, 1990; Moseley, 1990; Robic et al., 2009). However, this is also a consequence of a relative reduction in intracellular storage molecules of other carbohydrates and lipids or proteins, in relation to a more stable extracellular matrix. Few species have been shown to shift the monomer structure of extracellular polysaccharides; however, it is apparent that the degree of sulfation is adaptable in the extracellular matrix (Carmona, Vergara, Lahaye, & Niell, 1998) and we know that the rigidity of alginates for example depends on the anionic state of the water (Smidsrød & Grasdalen, 1984).

8.3.5 Extraction of Polysaccharides

The diversity of extraction processes and technologies of chemical engineering for diverse sources of biomass has expanded. These techniques are now being adopted for seaweed biomass, and the sequential biorefinery concept to extract a palette of seaweed compounds is growing to increase the value of existing biomass (Francavilla et al., 2013) and also to realise diverse values from new sources (Lorbeer et al., 2013). However, the biorefinery process has a middle step of handling between the source and the processing facilities. The treatment of biomass directly from harvest until extraction has been shown to affect the quality and yields of polysaccharides (Robic, Sassi, Sassi, & Lahaye, 2008). Thus, consistent and quality-controlled handling procedures are essential in achieving high-value polysaccharides.

The methods of extraction have been shown to deliver different results of product but also the integrity of the molecules. Ale et al. (2011) point out that in many instances, the integrity of the molecules is damaged and therefore the bioactivity as well. Thus, extraction method and standards are just as vital to the selection of species in delivering bioactive product. Therefore, high standards of quality control and traceability are required to maintain the bioactivity of sulfated polysaccharide products (Ale et al., 2011).

Further to extraction process and size fraction selection, post-processing of molecules to increase the degree of sulfation has been developed (Arlov, 2012). This can deliver increased degrees of sulfation, which as described above can deliver desirable outcomes for bioactivity (Ale et al., 2011). Conversely, desulfation has also been developed to achieve polysaccharide function that improves with less sulfated product (Navarro et al., 2007).

Finally, reducing the molecular weight while maintaining active sulfation and integrity of shorter chains has been developed by enzymatic processes aided by the discovery of novel enzymes (Sassi et al., 2011). It has been established that many polysaccharides evolved to provide resistance to degradation (Bobin-Dubigeon, Lahaye, Guillon, Barry, & Gallant, 1997). However, in the contrast to interpretation that such polysaccharides are not degradable (O'Sullivan et al., 2010), it more correct to consider this as resistance to common enzymes and a matter of not having found the more active enzymes; many of which exist in marine organisms. Marine bacteria that have been exposed to these polysaccharides for a long time have evolved enzymes to cleave them and are found in the gut of marine herbivores. Over time, such enzymes are also inherited in the human gut from the ingestion of seaweeds with bacteria that can degrade seaweed polymers (Hehemann et al., 2010). However, the differences in molecular weights of the natural sources of polysaccharides within a species are not simply shorter and longer chains of the same repeating monosaccharide patterns; they are in fact uniquely structured molecules at different sizes. For example, lower molecular weight fractions of fucoidan have different oligosaccharide ratios to higher molecular weight ones (Holtkamp et al., 2009). Therefore, the selection of natural fractions is important in consistency of bioactive function, and in addition to synthetic fractionation.

8.4 DISCUSSION

Following on from this review, a simple categorical statement can be made that there are clear distinctions between the categories of polysaccharide metabolites between red, green and brown seaweeds. These distinctions these have been successfully commercialised during the last century to deliver important industrial compounds in the food and fluid product industries that we enjoy today. However, we are looking at a world that goes beyond the mechanical hydrogel industries of agar, alginate and carrageen, although those industries have paved the way and will provide the solid foundations upon which to explore the immense diversity and applications in biologically functional applications. The worlds of seaweed colloidal chemistry and glycobiology are merging and the extensive expertise within each field will drive rapid progress into new and unexplored aspects of global and human health through applications in food production, human nutrition and food security, as well as medical and intelligent materials science.

Translating the chemical ecology of seaweeds to practical efficacy in an anthropogenic setting includes translating fundamental scientific knowledge into commercially relevant and practical solutions. Such fundamental knowledge includes understanding biological life histories, taxonomic specificity of metabolic processes and metabolites, appreciating the role of chemical ecology and the environmental influences on seaweed metabolites and applying this knowledge into cultivation systems. Commercial relevance on the other hand needs to address the challenge of scaling production systems as well as taking evidence of functionality, including clinical or intervention trials to markets. It is particularly challenging to address the need for evidence-based clinical trials, using seaweeds and extracts in competition with traditional nutritional and health research, when there is such complexity, diversity and variability of the source and the applications.

However, in terms of emerging trends and a focus of interest, the sulfated polysaccharides stand out as the unique and continuing feature of seaweeds that are of interest, especially for health and medical applications and thus higher value products. This is concurrent with the frontier field of glycobiological research; however, it also demonstrates that the diversity of seaweed-sulfated polysaccharides needs to be tamed and controlled with a focus on the application that is to be targeted.

The range of sulfated polysaccharides within seaweeds have evolved in a race to support species against the marine elements, in competition with other species, to protect against herbivory, and to protect against pathogens (Lobban & Harrison, 2000); not to mention the intraorganismal cellular communication systems. Thus, control of the production of sulfated polysaccharides from seaweeds must start at the genetic level with control of the species. There is pattern and consistency in the heteropolysaccharide diversity that can be controlled to a large degree by increasing species-specific production, as well as life history control. To a large extent, this can be controlled in cultivation over wild harvest, especially in terms of the capacity to scale up the consistency of these bioactive molecules.

Then the consequent controls over the environment, processing, handling and extraction, eventual chemical engineering and consistent evidence in application are then needed to achieve the potential of the many promised bioactive applications. The sophistications of traceable and reliable chains of seaweed biomass production for high value markets are not well established. However, having identified the components of control and traceability, this should contribute to progressing this agenda (Figure 8.4). Quality control systems are already at a stage of proof of concept and will be vital in the long-term respect for efficacy of these natural products for functional applications. For example, ulvan composition and structure quality control methods have been developed (Robic, Sassi, Dion, Lerat, & Lahaye, 2009) and will be most important if biomass is sourced from the wild, and/or to qualify that high value cultivated biomass meets the standards for pharmaceutical or other high-grade products.

Of importance is the need for consistent and traceable sources of seaweed polysaccharides to be applied into clinical trials; something that is sorely lacking. Despite the increasing number of publications in medical literature on the bioactivity of seaweeds, there were only 26 human clinical trials related reported through MEDLINE in the last 2 decades (Table 8.3), prior to which there was only a couple of decades of research on reintroducing *Laminaria* as a tent, dilating device, in medical procedures which

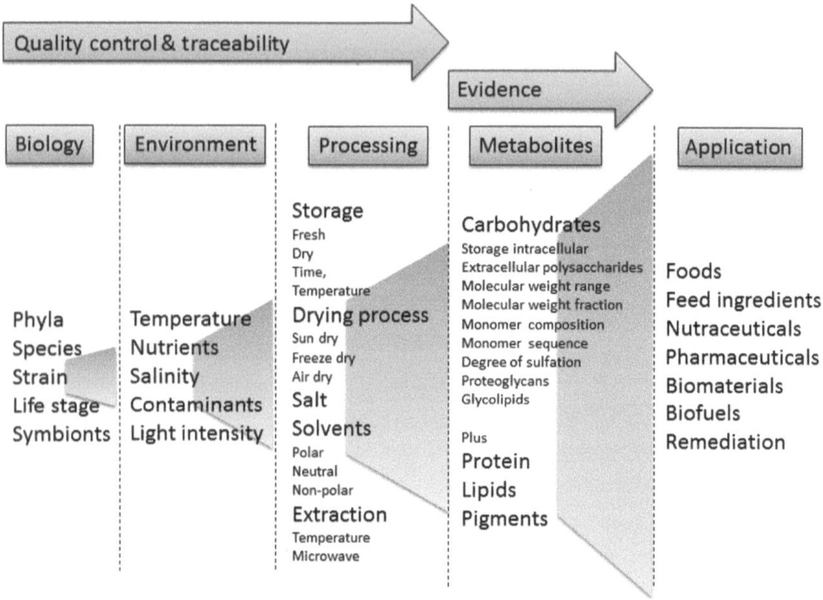

Figure 8.4 The expanding diversity of seaweed metabolite content, form and function as a result of the vertically integrated processes that will deliver an extraordinary plethora of human applications.

were later phased out. Thus, the urgency of the opportunities that seaweed biomass cultivation can contribute to requires investment in human intervention trials alongside other phytochemicals (Traka & Mithen, 2011).

There is an increasing interest to deliver more natural products from sulfated polysaccharides; and in the absence of consistency being delivered from the diverse metabolites in seaweeds, there are efforts to circumvent this issue of heterogeneity of seaweed SPS by sourcing marine SPS from more consistent molecular sources such as invertebrates (Pomin, 2009). The field of glycobiological research will continue to seek sources of these SPS and this may occur with a degree of synthetic manipulation from existing seaweed biomass (Arlov, 2012), the manufacturing of analogues from microbial production of polysaccharides (Redouan E., Emmanuel P., Christine B., et al., 2010; Redouan E., Emmanuel P., Michelle P., et al., 2010) or simply a synthetic chemistry approach (Sears & Wong, 2001). As the medical world of glycobiology is seeking sources and molecular diversity in form and function of SPS, a synthetic chemistry approach has been adopted to access new SPS. However, evolutionary history has already provided for diversity and abundance of SPS molecules in seaweeds that exceeds most other options as a viable source. Seaweed biomass production rates, the

Table 8.3 Summary of Clinical Trials that Involve Seaweed, as Found in the Journal Database MEDLINE

Year	Group	Genus	Species	Activity	Component	Dose	Comments
(Abe et al., 2013)	Brown	*Cladosiphon*		Anticholesterol	Fucoidan	4 g daily for 2 weeks	Safe high dose ingestion & favourable change in cholesterol
(Ruxton & Jenkins, 2013)	Brown	*Laminaria*	*digitata*	Anti-acne	Oligosaccharides		Reduced severity of acne
(Mori et al., 2012)	Brown	*Cladosiphon*		Antiviral	Fucoidan	0.83 g daily for 12 months	Safe and potential for use with antiviral hepatitis A
(Paradis et al., 2011)	Brown	*Fucus* *Ascophyllum*	*vesiculosis* *digitata*	Antidiabetic	Whole	500 mg	Increased insulin sensitivity
(Fazekas et al., 2012)	Red			Antiviral	Iota-carageenan		Reduced viral load and longevity of flu symptoms
(Myers, 2011)	Brown	*Fucus*	*vesiculosis*	Immunomodulation	Fucoidan	100/1000 mg daily for 12 weeks	Increased macrophage activity
(Ikeguchi, 2011)	Brown	*Cladosiphon*		Antitoxicity	Fucoidan	4 g daily	Reduced fatigue and increased chemo. cycles
(Araya, 2011)	Brown	*Cladosiphon*		Antiviral	Fucoidan	6 g daily for 6–13 months	Lowered proviral load in patients. No disease progression
(Teas et al., 2011)	Brown	*Alaria*	*esculenta*	Metabolic regulation	Whole	5 g	Reduced breast cancer associated growth hormone triggered by soy

Continued

Table 8.3 Summary of Clinical Trials that Involve Seaweed, as Found in the Journal Database MEDLINE—cont'd

Year	Group	Genus	Species	Activity	Component	Dose	Comments
(Myers et al., 2010)	Brown	*Laminaria*	*japonica*	Joint health	Fucoidan	100/1000 mg	Phase I & II reduced symptoms of osteo-arthritis
(Irhimeh et al., 2009)	Brown	*Fucus*	*vesiculosis*	Joint health	Fucoidan	100/1000 mg	"
	Brown	*Macrocystis*	*pyrifera*	Joint health	Fucoidan	100/1000 mg	"
	Brown	*Undaria*	*pinnatifida*	Anticoagulant	Fucoidan	3 g daily for 12 days	
(Miyai et al., 2008)	Brown	*Laminaria*	*japonica*	Thyroid function	Whole		>15 g dry kombu supressed thyroid function
(Irhimeh et al., 2007)	Brown	*Undaria*	*pinnatifida*	Cell mobility	Fucoidan	3 g daily for 12 days	Increased expression of CXCR4 on human CD34+ cells
(Juffrie et al., 2006)	Brown	*Cladosiphon*	*okanimuras*	Gut	Fucoidan	100 mg Daily for 3 weeks	Promotes gastric ulcer healing through fibroblast epidermal growth factors
(Skibola, 2004)	Brown	*Fucus*		Endocrine regulation	Whole	700 and 1400 mg/day	Reduced oestrogen, prolonged menstrual cycle
(Cooper et al., 2002)	Brown	*Undaria*	*pinnatifida*	Antiviral	Whole	560 mg/day	Reduced healing time for herpes symptoms
(Ladodo et al., 2002)	Brown			Nutritional	Whole		Russian article–addressed iodine deficiency in children

(Aquaron et al., 2002)	Brown	*Laminaria*	*hyperborea*	Nutritional	Whole		Overcame iodine deficiency
	Red	*Gracilaria*	*verrucosa*	Nutritional	Whole		Overcame iodine deficiency
(Fujimura et al., 2002)	Brown	*Fucus*	*vesiculosis*	Skin	Fucoidan	1% formulation	Reduced skin thickness and increased elasticity
(Fujita et al., 2000)				Osteoarthritis			Increased bone density
(Ohgitani et al., 1997)				Osteoarthritis			Increased calcium uptake
(Fujita, Ohue, Fujii, Miyauchi, & Takagi, 1996)				Osteoarthritis			Increased calcium uptake
(Krotkiewski et al., 1991)				Hypertension		6, 12 & 24 g/day	12 & 24 g reduced hypertension

high yields of polysaccharides and the opportunity that diversity presents are all reasons to try and increase the control of seaweed polysaccharides in the vertical chain of production as the source for biomedical and other glycobiological applications. Further to this are the sustainability, nutritional security and socioeconomic development opportunities that are features of seaweed production industries; all aspects of future industries that the world is looking for.

REFERENCES

Abe, S., Hiramatsu, K., Ichikawa, O., Kawamoto, H., Kasagi, T., Miki, Y., et al. (2013). Safety evaluation of excessive ingestion of mozuku fucoidan in human. *Journal of Food Science 78*(4), T648–T651.

Ale, M.T., Mikkelsen, J. D., & Meyer, A. S. (2011). Important determinants for fucoidan bioactivity: a critical review of structure-function relations and extraction methods for fucose-containing sulfated polysaccharides from brown seaweeds. *Marine Drugs*, *9*, 2106–2130.

Alexander, A. (2006). *The companion to tasmanian history*. Hobart, Center for Aboriginal Studies, Univeristy of Tasmania http://www.utas.edu.au/library/companion_to_tasmanian_history.

Allen, V. G., Pond, K. R., Saker, K. E., Fontenot, J. P., Bagley, C. P., Ivy, R. L., et al. (2001). Tasco: Influence of a brown seaweed on antioxidants in forages and livestock–a review. *Journal of Animal Science*, *79*, E21–E31.

Alper, J. (2001a). Saving lives with sugar. *Science*, *291*(5512), 2339.

Alper, J. (2001b). Searching for medicine's sweet spot. *Science*, *291*(5512), 2338–2343.

Alves, A., Sousa, R., & Reis, R. (2012). A practical perspective on ulvan extracted from green algae. *Journal of Applied Phycology*, 1–18. http://dx.doi.org/10.1007/s10811-012-9875-4.

Aquaron, R., Delange, F., Marchal, P., Lognone, V., & Ninane, L. (2002). Bioavailability of seaweed iodine in human beings. *Cell Molecular Biology*, *48*(6), 563–569.

Araya, N. (2011). Fucoidan therapy decreases the proviral load in patients with human T-lymphotropic virus type-1-associated neurological disease. *Antiviral Therapy*, *16*(1), 89–98.

Arlov, O. (2012). *Heparin analogs created by sulfation of alginates using a chemoenzymatic strategy*. Norwegian Univeristy of Science and Technology.

Bixler, H., & Porse, H. (2011). A decade of change in the seaweed hydrocolloids industry. *Journal of Applied Phycology*, *23*(3), 321–335.

Bobin-Dubigeon, C., Lahaye, M., Guillon, F., Barry, J.-L., & Gallant, D. J. (1997). Factors limiting the biodegrdataion of *Ulva* sp. cell-wall polysaccharides. *Journal of Science Food and Agriculture*, *75*, 341–351.

Briand, X., Cluzet S., Dumas B., Esquerre-Tugaye M.-T., Salamagne S. (2010). *Ulvans as activators of plant defense and resistance reactions against biotic or abiotic stresses.* U. S. Patent. U.S., TIMAC Agro International. USA 7820176 B2.

Carmona, R., Vergara, J., Lahaye, M., & Niell, F. X. (1998). Light quality affects morphology and polysaccharide yield and composition of *Gelidium sesquipedale* (Rhodophyceae). *Journal of Applied Phycology*, *10*(3), 323–332.

Chan, C.-X., Ho, C.-L., & Phang, S.-M. (2006). Trends in seaweed research. *Trends in Plant Science*, *11*(4), 165–166.

Chang, C.-F. (2012). *Carbohydrates–comprehensive studies on glycobiology and glycotechnology*. INTECH.

Chen, L. C. M., Craigie J. S., McCandless, E. L. McLachlan, J. L. Neish, A. C. Shacklock, P. F. Walter, J. A. (1975). Algal polysaccharide production. www.freepatentsonline.com/3879890.pdf. U. S. Patent. U.S., Canadian Patents and Development Ltd. Ontario, Canada. US Patent 3879890: 6.

Chen, W. Z., Li, Y. M., & Yu, M. H. (2008). Astragalus polysaccharides: an effective treatment for diabetes prevention in mice. *Experimental and Clinical Endocrinology and Diabetes, 116*(8), 468–474.

Chen, J. H., & Liu, J. (2005). Characterization of the structure of antithrombin-binding heparan sulfate generated by heparan sulfate 3-O-sulfotransferase 5. *Biochimica Et Biophysica Acta-general Subjects, 1725*(2), 190–200.

Chengkui, Z. (1984). Phycological research in the development of the Chinese seaweed industry. *Hydrobiologia, 116-117*(1), 7–18.

Chizhov, A. O., Dell, A., Morris, H. R., Haslam, S. M., McDowell, R. A., Shashkov, A. S., et al. (1999). A study of fucoidan from the brown seaweed *Chorda filum*. *Carbohydrate Research, 320*(1–2), 108–119.

Chopin, T., Hanisak, M. D., Koehn, F. E., Mollion, J., & Moreau, S. (1990). Studies on carrageenans and effects of seawater phosphorus concentration on carrageenan content and growth of *Agardhiella subulata* (C. Agardh) Kraft and Wynne (Rhodophyceae, Solieriaceae)*. *Journal of Applied Phycology, 2*, 3–16.

Cock, J. M., Sterck, L., Rouze, P., Scornet, D., Allen, A. E., Amoutzias, G., et al. (2010). The Ectocarpus genome and the independent evolution of multicellularity in brown algae. *Nature, 465*(7298), 617–621.

Collén, J., Porcel, B., Carré, W., Ball, S. G., Chaparro, C., Tonon, T., et al. (2013). Genome structure and metabolic features in the red seaweed *Chondrus crispus* shed light on evolution of the Archaeplastida. In *Proceedings of the National Academy of Sciences*.

Cooper, R., Dragar, C., Elliot, K., Fitton, J. H., Godwin, J., & Thompson, K. (2002). GFS, a preparation of Tasmanian *Undaria pinnatifida* is associated with healing and inhibition of reactivation of Herpes. *BMC Complementary & Alternative Medicine, 2*. http://www.biomedcentral.com/1472-6882/1472/1411.

Craigie, J. S. (2010). Seaweed extract stimuli in plant science and agriculture. *Journal of Applied Phycology, 23*(3), 371–393.

Dillehay, T D., C. Ramirex, M. Pino, M. B. Collins, J. Rossen, J. D. Pino-Navarro. (2008). Food, medicine, and the peopling of South America. *Science, 320*(784), 784–786.

Elisabeth Mann, Borgese (1980). *Seafarm: The story of aquaculture*. New York: Harry N. Abrams, Incorporated. ISBN 0-8109-1604-5.

FAO. (2010). *State of the world fisheries & aquaculture*. Rome: UN Fisheries and Agricultural Organisation.

FAO. (2011). *FIGIS global production statistics 1950–2011*. from http://www.fao.org/fishery/statistics/en.

Fazekas, T., Eickhoff, P., Pruckner, N., Vollnhofer, G., Fischmeister, G., Diakos, C., et al. (2012). Lessons learned from a double-blind randomised placebo-controlled study with a iota-carrageenan nasal spray as medical device in children with acute symptoms of common cold. *BMC Complementary & Alternative Medicine, 12*, 147.

Fitton, J. H. (2011). Therapies from fucoidan; multifunctional marine polymers. *Marine Drugs, 9*(10), 1731–1760.

Foffi, G., Pastore, A., Piazza, F., & Temussi, P. A. (2013). Macromolecular crowding: chemistry and physics meet biology (Ascona, Switzerland, June 10–14, 2012). *Physical Biology, 10*(4), 040301.

Francavilla, M., Franchi, M., Monteleone, M., & Caroppo, C. (2013). The red seaweed gracilaria gracilis as a multi products source. *Marine Drugs* (11), 3754–3776.

Fujimura, T., Tsukahara, K., Moriwaki, S., Kitahara, T., Sano, T., & Takema, Y. (2002). Treatment of human skin with an extract of *Fucus vesiculosus* changes its thickness and mechanical properties. *Journal of Cosmetic Science, 53*(1), 1–9.

Fujita, T., Fujii, Y., Goto, B., Miyauchi, A., & Takagi, Y. (2000). Peripheral computed tomography (pQCT) detected short-term effect of AAA Ca (heated oyster shell with heated algal ingredient HAI): a double-blind comparison with $CaCO_3$ and placebo. *Journal of Bone Mineral Metabolism, 18*(4), 212–215.

Fujita, T., Ohue, T., Fujii, Y., Miyauchi, A., & Takagi, Y. (1996). Heated oyster shell-seaweed calcium (AAA Ca) on osteoporosis. *Calcified Tissue International, 58*(4), 226–230.

Godfray, H. C. J., Beddington, J. R., Crute, I. R., Haddad, L., Lawrence, D., Muir, J. F., et al. (2010). Food security: the Challenge of feeding 9 billion people. *Science, 327*(327), 812–818.

Gomez, I., & Huovinen, P. (2012). Morphofunctionality of carbon metabolism in seaweeds. In C. Wiencke, & K. Bischof (Eds.), *Seaweed biology* (p. 219). Berlin-Heidelberg: Springer-Verlag.

Gordon-Mills, E. M., & McCandless, E. L. (1975). Carrageenans in the cell walls of *Chondrus crispus* Stack. (Rhodophyceae, Gigartinales). I. Localization with fluorescent antibody. *Phycologia, 14*(4), 275–281.

Hafting, J. T., Critchley, A., Cornish, M. L., Hubley, S. A., & Archibald, A. F. (2011). On-land cultivation of functional seaweed products for human usage. *Journal of Applied Phycology, 24*(3), 385–392.

Hallak, L. K., Collins, P. L., Knudson, W., & Peeples, M. E. (2000). Iduronic acid-containing glycosaminoglycans on target cells are required for efficient respiratory syncytial virus infection. *Virology, 271*(2), 264–275.

Halling, C., Wikström, S., Lilliesköld-Sjöö, G., Mörk, E., Lundsør, E., & Zuccarello, G. (2012). Introduction of Asian strains and low genetic variation in farmed seaweeds: indications for new management practices. *Journal of Applied Phycology*, 1–7.

Hanbo, F., Xuan, M., Bin, C., & Chunlan, X. (2014). Influence of gelling properties on protein imprinted agarose gel membrane. *Journal of Applied Polymer Science, 131*(11). http:/ dx.doi.org/ 10.1002/app.40323.

Harris, C., Matsuda, K., & Sattelle, D. B. (2013). Dr. Kathleen Drew-Baker, "Mother of the Sea", a Manchester scientist celebrated each year for half a century in Japan. *Bioessays, 35*(9), 838–839.

Hehemann, J.-H., Correc, G., Barbeyron, T., Helbert, W., Czjzek, M., & Michel, G. (2010). Transfer of carbohydrate-active enzymes from marine bacteria to Japanese gut microbiota. *Nature, 464*, 908–912.

Henderson, J. (2004). *The roman book of gardening*. London, New York: Routledge.

Hidari, K. I. P.J., Abe, T., & Suzuki, T. (2013). Carbohydrate-related inhibitors of dengue virus entry. *Viruses, 5*, 605–618.

Hill, D. R., Rho, H. K., Kessler, S. P., Amin, R., Homer, C. R., McDinal, C., et al. (2013). Human milk hyaluronan enhances innate defense of the intestinal epithelium. *Journal of Biological Chemistry, 288*, 29090–29104.

Holdt, S. L., & Kraan, S. (2011). Bioactive compound in seaweed: functional food applications and legislation. *Journal of Applied Phycology, 23*(3), 543–597.

Holtkamp, A., Kelly, S., Ulber, R., & Lang, S. (2009). Fucoidans and fucoidanases–focus on techniques for molecular structure elucidation and modification of marine polysaccharides. *Applied Microbiology and Biotechnology, 82*(1), 1–11.

Hurtado, A., Montaño, M., & Martinez-Goss, M. (2013). Commercial production of carrageenophytes in the Philippines: ensuring long-term sustainability for the industry. *Journal of Applied Phycology, 25*(3), 733–742.

Ikeguchi, M. (2011). Fucoidan reduces the toxicities of chemotherapy for patients with unresectable advanced or recurrent colorectal cancer. *Oncol Letters, 2*(2), 319–322.

Irhimeh, M. R., Fitton, J. H., & Lowenthal, R. M. (2007). Fucoidan ingestion increases the expression of CXCR4 on human CD34+ cells. *Experimental Hematology, 35*(6), 989–994.

Irhimeh, M. R., Fitton, J. H., & Lowenthal, R. M. (2009). Pilot clinical study to evaluate the anticoagulant activity of fucoidan. *Blood Coagulation & Fibrinolysis, 20*(7), 607–610.

Jaulneau, V., Lafitte, C., Jacquet, C., Fournier, S., Salamagne, S., Briand, X., et al. (2010). Ulvan, a sulfated polysaccharide from green algae, activates plant immunity through the jasmonic acid signaling pathway. *Journal of Biomedicine and Biotechnology, 2010*(525291), PMC2860583.

Jiang, Z., Fang, J. G., Mao, Y. Z., Han, T. T., & Wang, G. H. (2012). Influence of seaweed aquaculture on marine inorganic carbon dynamics and sea-air CO_2 flux. *Journal of the World Aquaculture Society*, *44*(1), 133–140.

Jiao, G., Yu, G., Zhang, J., & Ewart, H. S. (2011). Chemical structures and bioactivities of sulfated polysaccharides from marine algae. *Marine Drugs*, *9*, 196–223.

Johnson, N. (1990). Seaweed and its synthetic analogues in obstetrics and gynaecology 450BC–1990AD. *Journal of the Royal Society of Medicine*, *83*, 387–389.

Juffrie, M., Damayanti, W., Djumhana, A., & Ahmad, H. (2006). The efficacy of fucoidan on gastric ulcer. *Indonesian Journal of Biotechnology*, *11*(2), 908–913.

Khan, W., Rayirath, U. P., Subramanian, S., Jithesh, M. N., Rayorath, P., Hodges, D. M., et al. (2009). Seaweed extracts as biostimulants of plant growth and development. *Journal of Plant Growth Regulation*, *28*(4), 386–399.

Khan, W., Zhai, R., Souleimanov, A., Critchley, A. T., Smith, D. L., & Prithiviraj, B. (2012). Commercial extract of *Ascophyllum nodosum* improves root colonization of alfalfa by its bacterial symbiont *Sinorhizobium meliloti*. *Communications in Soil Science and Plant Analysis*, *43*(18), 2425–2436.

Kim, W., Koo, Y.-K., Jung, M.-K., Moon, H., Kim, S., Synytsya, A., et al. (2010). Anticoagulating activities of low-molecular weight fuco-oligosaccharides prepared by enzymatic digestion of fucoidan from the sporophyll of Korean *Undaria pinnatifida*. *Archives of Pharmacal Research*, *33*(1), 125–131.

Kirkendale, L., Saunders, G., & Winberg, P. C. (2013). A molecular survey of *Ulva* (Chlorophyta) in temperate Australia reveals enhanced levels of cosmopolitanism. *Journal of Phycology*, *49*(1), 69–81.

Kraan, S. (2012). Algal polysaccharides, novel applications and outlook. In C.-F. Chang (Ed.), *Carbohydrates—comprehensive studies on glycobiology and glycotechnology*. INTECH.

Krotkiewski, M., Aurell, M., Holm, G., Grimby, G., & Szczepanik, J. (1991). Effects of a sodium-potassium ion-exchanging seaweed preparation in mild hypertension. *American Journal of Hypertension*, *4*(6), 483–488.

Kumari, P., Kumar, M., Reddy, C. R. K., & Jha, B. (2013). Algal lipids, fatty acids and sterols. In H. Dominguez (Ed.), *Functional ingredients from algae for foods and nutraceuticals* (pp. 87–134). Cambridge: Woodhead Publishing.

Ladodo, K. S., Rybakova, E. P., Stefanova, I. L., Mokshantseva, I. V., & Timoshenko, N. V. (2002). Clinical evaluation of the usage of iodine-containing food products in the diet of children. *(Russain) Vopr Pitan*, *71*(5), 10–13.

Lahaye, M. (1998). NMR spectroscopic characterisation of oligosaccharides from two *Ulva rigida* ulvan samples (Ulvales, Chlorophyta) degraded by a lyase. *Carbohydrate Research*, *314*(1–2), 1–12.

Leiro, J. M., Castro, R., Arranz, J. A., & Lamas, J. (2007). Immunomodulating activities of acidic sulfated polysaccharides obtained from the seaweed *Ulva rigida* C. Agardh. *International Immunopharmacology*, 7(7), 879–888.

Liu, W., Wu, K., Mao, W., Wu, Y., & Ouyang, J. (2010). *Astragalus* polysaccharide improves insulin sensitivity in KKAy mice: regulation of PKB/GLUT4 signaling in skeletal muscle. *Journal of Ethnopharmacology*, *127*(1), 32–37.

Lluisma, A. O., & Ragan, M. A. (1997). Expressed sequence tags (ESTs) from the marine red alga *Gracilaria gracilis*. *Journal of Applied Phycology*, *9*, 287–293.

Lobban, C., & Harrison, P. J. (Eds.). (2000). *Seaweed ecology and physiology*. New York: Cambridge Univeristy Press.

Lorbeer, A. J., Tham, R., & Zhang, W. (2013). Potential products from the highly diverse and endemic macroalgae of Southern Australia and pathways for their sustainable production. *Journal of Applied Phycology*, 1–16.

Miranda, L. N., Hutchison, K., Grossman, A., & Brawley, S. H. (2013). Diversity and abundance of the bacterial community of the red macroalga porphyra umbilicalis: did bacterial farmers produce macroalgae? *PLoS ONE*, *8*(3), e58269.

Miyai, K., Tokushige, T., & Kondo, M. (2008). Suppression of thyroid function during ingestion of seaweed "Kombu" (*Laminaria japonica*) in normal Japanese adults. *Endocrine Journal, 55*(6), 1103–1108.

Mori, N., Nakasone, K., Tomimori, K., & Ishikawa, C. (2012). Beneficial effects of fucoidan in patients with chronic hepatitis C virus infection. *World Journal of Gastroenterology, 18*(18), 2225–2230.

Moseley, C. M. (Ed.). (1990). *The effect of cultivation conditions on the yield and quality of carageenans in Chondrus crispus. Introduction to applied phycology*. The Hague, The Netherlands: SPB Academic Publishing.

Myers, S. P. (2011). A combined phase I and II open-label study on the immunomodulatory effects of seaweed extract nutrient complex. *Biologics: Targets & Therapy* (5), 45–60.

Myers, S. P., O'Connor, J., Fitton, J. H., Brooks, L., Rolfe, M., Connellan, P., et al. (2010). A combined phase I and II open label study on the effects of a seaweed extract nutrient complex on osteoarthritis. *Biologics: Targets & Therapy, 4*, 33–44.

Navarro, D. A., Flores, M. L., & Stortz, C. A. (2007). Microwave-assisted desulfation of sulfated polysaccharides. *Carbohydrate Polymers, 69*(4), 742–747.

Ohgitani, S., Fujii, Y., & Fujita, T. (1997). Effects of calcium supplementation using AAA Ca or milk on nocturnal bone resorption in young women. *Nihon Ronen Igakkai Zasshi, 34*(9), 743–747.

O'Sullivan, L., Murphy, B., Mc Loughlin, P., Duggan, P., Lawlor, P. G., Hughes, H., et al. (2010). Prebiotics from marine macroalgae for human and animal health applications. *Marine Drugs, 8*, 2038–2064.

Paradis, M.-E., Couture, P., & Lamarche, B. (2011). A randomised crossover placebo-controlled trial investigating the effect of brown seaweed (*Ascophyllum nodosum* and *Fucus vesiculosus*) on postchallenge plasma glucose and insulin levels in men and women. *Applied Physiology Nutrition and Metabolism, 36*(6), 913–919.

Paradossi, G., Cavalieri, F., & Chiessi, E. (2002). A Conformational study on the algal polysaccharide ulvan†. *Macromolecules, 35*(16), 6404–6411.

Paradossi, G., Cavalieri, F., Pizzoferrato, L., & Liquori, A. M. (1999). A physico-chemical study on the polysaccharide ulvan from hot water extraction of the macroalga *Ulva. International Journal of Biological Macromolecules, 25*(4), 309–315.

Park, S. B., Chun, K. R., Kim, J. K., Suk, K., Jung, Y. M., & Lee, W. H. (2010). The differential effect of high and low molecular weight fucoidans on the severity of collagen-induced arthritis in mice. *Phytotherapy Research, 24*, 1384–1391.

Paulert, R., Ebbinghaus, D., Urlass, C., & Moerschbacher, B. M. (2010). Priming of the oxidative burst in rice and wheat cell cultures by ulvan, a polysaccharide from green macroalgae, and enhanced resistance against powdery mildew in wheat and barley plants. *Plant Pathology, 59*(4), 634–642.

Pomin, V. H. (2009). An overview about the structure-function relationship of marine homopolysaccharides with regular chemical structures. *Biopolymers, 91*(8), 601–609.

Qi, H., Zhang, Q., Zhao, T., Chen, R., Zhang, H., Niu, X., et al. (2005). Antioxidant activity of different sulfate content derivatives of polysaccharide extracted from *Ulva pertusa* (Chlorophyta) in vitro. *International Journal of Biological Macromolecules, 37*(4), 195–199.

Qi, H., Zhao, T., Zhao, T., Chen, R., Zhang, H., Niu, X., et al. (2005). Antioxidant activity of different molecular weight sulfated polysaccharides from *Ulva pertusa* Kjellm (Chlorophyta). *Journal of Applied Phycology, 17*, 527–534.

Reddy, C. R. K., Jha, B., Fujita, Y., & Ohno, M. (2008). Seaweed micro-propagation techniques and their potential: an overview. *Journal of Applied Phycology, 20*, 609–617.

Redouan, E., Emmanuel, P., Christine, B., Bernard, C., Josiane, C., & Cédric, D. (2010). Development of new ulvan-like polymer by regioselective oxidation of gellan exopolysaccharide using TEMPO reagent. *Carbohydrate Polymers, 80*(2), 485–490.

Redouan, E., Emmanuel, P., Michelle, P., Bernard, C., Josiane, C., & Cedric, D. (2010). Evaluation of antioxidant capacity of ulvan-like polymer obtained by regioselective oxidation of gellan exopolysaccharide. *Food Chemistry, 127*, 967–983.

Rindi, F., Soler-Vila, A., & Guiry, M. D. (2012). Taxonomy of marine macroalgae used as sources of bioactive compounds. In M. Hayes (Ed.), *Marine bioactive compounds* (pp. 1–51). New York, Dordrecht, Heidelberg, London: Springer.

Robic, A., Sassi, J.-F., Dion, P., Lerat, Y., & Lahaye, M. (2009). Seasonal variability of physicochemical and propertiesof ulvan in two *Ulva* species (Chlorophyta) from the Brittany Coast. *Journal of Phycology, 45*, 962–973.

Robic, A., Sassi, J. F., Sassi, J. F., & Lahaye, M. (2008). Impact of stabilization treatments of the green seaweed *Ulva rotundata* (Chlorophyta) on the extraction yield, the physico-chemical and rheological properties of ulvan. *Carbohydrate Polymers, 74*(3), 344–352.

Robinson, N. A., Winberg, P., & Kirkendale, L. (2012). Genetic improvement of macroalgae: Status to date and needs for the future. *Journal of Applied Phycology, 25*, 703–716.

Rudd, P. M., Elliott, T., Cresswell, P., Wilson, I. A., & Dwek, R. A. (2001). Glycosylation and the immune system. *Science, 291*(5512), 2370–2376.

Rupérez, P., Ahrazem, O., & Leal, J. A. (2002). Potential antioxidant capacity of sulfated polysaccharides from the edible marine brown seaweed *Fucus vesiculosus*. *Journal of Agricultural and Food Chemistry, 50*(4), 840–845.

Ruxton, C. H. S., & Jenkins, G. (2013). A novel topical ingredient derived from seaweed significantly reduces symptoms of acne vulgaris: a general literature review. *Journal of Cosmetic Science, 64*(3), 219–226.

Sassi, J. F., Nyvall-Collen, P., Helbert, W., Marfaing, H., Lerat, Y., Yvergnaux, F., et al. (2011). Design and upscaling of an enzymatic process for producing bioactive oligosaccharides from ulvan, a polysaccharide extracted from green seaweeds. In M. A. Borowitzka (Ed.), *4th congress of the international society for applied phycology*. .

Sears, P., & Wong, C.-H. (2001). Toward automated synthesis of oligosaccharides and glycoproteins. *Science, 291*(5512), 2344–2350.

Shanmugam, M., & Mody, K. H. (2000). Heparinoid-active sulfated polysaccharides from marine algae as potential blood anticoagulant agents. *Current Science, 79*(12), 1672–1683.

Silva, T. H., Alves, A., Popa, E. G., Reys, L. L., Gomes, M. E., Sousa, R. A., et al. (2012). Marine algae sulfated polysaccharides for tissue engineering and drug delivery approaches. *Biomatter, 2*(4), 278–289.

Silva, P. C., Basson, P. W., & Moe, R. L. (1996). *Catalogue of the marine benthic algae of the Indian Ocean*. Berkeley: University of Califormia Press.

Singh, R. S., Thakur, S. R., & Bansal, P. (2013). Algal lectins as promising biomoleculaes for biomedical research. *Critical Reviews in Microbiology*, 1–12. early online.

Skibola, C. F. (2004). The effect of *Fucus vesiculosus*, an edible brown seaweed, upon menstrual cycle length and hormonal status in three pre-menopausal women: a case report. *BMC Complementary and Alternative Medicine, 4*(10).

Skriptsova, A., Shevchenko, N., Zvyagintseva, T., & Imbs, T. (2010). Monthly changes in the content and monosaccharide composition of fucoidan from *Undaria pinnatifida* (Laminariales, Phaeophyta). *Journal of Applied Phycology, 22*(1), 79–86.

Smidsrød, O., & Grasdalen, H. (1984). Polyelectrolytes from seaweeds. *Hydrobiologia, 116–117*(1), 19–28.

Subramanian, S., Sangha, J., Gray, B., Singh, R., Hiltz, D., Critchley, A., et al. (2011). Extracts of the marine brown macroalga, *Ascophyllum nodosum*, induce jasmonic acid dependent systemic resistance in *Arabidopsis thaliana*; against *Pseudomonas syringae* pv. *tomato* DC3000 and *Sclerotinia sclerotiorum*. *European Journal of Plant Pathology*, 1–12.

Sutherland, J. E., Lindstrom, S. C., Nelson, W. A., Brodie, J., Lynch, M. D. J., Hwang, M. S., et al. (2011). A new look at an ancient order: generic revision of the Bangiales (Rhodophyta). *Journal of Phycology, 47*, 1131–1151.

Teas, J., Irhimeh, M. R., Druker, S., Hurley, T. G., Hebert, J. R., Savarese, T. M., et al. (2011). Serum IGF-1 concentrations change with soy and seaweed supplements in healthy postmenopausal American women. *Nutrition and Cancer, 63*(5), 743–748.

Traka, M. H., & Mithen, R. F. (2011). Plant science and human nutrition: challenges in assessing health-promoting properties of phytochemicals. *Plant Cell, 23*(7), 2483–2497.

Tseng, C. K. (1944). Agar: a valuable seaweed product. *Scientific Monthly, 58,* 24–32.

Usov, A. I., & Zelinsky, N. D. (2013). *Chemical strutrues of algal polysaccharides. Funcitonal Ingredients from Algae for Food and Nutraceuticals. H. Dominguez.* Cambridge: Woodhead Publishing.

Venugopal, V. (2011). *Marine polysaccharides: Food applications.* Florida: CRC Press.

Villanueva, R. D., Montaño, N. E., Romero, J. B., Aliganga, A. K. A., & Enriquez, E. P. (2005). Seasonal variations in the yield, gelling properties, and chemical composition of agars from *Gracilaria eucheumoides* and *Gelidiella acerosa* (Rhodophyta) from the Philippines. *Botanica Marina, 42,* 175.

Volpe, R., & Sawin, C. (2011). The History and iconography relating to the thyroid gland. In J. A. H. Wass, & P. M. Stewart (Eds.), *Oxford textbook of endocrinology and diabetes.* London: Oxford University Press.

Wally, O. D., Critchley, A., Hiltz, D., Craigie, J., Han, X., Zaharia, L. I., et al. (2012). Regulation of phytohormone biosynthesis and accumulation in *Arabidopsis* following treatment with commercial extract from the marine macroalga *Ascophyllum nodosum. Journal of Plant Growth Regulation,* 1–16.

Wijesekara, I., Pangestuti, R., & Kim, S.-K. (2011). Biological activities and potential health benefits of sulfated polysaccharides derived from marine algae. *Carbohydrate Polymers, 84*(1), 14–21.

Witvrouw, M., & De Clerck, E. (1997). Sulfated polysaccharides extracted from sea algae as potential antiviral drugs. *General Pharmacology, 4,* 497–511.

Yung, P. Y., Burke, C., Lewis, M., Kjelleberg, S., & Thomas, T. (2011). Novel antibacterial proteins from the microbial communities associated with the sponge *Cymbastela concentrica* and the Green alga *Ulva australis. Applied and Environmental Microbiology,* 77(4), 1512–1515.

Zhao, X., Li, B., Xue, C., & Sun, L. (2011). Effect of molecular weight on the antioxidant property of low molecular weight alginate from *Laminaria japonica. Journal of Applied Phycology, 24,* 295–300.

CHAPTER NINE

Polysaccharides from *Codium* Species: Chemical Structure and Biological Activity. Their Role as Components of the Cell Wall

Paula Virginia Fernández*, Paula Ximena Arata* and Marina Ciancia*,†,1

*Cátedra de Química de Biomoléculas, Departamento de Biología Aplicada y Alimentos, Facultad de Agronomía, Universidad de Buenos Aires, Argentina

†CIHIDECAR-CONICET, Departamento de Química Orgánica, Facultad de Ciencias Exactas y Naturales, Universidad de Buenos Aires, Argentina

[1]Corresponding author: e-mail address: ciancia@agro.uba.ar

Contents

Abstract

Polysaccharides from *Codium* species have drawn attention due to their high anticoagulant activity. However, many different mechanisms of action have been proposed and the structure of the active compounds was usually not elucidated. Recently, structural determination has been carried out using purified fractions showing the presence of at least three different polysaccharide types, namely: (1) 3-linked β-D-galactans with ramifications on C-6 and terminal β-D-galactopyranose units with pyruvic acid ketal

ISSN 0065-2296
http://dx.doi.org/10.1016/B978-0-12-408062-1.00009-3

linked to *O*-3 and *O*-4 (*S*-configuration). Sulphation occurs mainly on C-4 and some of the 3-linked galactose units are substituted by pyruvic acid ketals linked to *O*-4 and *O*-6 (*R*-isomer). (2) 3-linked linear pyranosic β-L-arabinans sulphated on C-2 and/or C-4. (3) 4-linked β-D-mannans, partially sulphated on C-2 or on C-6, but to a lesser extent. The backbone of this polysaccharide is similar to that of the fibrillar component of the cell wall. Antiherpetic activity and activity on the immune system were determined for sulphated galactans from *Codium fragile*, which do not have significant anticoagulant effect, while highly sulphated arabinans have an important anticoagulant activity by mechanisms different to those described for heparin. The utricle cell wall of *C. fragile* has a sandwich structure of two fibrillar-like layers delimiting a middle amorphous-like zone. Mannans and hydroxyproline-rich glycoprotein-like epitopes are located in two distinct cell-wall layers, whereas sulphated polysaccharides are distributed in the middle area of the wall. The overall cell-wall polymer arrangement in the utricles of different species proved to be somewhat different.

9.1 INTRODUCTION

The genus *Codium* (Codiaceae, Bryopsidales, Chlorophyta) currently comprises around 125 species widely distributed through the world's seas with the exception of the polar regions, being mainly found in temperate and subtropical areas (Oliveira-Carvalho, Oliveira, Barreto-Pereira, & Verbruggen, 2012). Seaweeds of this genus show a broad variation of forms and occur in various habitats, and, as the majority of those belonging to the suborder Bryopsidineae, they exhibit no calcification. *Codium* thalli can spread out over hard surfaces as mats, form spheres or grow upright, either unbranched and finger-like, or branched, with cylindrical or flattened branches. Anatomically, a *Codium* thallus is composed of a single, giant, branched tubular cell containing multiple nuclei, the branches are usually called siphons. The centre of the thallus consists of an entangled mesh of siphons, whereas in the surrounding cortex, siphons are closely adjoined and swollen into utricles. The utricles occur in a wide array of forms, varying in size, shape and composition, with gametangia and/or hairs borne along their sides (Figure 9.1). *Codium* is found in marine habitats ranging from rocky coasts exposed to full wave-forces to calm lagoons, from intertidal habitats to deep reefs, from arctic to tropical waters and from eutrophic estuaries to nutrient-depleted coral reefs (Verbruggen et al., 2007).

For the last two decades, *Codium* species have been of public and scientific concern because of the invasive, bloom-forming nature of some of them. *Codium fragile* subspecies *tomentosoides* is the most invasive seaweed in the world, and it is believed to be native to Japan and was then unintentionally spread around the world. Another species, *Codium isthmocladum*, forms

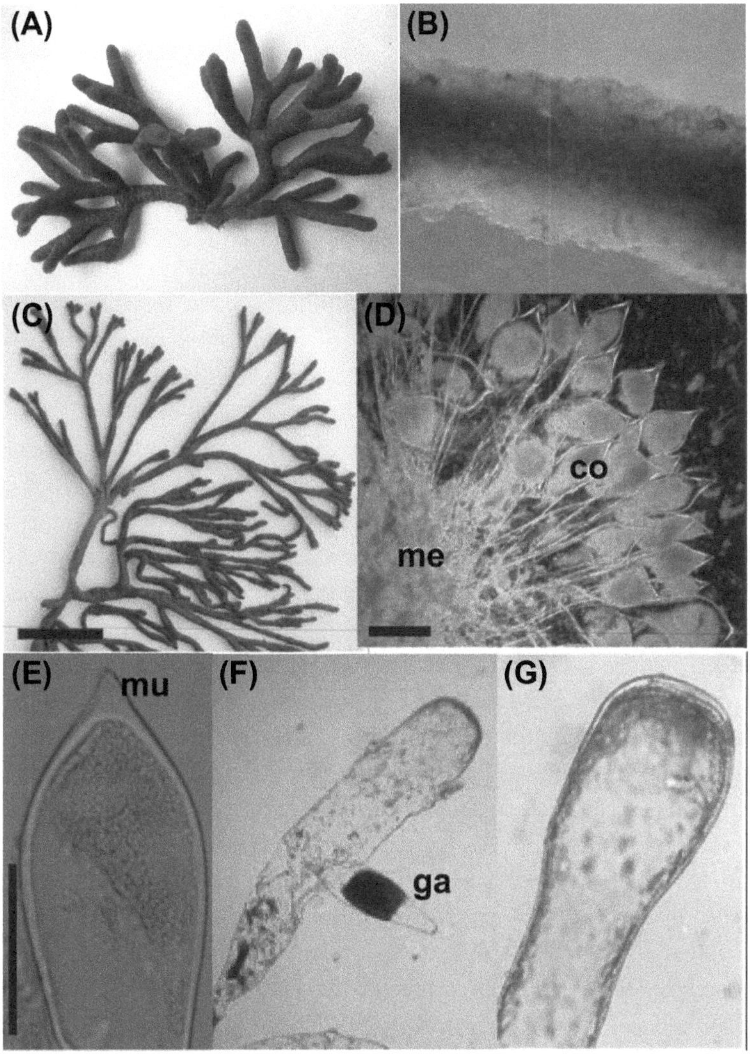

Figure 9.1 (A) General aspect of the Thallus of *Codium vermilara*. (B) Detail of the thallus of *Codium decorticatum*. (C) General aspect of the dry thallus of *Codium fragile*. Scale bar = 1 cm. (D) Inner structure of the thallus of *C. fragile*. Scale bar = 250 μm (me: medullar region; co: photosynthetic cortex). (E) Detail of an utricle of *C. fragile*. Scale bar = 250 μm (mu: mucron). (F) Detail of an utricle of *C. decorticatum* (ga: gametangium). (G) Detail of an utricle of *C. vermilara*. *(C), (D), and (E) from Estevez et al., 2009; (G) from Fernández et al., 2010; (B) and (F) courtesy from María Paula Raffo (Cenpat-Conicet, Puerto Madryn, Argentina).* (See the colour plate.)

harmful blooms on reefs, consequence of increased eutrophication. On the other hand, *Codium* species are used as food for cultured abalone, they are consumed by humans, and they are a source of bioactive compounds (Verbruggen et al., 2007).

Cell walls from marine algae, like red and brown seaweeds, together with marine angiosperms, biosynthesise sulphated polysaccharides, mainly galactan sulphates (carrageenans, agarans) (Usov, 2011) or sulphated fucans (Usov & Bilan, 2009). Although less studied, cell walls from green seaweeds also comprise sulphated polysaccharides. The fact that seaweeds, as well as sea invertebrates, like sea urchins (Vilela-Silva, Hirohashi, & Mourão, 2008) and echinoderms (Pereira, Mulloy, & Mourão, 1999), possess structurally related sulphated polysaccharides in their cell walls-intercellular matrices might be the result of convergent and extreme directional selection forces in their marine environments to high saline conditions (Aquino & Landeira-Fernández, 2005). These sulphated polysaccharides in high ionic strength media such as seawater would remain strongly charged providing mechanical stress resistance, hydration by gel formation, and they would act as ionic and osmotic regulators (Kloareg & Quatrano, 1988).

Green seaweeds biosynthesise matrix-sulphated polysaccharides with a great variety of chemical structures. They have been classified into two main groups (Percival & McDowell, 1981): (1) sulphated glucuronoxylorhamnans and glucuronoxylorhamnogalactans, with negative optical rotation, and (2) sulphated xyloarabinogalactans, which have positive optical rotation. As examples of the first group, the major acidic polysaccharide structure of *Ulva* and *Enteromorpha* species (Ulvales) is a linear sulphated polymer comprising alternating 4-linked β-D-glucuronic acid and α-L-rhamnose units (Lahaye & Robic, 2007), while glycuronorhamnans/rhamnans from *Monostroma* and *Gayralia* (Ulotrichales) have a backbone consisting in 2- and 3-linked α-L-rhamnose units with different substitution patterns (Cassolato et al., 2008; Lee, Koizumi, Hayashi, & Hayashi, 2010; Li et al., 2011). On the other hand, seaweeds of the Bryopsidales (Chlorophyta) belong to the second group, and only small to trace amounts of uronic acids and rhamnose, if any, were detected in the polysaccharides.

Until the present century, the only detailed structural studies of sulphated polysaccharides from this order were those of Percival and coworkers (Percival & McDowell, 1981), including those carried out on the polysaccharides from *C. fragile* (Love & Percival, 1964a). From that time until a few years ago, most of the work published comprised the characterisation of some extracts or fractions derived from them with interesting anticoagulant activity (Hayakawa et al., 2000; Matsubara et al., 2001; Matsubara, Matsuura, Hori, & Miyazawa, 2000;Siddhanta, Shanmugam, Mody, Goswami, & Ramabat, 1999; Uehara, Takeshita, & Maeda, 1992).

In addition, there is also information available about the fibrillar components of the cell walls of the Bryopsidales. These seaweeds lack cellulose, or at least this polymer is not the major fibrillar component of the cell wall, and have β-(1→4)-mannans (Chanzy, Grosrenaud, Vuong, & Mackie, 1984; Ciancia et al., 2007; Huizing & Rietema, 1975; Kaihou, Hayashi, Otsuru, & Maeda, 1993; Love & Percival, 1964b; Mackie, 1969) or β-(1→3)-xylans (Ciancia et al., 2012; Fukushi, Otsuru, & Maeda, 1988; Mackie, 1969; Mackie & Percival, 1959; Maeda, Fukushi-Fujikura, & Otsuru, 1990; Yamagaki, Maeda, Kanazawa, Ishizuka, & Nakanishi, 1997) instead. In addition, it has been shown that the major fibrillar component of the cell wall can vary in the different life stages (Huizing & Rietema, 1975; Huizing, Rietema, & Sietsma, 1979; Wutz & Zetsche, 1976). Thus, gametophyte microthalli of *Derbesia* sps. biosynthesise xylans, while sporophyte macrothalli of the same genus produce mannans. Similarly, gametophyte macrothalli from the genus *Bryopsis* biosynthesise xylans, while sporophyte microthalli produce mannans. However, considerable amounts of glucans were detected in some cases (Huizing et al., 1979; Wutz & Zetsche, 1976). Moreover, cellulose was found to be a minor fibrillar component of the cell walls from *Bryopsis maxima* and *Bryopsis plumosa* (Ciancia et al., 2012; Fukushi et al., 1988; Maeda et al., 1990).

9.2 STRUCTURE OF CELL WALL POLYSACCHARIDES FROM *CODIUM* SPECIES

The first structural study about the sulphated polysaccharides from *Codium* was carried out by Love and Percival (1964a), who found that the water extracts from *C. fragile* contained galactose and then arabinose as major monosaccharide constituents, but also small amounts of xylose, rhamnose, glucose, and mannose. This product contained 22% of sulphate (as SO_3Na), and minor amounts of uronic acids. It also contained 25% of protein and it was supposed to be a proteoglycan. Glucose and mannose were eliminated by periodate oxidation and were considered to arise from contaminant polymers. The resulting product was fractionated by anion exchange chromatography obtaining a fraction with similar quantities of galactose and arabinose, and important amounts of sulphate that would be in part, linked to C-2 or C-3 of the arabinose units. Partial acid hydrolysis and fractionation of the oligosaccharides, allowed to isolate a neutral fraction containing 3-*O*-β-L-arabinopyranosyl-L-arabinose and 3-*O*-β-D-galactopyranosyl-D-galactose and an acidic fraction containing galactose 4-sulphate and galactose 6-sulphate. Results obtained from seaweeds collected in two distant locations were similar.

This pioneer study gave a first clear idea about the structure of these polysaccharides. At that stage, they were considered as arabinogalactans, being both sugars possibly in the same molecule.

9.2.1 Sulphated Galactans

During the following decades, no structural studies were published. The only reports correspond to very general characterisations of compounds with interesting biological properties, mostly anticoagulants, until the isolation and structural studies of the galactan obtained from *Codium yezoensis* (Bilan, Vinogradova, Shashkov, & Usov, 2006, 2007).

By fractionation of the water extracts of this seaweed a very complex sulphated galactan was isolated. Thus, its elucidation required several modifications of the polysaccharide, i.e. desulphation and depyruvylation, followed by methylation analysis and Smith degradations, these chemical data were supported by NMR spectroscopy. This galactan showed a highly ramified structure, which contained linear backbone segments of 3-linked β-D-galactopyranose residues connected by (1→6) linkages, about 40% of 3-linked residues being additionally substituted at C-6, possibly by short oligosaccharide residues also containing (1→3) and (1→6) linkages. Sulphate groups were found mainly at C-4 and in minor amounts at C-6.

The most unusual feature of this water-soluble galactans was the high pyruvate content (one pyruvate and two sulphate per every four galactose residues). Pyruvate was found to form mainly five-membered cyclic ketals with *O*-3 and *O*-4 of the nonreducing terminal galactose residues (*S*-isomer). However, a minor part of pyruvate formed six-membered cyclic ketals with *O*-4 and *O*-6 (*R*-configuration). This feature is common to red algal galactans, which often contain 3-linked 4,6-*O*-(1′-carboxy)ethylidene-D-galactopyranose units. Both six- and five-membered cyclic pyruvate ketals of different monosaccharide residues were detected in bacterial polysaccharides, but seldom in green algal polysaccharides. This galactan was the first algal polysaccharide containing five-membered cyclic pyruvate ketals studied. Its putative fragment containing all the main structural features is shown in Figure 9.2.

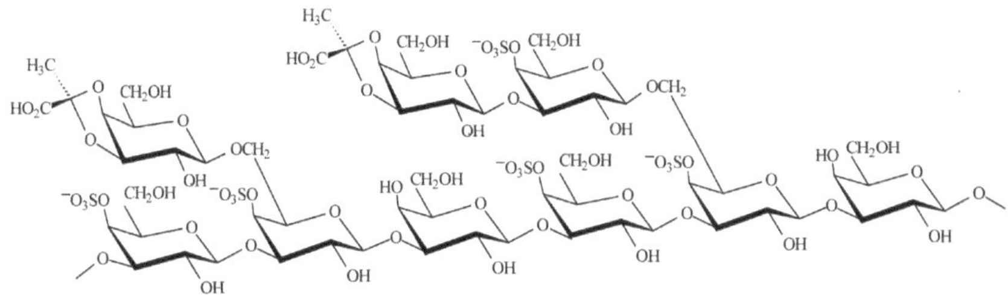

Figure 9.2 A possible sequence for the galactan from *Codium yesoense*.

Hence, the pyruvylated galactan sulphate isolated from *Codium yezoense* resembles red algal galactans in composition, but is quite different from these galactans in structure, having a ramified backbone similar to that of the 3,6-linked β-D-galactans of terrestrial plants, and bearing cyclic pyruvate ketals linked mainly to *O*-3 and *O*-4 of nonreducing terminal galactose residues, as in bacterial polysaccharides.

At the same time, the sulphated polysaccharides extracted with water from *C. fragile* and *Codium vermilara* from the Atlantic coast of Patagonia were studied (Ciancia et al., 2007). The first room temperature water extracts from both seaweeds (A1 and V1, respectively) showed galactose and arabinose as major sugar components. As no fractionation was carried out, it was not possible to deduce if this extracts contained arabinogalactans or a mixture of arabinans and galactans. However, in general terms, the galactan moieties in both extracts were similar, and similar to the galactan from *C. yezoense*. V1 was further fractionated by precipitation with potassium chloride and the product which remained soluble was a pure sulphated galactan (Fernández et al., 2013). This polysaccharide has a composition very similar to that of the galactan from *C. yezoense*; however, the structural units obtained by methylation and desulphation-methylation and NMR spectroscopy were somewhat different, mainly the presence of important amounts of 3-linked 4,6-disulphated units (~20%) and also the fact that the pyruvylated terminal units would form single stubs and not short chains. No appreciable pyruvylation was detected in the 3-linked β-D-galactose units of the backbone (Fernández, 2012).

Galactan sulphates from *C. fragile* of two different locations, Puerto Deseado (Santa Cruz), Argentina (F1 and F6), and Kamakura, Japan (FG) were studied at the same time by Estevez, Fernández, Kasulin, Dupree, & Ciancia (2009) and Ohta, Lee, Hayashi, & Hayashi (2009), respectively. Table 9.1 gives the structural units deduced from data published in these papers. These galactans showed the same structural units found for other galactans from *Codium* species, with the exception of F1 where structural data suggests

Table 9.1 Structural Units Found in Sulphated Galactans from *Codium fragile* from Two Different Locations

	***Codium fragile* from Argentina**[b]		***Codium fragile* from Japan**
Structural Units[a]	**F1**	**F6**	**FG**
→3)β-D-Gal*p*(1→	25	26	19
→3) β-D-Gal*p* 4S(1→	3	6	8
3,4Pyr-β-D-Gal*p*(1→	28	36	40
→3,6)β-D-Gal*p*(1→	24	12	7
→3,6)β-D-Gal*p* 4S(1→	–	10	25
→3)β-D-Gal*p*4,6-pyr(1→	8	11	–
3,4Pyr-β-D-Gal*p* 6S(1→[c]	15	–	–

[a]Nomenclature used: Gal*p*: galactopyranose; S: sulphate; Pyr: pyruvate.
[b]Small amounts of other sugars were considered to derive from contaminant polymers.
[c]Speculative unit, deduced from the desulphation pattern.

the presence of terminal 3,4-pyruvylated β-D-galactose 6-sulphate units. In addition, there are important differences in the percentages of the structural units: (1) most of the 3,6-linked galactose units are sulphated in FG, while they are not sulphated in F1 and in F6 only half of these units are sulphated; (2) there are not only 3,4-pyruvylated galactose terminal units, but also significant amounts of 3-linked 4,6-pyruvilated β-D-galactose units in F1 and F6, while the latter units are absent in FG.

These results show that although similar, galactans from different *Codium* species have important interspecific differences (Bilan et al., 2007; Ciancia et al., 2007; Farias et al., 2008; Ohta et al., 2009); moreover, also important differences between galactans of the same species but from different locations were observed. Besides, galactans F1 and F6, fractions from the same seaweed and location are also distinct (Table 9.1).

9.2.2 Sulphated Arabinans

Room-temperature water extraction of *Codium latum* and further fractionation produced a sulphated arabinan with 20.7% sulphate (molar ratio arabinose/sulphate 1/0.3–0.4), which, based on IR and ^{1}H NMR spectroscopy (anomeric proton δ 5.240 (D_2O)), was considered a sulphated furanosic 5-linked α-L-arabinan (Uehara et al., 1992).

In addition, after extensive fractionation of the cold-water extract of *Codium dwarkense*, a sulphated arabinan (A2a) with 41.45% sulphate (molar ratio arabinose/sulphate 1/1) and a sulphated arabinogalactan (A2b) with 31.85% sulphate were obtained in ~0.3% and ~0.4% yield, respectively

(Matsubara et al., 2001). A2a gave after extensive fractionation a product (Jia) with the highest anticoagulant activity (molar ratio arabinose/sulphate 1/1). The structure of this product was deduced from IR and ^{1}H NMR spectra as a furanosic α-L-arabinan without any specification of the position of the sulphate groups. Only the anomeric proton in the original product δ 5.21 (D_2O) and desulphated derivative δ 4.93 (DMSO) were informed (Siddhanta et al., 1999). In a screening of inhibition of thrombin by sulphated polysaccharides isolated from green algae, arabinans were extracted from *Codium divaricatum*, *Codium adhaerens*, *C. latum*, and *C. fragile* (molar ratio arabinose:sulphate, 0.6, 0.8, 0.8 and 0.5, respectively). No structural details were determined (Hayakawa et al., 2000). These papers give an idea that sulphated arabinans are usual components of *Codium* cell walls. However, there was a lot of confusion due to reports with very different structures, but no detailed information.

The room temperature water extracts from *C. fragile* and *C. vermilara* showed important amounts of arabinose. Methylation and desulphation-methylation analyses showed that 3-linked 2,4-disulphated arabinopyranose was the major structural unit. No evidence of the presence of arabinose in the furanose form was found. From the NMR spectra, it was suggested that these residues were β-anomers (Ciancia et al., 2007). On the other hand, hot-water extracts from *C. vermilara* indicated high amounts of non-sulphated arabinopyranose units. Later, arabinans from the same seaweed (Ab1 and Ab2) were isolated in a pure form (96–97% of arabinose, molar ratio arabinose:sulphate 1:1.8 and 1:1.5, respectively) and their structure was determined in detail by chemical and spectroscopic methods (Fernández et al., 2013) (Table 9.2). Both arabinans were extracted with water and obtained after separation from other polymers, mainly sulphated galactans.

Table 9.2 Linkage Analysis of Pyranosic Arabinans from *Codium vermilara*[a]

Monosaccharide	Structural Unit[b]	Ab1	Ab2
2,3,4-Ara[c]	β-L-Ara*p*(1→	1	6
2,4-Ara	→3) β- L-Ara*p* (1→	13	35
2-Ara	→3) β-L-Ara*p*4S (1→[d]	26	30
4-Ara	→3) β-L-Ara*p*2S (1→	10	7
–	→3) β-L-Ara*p*2,4S (1→	50	22
Sulphate content (as SO_3Na)		47	38

[a]All the methylated monosaccharides were derivatives of L-arabinose.
[b]β-configuration of these units was deduced by optical rotation and NMR spectroscopy.
[c]2,3,4-Ara indicates 2,3,4-tri-*O*-methyl-L-arabinose, etc.
[d]Nomenclature used, S: sulphate.

Initially, the room-temperature water extract was fractionated by anion-exchange chromatography, but separation was difficult and its reproducibility was low (Fernández, 2012). Then, taking into account previous reports (Siddhantha et al., 1999), separation by potassium chloride precipitation, like that used frequently for carrageenans, was successfully applied to the room temperature water extract, obtaining Ab1 through a clean separation. However, it was only possible to isolate Ab2 from the hot-water extract after elimination of α-glucans by treatment with amylase and further precipitation of the treated product with potassium chloride (Fernández et al., 2013).

Ab1 was isolated through a sharp gelification at 0.115 M KCl. No precipitation occurred when using sodium chloride, indicating that the insolubilisation is K^+-specific as in the case of kappa-carrageenans. Lower or higher concentrations of potassium chloride (upper limit 2 M) did not show any other precipitation. The mechanism of gel formation in dilute potassium chloride solutions has been extensively studied for κ-carrageenan where junction zones in gels are formed by quaternary interactions at the superhelical level (Nilsson & Piculell, 1991) between ordered tertiary structures (double helices), promoted by potassium ions, through the one-stage domain mechanism of aggregation (coil→double helix→gel) (Robinson, Morris, & Rees, 1980). A similar mechanism was suggested for gel formation of other polysaccharides in which the primary mode of interchain association is through multistranded helices (Aspinall, 1982; Robinson et al., 1980). Ab2, less sulphated than Ab1, also precipitated with potassium chloride, but at a higher concentration (0.5 M).

9.2.3 Mannans

Linear β-(1→4)-D-mannans are the major fibrillar component of cell walls from *Codium* species (Chanzy et al., 1984; Ciancia et al., 2007; Estevez et al., 2009; Kahiou et al., 1993; Love & Percival, 1964b; Mackie & Sellen, 1969; Percival & McDowell, 1981). These polymers are the predominant skeletal wall polymers in several other genera of green algae, including *Derbesia* (Bryopsidales), *Acetabularia*, *Halicoryne*, *Dasycladus*, *Neomeris*, *Cymopolia*, and *Batophora* (Dasycladales) (Dunn et al., 2007; Percival & McDowell, 1981). Substitution to even a small extent would hinder their ability to form high tensile strength microfibrils, but might enhance their ability to form a flexible network (Dunn et al., 2007).

The insoluble mannan obtained from *C. fragile* after exhaustive water extraction was studied by degradation with a β-mannanase and analysis of the resulting oligosaccharides; confirming the linear nature of this structure and the β-(1→4)-linkages, no evidence of side chains was found (Estevez et al., 2009). On the other hand, soluble mannan structures were found to be important components of the hot-water extracts from *C. fragile* and *C. vermilara*. The structure of the backbone was clearly established, being identical to that of the fibrillar component. The reason of the high solubility of these mannans in hot water was not clearly established. The number average molecular weight for those from *C. fragile* was 39–49 KDa and no side chains were detected. However, methylation analyses of mannans from *C. vermilara* suggested possible ramifications (Ciancia et al., 2007). Later, by anion exchange chromatography of the first hot-water extract from this seaweed, a fraction containing 94% of mannose and 12% of sulphate was isolated in low yield. The sulphate groups were linked to C-2 of 23% of the mannose units, while most of these units were not substituted. This degree of sulphation would explain the higher solubility of this β-(1→4)-D-mannans, compared to that of the nonsulphated fibrillar polymer. Other fractions also contain mannose in different proportions, but structures of these components were not studied (Fernández, Estevez, Cerezo, & Ciancia, 2012).

It has been shown that the unrelated green seaweed *Codiolum pusillum* (Chlorococcales, Endosphaeraceae) synthesizes a β-(1→4)-D-mannan branched at C-6 and carrying sulphate groups on C-2, although this is not the major sulphated polysaccharide biosynthesised by this unicellular seaweed (Carlberg & Percival, 1977).

Moreover, a β-(1→4)-D-mannan with similar substitution pattern to that of *C. pusillum* was isolated from red seaweeds *Chondrophycus papillosus* and *Chondrophycus flagelliferus* (Ceramiales) (Cardoso, Noseda, Fujii, Zibetti, & Duarte, 2007). This mannan was partially sulphated on C-2 of the mannose units and also substituted on C-6 with single stubs of β-D-xylose, as well as with β-D-mannose 2-sulphate units. On the other hand, the xylomannans from red seaweeds of the order Nemaliales have a linear backbone of α-(1→3)-D-mannopyranose units with different sulphation patterns (Pérez Recalde, Noseda, Pujol, Carlucci, & Matulewicz, 2009).

Linear β-D-mannans fulfill primarily structural functions not only in some green seaweeds, but also in the seeds of many plants, such as ivory nut, green coffee and coconut kernel. A low degree of α-(1→6)-D-galactose substitution results in (1) linear mannan chains with a degree of polymerisation of ~15 compacted into dense granular and crystalline structures (mannan I), or in (2) microfibrils, similar to those of cellulose that are less crystalline and with a higher degree of polymerisation of ~80 (mannan II). Both forms are insoluble and provide rigidity and protection to the endosperm. Once germination takes place, these mannans can be mobilised as nonstarch storage polysaccharides. Water retention by galactomannans is particularly important to the long-term survival of legume seeds in arid areas (Van Zyl, Rose, Trollope, & Gorgens, 2010).

Recently, a very different structure was proposed for a mannan isolated from a water extract of *C. fragile* collected from the coast of Sokcho, Gangwon province, Korea, comprising 3-linked α-D-mannose units (Tabarsa, Karnjanapratum, Cho, Kim, & You, 2013). This structure is completely different to those isolated and characterised by many other researches for the same species from various locations (Chanzy et al., 1984; Ciancia et al., 2007; Estevez et al.,2009; Love & Percival, 1964b, Mackie & Sellen, 1969). Evidence presented by the authors could be subject of different interpretations.

9.2.4 Hydroxyproline-Rich Glycoproteins

Arabinogalactan proteins (AGPs) and extensins are hydroxyproline-rich glycoproteins (HRGPs) usually present in plant cell walls (Albersheim, Darvill, Roberts, Sederoff, & Staehelin, 2011).

There are contradictory data about the exact functions of AGPs in a broad variety of processes as plant growth and development, plant defence, cell proliferation, cell expansion, cell differentiation, cell extension and somatic embryogenesis. The group of 'classical' AGPs consists of hyper-glycosylated polypeptides with a C-terminal hydrophobic sequence that directs the addition of a glycosylphosphatidylinositol (GPI) anchors and tethers them to the plasma membrane. Soluble AGP monomers are released through or into the middle lamella and the intercellular space. The continuous release of AGPs is characteristic of the rapid growing plant cells.

On the other hand, extensins are secreted as rod-like monomers with limited flexibility in an extended helical conformation. Extensin peroxidase catalyses polymerisation of extensin monomers via diisodityrosine and pulcherosine (tetramer and trimer of tyrosine, respectively) to yield cross-linked networks. Recently, it has been postulated that a positively charged extensin scaffold reacts with acidic pectin to form extensin pectate, which further templates orderly assembly of a nascent cell wall, changing the focus on extensin, from its role exclusively in the cessation of growth to an essential role in its initiation (Cannon et al., 2008).

Although they have also been detected in some Chlorophycean and Charophyceae green algae (Domozych et al., 2012), the first seaweed that was found to contain HRGPs was *C. fragile* (Estevez et al., 2009) by immunolabeling using antibodies against specific cell wall HRGP epitopes (AGPs and extensins) and by reaction with the β-glucosyl Yariv reagent.

In addition, a unique furanosic α-arabinosyl structure was detected in an arabinose-rich fraction obtained by fractionation of a room temperature water extract of the same seaweed, consisting in a 5-linked arabinan branched at C-3. 3-Linked and 3,6-linked β-D-galactose units were also detected. Although the latter units could be part of the major galactan biosynthesised by this seaweed, altogether, these results showed the presence in this fraction of AGP-like structures. Cell walls from *C. vermilara* also showed to have similar HRGPs by immunolabeling and by reaction with the Yariv reagent (Fernández, Ciancia, Miravalles, & Estevez, 2010); however, no furanosic arabinan structures were detected.

9.2.5 The Cell Wall Polysaccharides from *Codium*

Cell walls in *C. vermilara* are composed of ~32% (w/w) β-(1→4)-D-mannans, ~12% sulphated polysaccharides, and small amounts of hydroxyproline-rich glycoproteins (Fernández et al., 2010). Similar quantities of mannans and sulphated polysaccharides were reported previously in the related seaweed *C. fragile* (Estevez et al., 2009). Overall, both seaweed cell walls comprise ~40–44% of their dry weights. No other polysaccharide system from *Codium* was studied as a whole, although most of the data published until now suggest the possibility of similar polysaccharide systems. Within the sulphated polysaccharides group, a variety of polysaccharide structures were isolated and characterised, comprising (1) highly ramified pyruvylated sulphated galactans, with 3,6-, and 3,6-linkages, (2) linear sulphated pyranosic 3-linked arabinans and (3) sulphated 4-linked β-mannans. In spite of the fact that certain other structures were proposed (Kahiou et al., 1993; Tabarsa et al., 2013), it appears that this outline would be repeated for all the systems of sulphated polysaccharides from *Codium* species, with small differences. Results obtained so far for the cell wall components of *Codium decorticatum* (Fernández, Raffo, Alberghina, Ciancia, unpublished results) supports this generalisation.

9.3 POLYSACCHARIDE DISTRIBUTION ON THE CELL WALL

Despite the extensive knowledge of the chemical structure of most of the cell-wall polymers, the ways in which they interact to assemble into the cell-wall architecture remain elusive for most groups of plants. There are different analytical techniques available to achieve the understanding of the spatial arrangement of the cell-wall components. The use of *in situ* methods, such as the current library of monoclonal antibodies and carbohydrate binding modules (CBMs) against several cell-wall carbohydrate epitopes (Moller et al., 2007), together with noninvasive Fourier transform infrared spectroscopy coupled to a microscope (FTIR microspectroscopy), allows us to perform chemical imaging in plant cells and reveals intrinsic heterogenities in a single cell wall or in many different cell-wall types within a tissue (Carpita et al., 2001; McCartney et al., 2006).

In green algae, spatial arrangement of the cell-wall polymers was characterised only in simple models, such as volvocacean and desmid algae (Domozych & Domozych, 2003; Domozych, Lambiasse, Kiemle, & Gretz, 2009; Ender, Godl, Wenzl, & Sumper, 2002; Hallman, 2006) and also in the more complex wall of *Ulva* (Bobin-Dubigeon, Lahaye, Guillon, Barry, & Gallant, 1997). Lately, *in situ* methods were applied to study spatial arrangement of well characterised cell-wall polymers in *Codium* species (Estevez et al., 2009; Fernández et al., 2010).

In situ distribution of structural β-(1→4)-D-mannans in cell walls at the utricle tip from *C. fragile* was analysed using SR-FTIR microspectroscopy, Calcofluor White staining, birefringence detected by phase contrast and antibody labelling with anti-mannan mAb. These techniques showed a similar localisation pattern, with a central layer of the cell wall rich in fibrillar mannans. In addition, after an *in situ* enzymatic treatment with mannanase performed on cross-sections, cell-wall integrity was severely damaged, showing the important role of this polymer in cell-wall assembly and stability. Distribution of sulphated polysaccharides was determined using Toluidine Blue O (at pH = 1) and ruthenium red staining. The latter reacts not only with sulphate groups but also with the carboxylate groups of pyruvic acid. These polymers localise mostly in the tip of the mucron area and close to the plasma membrane. Finally, the *in situ* distribution of AGPs and extensin carbohydrate epitopes was assayed by several probes. AGP epitopes were localised in the inner part of the cell wall, close to the plasma membrane, and also in the outer faces of the utricle cell wall. Extensin epitopes were detectable mostly in the tip of the mucron, showing that the distribution of both types of glycoproteins only overlaps in the apical region of the utricle cells, but not in the cell-wall region close to the plasma membrane. Coomasie Blue staining and characteristic FTIR protein absorbance bands in SR-FTIR map are in good agreement with these results.

A sandwich structure was proposed for the utricle cell walls from *C. fragile*, with two boundary zones of HRGPs and a central layer rich in β-(1→4)-D-mannans and sulphated polysaccharides. The presence of fibrillar mannans and charged polysaccharides in the same region suggests

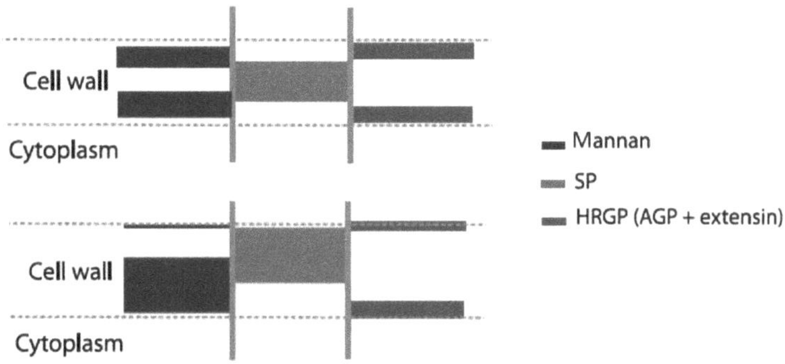

Figure 9.3 Proposed models for the utricule cell wall of *Codium vermilara* (above) and *Codium fragile* (below), including the *in situ* distribution of fibrillar mannan, sulphated polysaccharides (SP) and hydroxyproline-rich glycoproteins (HRGP), including arabinogalactan proteins (AGP) and extensins.

an important degree of interaction between both macromolecular types (Figure 9.3).

In *C. vermilara*, the utricle cell wall shows an external cuticle and two fibrillar-like layers delimiting a middle amorphous layer. *In situ* localisation of the main polymers was assayed by immunolabelling and chemical imaging. β-(1→4)-D-mannans are localised in two well-defined layers with similar development. *In situ* enzymatic treatment with mannanase caused not only an important loss of CW staining and birefringence, but also severe damage in some cell-wall areas, as it was shown previously for *C. fragile*. SR-FTIR microspectroscopy showed that sulphated polysaccharides concentrate in the amorphous central layer of the cell wall. Toluidine Blue O staining showed a similar pattern. AGPs labelling with β-glucosyl Yariv reagent showed that these glycoproteins are present in two marginal cell-wall layers. This two-layered distribution of AGPs was confirmed by immunolabelling, while extensin epitopes (also detected with specific antibodies) followed a similar pattern.

Based on these results, cell wall models for *C. fragile* and *C. vermilara* were proposed. The overall cell-wall polymer arrangement is different in both species, in spite of being phylogenetically very close. In *C. fragile*, mannans and HRGPs are distributed in a very wide internal layer, and only a narrow external layer was observed. On the other hand, in *C. vermilara*, two boundary zones of fibrillar mannans and HRGPs were detected, whereas in both seaweeds sulphated polysaccharides form an amorphous middle layer. The apical high concentration of AGP epitopes in *C. fragile* and *C. vermilara* suggests a relationship with apical expansion in the tip-growing cells, as it was shown in the moss protonemal growth and in the tip growth of pollen tubes. Extensins have been proposed to contribute to wall architecture, increased tensile strength, and to participate in a cell-plate formation interpenetrating cross-linked networks in the wall. Since *Codium* and all Bryopsidales are composed of continuous siphons lacking cross-cell walls and cell-plate formation, it is possible that the apical extensin scaffold would provide mechanical support just after or during apical expansion.

9.4 BIOLOGICAL ACTIVITY

Sulphated polysaccharides are known to have a number of biological activities including anticoagulant, antiviral, and immuno–inflammatory activities that could be used in nutraceutical/functional food, cosmetic/cosmeceutical and pharmaceutical applications.

Marine environment offers a great biodiversity and original sulphated polysaccharides, which have different chemical structures that are species-specific. The study of the biological properties of these polysaccharides could provide a valid alternative to traditional polysaccharides, such as glycosaminoglycans. Marine polysaccharides have a real potential for natural product drug discovery and for the delivery of new marine-derived products for therapeutic applications (Jiao, Yu, Zhan, & Ewart, 2011).

The fine structure of these sulphated polysaccharides allows them to behave in a similar way to the biologically active and informative oligosaccharide motifs of human glycoproteins, proteoglycans and glycolipids. Overall, the specific bindings of sulphated polysaccharides with basic proteins, or domains in viral particles, and thus binding affinities in molecular interactions account in conjunction to the resultant pharmacological actions and effectiveness of these glycans. The nature of these interactions has proved to be not a mere consequence of the net charge originated from sulphate content, but mostly driven by stereospecific features like monosaccharide composition, sulphation and glycosylation sites, anomericity and conformational preference (Mulloy, Ribeiro, Alves, & Mourão, 1994).

Although most of the research was carried out on sulphated polysaccharides from red and brown seaweeds, in the last few years promising results in this field were also found about sulphated rhamnans or glucuronorhamnans from green seaweeds of the first group (see Introduction). Besides, there are many reports about different biological activities of sulphated polysaccharides from *Codium*. However, only a few have been carried out with well purified and defined structures, and we will focus on these studies.

9.4.1 Anticoagulant Activity

Antithrombotic agents have been extensively used as a systemic therapy in cardiovascular diseases and heparin is the initial choice; nevertheless, heparin can induce several side effects, such as development of thrombocytopenia, arterial embolism, bleeding complications and so on. Furthermore, the incidence of prion-related diseases in mammals and the increasing requirements of anticoagulant therapy increase the need to look for alternative sources of anticoagulant and antithrombotic compounds.

The anticoagulant activity of sulphated polysaccharides is mainly attributed to thrombin inhibition mediated by antithrombin and/or heparin cofactor II, with different effectiveness depending of the compound. Other mechanisms were also proposed and these differences could be attributed to the diversity of structures of the polysaccharides evaluated and to the fact that one compound may have more than one target protease. There are some previous reviews about anticoagulant activity of sulphated polysaccharides from seaweeds (Ciancia, Quintana, & Cerezo, 2010; Matsubara, 2004; Shanmugam & Mody, 2000).

An exhaustive revision of the work carried out on the anticoagulant properties of sulphated polysaccharides from seaweeds concluded that the driving force for the formation of the sulphated polysaccharide/protein complex is the nonspecific polar interaction between the negatively and positively charged groups in the polysaccharide and protein, respectively and that the complex is further stabilised by short-range interactions. The polysaccharide binding site should be able to go through the following conformational steps in the formation of the complex: random coil→ordered conformation→low distortion of this conformation to form a complementary fitting structure with the protein backbone. The sulphated monosaccharide units with the highest potential for anticoagulant activity should have two sulphate groups and a glycosidic linkage on the pyranose ring with C-2, C-3 and C-4 in 2S, 3R, 4R or 2R, 3S, 4S configurations for galactose, fucose and arabinose and 2S, 3S, 4R, for rhamnose. Three distributions of these substituents appear: 3-linked 2,4-disulphated units, 4-linked 2,3-disulphated units and 2-linked 3,4-disulphated residues. These types of units have the possibility, through the equilibrium of the chair conformations, to place their sulphate groups in adequate spatial positions to interact with basic groups of the protein (Ciancia et al., 2010).

These structural conditions are in agreement with the fact that pure sulphated galactans from *Codium* usually did not show high anticoagulant activity. Although there are some reports of samples rich in galactose with important anticoagulant activity, this activity could be due to other structures present in the sample (Ciancia et al., 2010; Ciancia et al., 2007). However, a sulphated polysaccharide was obtained from *Codium cylindricum* by an extraction and purification procedure (Matsubara, 2004; Matsubara et al., 2001); it was composed by galactose as major monosaccharide, but also glucose (11%) and sulphate (13.1%). The galactan structure would be similar to the other galactans obtained from *Codium* species, although the presence of pyruvic acid was not investigated (Matsubara, 2004). It prolonged thrombin time (TT) and activated partial thromboplastin time (APTT), but did not affect prothrombin time (PT) (APtt and TT ratios >10 for a concentration of the sample of 15 μg/ml). It did not exert an indirect effect on thrombin by potentiating antithrombin (AT) or heparin cofactor II (HCII), and thus it would inhibit thrombin directly or fibrin polymerisation. It also showed direct inhibition of factor Xa, but at high concentrations (Matsubara et al., 2001).

Highly sulphated arabinans from *C. dwarkense* showed very high anticoagulant activity by global coagulation tests, i.e. a TT ratio >10 was achieved with a 5 μg/ml solution with respect to the control (Siddhanta et al., 1999).

In a screening of inhibition of thrombin by sulphated polysaccharides isolated from green algae, the arabinans from several *Codium* species were analysed. Anticoagulant mechanism proved to be by an HCII-dependent pathway with higher effectiveness than that of heparin or dermatan sulphate. It was suggested that binding of sulphated polysaccharides to HCII mainly induces conformational changes in the reactive site of this serpin, thereby optimising its interaction with thrombin, its target protease (Hayakawa et al., 2000).

On the contrary, proteoglycan from *Codium pugniformis* containing major amounts of glucose (72.4%), but also arabinose (17.3%) and galactose (10.0%), showed both direct and antithrombin mediated inhibition for thrombin activity, but it did not potentiate HCII (Matsubara et al., 2000). We speculate that the activity would be due to the arabinogalactan or to the arabinan moiety and that possibly glucose derives from contaminant glucans that were not eliminated by the fractionation procedure carried out.

The anticoagulant properties of extracts from *C. fragile* and *C. vermilara* were evaluated by general coagulation assays. All the fractions significantly prolonged APTT and TT regards to control saline solution, in dose-dependent manner, being this effect higher in those with higher arabinose and sulphate content. As fibrinogen levels were in normal range and no effect was demonstrated by Reptilase® time, neither the fibrinogen molecule, nor the polymerisation process would be altered. Anticoagulant activity was intermediate between that of DS and heparin. Besides, several fractions induced platelet aggregation by dose-dependent manner. In order to understand a possible aggregation mechanism, the extracts were tested on thrombastenic platelets that have GPIIbIIIa deficiency. This glycoprotein is the main receptor necessary to provide adequate support for platelet aggregation. The GPIIbIIIa receptors bind fibrinogen and other adhesive proteins to form cross-bridges between adjacent platelets. As no induction of aggregation was observed, it was concluded that GPIIbIIIa was involved in the aggregation mechanism when the seaweed extracts were used as agonists (Ciancia et al., 2007).

The highly sulphated 3-linked β-arabinan (Ab1) obtained from green seaweed *C. vermilara* (Bryopsidales), comprising major amounts of units sulphated on C-2 and C-4, showed anticoagulant activity by global coagulation tests. Less sulphated arabinans with similar backbones have less or no activity. Ab1 exerts its activity through direct and indirect (antithrombin- and heparin cofactor II-mediated) inhibition of thrombin. Direct thrombin inhibition was studied in detail. By native PAGE, it was possible to detect formation of a complex between Ab1 and human thrombin (HT). Ab1 binding to HT was measured by fluorescence spectroscopy. CD spectra of the Ab1-complex suggested that ligand binding induced a small conformational change on HT. Ab1-thrombin interactions were studied by molecular dynamic simulations using the persulphated octasaccharide as model compound. Most carbohydrate-protein contacts were postulated to occur by interaction of sulphate groups with basic amino acid residues in the surface of the enzyme, being more than 60% of them performed by the exosite 2-composing residues. In these interactions, the sulphate groups on C-2 showed to interact more intensely with thrombin structure. In contrast, the disulphated oligosaccharide does not promote major conformational modifications at the catalytic site when complexed to exosite 1. This pyranosic sulphated arabinan exerts its anticoagulant activity by a mechanism different to that of heparin, which interacts with exosite 2 of thrombin, but it only inhibits the enzyme by an AT-dependent mechanism, meanwhile dermatan sulphate potentiates HCII without binding to thrombin. The unusual mechanism described for Ab1, direct thrombin inhibition by interaction with exosite 2 that alters active site without mediation of AT, could be of interest in the context of development of antithrombotic strategies targeting thrombin (Fernández et al., 2013).

9.4.2 Antiangiogenic Activity

Angiogenesis, the formation of new blood vessels, plays an important role in physiological processes. In addition, angiogenesis is involved in several pathological conditions, including tumour growth and metastasis, atherosclerosis and diabetic retinopathy (Folkman, 1995). Inhibition of angiogenesis suppresses the tumour growth and metastasis.

It has been shown that fucoidans (Guerra Dore et al., 2013; Liu et al., 2012; Matou, Helley, Chabut, Bros, & Fischer, 2002; Pomin, 2012) and sulphated galactans from red seaweeds (De Souza et al., 2012) have antiangiogenic activity. However, to the best of our knowledge, only the anticoagulant sulphated galactan isolated from *Codium cylindricum* was investigated in this sense (Matsubara, Mori, Matsumoto, Hori, & Miyazawa, 2003). This galactan suppressed microvessel formation in an *ex vivo* serum-free matrix culture model using rat aortic ring. It also inhibited human umbilical vein endothelial cells (HUVEC) tube formation on reconstituted basement membrane gel. These effects were already evident at a concentration of 5 μg/ml of the polysaccharide. These results show the interest of further studies in this area of research.

9.4.3 Antiviral Activity

Several sulphated seaweed polysaccharides show antiviral activity against enveloped viruses, including important human patogens as Herpex simplex virus (HSV), human immunodeficiency virus (HIV), dengue virus and others (Damonte, Matulewicz, & Cerezo, 2004). Their antiviral activity appears to be based on the ability to interfere with the initial attachment of the virus to the target cell, blocking the viral entry. This activity was mainly reported for polysaccharides from red seaweeds (carrageenans, agarans and xylomannans) and brown seaweeds (fucoidans). In 1999, Lee, Hayashi, Maeda, & Hayashi, 2004 described the anti-HSV-1, anti-HCMV and anti–HIV-1 activity of a rhamnnan from the green seaweed *Monostroma latissimum*. Later, the same group assayed for anti-HSV-1 activity 11 polysaccharides from green algae, including two arabinans from *C. latum* and *C. fragile* and an arabinoxylogalactan from *C. adhaerens*. Although no detailed structural information about these polysaccharides was given, differences in antiviral potency among arabinans appear to be related with sulphate content. Arabinan from *C. latum*, as well as other compounds evaluated, showed the highest activity when added to the medium at the time of viral infection, but it maintained its activity at high levels even when added to the medium 8 h post-infection.

Finally, a highly branched galactan from *C. fragile,* FG, inhibited the replication of HSV-2, possibly by interfering in the early steps such as virus adsorption and penetration into host cells. FG suppressed virus production most efficiently when added at the same time as virus infection and throughout the incubation thereafter. Pretreatment of host cells with FG showed no inhibitory effects, and less anti-HSV-2 effect was observed when added only during viral infection. FG maintained antiviral activity at higher levels when added to the medium even after 6 h of virus infection. In addition, FG was found to make the virion lose its infectivity. In contrast, FG did not show anti-influenza A virus effect (Ohta et al. 2009).

9.4.4 Immunostimulating Activity

Different cell types are involved in innate and adaptive immune responses. Among them, macrophages have a central role, as they act removing pathogens, presenting antigens to lymphocytes and releasing biologically active molecules that regulate other cell's activity (Leiro, Castro, Arranz, & Lamas, 2007). They also participate in inflammatory processes, promoting them, as well as releasing anti-inflammatory cytokines. Some sulphated polysaccharides are capable of mimicking the protein-binding characteristics of mammalian glycocongujates (Nika, Mulloy, Carpenter, & Gibbs, 2003), thus modifying the activity of macrophages and exerting immunomodulatory effects. Sulphated polysaccharides can also interact with several cytokines and exert direct or indirect modulating effects (Nika et al., 2003). In some cases, antitumour activity reported for seaweed polysaccharides was found to be associated with macrophages, by means of the release of certain cytokines such as tumour necrosis factor-α (TNF- α) (Leiro et al., 2007).

The same branched galactan from *C. fragile,* described above as having antiviral activity (Ohta et al., 2009), was found to stimulate the production of nitric oxide (NO) by activation of inducible nitric oxide syntase (iNOS), at the mRNA and protein levels (Lee, Ohta, Hayashi, & Hayashi, 2010). Nitric oxide may inhibit an early stage in viral replication, and thus prevent viral spreads. In addition, this galactan also enhances the production of both pro-inflammatory (IL-1, IL-6, IL-12 and TNF-α) and anti-inflammatory (IL-10) cytokines.

Crude cell walls, as well as galactose and mannose-rich fractions from *C. fragile,* were reported to improve macrophage cells proliferation (Tabarsa et al., 2013). Besides, they also stimulate cytokines (COX-2, IL-1β, IL-6,

IL-10 and TNF-α) and NO release by these cells. Increased levels of NO were related with enhanced mRNA and protein expression of iNOS in the presence of the algal polysaccharides. This stimulation effect on macrophages could be due to the activation of NF-κB and MAPK pathways.

Nika et al. (2003) assayed the binding of a sulphated arabinogalactan from *C. fragile* to IL-2, IL-7 and gamma interferon (INF- γ). This polysaccharide appears to bind cytokines to a similar extent as λ-, κ- and ι-carrageenans. Sulphated polysaccharides could be useful for cytokine-based therapies, as they help to concentrate cytokine molecules close to their site of secretion and also protect them against proteolytic degradation (Nika et al., 2003).

9.5 CONCLUSION

In summary, *Codium* species biosynthesise sulphated polysaccharides with very distinct structural features, which have been elucidated in recent years. Some of them have different biological activities with a great potential in pharmaceutical applications. However, a lot of work is still pending for the development of drugs suitable for use in human health

ACKNOWLEDGEMENTS

This work was supported by grants from National Research Council of Argentina, CONICET (PIP 559–2009), and the National Agency for Promotion of Science and Technology, ANPCYT (PICT 2008-0500). Marina Ciancia is a member of the National Research Council of Argentina.

REFERENCES

Albersheim, P., Darvill, A., Roberts, K., Sederoff, R., & Staehelin, A. (2011). *Plant cell walls: from chemistry to biology*. New York, USA: Garland Science, Taylor and Francis Group.

Aquino, R. S., & Landeira-Fernández, A. M. (2005). Occurrence of sulphated galactans in marine angiosperms, evolutionary implications. *Glycobiology, 15*, 11–20.

Aspinall, G. O. (1982). Isolation and fractionation of Polysaccharides. In G. O. Aspinall (Ed.), *The polysaccharides* (Vol. 1) (pp. 19–34). Orlando, Florida,: Academic Press.

Bilan, M. I., Vinogradova, E.V., Shashkov, A. S., & Usov, A. I. (2006). Isolation and preliminary characterization of a highly pyruvylated galactan from *Codium yezoense* (Bryopsidales, Chlorophyta). *Botanica Marina, 49*, 259–262.

Bilan, M. I., Vinogradova, E. V., Shashkov, A. S., & Usov, A. I. (2007). Structure of a highly pyruvylated galactan sulphate from the pacific green alga *Codium yezoense* (Bryopsidales, Chlorophyta). *Carbohydrate Research, 342*, 586–596.

Bobin-Dubigeon, C., Lahaye, M., Guillon, F., Barry, J. L., & Gallant, D. J. (1997). Factors limiting the biodegradation of *Ulva* sp cell-wall polysaccharides. *Journal of the Science of Food and Agriculture, 75*, 341–351.

Cannon, M. C., Terneus, K., Hall, Q., Wang, L. T. Y., Wegenhart, B. L., Chen, L., et al. (2008). Self-assembly of the plant cell wall requires an extensin scaffold. *Proceedings of the National Academy of Sciences of the United States of America, 105*, 2226–2231.

Cardoso, M. A., Noseda, M. D., Fujii, M. T., Zibetti, R. G. M., & Duarte, M. E. R. (2007). Sulphated xylomannans isolated from red seaweeds *Chondrophycus papillosus* and *C. flagelliferus* (Ceramiales) from Brazil. *Carbohydrate Research, 342*, 2766–2775.
Carlberg, G. E., & Percival, E. (1977). Carbohydrates of the seaweeds *Urospora wormskioldii* and *Codiolum pusillum*. *Carbohydrate Research, 57*, 223–234.
Carpita, N. C., Defernez, M., Findlay, K., Wells, B., Shoue, D. A., Catchpole, G., et al. (2001). Cell wall architecture of the elongating maize coleoptile. *Plant Physiology, 127*, 551–565.
Cassolato, J. E. F., Noseda, M. D., Pujol, C. A., Pellizzari, F. M., Damonte, E. B., & Duarte, M. E. R. (2008). Chemical structure and antiviral activity of the sulphated heterorhamnan isolated from the green seaweed *Gayralia oxysperma*. *Carbohydrate Research, 343*, 3085–3095.
Chanzy, H. D., Grosrenaud, A., Vuong, R., & Mackie, W. (1984). The crystalline polymorphism of mannans in plant cell walls and after recrystallisation. *Planta, 161*, 320–329.
Ciancia, M., Alberghina, J., Arata, P. X., Benavides, H., Leliaert, F., Verbruggen, H., et al. (2012). Characterization of cell wall polysaccharides of the coencocytic green seaweed *Bryopsis plumosa* (Bryopsidaceae, Chlorophyta) from the Argentine coast. *Journal of Phycology, 48*, 326–335.
Ciancia, M., Quintana, I., & Cerezo, A. S. (2010). Overview of anticoagulant activity of sulphated polysaccharides from seaweeds in relation to their structures, focusing on those of green seaweeds. *Current Medicinal Chemistry, 17*, 2503–2529.
Ciancia, M., Quintana, I., Vizcarguenaga, M. I., Kasulin, L., Dios, A., Estevez, J. M., et al. (2007). Polysaccharides from the green seaweeds *Codium fragile* and *C. vermilara* with controversial effects on hemostasis. *International Journal of Biological Macromolecules, 41*, 641–649.
Damonte, E. B., Matulewicz, M. C., & Cerezo, A. S. (2004). Sulphated seaweed polysaccharides as antiviral agents. *Current Medicinal Chemistry, 11*, 2399–2419.
De Souza, L. A. R., Dore, C. M. P.G., Castro, A. J. G., De Azevedo, T. C. G., De Oliveira, M. T. B., Moura, M. F. V., et al. (2012). Galactans from the red seaweed *Amansia multifida* and their effects on inflammation, angiogenesis, coagulation and cell viability. *Biomedicine and Preventive Nutrition, 2*, 154–162.
Domozych, D. S., Ciancia, M., Fangel, J., Mikkelsen, M. D., Ulvskov, P., & Willats, W. G. T. (2012). The cell walls of green algae, a journey through evolution and diversity. *Frontiers in Plant Science, 3*, 1–7. article 82.
Domozych, D. S., & Domozych, C. R. (2003). Mucilage processing and secretion in the green alga *Closterium. II.* Ultrastructure and immunochemistry. *Journal of Phycology, 29*, 659–667.
Domozych, D. S., Lambiasse, L., Kiemle, S. N., & Gretz, M. R. (2009). Cell-wall development and bipolar growth in the desmid *Penium margaritaceum* (Zygnematophyceae, Streptophyta). Asymmetry in a symetric world. *Journal of Phycology, 45*, 879–893.
Dunn, E. K., Shoue, D. A., Huang, X., Kline, R. E., MacKay, A. L., Carpita, N. C., et al. (2007). Spectroscopic and biochemical analysis of regions of the cell wall of the unicellular 'mannan weed', *Acetabularia acetabulum*. *Plant Cell Physiology, 48*, 122–133.
Ender, F., Godl, K., Wenzl, S., & Sumper, M. (2002). Evidence for autocatalytic cross-linking of hydroxiproline-rich glycoproteins during extracellular matrix assembly in Volvox. *Plant Cell, 14*, 1147–1160.
Estevez, J. M., Fernández, P. V., Kasulin, L., Dupree, P., & Ciancia, M. (2009). Chemical and in situ characterization of macromolecular components of the cell walls from the green seaweed *Codium fragile*. *Glycobiology, 19*(3), 212–228.
Farias, E. H. C., Pomin, V. H., Valente, A. P., Nader, H. B., Rocha, H. A. O., & Mourão, P. A. S. (2008). A preponderantly 4-sulphated, 3-linked galactan from the green alga *Codium isthmocladum*. *Glycobiology, 18*, 250–259.
Fernández. P. V. (2012). *Polysaccharides from green seaweed* Codium *vermilara, fine structure, its arrangement in the cell wall. Anticoagulant Activity of a Pyranosic Sulphated Arabinan* (Ph.D. thesis University of Buenos Aires, Argentina).
Fernández P. V., Raffo M. P., Alberghina J., Ciancia M., unpublished results. Polysaccharides from the green seaweed *Codium decorticatum*. Structure and cell wall distribution.

Fernández, P.V., Ciancia, M., Miravalles, A. B., & Estevez, J. M. (2010). Cell-wall polymer mapping in the coenocytic macroalga *Codium vermilara* (Bryopsidales, Chlorophyta). *Journal of Phycology*, *46*, 456–465.

Fernández, P.V., Estevez, J. M., Cerezo, A. S., & Ciancia, M. (2012). Sulphated β-D-mannan from the green seaweed *Codium vermilara*. *Carbohydrate Polymers*, *87*, 916–919.

Fernández, P.V., Quintana, I., Cerezo, A. S., Caramelo, J. J., Pol-Fachin, L., Verli, H., et al. (2013). Anticoagulant activity of a unique sulphated pyranosic (1→3)-β-L-arabinan through direct interaction with thrombin. *Journal of Biological Chemistry*, *288*, 223–233.

Folkman, J. (1995). Angiogenesis in cancer, vascular, rheumatoid and other disease. *Nature Medicine*, *1*, 27–31.

Fukushi, Y., Otsuru, O., & Maeda, M. (1988). The chemical structure of the D-xylan from the main cell wall constituents of *Bryopsis maxima*. *Carbohydrate Research*, *182*, 313–320.

Guerra Dore, C. M. P., Faustino Alves, M. G. C., Santos, N. D., Cruz, A. K. M., Câmara, R. B. G., Castro, A. J. G., et al. (2013). Antiangiogenic activity and direct antitumor effect from a sulphated polysaccharide isolated from seaweed. *Microvascular Research*, *88*, 12–18.

Hallman, A. (2006). The pherophorins, common, versatile building blocks in the evolution of extracellular matrix architecture in Volvocales. *Plant Journal*, *45*, 292–307.

Hayakawa, Y., Hayashi, T., Lee, J.-B., Srisomporn, P., Maeda, M., Ozawa, T., et al. (2000). Inhibition of thrombin by sulphated polysaccharides isolated from green seaweeds. *Biochimica et Biophysica Acta*, *1543*, 86–94.

Huizing, H. J., & Rietema, H. (1975). Xylan and mannan as cell wall constituents of different stages in the life-histories of some siphoneous green algae. *British*, *10*, 13–16.

Huizing, H. J., Rietema, H., & Sietsma, J. H. (1979). Cell wall constituents of several siphoneous green algae in relation to morphology and taxonomy. *British Phycological Journal*, *14*, 25–32.

Jiao, G., Yu, G., Zhang, J., & Stephen Ewart, H. (2011). Chemical structures and bioactivities of sulphated polysaccharides from Marine algae. *Marine Drugs*, *9*, 196–223.

Kaihou, S., Hayashi, T., Otsuru, O., & Maeda, M. (1993). Studies on the cell-wall mannan of the siphonous green algae, *Codium* latum. *Carbohydrates Research*, *240*, 207–218.

Kloareg, B., & Quatrano, R. S. (1988). Structure of the cell wall of marine algae and eco-physiological functions of the matrix polysaccharides. *Oceanography and Marine Biology-an Annual Review*, *26*, 259–315.

Lahaye, M., & Robic, A. (2007). Structure and functional properties of Ulvan, a polysaccharide from Green seaweeds. *Biomacromolecules*, *8*, 1765–1774.

Lee, J. B., Hayashi, K., Maeda, M., & Hayashi, T. (2004). Antiherpetic activities of sulphated polysaccharides from green algae. *Planta Medica*, *70*, 813–817.

Lee, J. B., Koizumi, S., Hayashi, K., & Hayashi, T. (2010). Structure of rhamnan sulphate from the green alga *Monostroma nitidum* and its anti-herpetic effect. *Carbohydrate Polymers*, *81*, 572–577.

Lee, J. B., Ohta, Y., Hayashi, K., & Hayashi, T. (2010). Immunostimulating effects of a sulphated galactan from *Codium fragile*. *Carbohydrate Research*, *345*, 1452–1454.

Leiro, J. M., Castro, R., Arranz, J. A., & Lamas, J. (2007). Immunomodulating activities of acidic polysaccharides obtained from the seaweed *Ulva rigida* C. Agardh. *International Immunopharmacology*, 7, 879–888.

Li, H., Mao, W., Zhanga, X., Qia, X., Chena, Y., Chena, Y., et al. (2011). Structural characterization of an anticoagulant-active sulphated polysaccharide isolated from green alga *Monostroma latissimum*. *Carbohydrate Polymers*, *85*, 394–400.

Liu, F., Wang, J., Chang, A. K., Liu, B., Yang, L., Li, Q., et al. (2012). Fucoidan extract derived from *Undaria pinnatifida* inhibits angiogenesis by human umbilical vein endothelial cells. *Phytomedicine*, *19*, 797–803.

Love, J., & Percival, E. (1964a). The polysaccharides of the green seaweed *Codium fragile*, part II. the water-soluble sulphated polysaccharides. *Journal of the Chemical Society*, 3338–3345.

Love, J., & Percival, E. (1964b). The polysaccharides of the green seaweed *Codium fragile*, part III. A β-1,4-linked mannan. *Journal of the Chemical Society*, 3345–3349.
Mackie, W. (1969). The degree of polymerization of xylan in the cell wall of the green seaweed *Penicillus dumetosus*. *Carbohydrate Research*, *9*, 247–249.
Mackie, I. M., & Percival, E. (1959). The constitution of xylan from the green seaweed *Caulerpa filiformis*. *Journal of the Chemical Society*, 1151–1156.
Mackie, W., & Sellen, D. B. (1969). The degree of polymerization and polydispersity of mannan form the cell wall of the green seaweed *Codium fragile*. *Polymer*, *10*, 621–632.
Maeda, M., Fukushi-Fujikura, Y., & Otsuru, O. (1990). Cellulose in the cell wall of the siphonous green alga, *Bryopsis maxima*. *Carbohydrate Research*, *201*, 91–99.
Matou, S., Helley, D., Chabut, D., Bros, A., & Fischer, A.-M. (2002). Effect of fucoidan on fibroblast growth factor-2-induced angiogenesis in vitro. *Thrombosis Research*, *106*, 213–221.
Matsubara, K. (2004). Recent advances in marine algal anticoagulants. *Current Medicinal Chemistry - Cardiovascular & Hematological Agents*, *2*, 13–19.
Matsubara, K., Matsuura, Y., Bacic, A., Liao, M. L., Hori, K., & Miyazawa, K. (2001). Anticoagulant properties of a sulphated galactan preparation from a marine green alga, *Codium cylindricum*. *International Journal of Biological Macromolecules*, *28*, 395–399.
Matsubara, K., Matsuura, Y., Hori, K., & Miyazawa, K. (2000). An anticoagulant proteoglycan from the marine green alga, *Codium pugniformis*. *Journal of Applied Phycology*, *12*, 9–14.
Matsubara, K., Mori, M., Matsumoto, H., Hori, K., & Miyazawa, K. (2003). Antiangiogenic properties of a sulphated galactan isolated from a marine green alga, *Codium cylindricum*. *Journal of Applied Phycology*, *15*, 87–90.
McCartney, L., Blake, A. W., Flint, J., Bolam, D. N., Boraston, A. B., Gilbert, H. J., et al. (2006). Differential recognition of plant cell walls by microbial xylan-specific carbohydrate-binding modules. *Proceedings of the National Academy of Sciences of the United States of America*, *103*, 4765–4770.
Moller, I., Sørensen, I., Bernal, A. J., Blaukopf, C., Lee, K., Øbro, J., Pettolino, F., Roberts, A., Mikkelsen, J. D., Knox, J. P., Bacic, A., Willats, W. G. T. (2007). High-throughput mapping of cell-wall polymers within and between plants using novel microarrays, *Plant Journal*, *50*(6), 1118–1128.
Mulloy, B., Ribeiro, A. C., Alves, A. P., Vieira, R. P., & Mourão, P. A. (1994). Sulphated fucans from echinoderms have a regular tetrasaccharide repeating unit defined by specific patterns of sulfation at the 0-2 and 0-4 positions. *Journal of Biological Chemistry*, *269*(35), 22113–22123.
Nika, K., Mulloy, B., Carpenter, B., & Gibbs, R. (2003). Specific recognition of immune cytokines by polysaccharides from marine algae. *European Journal of Phycology*, *38*, 257–264.
Nilsson, S., & Piculell, L. (1991). Helix-coil transitions of ionic polysaccharides analyzed within the Poisson-Boltzmann cell model. 4. Effects of site-specific counterion binding. *Macromolecules*, *24*, 3804–3811.
Ohta, Y., Lee, J. B., Hayashi, K., & Hayashi, T. (2009). Isolation of sulphated galactan from *Codium fragile* and its antiviral effect. *Biological and Pharmaceutical Bulletin*, *32*, 892–898.
Oliveira-Carvalho, M. F., Oliveira, M. C., Barreto-Pereira, S. M., & Verbruggen, H. (2012). Phylogenetic analysis of Brazilian *Codium* species, including the new species *C. Pernambucensis*. *European Journal of Phycology*, *47*, 355–365.
Percival, E., & McDowell, R. H. (1981). Algal walls. Composition and biosynthesis. *Encyclopedia of Plant Physiology*, *13B*, 277–316.
Pereira, M. S., Mulloy, B., & Mourão, P. A. (1999). Structure and anticoagulant activity of sulphated fucans. Comparison between the regular, repetitive, and linear fucans from echinoderms with the more heterogeneous and branched polymers from brown algae. *Journal of Biological Chemistry*, *274*(12), 7656–7667.
Pérez Recalde, M., Noseda, M. D., Pujol, C. A., Carlucci, M. J., & Matulewicz, M. C. (2009). Sulphated mannans from the red seaweed *Nemalion helminthoides* of the South Atlantic. *Phytochemistry*, *70*, 1062–1068.

Pomin, V. H. (2012). Fucanomics and galactanomics, current status in drug discovery, mechanisms of action and role of the well-defined structures. *Biochimica et Biophysica Acta, 1820*, 1971–1979.

Robinson, G., Morris, E. R., & Rees, D. A. (1980). Role of double helices in carrageenan gelation, the domain model. *Journal of the Chemical Society, Chemical Communications*, 152–153.

Shanmugam, M., & Mody, K. H. (2000). Heparinoid-active sulphated polysaccharides from marine algae as potential blood anticoagulant agents. *Current Science, 79*, 1672–1983.

Siddhanta, A. K., Shanmugam, M., Mody, K. H., Goswami, A. M., & Ramabat, B. K. (1999). Sulphated polysaccharides of *Codium dwarkense Boergs.* from the west coast of India, chemical composition and blood anticoagulant activity. *International Journal of Biological Macromolecules, 26*, 151–154.

Tabarsa, M., Karnjanapratum, S., Cho, M. L., Kim, J. K., & You, S. G. (2013). Molecular characteristics and biological activities of anionic macromolecules from *Codium fragile*. *International Journal of Biological Macromolecules, 59*, 1–12.

Uehara, T., Takeshita, M., & Maeda, M. (1992). Studies on anticoagulant-active arabinan sulphates from the green alga, *Codium latum*. *Carbohydrate Research, 23*, 309–311.

Usov, A. I. (2011). Polysaccharides of the red algae. *Advances in Carbohydrate Chemistry and Biochemistry, 65*, 115–217.

Usov, A. I., & Bilan, M. I. (2009). Fucoidans-sulphated polysaccharides of brown algae. *Russian Chemical Reviews, 78*, 785–799.

Van Zyl, W. H., Rose, S. H., Trollope, K., & Gorgens, J. F. (2010). Fungal β-mannanases, mannan hydrolysis, heterologous production and biotechnological applications process. *Biochemistry, 45*, 1203–1213.

Verbruggen, H., Leliaert, F., Maggs, C. A., Shimada, S., Schils, T., Provan, J., et al. (2007). Species boundaries and phylogenetic relationships within the green algal genus *Codium* (Bryopsidales) based on plastid DNA sequences. *Molecular Phylogenetics and Evolution, 44*, 240–254.

Vilela-Silva, A. C., Hirohashi, N., & Mourão, P. A. (2008). The structure of sulphated polysaccharides ensures a carbohydrate-based mechanism for species recognition during sea urchin fertilization. *International Journal of Developmental Biology, 52*(5–6), 551–559.

Wutz, M., & Zetsche, K. (1976). Zur biochimie und regulation des heteromorphen generationswechsels der grünalge. *Planta, 129*, 211–216.

Yamagaki, T., Maeda, M., Kanazawa, K., Ishizuka, Y., & Nakanishi, H. (1997). Structural clarification of *Caulerpa* cell wall β-(1→3)-xylan by NMR spectroscopy. *Bioscience Biotechnology Biochemistry, 61*, 1077–1080.

CHAPTER TEN

Enzymatic Recovery of Metabolites from Seaweeds: Potential Applications

Kevin Hardouin*,†,1, Gilles Bedoux*, Anne-Sophie Burlot*, Pi Nyvall-Collén‡ and Nathalie Bourgougnon*

*Université de Bretagne-Sud, Laboratoire de Biotechnologie et Chimie Marine, EA 3884, Institut Universitaire Européen de la Mer, Vannes, France
†OLMIX SA, Bréhan, France
‡Amadéite SAS, Bréhan, France
[1]Corresponding author: e-mail address: kevin.hardouin@univ-ubs.fr

Contents

Abstract

Seaweeds are the basis of a multibillion-dollar economy with an impact on highly diverse sectors, including food, feed, textile, pharmaceutical, nutraceutical, cosmetic, chemistry, bioactive compounds, and biotechnological sectors like bioenergy. Molecules from seaweed have provided promising drug leads, offered new targets for synthetic chemists, and provided opportunities for the elucidation of unusual biosynthetic pathways. Seaweed farming has developed all over the world. Simultaneously,

Advances in Botanical Research, Volume 71
ISSN 0065-2296
http://dx.doi.org/10.1016/B978-0-12-408062-1.00010-X

the increasing report of outbreaks of 'green tides' ('brown or red tides'), fouling species and large piles of decomposing biomass which appear along shallow sandy bays represent a true economic constraint for the affected communities. Harvesting of invasive and proliferative species is an opportunity for collecting an important biomass for research of new compounds of interest. This review describes enzymatic processing of seaweeds as reported for the last 20 years. At this time when sustainable development and environmental protection are regarded as key factors, studies about innovative extraction processes of active compounds from natural products have attracted special attention. The use of enzyme treatment as a tool to improve the extraction efficiency of bioactive compounds from seaweeds shows several benefits. This chapter presents a synthesis of the conditions applied for seaweed cracking. Three different types of hydrolysis are used depending on the process design and the nature of the enzymes. Biochemical composition of hydrolysates, enzymatic hydrolysis efficiency, and biological activities depend on the nature of the macroalgae, the enzyme activity, the experimental conditions and the design process. Among the biological properties, the hydrolysates show various activities: antioxidant, antiviral, antiproliferative, antihypertensive and anticoagulative. These data demonstrate some potential applications of seaweed metabolites obtained by enzymatic hydrolysis and of the prospects for the future.

10.1 INTRODUCTION

Algae is a functional definition including aquatic, photosynthetic, organisms of prokaryotic or eukaryotic origin. The group include unicellular and multicellular organisms of different phylogenetic origins as Archaeplastida (Chlorophyta, Rhodophyta and Glaucophyta), Rhizaria, Excavata, Chromista (Heterokonts, including Phaeophyceae, Dinoflagellates, Cryptophyta and Haptophyta). Algae are separated from Plantae by their lack of true roots, stems leaves and embryos. There are approximately 25–30,000 species of algae on earth and 100,000 including diatoms. The term 'seaweed' is used to describe multicellular algae and include members of red, green and brown algae, which are the only groups of algae that have developed complex multicellularity. The rest of this manuscript will concern seaweed. The three different groups of seaweed were empirically distinguished in the mid-nineteenth century on the basis of thallus colour, green (Chlorophyceae), red (Rhodophyceae) and brown (Phaeophyceae), which have since been shown to be monophyletic groups. Seaweeds occur in different shapes and sizes, ranging from single-cell monolayers or filaments to complex structures that approximate the structure of plants.

Seaweeds are traditionally used as sea vegetables in Asian countries such as China, Japan and Korea since thousands of years.

Around 500 species of marine macroalgae have been used for centuries for human food and medicinal purposes, directly mostly in Asia or indirectly, mainly by the production of fertilizers and the more recent phycocolloid industry (agars, carrageenans and alginates). Marine algae are rich sources of diverse bioactive compounds with various biological activities. Their importance as a source of novel bioactive substances is growing rapidly with numerous publications on various biological activities of marine algal origin (Bhakuni & Rawat, 2005; Holdt & Kraan, 2011; Samarakoon & Jeon, 2012; Stengel, Connan, & Popper, 2011; Wijesekara, Pangestuti, & Kim, 2011). Excluding the bioactive secondary metabolites that target the pharmaceutical market, the most common compounds from algae with existing commercial applications are, besides phycocolloids, polyunsaturated fatty acids, steroids, carotenoids and lectins.

In the world, 94% of the seaweeds produced come from aquaculture and only 6% from harvesting of wild biomass (Buchholz, Krause, & Buck, 2012). Exportations represent approximately two million tons per year for a trade of 680 M₡. The producers are Indonesia 32%, Chile 19%, China 11%, Ireland 8%, Korea 5%, Peru 4%, Philippines 4% and other countries 19%. France represents less than 1% of the world seaweed production (Chopin & Sawhney, 2009; McHugh and Food and Agriculture Organization of the United Nations, 2003). 75% of the seaweed production is used for food, 12% for agriculture and 13% for the colloids industry. The worldwide increase in the demand for raw materials for the sectors of food, cosmetics, bioactive molecules and more recently for chemistry and bioenergy, raises the issue of the sustainability of the French and European industries. In Europe, only 1% is used as vegetable, 24% in agriculture and 75% for colloids industry.

Native or introduced macroalgae can become proliferative and have profound adverse ecological impacts including the alteration of ecosystem structure, reduction of indigenous biodiversity and economic losses. Excessive growth of some opportunistic seaweed in reaction to medium disturbance is observed with increasing frequency in coastal ecosystems close to agricultural or strongly urbanised and industrialised zones. These massive proliferations can cause different inconveniences to the population, for example on health and tourism. However, this important bioresource also represents an interesting opportunity (Smetacek & Zingone, 2013) to explore the production of new biomolecules for applications in food, feed, agriculture, pharmaceutics, cosmetics, biofuels...etc. (Cardozo et al., 2007; Fleurence, 1999a; Kapraun, 1999; Stengel et al., 2011). Seaweeds are among the first marine organisms chemically studied, with more than 3600 articles published describing 3300 secondary metabolites from marine plants and algae, and they still remain an almost endless source of new bioactive compounds (Ioannou & Roussis, 2009, pp. 51–81). Due to the wide range of competitive environments they live in, their divergent phylogenetic origins and their respective evolutions, marine organisms have developed unique defence strategies and bioactive compounds that, in some cases, are unparalleled by their terrestrial counterparts (Kornprobst, 2005). Secondary metabolic pathways evolved as a result of complex interactions between organisms in their native habitats, but the role of natural products in mediating such interactions remains poorly understood in the majority of cases (Ioannou & Roussis, 2009, pp. 51–81).

Many different procedures have been studied for the extraction of metabolites from seaweeds. For industrial applications, ideal extraction method should be quantitative, nondestructive and solvent-free, characterised by an increased extraction rate allowing shorter extraction times and reduced energy and financial costs. These extractions can be performed by chemical means like organic solvents, alkaline and acid conditions, hot water or mechanic treatments like, pressure, sonication, grinding or freezing. The extraction of metabolites from seaweed is usually performed by coupling a chemical and a physical treatment. Over the past decade, the enzyme-assisted extraction method has been studied many times (Samarakoon & Jeon, 2012; Wijesinghe & Jeon, 2012) showing the interest of this eco-friendly method compared to the conventional methods, improving both quality and yield. In most studies, the enzymes, or multienzyme mixes, used are commercial preparations that are already used in the food industries for the hydrolysis of starch, cellulose, proteins and generally for molecules also present in terrestrial plants. The identification of bacterial degradation pathways or processes controlling biofilm could be also used for production of novel enzymes (Correc, Hehemann, Czjzek, & Helbert, 2011; Courtois, 2009; Lundqvist, Jam, Barbeyron, Czjzek, & Sandström, 2012) to improve seaweed biomass processing.

The aim of this chapter is a discussion about the use of enzyme-assisted extraction as an alternative method to improve the recovery of bioactive compounds from seaweeds.

10.2 SEAWEED RESOURCES

10.2.1 Harvesting and Aquaculture

The seaweed biorefinery requires a significant amount of biomass. The harvesting of abundant native species, invasive species and seaweeds from aquaculture constitute the three types of biomasses.

Abundant native biomasses constitute large fields of seaweeds such as kelp *Laminaria digitata* and *Laminaria saccharina* (Steneck et al., 2003). *Laminaria* species or *Macrocystis pyrifera* 'giant kelp' forms dense submarine 'forests' on hard substrata or occasionally on coarse sand at depths less than 15–40 m off the coast of Brittany (France), southern California (USA) or Tasmania. These forests are some of the most productive communities on earth with much of the annual production exported as large floating rafts. The kelp harvest increased annually from 10,000 tons in the early 1930s to a high of 170,000 tons in the late 1970s for alginate extraction and abalone feed. The harvesting for alginates ceased in 2005, but the kelp is still harvested as feed for abalone aquaculture. These biomasses represent a very important resource of seaweed. However, the harvesting represents only 6% of the total tonnage of seaweeds that are industrially used in the world (Buchholz et al., 2012).

Currently, the majority (94%) of seaweeds that are industrially processed comes from aquaculture. As demand increases, natural populations frequently become overexploited and the need for the cultivation of the appropriate species emerges. Mariculture is mainly developed in Asia where China produced 63%, Indonesia 14% and Philippines 11%, by quantity, of the global international production (Chopin & Sawhney, 2009; Douglas-Westwood, 2005; FAO, 2011; McHugh and Food and Agriculture Organization of the United Nations, 2003). In 2008, the main cultivated seaweeds were *Saccharina japonica* (formerly *Laminaria japonica*), *Eucheuma* sp., *Kappaphycus* sp., *Undaria pinnatifida* (Wakame), *Gracilaria* spp. and *Porphyra* spp. (Nori). *Eucheuma* sp. and *Kappaphyccus* are cultivated for colloids industry (carrageenans) and the others for human consumption (FAO, 2011). Seaweeds are cultivated on floating bamboo modules in deep water and on mangrove nets or ropes in shallow waters (Pereira & Yarish, 2008). With these installations, seaweeds are always exposed to the sunlight and the growth rate is increased (Figure 10.1). Seaweeds can be cultivated in single-species culture or in integrated multitrophic aquaculture (Chopin et al., 2008). Historically, seaweeds were harvested by hand cutting with sickles or scythes on rocks at low tides or with dragnets. But in recent year mechanical method, by boat or trucks, had been developed (Titlyanov & Titlyanova, 2010).

These fields of seaweeds can also be due to the presence of introduced species that have colonised some ecosystems. Introduced species concerns seaweed that had settled in a non-native area by human activity (aquarium introduction, hull fouling, intentional aquaculture, etc.) (Ruiz, Fofonoff, Carlton, Wonham, & Hines, 2000) and successfully installed in this new

Figure 10.1 (A) *Kappaphyccus* sp. farming on ropes and (B) open air drying in Bali, Indonesia, April 2013. (See the colour plate.) © *Balakrishnan Prithiviraj and Kevin Hardouin.*

environment. Williams and Smith (2007) reviewed over 407 global seaweed introduction events caused by 277 seaweed species, including 165 Rhodophytes (60%), 66 Ochrophytes (24%), 45 Chlorophytes (16%) and one Charophyte (<0.5%). Some of these species, both native and invasive, causes every year very important blooms or 'tides' of seaweed. As examples, *Solieria chordalis* (native) in France (Tavennec, 2009, 107 pp.), *Ulva prolifera* (native) in China (Hu et al., 2010; Zhang et al., 2013), *Ulva rotundata* and *Ulva armoricana* (native) in France (Coat et al., 1998; Smetacek & Zingone, 2013), *Codium fragile* (invasive) in Canada (D'Amours & Scheibling, 2007), *Caulerpa* sp. (invasive) in Mediterranean Sea and Florida (Cebrian & Ballesteros, 2009; Lapointe & Bedford, 2010), *Grateloupia turuturu* (invasive) in North America (Mathieson, Dawes, Pederson, Gladych, & Carlton, 2007) and in France (Denis et al., 2010), *Undaria pinatifida* (invasive) in New Zealand (Schiel & Thompson, 2012), in Argentina (Casas, Scrosati, & Piriz, 2004) and in Europe (Sfriso & Facca, 2013), *Sargassum muticum* (invasive) in Canada and in Europe from Norway to Portugal (Engelen & Santos, 2009; Sfriso & Facca, 2013) are responsible of these blooms. These blooms represent a very important biomass that can be easily harvested in shallow sandy bays or directly on beaches (Avila & Seguel, 1993; Cardozo et al., 2007; Charlier, Morand, & Finkl, 2008; Charlier, Morand, Finkl, & Thys, 2007; Hu et al., 2010; Kirkman & Kendrick, 1997). Specific techniques have been developed in order to limit the impact of harvesting on ecosystems (Stagnol, Renaud, & Davoult, 2013).

10.2.2 Focus on French Seaweed Biomass

Marine algae from the French Brittany coasts (North West of France) constitute a significant and diversified natural vegetable production. The diversity of macroalgal species occurring in Brittany is due to abiotic factors (substrate, seawater temperature, hydrodynamism, light conditions), which are optimal for the development of macroalgae. Nearly 700 species of algae are currently listed on the Breton littoral. France is the 7th producer of marine algae in the world and the 1st in Europe. Moreover, 90% of the French production is harvested in Brittany. Among macroalgae present in Brittany, brown algae are the main group, which is harvested, following by red and green algae. Among brown algae, *Ascophyllum nodosum*, *L. digitata* and *Fucus serratus* were the most harvested species. As for red and green algae, harvesters collected mainly *Palmaria palmata*, the two allied morphological species *Chondrus crispus*/*Mastocarpus stellatus* and *Ulva* sp., 80,000 tons of collected algae each year in Brittany (France) are transformed into colloids, are intended for agri-food or cosmetic industries, or are employed as organic compounds (Bourgougnon & Stiger-Pouvreau, 2011).

Among the flora encountered in Brittany, native or introduced macroalgae can become invasive or proliferative and have profound adverse ecological impacts including the alteration of ecosystem structure, reduction of indigenous biodiversity and economic losses. Excessive growth of some opportunist seaweeds in reaction to medium disturbance is observed with increasing frequency in coastal ecosystems close to agricultural or strongly urbanised and industrialised zones (Hardouin et al., 2013). Approximately 70 species of invasive macroalgae are listed in France, sometimes developing important blooms giving rise to important landings of seaweed. The majority of these species just settles in the environment but some of it spreads on the seabed, unbalances the ecosystems and causes important landings (Williams & Smith, 2007).

10.2.2.1 Ulva *sp. (Chlorophyceae)*

These edible seaweeds, commonly called 'sea lettuce' for appearance and colour like lettuce, belong to the Ulvaceae (Figure 10.2). One hundred *Ulva* species are currently listed in the world (Algaebase). *Ulva* sp. attached to rocks during the first step of life grows in flotation in water (De Casabianca, Barthelemy, Serrano, & Sfriso, 2002). Periodically, particular species overgrow as 'green-tides' in eutrophicated coastal and lagoon waters (Venice lagoon, China) and create ecological and economic problems. In Brittany in France, the opportunistic growth ability of *Ulva* is associated with proliferation in eutrophicated coastal waters or with contamination of algal closed cultures (Lahaye & Robic, 2007). *Ulva* blooms are significant nuisances to

Taxonomy:	
Empire:	Eukaryota
Kingdom:	Plantae
Phylum:	Chlorophyta
Class:	Ulvophyceae
Order:	Ulvales
Family:	Ulvaceae
Genus:	Ulva
Specie:	*Ulva sp.*

Figure 10.2 Taxonomy and picture of *Ulva* sp. (See the colour plate.) © *Kristian Peters.*

make harvesting essential in some places (Ducreux, 2010; Smetacek & Zingone, 2013). In Brittany, each year 100,000 tons of landed algae are removed and disposed. The cost is estimated to be US$10–150 per ton (Charlier et al., 2008).

Seasonal fluctuations of protein, neutral sugars or uronic acid's content in seaweeds were previously reported in relation with species, geographic areas with values between 11 and 55% ash, 15–65% carbohydrates, 4–44% proteins and 0.3–1.6% lipids (%DW) (Holdt & Kraan, 2011). The skeletal phase of cell wall consists in cellulose (β-1,4-D-anhydroglucopyranose units), xyloglucan (cellulose-like material composed of D-xylose units) and glucuronans (Lahaye, Jegou, & Buleon, 1994; Lahaye & Robic, 2007; Popper et al., 2011). The cell-wall polysaccharides are essentially represented by ulvan, which is a complex sulphated polysaccharide representing 8–29% of the algae dry weight (Robic, Sassi, & Lahaye, 2008). The two major repeating disaccharides are aldobiuronic acids designated as type A: ulvanobiuronic acid 3-sulphate (A 3s) and type B: ulvanobiuronic acid 3-sulphate (B 3s). Xylose partially sulphated residues at O-2 can also occur as in place of uronic acids. In addition, glucuronic acid can branch at O-2 of rhamnose 3-sulphate (Lahaye & Robic, 2007; Popper et al., 2011).

10.2.2.2 Solieria chordalis *(Rhodophyceae)*

Only nine species of *Solieria* are listed in the world belonging to the family Solieriaceae (Figure 10.3), in which are also found the genera *Eucheuma, Kappaphycus* or *Sarcodiotheca. S. chordalis* is found in the Mediterranean Sea and Atlantic Ocean from Morocco to Southwest England. Little is known about its ecology, and Floc'h et al. (1987) showed that *S. chordalis* develops in the upper sublittoral, commonly in sheltered areas where salinities ranged between 26 and 35‰, and on a variety of substrates including gravel, shells and other algae. *S. chordalis* is observed in the Gulf of Morbihan (France) where strandings have become more abundant between July and October (Tavennec, personal communication, 2009). During the summertime, it is then a question of removing more than 2000 tons of red algae mainly of *S. chordalis* from the high part of the beaches.

Much less information is available in the literature regarding to the chemical composition of *S. chordalis.* According to Hardouin et al. (2013), the chemical composition of *S. chordalis* contains 27.4% ash, 9.8% neutral sugars, 8.0% uronic acids, 5.6% sulphate groups, 29.3% proteins and 2.1% total phenol. The genus *Solieria* contains iota-carrageenan as the main component. The structural components of this polysaccharide are mainly

Taxonomy:	
Empire:	Eukaryota
Kingdom:	Plantae
Phylum:	Rhodophyta
Class:	Florideophyceae
Order:	Gigartinales
Family:	Solieriaceae
Genus:	Solieria
Specie:	*Solieria chordalis*

Figure 10.3 Taxonomy and picture of *Solieria chordalis* collected in November 2013, in St-Gildas de Rhuys, Brittany (France). © *Anne-Sophie Burlot.*

a (D-anhydrogalactose-2-sulphate-galactose-4-sulphate)-type structure in association with methylated-ι-carrageenan, pyruvated α-carrageenan and the minor precursor, ν-carrageenan, in small amounts. The relative molecular weight of the native polysaccharide was estimated by LP-GPC as 913 kDa. (Estevez, Ciancia, & Cerezo, 2008; Bondu, Deslandes, Fabre, Berthou, & Guangli, 2010; Popper et al., 2011). The differentiation of each type of carrageenan is based on the nature of the units and the amount and the position of sulphate groups (Table 10.1). In *S. chordalis*, the flux of carbon issued from the photosynthesis is directed towards low-molecular weight carbohydrates of D-galactose and D-mannose floridoside (2-O-glycerol-α-D-galactopyranoside) and digeneaside (2-O-a-D-mannopyranosyl-D-glyceric acid) or floridean starch (polymer of 1,4-α-glucose) or cell-wall polysaccharides according to the external conditions (Bondu et al., 2010).

10.2.2.3 Sargassum muticum *(Phaeophyceae)*

More than 300 *Sargassum* species are currently listed in the world. The macroalga *S. muticum* (Yendo) Fensholt (Phaeophyceae, Fucales, Figure 10.4) is one of the most aggressive marine invaders (White & Shurin, 2011). Native from South East Asia, *S. muticum* was introduced in the Pacific Northwest around the 1940s with Japanese oysters (*Crassostrea gigas*) imported for aquaculture. *S. muticum* presently is distributed along Atlantic coasts, from Portugal to Norway.

Alginates, ascophyllans and fucoidans constitute the matrix phase of the cell wall of *Sargassum*. Alginate is composed of mannuronic and guluronic acid blocks (Kloareg & Quatrano, 1988; Popper et al., 2011). The

Table 10.1 Classification and Composition of the Different Types of Carrageenans

Carrageenans	Unit A: α-1,4 Linked	Unit B: β-1,3 Linked
κ-Carrageenans Group		
Kappa κ	3,6-anhydrogalactose	Galactose-4-sulphate
Iota ι	3,6-anhydrogalactose-2-sulphate	Galactose-4-sulphate
Mu μ	Galactose-6-sulphate	Galactose-4-sulphate
Nu ν	Galactose-2,6-disulphate	Galactose-4-sulphate
Omicron O	Galactose-2-sulphate	Galactose-4-sulphate
λ-Carrageenans Group		
Xi ξ	Galactose-2-sulphate	Galactose-2-sulphate
Lambda λ	Galactose-2,6-disulphate	Galactose-2-sulphate
Pi π	4,6-O-(1-carboéthylidène)-galactose-2-sulphate	Galactose-2-sulphate
Delta δ	Galactose-2,6-disulphate	Galactose
Alpha α	3,6-anhydrogalactose-2-sulphate	Galactose
Thêta θ	3,6-anhydrogalactose-2-sulphate	Galactose-2-sulphate
β-Carrageenans Group		
Bêta β	3,6 anhydrogalactose	Galactose
Gamma γ	Galactose-6-sulphate	Galactose
Psi ψ	Galactose-5-sulphate	Galactose-5-sulphate
Omega ω	3,6-anhydrogalactose	Galactose-6-sulphate

Taxonomy:	
Empire:	Eukaryota
Kingdom:	Chromista
Phylum:	Heterokontophyta
Class:	Phaeophyceae
Order:	Fucales
Family:	Sargassaceae
Genus:	Sargassum
Specie:	*Sargassum muticum*

Figure 10.4 Taxonomy and picture of *Sargassum muticum*. May 2009, Pointe du Diable, Brittany (France). © *Klervi Le Lann.*

other family of acids and soluble polysaccharides are represented by fucans, which are sulphated and ramified polysaccharides constituted by a large variety of sugars, as L-fucose and D-galactose. Laminarin, a sequence of 1,3-β-glucopyranose, constitute the energy storage sugar (Kloareg & Quatrano,

1988; Popper et al., 2011). The D-mannitol, simple sugar free in the cells, presents a role in the regulation of osmotic pressure.

The chemical composition of *Sargassum* sp. is 14–44% ash, 4–68% carbohydrates, 9–20% proteins and 0.5–3.9% lipids (%DW) (Holdt & Kraan, 2011). According to Hardouin et al. (2013), the chemical composition of French-dried *S. muticum* sample contained 25.5 ± 1.1% ash, 39.7 ± 1.3% neutral sugars, 1.8 ± 0.1% uronic acids, 8.8 ± 0.6% sulphated groups, 22.1 ± 0.6% proteins, 0.9% total phenol.

10.2.3 Spatiotemporal Variability of Seaweeds

The seaweed growth is directly linked to their ecosystems. The rate and type of available nutrients and the temperature have a direct impact on the composition of seaweed. Therefore, there is an important variability in the composition depending on the areas and the seasons. Stiger and Payri, 1999 highlighted the variations in density, maturity index, mean size, sex distribution and fertility in populations of *Sargassum mangarevense* and *Turbinaria ornata* from Tahitian reefs over 16 months. Floc'h et al. (1987) showed that the carbohydrate and proteins amounts in *Gracilaria cervicornis* are linked to the life cycle of seaweed. The highest amount of carbohydrate is observed during the maximum growth period, which also corresponds to a reduction of protein content and an increasing in photosynthetic activity. The relationship between the environmental parameters and the composition of *G. cervicornis* and *Sargassum vulgare* was established by Marinho-Soriano, Fonseca, Carneiro, and Moreira (2006). The study shows that proteins and carbohydrates are the most affected by these variations. The amount of proteins oscillates in a range of 142.92 g/kg in February to 227.08 g/kg in June in *G. cervicornis* and in a range of 91.88 g/kg in February to 199.38 g/kg in August in *S. vulgare*. The amount of carbohydrate oscillates in a range of 577.14 g/kg in February to 682.87 g/kg in April and in a range of 526.19 g/kg in November and 685.36 g/kg in December, respectively in *G. cervicornis* and *S. vulgare*. The study also correlates these variations and the environmental conditions as demonstrated by Rosenberg and Ramus (1982). The study showed that this period coincides with an increasing of temperature, salinity and sunlight intensity, which confirmed influence on carbohydrate synthesis. Spatiotemporal variations in the composition of seaweeds have widely been identified (Alves de Guimaraens & Coutinho, 2000; Duke, Litaker, & Ramus, 1987; Kaimoussi, Mouzdahir, & Saih, 2005; Le Lann, Connan, & Stiger-Pouvreau, 2012; Plouguerné et al., 2006; Rindi & Guiry, 2004; Stiger, Deslandes, & Payri, 2004; Villares, Puente, & Carballeira, 2002; Villares &

Carballeira, 2003). The variability of seaweeds has to be studied before a potential industrial application because it will have a direct impact on the efficiency of the enzymatic hydrolysis and the composition of hydrolysates.

10.3 ENZYME-ENHANCED PROCESSES FOR SEAWEED BIOREFINERY

This promising biotechnological procedure has been widely used to improve the extraction efficiency of bioactive components from land plants. The successful use of this technique to enhance the recovery of polyphenols from citrus peel, grape skin, apple skin, black currant or *Ginkgo biloba* leaves has been reported (Wijesinghe & Jeon, 2012). At the same time, studies about the extraction of active compounds from marine natural products have attracted special attention in the last recent years. The application of the enzyme-assisted extraction (EAE) method on seaweed materials was also reported (Wijesinghe & Jeon, 2012).

Seaweeds represent an attractive and sustainable biomass for simultaneous carbon capture, bioremediation of wastewater, and recycling in biofuel production. Indeed, with increased demands for renewable biofuels, the development of technological enzymatic processes for the production of third generation biofuel from invasive algal biomass will contribute to give information about composition of biomass before and after hydrolysis (Wang, Liu, & Wang, 2011). Biomass-derived fuels also would reduce the reliance on carbon extraction (*petroleum*) for combustion. The industrialisation of enzymatic hydrolysis process for the production of biofuels leads to the publication of few patents (Suryanarayan et al., 2013; Yoon et al., 2010).

The capacity of microorganisms to degrade the walls of microalgae and macroalgae has been isolated (Jung, Lim, Kim, & Park, 2013) and their ability, for example, to convert cellulose into glucose for the bioethanol production has been proven (Trivedi, Gupta, Reddy, & Jha, 2013). The glucose is then used for the production of biofuels by fermentation under the same systems used for the production of fuels from sugarcane. For example, study of Borines, de Leon, and Cuello (2013) focused on the pretreatment, enzyme saccharification and fermentation of *Sargassum* spp. for bioethanol production. Proximate analysis of dried *Sargassum* spp. gave 10.2% of protein, 9.8% of crude fibre, 26.2% of ash and 41.8% of total carbohydrates. After hydrolysis, amount of carbohydrate in the sample was 47.06%, which is within the reported range of carbohydrate content of *Sargassum* species. These are the 3rd generation biofuels (Alam et al., 2012).

Traditional extraction techniques to obtain bioactive materials from seaweeds have used different organic solvents or water. However, there have been several controversial points such as extremely low recovery rate and strict regulations in use of organic solvents in the food industry, and limited yield in the recovery of water-soluble components in water extracts. The advantages of the enzyme-assisted extraction of algal metabolites or the production of low molecular weight fractions by enzymatic hydrolysis of macromolecules from seaweeds have been amply demonstrated (Wijesinghe, Athukorala, & Jeon, 2011; Jeon, Wijesinghe, & Kim, 2011). Extraction techniques using digestive commercial enzymes such as carbohydrases and proteases were employed in order to degrade seaweed tissues and to help releasing a variety of bioactive compounds from the seaweeds. These kinds of enzymes can convert water-insoluble seaweeds into water-soluble materials. Additionally, it has a high yield (approx. 50%) and higher radical scavenging activity than organic solvent extracts. Presently, studies show that the use of commercial enzymes preparations or enzyme mixes will often be preferred. In this case, the efficiency of the enzyme and a low production cost will be chosen at the expense of specificity. However, when it comes to extract specific active molecules or produce specific functional molecules, not from biomass directly, but from a raw component, hydrolysis by specific enzymes will be used. The productions of bioactive oligosaccharides from nonactive parietal polysaccharides or active peptides from proteins are good examples (El Boutachfaiti et al., 2009; Jeon et al., 2011; Wijesinghe et al., 2011). The algal material thus undergoes a preliminary treatment to remove the molecules that serve as substrate to the enzymes. The use of enzymes in industry presents many advantages (Table 10.2). For example, the enzymes have the advantage of being a catalyst, which means that it will be intact at the end of the process. Any application of immobilised enzymes has already been tested on seaweeds or seaweeds extracts but it could be developed. Because they are biological molecules, enzymes are also nontoxic, active under soft conditions, biodegradable and capable of high specificity for their substrate conditions. Enzymes produced in most cases by bacterial or fungal fermentation provides the opportunity for a large-scale production. However, they still show some disadvantages. As biological molecules, they will be less stable than chemical catalysts and often has a higher cost. Furthermore, the enzymatic hydrolysis reaction is a homogeneous catalytic phase, which in some cases can present problems in some manufacturing processes.

However, several strategies can be implemented and designed to achieve these objectives:

- The first one, the *enzyme-assisted extraction* aims to improve extraction of molecules other than those affected by the enzyme. This type of extraction often calls for use of different commercial enzyme preparations not specifically developed for the use on seaweed.
- The second one, the *specific enzymatic hydrolysis* concerns the molecules production of interest. In this case, the use of specific enzymes is needed. This technique can be accomplished directly on raw material or after a first extraction releasing molecules that become a substrate.
- The third one, the *assisted enzymatic hydrolysis*, consists in the assistance of the hydrolysis by mechanical, physical or biological methods, which can be carried out before or during the enzymatic hydrolysis in order to facilitate the release of the product.

10.3.1 Enzyme-Assisted Extraction

In most studies on the enzymatic hydrolysis of algae, enzymes are used as processing additives. Therefore, most articles deal with enzyme-assisted extraction and not specific enzymatic hydrolysis. Due to the nature of the cell walls of algae, the most commonly used enzymes are proteases and carbohydrases. The aim of the enzymatic treatment is to degrade the structure of the cell wall and solubilise the desired molecules. For certain molecules, this type of processing may significantly improve yields. Primary and secondary metabolites are extracted in most cases in their native forms. The enzymes are mainly commercial preparations from the food industry and obtained by fermentation of bacteria or fungi. These commercial enzyme preparations may have one or more declared activities, by the manufacturer. The declared enzyme activities are often the mains but these preparations may also contain secondary activities, which are minor but may still be of interest. Table 10.3 summarises studies and gives names and types of enzyme used and the extraction conditions (duration, temperature, pH, etc.).

Table 10.2 Advantages and Disadvantages of Enzymatic Hydrolysis

Advantages	Disadvantages
Not-consumed during reaction	
High conversion yield	Stability
Nontoxic and biodegradable	
High selectivity	Cost
High specificity	
Utilisation in soft conditions	
Large-scale production	

Most enzyme-assisted extractions have been performed on the red and brown algae, while tests on green algae are rare. In these studies, only few enzyme preparations have been used. It is, for carbohydrases, cellulase (EC 3.2.1.4), β-xylanase (EC 3.2.1.8), β-glucanase (EC 3.2.1.6), α-amylase (EC 3.2.1.1) or exo-1,4-α-glucosidase (EC 3.2.1.3) and proteases, alkaline endo-proteases (EC 3.4.21.62), neutral (EC 3.4.24.28) and the carboxy- and amino-peptidase. Coming from the food industry, these enzymes have been selected for their activities under mild conditions of pH and temperature. Carbohydrases are used at temperatures between 40 and 60 °C and at slightly acidic pH, between 4.5 and 7.0. Proteases, in turn, are used at similar temperatures, 40 °C or 50 °C but at higher pH value, 6.0 to 9.0. To increase the amount of extracted material, they are used over relatively long durations, from 12 to 24 h or 48 h.

In most studies, similar protocols are used. Seaweeds are dried and pulverised before being homogenised in water or a buffer and then subjected to enzymatic hydrolysis. The enzyme is denatured by a brief heat treatment and the extract is recovered by centrifugation. Many commercial enzyme preparations, some carbohydrases (Viscozyme L, Celluclast 1.5 L FG, AMG 300 L, 120 L Termamyl, Ultraflo L) and proteases (Neutrase 0.8 L, Protamex, Flavourzyme 500 MG, Alcalase 2.4 L FG, Kojizyme 500 MG) were tested, on several species of brown algae. In these studies, the objectives were to test the antiradical, antioxidant and anticoagulant potentials of hydrolysates (Ahn et al., 2004; Athukorala, Lee, Kim, Jeon, 2007; Athukorala et al., 2006; Heo, Jeon, Lee, Kim, & Lee, 2003; Heo, Park, Lee, Jeon, 2005; Jeon et al., 2011; Park, Shahidi, & Jeon, 2004; Wang et al., 2010; Wijesinghe & Jeon, 2012; Wijesinghe et al., 2011). Lahaye and Vigouroux, (1992) tested the effect of a cellulase and endo-1,4-β-xylanase on the liquefaction of *P. palmata*. Hydrolyses were carried out at 40 °C for 24 h in acetate buffer, sea water and distilled water. In the various extraction systems, the rate of liquefaction of the seaweeds obtained by the enzyme is significantly higher than that obtained for the control (buffer alone). They pass, respectively, for the buffer and the enzyme, from 72.3% to 87.1% in the acetate buffer, from 66.9% to 81.5% in distilled water and from 62.6% to 82.5% in sea water, on dried seaweed. As a final example, Dumay, Clément, Morançais, and Fleurence (2013) have studied the optimisation of the parameters of hydrolysis of *P. palmata* for the extraction of R-phycoerythrin with a xylanase (EC 3.2.1.8) from *Trichoderma longibrachiatum*. In this study, a number of parameters were investigated such as the pretreatment of the raw material (cut, ground or freeze-ground on dried or fresh seaweed), the duration

Table 10.3 Enzymatic-Assisted Extraction of Components from Seaweeds

Enzymes	Type	Microorganism	Temperature	Duration	pH	Algae	References
Carbohydrases							
AMG 300 L	Exo-1,4-α-D-glucosidase	*Aspergillus niger*	60 °C	12–24 h	4.5	Brown, green	Ahn, Jeon, Kang, Shin, and Jung (2004), Heo, Park, Lee, and Jeon (2005), Wijesinghe et al. (2011), Athukorala, Jung, Vasanthan, and Jeon (2006), Wang et al. (2010), Choi, Nguyen, and Sim (2010)
Celluclast 1.5 L FG	Cellulase	*Trichoderma reesei ATCC 26921*	50 °C	12–24 h	4.5	Brown, red	Ahn et al. (2004), Heo et al. (2005), Wijesinghe et al. (2011), Athukorala et al. (2006), Wang et al. (2010), Fleurence (1999a), Lahaye and Vigouroux (1992)
Cellulosin T2	Cellulase	*Trichoderma* sp.	37 °C	48 h	4.5	Green, red	Okuda et al. (2008)
Panlyve HCB 10	Xylanase	*A. niger*	–	–	–	Red	Fleurence (1999a)
Pentosanase	Pentosanase	*A. niger*	40 °C	24 h	4.5	Red	Lahaye and Vigouroux (1992)
Shearzyme 500 L	Endo-1, 4-xylanase	*Aspergillus orysae*	–	–	–	Red	Fleurence (1999a)
Termamyl 120 L	α-amylase	*Bacillus licheniformis*	60 °C	12–24h	6	Brown, green	Ahn et al. (2004), Heo et al. (2005), Wijesinghe et al. (2011), Athukorala et al. (2006), Wang et al. (2010), Choi et al. (2010)

Continued

Table 10.3 Enzymatic-Assisted Extraction of Components from Seaweeds—cont'd

Enzymes	Type	Microorganism	Temperature	Duration	pH	Algae	References
Ultraflo L	β-glucanase	*Humicola insolens*	60 °C	12–24 h	7	Brown, red	Ahn et al. (2004), Heo et al. (2005), Wijesinghe et al. (2011), Athukorala et al. (2006), Wang et al. (2010), Fleurence (1999a)
Viscozyme L	Arabanase, cellulase, β-glucanase, hemi-cellulase and xylanase	*Aspergillus accu-leatus*	50 °C	12–24 h	4.5	Brown	Ahn et al. (2004), Heo et al. (2005), Wijesinghe et al. (2011), Athukorala et al. (2006), Wang et al. (2010)
Proteases							
Alcalase 2.4 L FG	Alkalin endopro-tease	*B. licheniformis*	50 °C	12–24 h	8	Brown, red	Ahn et al. (2004), Heo et al. (2005), Wijesinghe et al. (2011), Athukorala et al. (2006), Wang et al. (2010), Cian, Martínez-Augustin, and Drago, (2012).
Flavourzyme 500 MG	Endoprotease and exopepti-dase	*A. oryzae*	50 °C	12–24 h	7	Brown	Ahn et al. (2004), Heo et al. (2004), Wijesinghe et al. (2011), Athukorala et al. (2006), Wang et al. (2010)
Kojizyme 500 MG	Amino-/car-boxypepti-dase	*A. oryzae*	40 °C	12–24 h	6	Brown	Ahn et al. (2004), Heo et al. (2005), Wijesinghe et al. (2011), Athukorala et al. (2006), Wang et al. (2010)

Neutrase 0.8L	Neutral endoprotease	*Bacillus amyloliquefaciens*	50 °C	12–24 h	6	Brown	Ahn et al. (2004), Heo et al. (2005), Wijesinghe et al. (2011), Athukorala et al. (2006), Wang et al. (2010)
Protamex	Alkalin and neutral endoproteases	*Bacillus* sp.	40 °C	12–24 h	6	Brown, red	Ahn et al. (2004), Heo et al. (2005), Wijesinghe et al. (2011), Athukorala et al. (2006), Wang et al. (2010)
Subtilisin DY	Alkalin serin protease	*Bacillus subtilis*	50 °C	4 h	8.5	Green	Tchorbanov and Bozhkova. (1988)
Trypsin	Endoprotéase	*Intestin*	50 °C	4 h	9	Red	Cian et al. (2012)
Umamizyme	Endo-/exo-peptidase	*Aspergillus oryzae*	50 °C	24 h	7	Red	Wang et al. (2010)

(0.5–6 h), temperature (10–40 °C) and the ratio enzyme/substrate (5–25 g/kg DW). The optimisation of hydrolysis parameters led to the improvement of the extraction yields of R-phycoerythrin from *P. palmata*, respectively, from 3.28 ± 0.064 g/kg DW before optimisation to 12.36 ± 0.37 g/kg DW after optimisation.

10.3.2 Specific Enzymatic Hydrolysis

As shown in the previous part, enzyme-assisted extraction is useful, but is a nonspecific action of the enzymes on the raw material. Another approach that has been tested in some studies is the specific enzymatic hydrolysis. It consists in the selection of enzymes to act specifically on target molecules and produce novel bioactive molecules. In this approach, the substrates can be whole algae or algal polysaccharides or proteins prior extracted by other methods. The mains objectives of these methods are to produce oligosaccharides, simple sugars, peptides or amino acids from algal biomass or algal extracts in a controlled and specific way. Very few specific enzymes for algal metabolites exist on the market so often the first step of these studies is the isolation, identification and purification of enzymes from organisms able to degrade algae constituents. These microorganisms can come from the intestinal flora of marine animals, which can digest algae or from symbiotic bacteria or fungi (Jung et al., 2013). This approach has the advantage to guide the hydrolysis to the production of small molecules of interest, mainly in the context of researching new molecules with biological activity or for applications in cosmetics, food, feed, etc. Researches are still at an early stage and only a few studies have isolated this type of enzymes.

As an example, Harshvardhan, Mishra, and Jha (2013) isolated a cellulase from a marine strain of *Bacillus* sp. (H1666). The characterisation of the enzyme showed that it has an optimum activity at 50 °C and pH 7.0. This enzyme was then tested for the saccharification of biomass of the green algae *Ulva lactuca*. The results showed the ability of the enzyme to degrade the starch, and to transform it into glucose (21 mg/g dry seaweed) directly on the whole seaweed.

In another study, an ulvan-lyase was produced from a strain of *Persicivirga ulvanivorans* (*Flavobacteriaceae*) (Barbeyron et al., 2010; Helbert & Nyvall-Collen, 2011) isolated from faeces of *Aplysia punctata* (*Mollusca*, *Gastropoda*), a marine gastropod whose intestinal flora is able to degrade *Ulva*. It has been shown that this ulvan-lyase acts by depolymerisation of ulvans by cleaving on glucuronic and iduronic acid sites. Similar enzymes had also been tested for the production of oligosaccharides from ulvan (Lahaye, 1998; Lahaye,

Brunel, & Bonnin, 1997). El Boutachfaiti et al. (2009) used a glucuronan-lyase (EC 4.2.2.14), isolated from a strain of *Trichoderma* sp., for the production of bioactive oligosaccharides from *U. lactuca*. They were first carried out the extraction and purification of ulvans and glucuronans before submitting samples to enzymatic hydrolysis. The results show that the enzymatic treatment with glucuronan-lyase is able to convert, after 72 h, glucuronans in mono-, di- and trisaccharides.

Fleurence, Massiani, Guyader, and Mabeau (1995) tested the use of specific polysaccharidases for the cell-wall destructuration in order to improve the extraction of proteins from various red algae species. These hydrolyses were carried out on couples algae/enzyme, *C. crispus*/κ-carrageenase; *P. palmata*/xylanase; *Gracilaria verrucosa*/β-agrarase, with or without a cellulase, and at the optimal activity of each enzyme/substrate couple. The study showed that, depending on the system used, the protein extraction yields vary and all enzymes do not provide the same benefit. On *C. crispus*/κ-carrageenase and *G. verrucosa*/β-agarase models, the specific enzyme do not significantly improve the protein extraction. However, these yields are greatly increased with the addition of the cellulase. A clear synergistic effect was observed between the two enzymes. The action of xylanase and cellulase (separately) on the cell wall of *P. palmata* increases four to five times the protein extraction yield but their coupled action shows no increasing compared to the control. An antagonistic effect between the two enzymes is clearly demonstrated. Menshova et al. (2014) used an endo-1,3-β-D-glucanase (LIV) from *Pseudocardium sacchalinensis*, and an exo-β-1,3-D-glucanase from *Chaetomium indicum*, to hydrolyse the laminarin of the brown algae *Eisenia bicyclis*, for the production of laminarioligosaccharides, in order to evaluate their anticancer potential. They showed an increase in the amount of di-, tri-, tetra- and penta-saccharides in hydrolysates by exoglucanase, with majority production of the disaccharides. Furthermore, the endoglucanase seems to have a specific action because only the trisaccharides forms show an increase in the hydrolysates.

The results obtained in these few, nonexhaustive, studies, confirm the interest and the potential associated with the use of specific enzymes for hydrolysis of algal metabolites and possibilities of oriented production of molecules of interest. Indeed, the selectivity of enzymes makes possible the targeted hydrolysis of seaweeds cell-wall molecules.

10.3.3 Improvement of Hydrolysis Efficiency

For optimal extraction the choice of the enzyme is important but results show that the pretreatment of the raw material have a significant effect on the final extraction. The purpose of the pretreatment is to facilitate the extraction of the compounds by destructuring the cell wall to increase the accessibility of the enzymes to their substrate. The destructuration can be achieved by different physical treatments (grinding, freezing, drying, pressing, temperature, ultrasound, etc.), chemical extractions (organic solvents, bases, acids, other chemicals, etc.) or biological processes (enzymes or microorganisms). For example, prior the extraction, a freezing step can be used to destructure the membranes and release the intracellular content directly upon thawing. A thermal shock has the effect of breaking the bonds between the parietal molecules and of denaturing proteins. It results in an increasing of the fragility of the cell wall, thereby facilitating the action of enzymes, and the release of molecules.

The storage of the raw material is also an important parameter for the quality of the extracts. The storage conditions and/or the pretreatment depend on the seaweed species because their compositions and cell-wall structures are different. Fresh algae, once harvested, are fragile and become an easily putrescible biomass. Paull and Chen (2008) showed a gradual deterioration of a material postharvest, at different temperatures and on several aspects (appearance, colour, phycobilin content and liquefaction). Romero, Villanueva, and Montaño (2008) found that the shelf life had an important impact on agars from *Gracilaria eucheumatoides*. After 31 months of storage, the sulphation rate and the 3,6-anhydrogalactose content did not change but the gel strength of agars was decreased by 3, the viscosity by 10 and the molecular weight by 5. Therefore, the conservation of raw material will have a direct impact on the quality of the fractions produced. Freezing, heat drying, or freeze-drying, soft acids or salts stabilisations or pasteurisation methods are commonly used to stabilise the raw material. All these methods have an effect on the algal components, and cause variations in the extraction yields and the quality of the products (Robic et al., 2008) and may therefore be considered as preliminary treatment to enzymatic hydrolysis. It is therefore important to be considered for interpreting the results because the differences can be significant.

Robic et al. (2008) studied the effect of the stabilisation conditions on the organic matter extraction yields and the quality of ulvan. The extraction yields of organic matter of fresh, frozen, dried and freeze-dried algae are little increased but there is an important increasing for the brined and salted seaweeds. They also showed that the extraction yield of ulvan significantly decreased after pretreatment. These results confirmed the effect of the treatment on the extraction yields. As an example of pretreatment of the raw material, Trivedi et al. (2013) studied the effect of several types of pretreatment on the reducing sugars extraction yields obtained by enzymatic hydrolysis of *Ulva fasciata*. The study shows that the use of dilute sulphuric acid added to high temperature (100 °C) gives low extraction yields of reducing sugars. However, the use of hot buffer (sodium acetate, 120 °C), dry heat (120 °C) and liquid ammonia increase yields about 3–8 times compared to sulphuric acid extraction. No effect of the duration of pretreatment was observed. The results after 30 and 90 min were identical. A similar study has shown that the hydrothermal pretreatment of *Solieria pacifica* (red algae) and *Monostroma nitidum* (green algae) increases the conversion yields of cellulose to glucose by enzymatic hydrolysis (Okuda et al., 2008). The maximum ratios were, respectively, 87.8% for *S. pacifica* and 79.9% for *M. nitidum* compared to a ratio of 27% for the unhydrothermally pretreated seaweeds. The use of enzymatic hydrolysis pulsed by ultrasound was compared with the conventional method for the acid extraction of metals in *Palmaria* (Dulse). Trypsin, pepsin and α-amylase were tested and the collection of extracts relative to the total metal content was determined (Peña-Farfal et al., 2005). The study shows that the amount of metal ions extracted depends on the type of enzyme used and the assistance of ultrasound recovers nearly 100% with pepsin. The yields obtained by single enzymatic hydrolysis give results inferior of the reference values for different metals. The sonication increases these yields and compensates these differences. The ultrasonic assisted enzymatic hydrolysis was also tested by Romarís-Hortas, Bermejo-Barrera, and Moreda-Piñeiro (2013) in order to establish a method to extract iodine in edible seaweeds (Kombu, Wakame, Nori, Dulse, Sea Lettuce). The method showed an increased recovery from 76% for alkaline extraction to 96% for the ultrasonic assisted extraction.

10.4 BIOLOGICAL ACTIVITIES OF SEAWEEDS HYDROLYSATES AND POTENTIAL APPLICATIONS

Seaweeds represent a valuable source of novel bioactive compounds. Over the years, many active components have been isolated from seaweeds with diverse chemical nature. Enzyme-assisted extraction of bioactive compounds from seaweeds could be a useful technique and the process may provide a valuable alternative to overcome physical and chemical extraction barriers. Novel extraction technique using digestive enzymes such as carbohydrases and proteases was employed in order to degrade seaweed tissues and to help the releasing of a variety of bioactive compounds from seaweeds (Ahn et al., 2004; Athukorala et al., 2006; Cian et al., 2012; Hardouin et al., 2013; Heo et al., 2005; Park et al., 2004; Siriwardhana et al., 2008; Wang et al., 2010). The enzymatic extraction can provide solutions both on the extraction yields and for the quality of extracts (conservation of biological activities, low cytotoxicity). As described above, they were tested on algae and demonstrated that it preserve the biological activities of molecules extracted while working under soft extraction conditions. Enzymatic hydrolysis could release molecules normally extracted with organic solvents.

Table 10.4 summarises the biological activities or interesting components present in hydrolysates. This table, that references a nonexhaustive list of hydrolysis works, still shows however that research has been done on all types of seaweeds, whether green, red or brown.

10.4.1 Antioxidant Activities

Recently, much attention has been paid on the antitumor, anticholesterolemic and antioxidant activities of seaweed components. Consequently, antioxidant activity is intensively focused due to the currently growing demand from the pharmaceutical industry in which there are interests in anti-aging and anticarcinogenic natural bioactive compounds, which possess health benefits. The antioxidant and free radical scavenging activities have essentially been identified in brown and red seaweeds (Ahn et al., 2004; Athukorala et al., 2006; Cian et al., 2012; Heo et al., 2005; Park et al., 2004; Siriwardhana et al., 2008; Wang et al., 2010). The potential antioxidant compounds were identified as some pigments (e.g. fucoxanthin, astaxanthin, carotenoid) and polyphenols (e.g. phenolic acid, flavonoid, tannins). Use of organic solvents such as ethanol or methanol is the common method for the extraction of polyphenolic compounds from plant materials. In addition, aqueous mixtures of the particular organic solvents also improve the extraction efficiency. But, some plant polyphenols appear to be twisted together with cell-wall polysaccharides and proteins via tight hydrophilic and hydrophobic bonds. Moreover, the distribution of polyphenols is not homogeneous inside the plant tissues (Wijesinghe & Jeon, 2012). The release of polyphenolic compounds can be enhanced by enzyme-enhanced degradation of cell-wall polysaccharides. For example, enzymatic extracts of the brown seaweeds might be valuable antioxidant sources. In 2003, Heo et al. evaluated the potential antioxidant activity of water-soluble enzymatic hydrolysates from a kelp *Ecklonia cava* by free radical scavenging and lipid peroxidation assays. Five polysaccharidases (Viscozyme, Celluclast, AMG, Termamyl and Ultraflo) and five proteases (Protamex, Kojizyme, Neutrase,

Table 10.4 Biological Activities and/or Bioactive Components Isolated from Seaweeds by Enzymatic Hydrolysis

Seaweeds	Enzymes	Component/Activity/Application	References
Green Seaweed			
Chlamydomonas reinhaardtii	β-glucosidase, α-amylase	–	Choi et al. (2010)
Chlorella sp.	Protease	Proteins/food	Tchorbanov and Bozhkova (1988)
Codium fragile	Glucosidase, cellulase, α-amylase, glucanase, protease, peptidase, multi-carbohydrase complex	Anticoagulant	Athukorala et al. (2007)
Enteromorpha linza	Cellulase, pectinase, hemicellulase	Proteins	Amano and Noda (1992)
Monostroma nitidum	Cellulase, pectinase, hemicellulase	Glucose	Amano and Noda (1992)
Monostroma nitidum	Cellulase	Glucose	Okuda et al. (2008)
Scenedesmus incrassatulus	Protease	Proteins/food	Tchorbanov and Bozhkova, (1988)
Ulva fasciata	Cellulase, pectinase, hemicellulase	Proteins	Amano and Noda (1992)
Ulva pertusa	Cellulase, pectinase, hemicellulase	Proteins	Amano and Noda (1992)
Ulva rigida	Cellulase	Proteins	Fleurence et al. (1995)
Ulva rotundata	Cellulase	Proteins	Fleurence et al. (1995)
Ulva sp.	Cellulase, xylanase, β-glucanase, arabanase, endopeptidase, endoprotease	Proteins, sulphated polysaccharides	Hardouin et al. (2013)
Red Seaweed			
Callymenia perforate	Pectinase, hemicellulase	Proteins	Amano and Noda (1992)
Chondrus crispus	κ-carrageenase, xylanase, cellulase, β-agarase	Protein accessibility	Fleurence et al. (1995)
Gracilaria verrucosa	κ-carrageenase, xylanase, cellulase, β-agarase	Protein accessibility	Fleurence et al. (1995)

Continued

Table 10.4 Biological Activities and/or Bioactive Components Isolated from Seaweeds by Enzymatic Hydrolysis—cont'd

Seaweeds	Enzymes	Component/Activity/Application	References
Palmaria palmata	Endo- and exopeptidase, protease complex, amino- and carboxypeptidase, metalloendoprotease, multi-carbohydrase complex, glucanase, glucosidase, cellulsae, α-amylase	Antioxidant	Wang et al. (2010)
Palmaria palmata	κ-carrageenase, xylanase, cellulase, β-agarase	Protein accessibility	Fleurence et al. (1995)
Palmaria palmata	Endo-β-1,4-D-xylanase, cellulase	–	Lahaye and Vigouroux (1992)
Porphyra columbina	Trypsin, alkalin protease	Antioxidant, Antitensin, antihypertensive, Immunomodulation	Cian et al. (2012)
Porphyra yezoensis	Pectinase, hemicellulase	Proteins	Amano and Noda (1990)
Solieria chordalis	Cellulase, xylanase, β-glucanase, arabanase, endopeptidase, endoprotease	Sulphated polysaccharides/ antiviral,	Hardouin et al. (2013)
Solieria pacifica	Cellulase	Glucose	Okuda et al. (2008)
Brown seaweed			
Ecklonia cava	Cellulase, α-amylase, β-glucanase, exo-1,4-α-D-glucosidase, multi-carbohydrase complex, alkalin endoprotease, neutral endoprotease, amino- and carboxypeptidase	Antioxidant	Heo et al. (2005)
Ecklonia cava	Exo-1,4-α-D-glucosidase, cellulase, α-amylase, β-glucanase, multi-carbohydrase complex	Anticoagulative, antiproliferative, antioxidant	Wijesinghe et al. (2011), Athukorala et al. (2006)
Hizika fusiformis	β-glucanase, peptidase	Antioxidant	Siriwardhana et al. (2008)
Ishige okamurae	Cellulase, α-amylase, β-glucanase, exo-1,4-α-D-glucosidase, multi-carbohydrase complex, alkalin endoprotease, neutral endoprotease	Antioxidant	Heo et al. (2005)

Sargassum coreanum	Cellulase, α-amylase, β-glucanase, exo-1,4-α-D-glucosidase, multi-carbohydrase complex, alkalin endoprotease, neutral endoprotease, amino- and carboxypeptidase	Antioxidant	Heo et al. (2005)
Sargassum fullvelum	Cellulase, α-amylase, β-glucanase, exo-1,4-α-D-glucosidase, multi-carbohydrase complex, alkalin endoprotease, neutral endoprotease, amino- and carboxypeptidase	Antioxidant	Heo et al. (2005)
Sargassum horneri	Cellulase, α-amylase, β-glucanase, exo-1,4-α-D-glucosidase, multi-carbohydrase complex, alkalin endoprotease, neutral endoprotease, amino- and carboxypeptidase	Antioxidant	Heo et al. (2005), Park et al. (2004)
Sargassum muticum	Cellulase, xylanase, β-glucanase, arabanase, endopeptidase, endoprotease	Proteins, polysaccharides, polyphenols	Hardouin et al. (2013)
Sargassum thunbergii	Cellulase, α-amylase, β-glucanase, exo-1,4-α-D-glucosidase, multi-carbohydrase complex, alkalin endoprotease, neutral endoprotease, amino- and carboxypeptidase	Antioxidant	Heo et al. (2005)
Scytosiphon lomentaria	Cellulase, α-amylase, β-glucanase, exo-1,4-α-D-glucosidase, multi-carbohydrase complex, alkalin endoprotease, neutral endoprotease, amino- and carboxypeptidase	Antioxidant	Heo et al. (2005), Ahn et al. (2004)
Undaria pinatifida	Cellulase, pectinase, hemicellulase	Proteins	Amano and Noda (1992)

Flavourzyme and Alcalase) commercial and inexpensive enzymes were used. Among all the hydrolysates, Celluclast hydrolysate effectively scavenged free radicals released from DPPH (1,1-diphenyl-2-pricrylhydrazyl) and recorded around 73% scavenging activity at the concentration of 4 mg/ml. These hydrolysates were thermally stable and DPPH radical scavenging activity remained 80% or higher at heating temperatures of 40 and 60 °C up to 12 h and around 80% at 100 °C up to 8 h. AMG and Ultraflo hydrolysates inhibited the lipid peroxidation of fish oil as that of α-tocopherol.

In the study of Heo et al. (2005), seven brown seaweeds were enzymatically hydrolysed to prepare water-soluble extracts by using the same enzymes. The enzymatic extracts exhibited more prominent effects in hydrogen peroxide scavenging activity (approximately 90%) compared to the other scavenging activities and the activity of enzymatic extracts was even higher than that of the commercial antioxidants. Moreover, the two enzymatic extracts strongly inhibited DNA damage (approximately 50%). Those extracts showed significantly ($p < 0.05$) remarkable scavenging effects in DPPH free radical scavenging assay and the activity indicated a marked correlation with phenolic contents. These studies have shown that the amount of active molecules in the hydrolysates is increased compare to a blank, but it does not depend on the nature of enzyme. In fact, proteases and carbohydrases could improve extraction of antioxidant active molecules. This point confirms that an enzyme-assisted extraction, involving commercial enzyme preparations, constitute a massive extraction and that the composition of the extract is not or not much affected.

Cian et al. (2012) demonstrated the antioxidant capacity of hydrolysates from *Porphyra columbina* (Rhodophyta). These authors obtained a crude protein fraction (nonhydrolysed), an alkalin protease hydrolysate and a trypsin hydrolysate. They have also done a sequential hydrolysis with these two enzymes. The antioxidant activity was evaluated by the DPPH radical scavenging, Trolox equivalent antioxidant capacity (TEAC), β-carotene oxidation assay and the oxygen radical absorbance capacity. The DPPH scavenging test shows that proteins hydrolysates present an activity significantly higher than the crude protein extract (nonhydrolysed) but no differences appears between the different hydrolysates for the four tests. The antioxidant activity is influenced by the proteins and peptides molecular weight and the amino acids composition. Bioactive extracts are low molecular weight and rich in aspartic, glutamic and alanine.

Antiradical activities were identified (Hardouin et al., in press) in crude hydrolysates obtained from French *Ulva* sp., *S. chordalis* and *S. muticum*. These extracts were obtained with commercial carbohydrases and proteases after a treatment with 0.5% (w:w) of enzyme during 5 h at 50 °C. The IC_{50} was in a range of 2.0–12.0 mg/ml for *Ulva* sp. extracts, 9.0–11.0 mg/ml in *S. chordalis* extracts and 5.0–9.0 mg/ml in *S. muticum* extracts. The total phenol content (TPC) of hydrolysates was also measured by DPPH scavenging test. Both in carbohydrases and proteases hydrolysates the TPCs were comprised between 4.4 and 5.7 for *S. muticum*, 0.6–1.0 %DW for *S. chordalis* and 0.4 and 1.0 for *Ulva* sp. The TPC of hydrolysate do not depend on the nature enzyme used. In this study, the relation between TPCs and antiradical activities was highlighted.

10.4.2 Antiviral Activities

Hardouin et al. (2013) propose the use of two commercial proteases and four carbohydrases treatments as a tool to improve the extraction efficiency of bioactive compounds from three abundant French seaweeds *S. chordalis*, *Ulva* sp. and *S. muticum*. In comparison with water extraction at 50 °C for the same time of treatment, enzymatic hydrolysis increased the extraction yields. The data from this study suggest the potential of the enzymatic hydrolysis for producing active fractions in function of the algal biomass, the behaviour of cell wall, the selectivity and the action of enzyme. Enzymatic hydrolysis appeared less effective for polyphenol but a promising softer technique for recovering proteins, neutral sugars, uronic acids and sulphate groups. Some antiviral activities had been characterised by cell viability in *S. chordalis* extracts. These authors have shown that *S. chordalis* hydrolysates present an effective anti-herpetic activity with EC_{50} between 23.0 and 101.1 μg/ml at a Multiplicity of Infection of 0.001 ID_{50}/cells. 100% and 98% cellular protection were obtained for 500 μg/ml of carbohydrase hydrolysate and hot water respectively. After 3 days experiment, no microscopically visible alteration of normal cell morphology was observed even at 500.0 μg/ml. The activity corresponds to fractions rich in sulphates groups (4.7 ± 0.2% and 5.0 ± 0.3% respectively). The importance of red macroalgae as a source of novel anti-HSV agents has been recognised and reported by many researchers (Bouhlal, Riadi, & Bourgougnon, 2010; Bouhlal et al., 2011; Vo, Ngo, Ta, & Kim, 2011). It was revealed that the aqueous extracts rich in sulphate groups (Bourgougnon, 2003; Bourgougnon, Lahaye, Chermann, & Kornprobst, 1993; Damonte, Matulewicz, & Cerezo, 2004) were capable of inhibiting the replication of HSV-1 at an EC_{50} range of 2.5–80 μg/ml without cytotoxic effect.

10.4.3 Other Activities

Athukorala et al. (2007) proposed to screen seaweeds for their potential anticoagulant activities, which might expand the possibility to find better anticoagulant drug. Twenty-two algal species were evaluated for their potential anticoagulant activities. In this study, hot water extracts from selected species, *C. fragile* and *Sargassum horneri* showed high activated partial thromboplastin time (APTT). But both Ultraflo extracts of *C. fragile* and *S. horneri* showed much better activities than other respective enzymatic hydrolysates. The polysaccharidase, Ultraflo, produced by *Humicola insolens* (Fungi), was the best enzyme preparation in releasing arabinose, from both soluble and insoluble arabinoxylan and its derivatives to the extraction media (ferulic acid from insoluble arabinoxylan). Furthermore, in both algal species, active compounds were mainly concentrated in >30 kDa fraction. In 2006, the same authors evaluated potential anticoagulant activity of seven brown algal species (*Ecklonia cava*, *Ishige okamurae*, *Sargassum fulvellum*, *Sargassum horneri*, *Sargassum coreanum*, *Sargassum thunbergii* and *Scytosiphon lomentaria*) after enzymatic hydrolysis using five carbohydrases (Viscozyme, Celluclast, AMG, Termamyl and Ultraflo). The active polysaccharide fraction of the AMG enzymatic extract of *E. cava* was purified using anion-exchange chromatography on DEAE-cellulose and gel filtration chromatography on Sepharose 4B. The anticoagulant compound studied was detected by prolongation of APTT, thrombin time and prothrombin time. The pure sulphated polysaccharide mainly composed of fucose and small amount of galactose, from *E. cava* showed almost similar anticoagulant activity to that of heparin and strongly interferes with coagulation cascade by inhibiting biological activity of serine proteases II, VII and X.

In addition, other recent reports revealed the immunomodulatory effects of enzymatic extracts from *E. cava* (Ahn et al., 2008, 2010). Kang et al. (2011) reported another successful use of EAE on purification of sulphated polysaccharide from *E. cava*. The study demonstrated the potential anti-inflammatory activity of a sulphated polysaccharide purified from amyloglucosidase (AMG)-assistant extract of *E. cava*. Therefore, the extraction of this active polysaccharide was specifically performed. This is a good example of the selectivity and specificity of enzymes. However, at this stage of researches, example of specific production of active metabolites from seaweeds by enzymatic hydrolysis has rarely been highlighted. Nevertheless, because of the very large number of algae metabolites known for their biological activities, it is certain that other active molecules will soon be isolated from hydrolysates and others may be produced from native metabolites present in algae. The presence of various polysaccharides of large quantities in the cell wall in brown seaweeds could strongly reduce the extraction efficiency during application of classical extraction methods. In addition, there are compounds attached to the cell wall (cell-wall-bound compounds), which cannot be easily extracted using typical extraction methods with aqueous solvents. Commonly, these polysaccharides have been extracted using water or aqueous organic solvents. However, the cell wall consists in complex polymers; it is not easy to extract active polysaccharides using solvent extraction processes. Moreover, this might limit the study and potential industrial applications of active components derived from seaweed. Interestingly, enzymatic digestion of algae gains high bioactive yield and shows enhanced biological activity in comparison with water and organic extract counterparts. The technique has impressive effects with characteristics of high catalytic efficiency, high specificity, mild reactive conditions and preserving the original efficacy of active compounds to the maximum (Wijesinghe & Jeon, 2012).

The antihypertensive activity of hydrolysates from *P. columbina* was observed (Cian et al., 2012) by the measurement of angiotensin-converting enzyme (ACE) inhibition of proteins hydrolysates. In general, food protein-based antihypertensive peptides can be produced using autolysis, fermentation or enzymatic hydrolysis methods. In view of the advantages of moderate environment, safety, directional production, and high efficiency, the enzymatic hydrolysis method was chosen to produce antihypertensive peptides from *Porphyra yezoensis* by these researchers. *P. yezoensis* is an important alga mainly cultivated in China, Japan and Korea. It is mainly processed into juice and sauce with low economic value. *P. yezoensis* enriched in proteins (25–50%) and polysaccharides (20–40%) has shown medicinal benefits such as inhibiting tumours, viruses, anti-ulcer, etc., which are attributed to bioactive polysaccharides. In this study, seven commercial enzymes were screened and then enzymatic hydrolysis conditions were optimised. The results showed that Alcalase was the most effective for hydrolysis and its optimum conditions to achieve the highest antihypertensive activity of peptide. The antihypertensive peptide produced under the optimum conditions had a high ACE inhibition rate and remained a high stability at different temperatures, pH values of 2.0 and 8.0, and after pepsin and trypsin treatments. *P. yezoensis* mainly contained albumin, gliadin and glutelin. The glutelin as the major component of the protein (77.1%) had the highest extraction yield (28.3%) and content (38.6%).

10.4.4 Proteins and Polysaccharides

Numerous macroalgal species, in particular the red seaweeds have been shown to contain significant levels of protein. In some cases, they contain higher quantities than some conventional protein rich foods such as soybean, cereals, eggs and fish (Fleurence, 1999b) as well as for the bioactive potentials of specific lectins, enzymes and protein derivatives such as peptides. Seaweeds traditionally consumed in Asian countries as sea vegetables seem to be a suitable source of proteins for human nutrition. At the same time, currently aquaculture consumes an important amount of 'wild' fishes and by-products from fish industry (Naylor et al., 2000). These fishes are processed into flour or hydrolysed to extract proteins (Nilsang et al., 2005) and used as feed. Actually, the main plant proteins used to animal nutrition concern soybean with different caused problems like digestibility, non-adapted enzymatic system for the degradation of this proteins or environmental and ecological problem like the request of a large quantity of fresh water, ethical food value with genetically modified soybean (Fleurence et al., 2012). In a context of economic and environmental difficulties, it seems advantageous to replace fishmeal by proteins from algal biomass. This type of application is also possible for terrestrial animals, although there is an additional constraint that the digestibility of proteins and polysaccharides by animals intestines.

The role of proteins as physiologically active components in the diet is being increasingly acknowledged (Fleurence et al., 2012; Harnedy & FitzGerald, 2011; Stengel et al., 2011). However, extraction of protein from most seaweed is difficult due to the presence of a large amount of cell wall polysaccharides, such as alginates or the carrageenans. The extraction of macroalgal proteins is hindered, in part, due to high viscosity and ionic interactions arising from the presence of cell wall and intracellular polysaccharides. In this case, enzymatic hydrolysis can bring profits for the hydrolysis of proteins into peptides increasing, their accessibility and assimilation by the organism (Fleurence, 1999b; Fleurence et al., 1995). The cooperation of polysaccharidases is an attractive and efficient alternative method, which is facilitating access to the algal protein fraction. Moreover, the initial nonspecific breakdown or the depolymerisation of structural and storage polysaccharides could enhance the easy access of proteases to their respective substrates, which are located inside the cells. Although more difficult to implement, this principle can also naturally be applied to human nutrition.

Amano and Noda (1990) and Wijesinghe and Jeon (2012) reported the use of algal cell-wall degradation enzymes to facilitate the extraction of proteins from the red alga *Porphyra yezoensis*. Moreover, they demonstrated the use of an enzymatic mixture, including digestive enzymes to improve the protein accessibility. In addition, the technique was tested on several algae including *Ulva pertusa*, *L. japonica* and *Kallymenia perforata* and significant differences were obtained in protein composition for all the treated seaweeds

(Amano & Noda, 1992). Fleurence et al. (1995) reported that the simultaneous application of carrageenase and cellulase activities on the red alga *C. crispus* led to a 10-fold increase in extraction efficiency of the proteins. Further, they demonstrated that the combined use of agarase and cellulase on another red alga *G. verrucosa*, led to a 3-fold increase in protein yield. Cian et al. (2012) showed that several hydrolysates from *P. columbina* obtained by using trypsin and Alcalase were enriched in peptides. Fleurence (1999b) and Castro et al. (2006) showed that enzymatic transformation of Ulvales biomass increases both the algal protein availability and digestibility. Ulvan degradation stimulates animal defenses. Ulvan, from *Ulva rigida,* enhances macrophage activities against pathogens and contributes to disease resistance in fish. A patent describing the capacity of ulvan to intercalate into clay was described with antimycotoxin applications in animal feed (Demais et al., 2006b; Demais, Brendle, Jaber, & Laza, 2006a). This characteristic opens the way for the preparation of novel interesting nanocomposites for applications in many different domains.

Harnedy and FitzGerald (2013) used polysaccharidases as prior treatment of Rhodophyta *P. palmata* cells for protein extraction during 30 min at 40 °C. They studied the effect of NaOH and N-acetyl-L-cysteine (NAC) concentration (osmotic shock), temperature, stirring and weight/volume ratio. They showed that the optimal alkaline soluble protein extraction occurred when using 0.12 M NaOH and 0.1 g/l of NAC with respective protein recovery of 5.76 and 6.18 g/100 gDW. They also showed that the pretreatment of the cell wall by carbohydrases (Shearzyme 500 L and Celluclast 1.5 L) resulted in a recovery of 8.39 g/100 gDW. But these conditions cannot be applied at a larger scale due to the high enzyme:substrate ratio required.

Proteins are essential molecules for animal nutrition but algal polysaccharides also appear to be very interesting molecules in the diet of livestock. Study of Moroney, O'Grady, O'Doherty, and Kerry (2012) showed that the addition of cell wall polysaccharides, laminarin and fucoidans, from the brown alga *L. digitata* can increase the quality and shelf life of pigs. Other applications, more ancestral (Vincent, 1924), ever exist such as the use as fertilizer, soil conditioner (Kapraun, 1999). The enriched composition of algae makes it an interesting source of minerals. It has been shown that a foliar application of carrageenans, on plants of *Nicotiana tabacum*, stimulates plant defense (Mercier et al., 2001).

In agricultural fields, new strategies of crop protection are needed urgently in the objective to minimise the chemical quantities of pesticides in the soil and their residues in food products. Among the new biological approaches, the stimulation of natural plants defenses is considered as one of the most promising alternative strategy for crop protection. The induction of plant defense mechanisms was associated with the production of elicitors by the pathogen (exogenous elicitor) or the plant–host (endogenous elicitor). Glucidic elicitors were mainly oligosaccharides though some of them were polysaccharides (El Modafar et al., 2012; Briand, WO Patent No. 094581 A1, 2005).

Macroalgae were considered as a natural resource of bioelicitors, which could be valourised into biopesticides. Polysaccharides from red and brown algae, particularly the laminarin, as well as the carrageenan and the alginate, were extensively studied due to their elicitor properties on plants natural defenses (Mercier et al., 2001; Courtois, 2009; El Modafar et al., 2012). El Modafar studied activities of glucuronan, ulvan- and oligoulvans isolated from *U. lactuca* for their elicitor capacity as natural defenses in tomato seedlings. The specific depolymerisation of the ulvan and the glucuronan by an ulvan-lyase and a glucuronan-lyase, produced by a bacterial strain allowed the production of different oligorhamnoglucuronans and oligo-glucuronans (Elboutachfaiti et al., 2009). The treatments of tomato seedlings with *U. lactuca* elicitors significantly reduced wilt development by pathogens, particularly the oligoulvans, which strongly reduce the wilting and the mortality of tomato plants. Ulvan and the oligoulvans induced in tomato seedlings the natural systemic defenses and the salicylic acid-dependent systemic acquired resistance. The high elicitor capacity of these bioelicitors seemed to be related to sulphate groups and the rhamnose residues (El Modafar et al., 2012).

Jaulneau et al. (2010) showed that ulvan induced a gene expression signature similar to that observed upon methyl jasmonate treatment. Involvement of jasmonic acid in ulvan response was confirmed by detecting induction of protease inhibitor activity and by hormonal profiling of JA, salicylic acid and abscisic acid. Ulvan activity on the hormonal pathway was further consolidated by using Arabidopsis hormonal mutants. Altogether, the results demonstrate that green algae are a potential reservoir of ulvan elicitor, which acts through the JA pathway.

All these existing and potential applications of seaweeds allow, by fractionation, the enhancing of the total biomass, reaching a zero-waste policy. Although the range of applications of algae in the world is numerous, this is also a new and undeveloped domain and the legislative aspect is not yet clear. Indeed, each country has its own legislation, which makes it difficult to transfer. Legislation should firstly be defined precisely and globally for the use of algae or algal extracts in food additives, but also define the potential risks and consequences. All these aspects make the development of application fields of algae difficult to establish (Holdt & Kraan, 2011).

10.5 CONCLUSION

Invasive seaweeds represent an important biomass and harvesting for refinery has a significant economic interest. Biological activities have been demonstrated for molecules from macroalgae and many others applications can certainly be highlighted. The use of enzymes for the recovery of primary and secondary metabolites permits differentiation of algae extracts obtained compared to those obtained by conventional methods. Enzymatic hydrolysis can improve extraction yields, allow release of molecules inaccessible by other methods and offers the possibility of producing novel biologically active molecules, such as peptides or oligosaccharides from high molecular weight compounds. Although at present little used, the enzymatic hydrolysis of algae is a technique that will certainly increase in the coming years, be developed on a larger scale by industries and used to its full capacities. Added to this, a very important seaweed resource exists and is enriched in metabolites. In France, this proliferating biomass has consequences for coastal life but its harvesting and destruction are expensive. Therefore, the establishment of a valuation chain would both reduce the cost of treatment but also develop a potential highly productive market. This sector is currently under construction. In 2012, a project named Ulvans ('Ulva VAlorisation Nutrition Santé'), led by Breton companies, was launched for the collection and refinery by enzymatic hydrolysis of invasive green algae of the genus *Ulva* sp. for applications in food and health for animal and plant (Algae Industry Magazine, 2013). Other projects will certainly follow and allow development of industrial applications.

REFERENCES

Ahn, G., Hwang, I., Park, E., Kim, J., Jeon, Y.-J., Lee, J., et al. (2008). Immunomodulatory effects of an enzymatic extract from *Ecklonia cava* on murine splenocytes. *Marine Biotechnology*, *10*(3), 278–289.

Ahn, C. B., Jeon, Y. J., Kang, D. S., Shin, T. S., & Jung, B. M. (2004). Free radical scavenging activity of enzymatic extracts from a brown seaweed *Scytosiphon lomentaria* by electron spin resonance spectrometry. *Food Research International*, *37*(3), 253–258.

Ahn, G., Park, E., Lee, W.-W., Hyun, J.-W., Lee, K.-W., Shin, T., et al. (2010). Enzymatic extract from *Ecklonia cava* induces the activation of lymphocytes by IL-2 production through the classical NF-κB pathway. *Marine Biotechnology*, *13*(1), 66–73.

Alam, F., Date, A., Rasjidin, R., Mobin, S., Moria, H., & Baqui, A. (2012). Biofuel from algae- is it a viable alternative? *Procedia Engineering, 49*, 221–227.

Algae Industry Magazine. (2013). *Olmix opens algae biorefinery in Brittany*. Algae Industry Magazine. http://www.algaeindustrymagazine.com/olmix-opens-algae-biorefinery-brittany/.

Alves de Guimaraens, M., & Coutinho, R. (2000). Temporal and spatial variation of *Ulva* spp. and water properties in the Cabo Frio upwelling region of Brazil. *Aquatic Botany, 66*(2), 101–114.

Amano, H., & Noda, H. (1990). Proteins of protoplasts from red alga *Porphyra yezoensis*. *Nippon Suisan Gakkaishi, 56*, 1859–1864.

Amano, H., & Noda, H. (1992). Proteins of protoplasts from several seaweeds. *Nippon Suisan Gakkaishi, 58*, 291–299.

Athukorala, Y., Jung, W. K., Vasanthan, T., & Jeon, Y. J. (2006). An anticoagulative polysaccharide from an enzymatic hydrolysate of *Ecklonia cava*. *Carbohydrate Polymers, 66*, 184–191.

Athukorala, Y., Lee, K. W., Kim, S. K., & Jeon, Y. J. (2007). Anticoagulant activity of marine green and brown algae collected from Jeju Island in Korea. *Bioresource Technology, 98*, 1711–1716.

Avila, M., & Seguel, M. (1993). An overview of seaweed resources in Chile. *Journal of Applied Phycology, 5*(2), 133–139.

Barbeyron, T., Lerat, Y., Sassi, J.-F., Le Panse, S., Helbert, W., & Collen, P. N. (2010). *Persicivirga ulvanivorans* sp. nov., a marine member of the family Flavobacteriaceae that degrades ulvan from green algae. *International Journal of Systematic and Evolutionary Microbiology, 61*(8), 1899–1905.

Bhakuni, D. S., & Rawat, D. S. (2005). *Bioactive marine natural products*. New York: Springer.

Bondu, S., Deslandes, E., Fabre, M. S., Berthou, C., & Guangli, Y. (2010). Carrageenan from *Solieria chordalis* (Gigartinales): structural analysis and immunological activities of the low molecular weight fractions. *Carbohydrate Polymers, 81*(2), 448–460.

Borines, M. G., de Leon, R. L., & Cuello, J. L. (2013). Bioethanol production from the macroalgae *Sargassum* spp. *Bioresource Technolology, 138*, 22–29.

Bouhlal, R., Haslin, C., Chermann, J. C., Colliec-Jouault, S., Sinquin, C., Simon, G., et al. (2011). Antiviral activities of sulfated polysaccharides isolated from *Sphaerococcus coronopifolius* (Rhodophytha, Gigartinales) and *Boergeseniella thuyoides* (Rhodophyta, Ceramiales). *Marine Drugs*, 7, 1187–1209.

Bouhlal, R., Riadi, H., & Bourgougnon, N. (2010). Antiviral activities of Morocco seaweeds extracts. *African Journal of Biotechnology, 9*, 7968–7975.

Bourgougnon, N. (2003). Anti-HIV compounds from red seaweeds. Review. In *Bioorganisms and medicinal chemistry Letter* (Vol. 9). New Hampshire, USA and Plymouth, UK: Recent Advances in Marine Biotechnology published by Science Publishers of Enfield. 40 pp.

Bourgougnon, N., Lahaye, M., Chermann, J. C., & Kornprobst, J. M. (1993). Composition and antiviral activities of sulfated polysaccharide from *Schizymenia dubyi* (Rhodophyta, Gigartinales). *Bioorganic and Medicinal Chemistry Letter, 3*, 1141–1146.

Bourgougnon, N., & Stiger-Pouvreau, V. (2011). Chemodiversity and bioactivity within red and brown macroalgae along the french coasts, Metropole and Overseas Departments and Territories. In S.-K. Kim (Ed.), *Handbook of marine macroalgae* (pp. 58–105). John Wiley & Sons, Ltd.

Briand, X., Cluzet, S., Dumas, B., Esquerre-Tugaye, M. T. & Salamagne, S. (October, 2005). WO. Patent No. 2005/094581. Use of ulvans as elicitors of mechanisms for nitrogen absorption and protein synthesis.

Buchholz, C. M., Krause, G., & Buck, B. H. (2012). Seaweed and man. In C. Wiencke, & K. Bischof (Eds.), *Seaweed Biology* (Vol. 219) pp. 471–493). Berlin, Heidelberg: Springer Berlin Heidelberg.

Cardozo, K. H. M., Guaratini, T., Barros, M. P., Falcão, V. R., Tonon, A. P., Lopes, N. P., et al. (2007). Metabolites from algae with economical impact. *Comparative Biochemistry and Physiology Part C: Toxicology & Pharmacology, 146*(1–2), 60–78.

Casas, G., Scrosati, R., & Piriz, M. L. (2004). The invasive kelp *Undaria pinnatifida* (Phaeophyceae, Laminariales) reduces native seaweed diversity in Nuevo Gulf (Patagonia, Argentina). *Biological Invasions, 6*(4), 411–416.

Castro, R., Piazzon, M. C., Zarra, I., Leiro, J., Noya, M., & Lamas, J. (2006). Stimulation of turbot phagocytes by *Ulva rigida* C. Agardh polysaccharides. *Aquaculture, 254*, 9–20.

Cebrian, E., & Ballesteros, E. (2009). Temporal and spatial variability in shallow- and deep-water populations of the invasive *Caulerpa racemosa* var. cylindracea in the Western Mediterranean. *Estuarine, Coastal and Shelf Science, 83*(4), 469–474.

Charlier, R. H., Morand, P., & Finkl, C. W. (2008). How Brittany and Florida coasts cope with green tides. *International Journal of Environmental Studies, 65*(2), 191–208.

Charlier, R. H., Morand, P., Finkl, C. W., & Thys, A. (2007). Green tides on the Brittany coasts. *Environmental Research Engineering Management, 3*(41), 52–59.

Choi, S. P., Nguyen, M. T., & Sim, S. J. (2010). Enzymatic pretreatment of *Chlamydomonas reinhardtii* biomass for ethanol production. *Bioresource Technology, 101*(14), 5330–5336.

Chopin, T., Robinson, S. M. C., Troell, M., Neori, A., Buschmann, A. H., & Fang, J. (2008). Multitrophic integration for sustainable marine aquaculture. In S. E. Jørgensen, & B. D. Fath (Eds.), *Encyclopedia of ecology* (pp. 2463–2475). Oxford: Academic Press.

Chopin, T., & Sawhney, M. (2009). Seaweeds and their Mariculture. In J. H. Steele (Ed.), *Encyclopedia of ocean Sciences* (2nd ed.) (pp. 317–326). Oxford: Academic Press.

Cian, R. E., Martínez-Augustin, O., & Drago, S. R. (2012). Bioactive properties of peptides obtained by enzymatic hydrolysis from protein byproducts of *Porphyra columbina*. *Food Research International, 49*(1), 364–372.

Coat, G., Dion, P., Noailles, M.-C., De Reviers, B., Fontaine, J.-M., Berger-Perrot, Y., et al. (1998). *Ulva armoricana* (Ulvales, Chlorophyta) from the coasts of Brittany (France). II. Nuclear rDNA ITS sequence analysis. *European Journal of Phycology, 33*(1), 81–86.

Correc, G., Hehemann, J. H., Czjzek, M., & Helbert, W. (2011). Structural analysis of the degradation products of porphyran digested by Zobellia galactanivorans β-porphyranase A. *Carbohydrate Polymers, 83*(1), 277–283.

Courtois, J. (2009). Oligosaccharides from land plants and algae: production and applications in therapeutics and biotechnology. *Current Opinion in Microbiology, 12*(3), 261–273.

Damonte, E. B., Matulewicz, M. C., & Cerezo, A. S. (2004). Sulfated seaweed polysaccharides as antiviral agents. *Current Medicinal Chemistry, 11*, 2399–2419.

De Casabianca, M. L., Barthelemy, N., Serrano, O., & Sfriso, A. (2002). Growth rate of Ulva rigida in different Mediterranean eutrophicated sites. *Bioresource Technology, 82*(1), 27–31.

Demais, H., Brendle, J., Jaber, M., & Laza, A. L. (September 2006a). FR. Patent No. 2006/2882997. Exfoliating an intercalated clay, useful in e.g. animal/human feeds, plastification, surface coatings and in nanocomposite, comprises preparing intercalated clay from clay and intercalating compound and lyophilizing in presence of water.

Demais, H., Brendle, J., Le Deit, H., Laza A. L., Lurton, L., Brault, D. (March 2006b). WO. Pantent No. 2006/030075. (Interspersed clay).

Denis, C., Morançais, M., Li, M., Deniaud, E., Gaudin, P., Wielgosz-Collin, G., et al. (2010). Study of the chemical composition of edible red macroalgae *Grateloupia turuturu* from Brittany (France). *Food Chemistry, 119*(3), 913–917.

Douglas-Westwood. (2005). *Marine industries global market analysis*.

Ducreux, P. (2010). Le risque d'intoxication aiguë par l'hydrogène sulfuré (H_2S) chez les professionnels chargés du ramassage, du transport et du traitement des "algues vertes". Unpublished Diplôme Médecine Agricole. Institut National de Médecine Agricole, Tours, France. Retrieved from http://www.fichier-pdf.fr/2011/01/07/memoire-algues-vertes-dr-ducreux/memoire-algues-vertes-dr-ducreux.pdf.

Duke, C. S., Litaker, R. W., & Ramus, J. (1987). Seasonal variation in RuBPcase activity and N allocation in the chlorophyte seaweeds *Ulva curvata* (kutz.) de toni and Codium decorticatum (Woodw.) Howe. *Journal of Experimental Marine Biology and Ecology, 112*(2), 145–164.

Dumay, J., Clément, N., Morançais, M., & Fleurence, J. (2013). Optimization of hydrolysis conditions of *Palmaria palmata* to enhance R-phycoerythrin extraction. *Bioresource Technology, 131*, 21–27.

D'Amours, O., & Scheibling, R. E. (2007). Effect of wave exposure on morphology, attachment strength and survival of the invasive green alga *Codium fragile* ssp. tomentosoides. *Journal of Experimental Marine Biology and Ecology, 351*, 129–142.

El Boutachfaiti, R., Delattre, C., Petit, E., El Gadda, M., Courtois, B., Michaud, P., et al. (2009). Improved isolation of glucuronan from algae and the production of glucuronic acids oligosaccharides using a glucuronan lyase. *Carbohydrate Research, 344*, 1670–1675.

El Modafar, C., Elgadda, M., El Boutachfaiti, R., Abouraicha, E., Zehhar, N., Petit, E., ... Courtois, J. (2012). Induction of natural defence accompanied by salicylic acid-dependant systemic acquired resistance in tomato seedlings in response to bioelicitors isolated from green algae. *Scientia Horticulturae, 138*, 55–63.

Engelen, A., & Santos, R. (2009). Which demographic traits determine population growth in the invasive brown seaweed *Sargassum muticum*? *Journal of Ecology, 97*(4), 675–684.

Estevez, J. M., Ciancia, M., & Cerezo, A. S. (2008). The system of sulfated galactans from the red seaweed *Gymnogongrus torulosus* (Phyllophoraceae, Rhodophyta): location and structural analysis. *Carbohydrate Polymers, 73*(4), 594–605.

Fleurence, J. (1999a). The enzymatic degradation of algal cell walls: a useful approach for improving the protein accessibility? *Journal of Applied Phycology, 11*, 313–314.

Fleurence, J. (1999b). Seaweed proteins: biochemical, nutritional aspects and potential uses. *Trends in Food Science & Technology, 10*(1), 25–28.

Fleurence, J., Massiani, L., Guyader, O., & Mabeau, S. (1995). Use of enzymatic cell wall degradation for improvement of protein extraction from *Chondrus crispus*, *Gracilaria verrucosa* and *Palmaria palmata*. *Journal of Applied Phycology*, 7(4), 393–397.

Fleurence, J., Morançais, M., Dumay, J., Decottignies, P., Turpin, V., Munier, M., et al. (2012). What are the prospects for using seaweed in human nutrition and for marine animals raised through aquaculture? *Trends in Food Science & Technology, 27*, 57–61.

Floc'h, J. Y., Deslandes, E., & Le Gall, Y. (1987). Evidence for vegetative propagation of the carrageenophyte *Solieria chordalis* (Solieriaceae, Rhodophyceae) on the coast of Brittany (France) and in culture. *Botanica Marina, 30*, 315–321.

Food and Agriculture Organization. (2011). *World aquaculture 2010*. Rome: Food and Agriculture Organization of the United Nations.

Hardouin, K., Burlot, A. S., Umami, A., Nyvall-Collen, P., Bedoux, G., & Bourgougnon, N. (December 2013). Enzyme-assisted extraction of bioactive metabolites from seaweeds. In M. Czjzek (Chair) (Ed.), *Cell biology, physiology and biochemistry. Journées de la Société Phycologique de France*. France: Station Biologique de Roscoff, in press.

Hardouin, K., Burlot, A.-S., Umami, A., Tanniou, A., Stiger-Pouvreau, V., Widowati, I., Bedoux, G., & Bourgougnon, N. (2013). Biochemical and antiviral activities of enzymatic hydrolysates from different invasive French seaweeds. *Journal of Applied Phycology. in press* http://dx.doi.org/10/1007/s10811-013-0201-6.

Harnedy, P. A., & FitzGerald, R. J. (2011). Extraction of protein from the macroalga *Palmaria palmata*. *Food Science and Technology, 51*, 375–382.

Harnedy, P. A., & FitzGerald, R. J. (2013). Extraction of protein from the macroalga *Palmaria palmata*. *LWT - Food Science and Technology, 51*(1), 375–382.

Harshvardhan, K., Mishra, A., & Jha, B. (2013). Purification and characterization of cellulase from a marine *Bacillus* sp. H1666: a potential agent for single step saccharification of seaweed biomass. *Journal of Molecular Catalysis B: Enzymatic, 93*, 51–56.

Helbert, W., and Nyvall-Collen, P. (December 2011). WO. Patent No. 2011/157966. Ulvan lyase, method for manufacturing same, and uses thereof.

Heo, S.-J., Jeon, Y.-J., Lee, J., Kim, H. T., & Lee, K.-W. (2003). Antioxidant effect of enzymatic hydrolysate from a Kelp, *Ecklonia cava*. *Algae, 18*(4), 341–348.

Heo, S., Park, E., Lee, K., & Jeon, Y. (2005). Antioxidant activities of enzymatic extracts from brown seaweeds. *Bioresource Technology*, *96*(14), 1613–1623.
Holdt, S. L., & Kraan, S. (2011). Bioactive compounds in seaweed: functional food applications and legislation. *Journal of Applied Phycology*, *23*(3), 543–597.
Hu, C., Li, D., Chen, C., Ge, J., Muller-Karger, F. E., Liu, J., et al. (2010). On the recurrent *Ulva prolifera* blooms in the Yellow Sea and East China Sea. *Journal of Geophysical Research*, *115*(C5).
Ioannou, E., & Roussis, V. (2009). *Natural products from seaweeds*. Berlin: Springer Science.
Jaulneau, V., Lafitte, C., Jacquet, C., Fournier, S., Salamagne, S., Briand, X., et al. (2010). Ulvan, a sulfated polysaccharide from green algae, activates plant immunity through the jasmonic acid signaling pathway. *Journal of Biomedicine and Biotechnology*, 525291, 11. http://dx.doi.org/10.1155/2010/525291.
Jeon, Y. J., Wijesinghe, W. A. J. P., & Kim, S. K. (2011). Enzyme-assisted extraction and recovery of bioactive components from seaweeds. In S. K. Kim (Ed.), *Handbook of Marine macroalgae* (pp. 221–228). John Wiley & Sons, Ltd.
Jung, K. A., Lim, S.-R., Kim, Y., & Park, J. M. (2013). Potentials of macroalgae as feedstocks for biorefinery. *Bioresource Technology*, *135*, 182–190.
Kaimoussi, A., Mouzdahir, A., & Saih, A. (2005). Seasonal variations of metal concentrations (Cd, Cu, Fe, Mn and Zn) in algae fucus spiralis from the coast of el Jadida City (Morocco). *Water Quality Research Journal of Canada*, *40*(1), 102–110.
Kang, S. M., Kim, K. N., Lee, S. H., Ahn, G., Cha, S. H., Kim, A. D., et al. (2011). Anti-inflammatory activity of polysaccharide purified from AMG-assistant extract of *Ecklonia cava* in LPS-stimulated RAW 264.7 macrophages. *Carbohydrate Polymers*, *85*(1), 80–85.
Kapraun, D. F. (1999). Red algal polysaccharide industry: economics and research status at the turn of the century. In *Sixteenth international seaweed Symposium* (pp. 7–14). Springer.
Kirkman, H., & Kendrick, G. A. (1997). Ecological significance and commercial harvesting of drifting and beach-cast macro-algae and seagrasses in Australia: a review. *Journal of Applied Phycology*, *9*(4), 311–326.
Kloareg, B., & Quatrano, R. S. (1988). Structure of the cell walls of marine algae and ecophysiological functions of the matrix polysaccharides. *Oceanographic Marine Biological Annual Review*, *26*, 259–315.
Kornprobst, J. M. (2005). *Substances naturelles d'origine marine: Chimiodiversité - Pharmacodiversité - Biotechnologies*. Editions Tec & Doc.
Lahaye, M., Brunel, M., & Bonnin, E. (1997). Fine chemical structure analysis of oligosaccharides produced by an ulvan-lyase degradation of the water-soluble cell-wall polysaccharides from *Ulva* sp. (Ulvales, Chlorophyta). *Carbohydrate Research*, *304*(3–4), 325–333.
Lahaye, M., Jegou, D., & Buleon, A. (1994). Chemical characteristics of insoluble glucans from the cell wall of the marine green alga *Ulva lactuca* (L.) Thuret. *Carbohydrate Research*, *262*(1), 115–125.
Lahaye, M., & Robic, A. (2007). Structure and functional properties of ulvan, a polysaccharide from Green seaweeds. *Biomacromolecules*, *8*(6), 1765–1774.
Lahaye, M., & Vigouroux, J. (1992). Liquefaction of dulse (*Palmaria palmata* (L.) Kuntze) by a commercial enzyme preparation and purified endo-β-1,4-D-xylanase. *Journal of Applied Phycology*, *4*, 329–337.
Lapointe, B. E., & Bedford, B. J. (2010). Ecology and nutrition of invasive *Caulerpa brachypus* f. parvifolia blooms on coral reefs off southeast Florida, U.S.A. *Harmful Algae*, *9*(1), 1–12.
Le Lann, K., Connan, S., & Stiger-Pouvreau, V. (2012). Phenology, TPC and size-fractioning phenolics variability in temperate Sargassaceae (Phaeophyceae, Fucales) from Western Brittany: native versus introduced species. *Marine Environmental Research*, *80*, 1–11.

Lundqvist, L. C. E., Jam, M., Barbeyron, T., Czjzek, M., & Sandström, C. (2012). Substrate specificity of the recombinant alginate lyase from the marine bacteria Pseudomonas alginovora. *Carbohydrate Research, 352*, 44–50.

Marinho-Soriano, E., Fonseca, P. C., Carneiro, M. A. A., & Moreira, W. S. C. (2006). Seasonal variation in the chemical composition of two tropical seaweeds. *Bioresource Technology, 97*(18), 2402–2406.

Mathieson, A. C., Dawes, C. J., Pederson, J., Gladych, R. A., & Carlton, J. T. (2007). The Asian red seaweed *Grateloupia turuturu* (Rhodophyta) invades the Gulf of Maine. *Biological Invasions, 10*(7), 985–988.

McHugh, D. J., & Food and Agriculture Organization of the United Nations (2003). *A guide to the seaweed industry*. Rome: Food and Agriculture Organization of the United Nations.

Menshova, R. V., Ermakova, S. P., Anastyuk, S. D., Isakov, V. V., Dubrovskaya, Y. V., Kusaykin, M. I., et al. (2014). Structure, enzymatic transformation and anticancer activity of branched high molecular weight laminaran from brown alga *Eisenia bicyclis*. *Carbohydrate Polymers, 99*, 101–109.

Mercier, L., Lafitte, C., Borderies, G., Briand, X., Esquerré-Tugayé, M.-T., & Fournier, J. (2001). The algal polysaccharide carrageenans can act as an elicitor of plant defense. *New Phytologist, 149*(1), 43–51.

Moroney, N. C., O'Grady, M. N., O'Doherty, J. V., & Kerry, J. P. (2012). Addition of seaweed (*Laminaria digitata*) extracts containing laminarin and fucoidan to porcine diets: Influence on the quality and shelf-life of fresh pork. *Meat Science, 92*(4), 423–429.

Naylor, R. L., Goldburg, R. J., Primavera, J. H., Kautsky, N., Beveridge, M. C., Clay, J., et al. (2000). Effect of aquaculture on world fish supplies. *Nature, 405*(6790), 1017–1024.

Nilsang, S., Lertsiri, S., Suphantharika, M., & Assavanig, A. (2005). Optimization of enzymatic hydrolysis of fish soluble concentrate by commercial proteases. *Journal of Food Engineering, 70*(4), 571–578.

Okuda, K., Oka, K., Onda, A., Kajiyoshi, K., Hiraoka, M., & Yanagisawa, K. (2008). Hydrothermal fractional pretreatment of sea algae and its enhanced enzymatic hydrolysis. *Journal of Chemical Technology & Biotechnology, 83*(6), 836–841.

Park, P. J., Shahidi, F., & Jeon, Y. J. (2004). Antioxidant activities of enzymatic extracts from an edible seaweed sargassum horneri using ESR spectrometry. *Journal of Food Lipids, 11*(1), 15–27.

Paull, R. E., & Chen, N. J. (2008). Postharvest handling and storage of the edible red seaweed Gracilaria. *Postharvest Biology and Technology, 48*(2), 302–308.

Peña-Farfal, C., Moreda-Piñeiro, A., Bermejo-Barrera, A., Bermejo-Barrera, P., Pinochet-Cancino, H., & de Gregori-Henríquez, I. (2005). Speeding up enzymatic hydrolysis procedures for the multi-element determination in edible seaweed. *Analytica Chimica Acta, 548*(1–2), 183–191.

Pereira, R., & Yarish, C. (2008). Mass production of marine macroalgae. In S. E. Jørgensen, & B. D. Fath (Eds.), *Encyclopedia of ecology* (pp. 2236–2247). Oxford: Academic Press.

Plouguerné, E., Le Lann, K., Connan, S., Jechoux, G., Deslandes, E., & Stiger-Pouvreau, V. (2006). Spatial and seasonal variations in density, maturity, length and phenolic content of the invasive brown macroalga *Sargassum muticum* along the coast of Western Brittany (France). *Aquatic Botany, 85*, 337–344.

Popper, Z. A., Michel, G., Hervé, C., Domozych, D. S., Willats, W. G. T., Tuohy, M. G., et al. (2011). Evolution and diversity of plant cell walls: from algae to flowering plants. *Annual Review of Plant Biology, 62*, 567–590.

Rindi, F., & Guiry, M. D. (2004). Composition and spatial variability of terrestrial algal assemblages occurring at the bases of urban walls in Europe. *Phycologia, 43*(3), 225–235.

Robic, A., Sassi, J.-F., & Lahaye, M. (2008). Impact of stabilization treatments of the green seaweed *Ulva rotundata* (Chlorophyta) on the extraction yield, the physico-chemical and rheological properties of ulvan. *Carbohydrate Polymers, 74*(3), 344–352.

Romarís-Hortas, V., Bermejo-Barrera, P., & Moreda-Piñeiro, A. (2013). Ultrasound-assisted enzymatic hydrolysis for iodinated amino acid extraction from edible seaweed before reversed-phase high performance liquid chromatography–inductively coupled plasma-mass spectrometry. *Journal of Chromatography a, 1309*, 33–40.

Romero, J. B., Villanueva, R. D., & Montaño, M. N. E. (2008). Stability of agar in the seaweed *Gracilaria eucheumatoides* (Gracilariales, Rhodophyta) during postharvest storage. *Bioresource Technology, 99*(17), 8151–8155.

Rosenberg, C., & Ramus, J. (1982). Ecological growth strategies in the seaweeds *Gracilaria foliifera* (Rhodophyceae) and *Ulva* sp. (Chlorophyceae): soluble nitrogen and reserve carbohydrates. *Marine Biology, 66*(3), 251–259.

Ruiz, G. M., Fofonoff, P.W., Carlton, J. T., Wonham, M. J., & Hines, A. H. (2000). Invasion of coastal marine communities in North America: apparent patterns, processes, and biases. *Annual Review of Ecology and Systematics*, 481–531.

Samarakoon, K., & Jeon, Y. J. (2012). Bio-functionnalities of proteins derived from marine algae. *Food Research International, 48*, 948–960.

Schiel, D. R., & Thompson, G. A. (2012). Demography and population biology of the invasive kelp *Undaria pinnatifida* on shallow reefs in southern New Zealand. *Journal of Experimental Marine Biology and Ecology, 434–435*, 25–33.

Sfriso, A., & Facca, C. (2013). Annual growth and environmental relationships of the invasive species *Sargassum muticum* and *Undaria pinnatifida* in the lagoon of Venice. *Estuarine, Coastal and Shelf Science, 129*, 162–172.

Siriwardhana, N., Kim, K. N., Lee, K.W., Kim, S. H., Ha, J. H., & Song, C. B. (2008). Optimisation of hydrophilic antioxidant extraction from Hizikia fusiformis by integrating treatments of enzymes, heat and pH control. *International Journal of Food Science And Technology, 43*, 587–596.

Smetacek, V., & Zingone, A. (2013). Green and golden seaweed tides on the rise. *Nature, 504*(7478), 84–88.

Stagnol, D., Renaud, M., & Davoult, D. (2013). Effects of commercial harvesting of intertidal macroalgae on ecosystem biodiversity and functioning. *Estuarine, Coastal and Shelf Science, 130*, 99–110.

Steneck, R. S., Graham, M. H., Bourque, B. J., Corbett, D., Erlandson, J. M., Estes, J. A., et al. (2003). Kelp forest ecosystems: biodiversity, stability, resilience and future. *Environmental Conservation, 29*(04).

Stengel, D. B., Connan, S., & Popper, Z. A. (2011). Algal chemodiversity and bioactivity: sources of natural variability and implications for commercial application. *Biotechnology Advances, 29*(5), 483–501.

Stiger, V., Deslandes, E., & Payri, C. E. (2004). Phenolic contents of two brown algae, *Turbinaria ornata* and *Sargassum mangarevense* on Tahiti (French Polynesia): interspecific, ontogenic and spatio-temporal variations. *Botanica Marina, 47*, 402–409.

Stiger, V., & Payri, C. E. (1999). Spatial and seasonal variations in the biological characteristics of two invasive brown algae, *Turbinaria ornata* (Turner) J. Agardh and *Sargassum mangarevense* (Grunow) Setchell (Sargassaceae, Fucales) spreading on the reefs of Tahiti (French Polynesia). *Botanica Marina, 42*(3), 295–306.

Suryanarayan, S., Nori, S. S., Kumar, S., Vadassery, N., Balendiran, S., Sayash Kumar, S. (June 2013). WO. Patent No. 2013/050860. Process of production of renewable chemicals and biofuels from seaweeds.

Tavennec, M. (2009). *La gestion d'une problématique algale dans un secteur hautement touristique, la presqu'île de Rhuys.* Rapport de master 2, Université de Nantes.

Tchorbanov, B., & Bozhkova, M. (1988). Enzymatic hydrolysis of cell proteins in green algae Chlorella and Scenedesmus after extraction with organic solvents. *Enzyme and Microbial Technology, 10*(4), 233–238.

Titlyanov, E. A., & Titlyanova, T. V. (2010). Seaweed cultivation: methods and problems. *Russian Journal of Marine Biology*, *36*(4), 227–242.

Trivedi, N., Gupta, V., Reddy, C. R. K., & Jha, B. (2013). Enzymatic hydrolysis and production of bioethanol from common macrophytic green alga *Ulva fasciata* Delile. *Bioresource Technology*, *150*, 106–112.

Villares, R., & Carballeira, A. (2003). Seasonal variation in the concentrations of nutrients in two green macroalgae and nutrient levels in sediments in the Rías Baixas (NW Spain). *Estuarine, Coastal and Shelf Science*, *58*(4), 887–900.

Villares, R., Puente, X., & Carballeira, A. (2002). Seasonal variation and background levels of heavy metals in two green seaweeds. *Environmental Pollution*, *119*(1), 79–90.

Vincent, V. (1924). *Les Algues Marines et leurs emplois agricoles, alimentaires, industriels, etc.*

Vo, T. S., Ngo, D. H., Ta, Q. V., & Kim, S. K. (2011). Marine organisms as a therapeutic source against herpes simplex virus infection. *European Journal of Pharmaceutical Science*, *44*, 11–20.

Wang, T., Jónsdóttir, R., Kristinsson, H. G., Hreggvidsson, G. O., Jónsson, J. Ó., Thorkelsson, G., et al. (2010). Enzyme-enhanced extraction of antioxidant ingredients from red algae *Palmaria palmata*. *Lwt-food Science and Technology*, *43*, 1387–1393.

Wang, X., Liu, X., & Wang, G. (2011). Two stage hydrolysis of invasive algal feedstock for ethanol fermentation. *Journal Integrative Plant Biology*, *53*, 246–253.

White, L. F., & Shurin, J. B. (2011). Density dependent effects of an exotic marine macroalga on native community diversity. *Journal of Experimental Marine Biology and Ecology*, *405*(1-2), 111–119.

Wijesekara, I., Pangestuti, R., & Kim, S. K. (2011). Biological activities and potential health benefits of sulfated polysaccharides derived from marine algae. *Carbohydrate Polymers*, *84*, 14–21.

Wijesinghe, W. A. J.P., Athukorala, Y., & Jeon, Y. J. (2011). Effect of anticoagulative sulfated polysaccharide purified from enzyme-assistant extract of a brown seaweed *Ecklonia cava* on Wistar rats. *Carbohydrate Polymers*, *86*(2), 917–921.

Wijesinghe, W. A. J.P., & Jeon, Y.-J. (2012). Enzyme-assistant extraction (EAE) of bioactive components: a useful approach for recovery of industrially important metabolites from seaweeds: a review. *Fitoterapia*, *83*(1), 6–12.

Williams, S. L., & Smith, J. E. (2007). A global review of the distribution, taxonomy, and impacts of introduced seaweeds. *Annual Review of Ecology, Evolution, and Systematics*, *38*(1), 327–359.

Yoon, Y. S., Goh, G. H., Song, J. H., Oh, S. H., Kang, S. Y., Park, C. H. (February 24, 2010). WO. Patent No. 2010/098585 A2. Method for producing biofuels via hydrolysis of seaweed extract using heterogeneous catalyst.

Zhang, J. H., Huo, Y. Z., Zhang, Z. L., Yu, K. F., He, Q., Zhang, L. H., et al. (2013). Variations of morphology and photosynthetic performances of *Ulva prolifera* during the whole green tide blooming process in the Yellow Sea. *Marine Environmental Research*, *92*, 35–42.

CHAPTER ELEVEN

Phycoerythrins: Valuable Proteinic Pigments in Red Seaweeds

Justine Dumay[1], Michèle Morançais, Mathilde Munier, Cécile Le Guillard and Joël Fleurence
LUNAM, Université de Nantes, Nantes, France
[1]Corresponding author: e-mail address: justine.dumay@univ-nantes.fr

Contents

Advances in Botanical Research, Volume 71
ISSN 0065-2296
http://dx.doi.org/10.1016/B978-0-12-408062-1.00011-1

Abstract

Seaweeds are traditionally used as sea vegetables in Asian countries but their consumption by western people is marginal. Species used as sea vegetables are mainly red and brown algae. The biochemical content of marine red algae provides a high nutritional value, similar to that found in traditional vegetable food sources. Nowadays, the use of red seaweeds as sources of proteins, pigments, or minerals appears to be an interesting opportunity for the valorization of a marine resource. Phycobiliproteins, which are the main light-harvesting pigments of photosystem II in Cyanobacteria and Rhodophyta, are the only water-soluble algal pigments. Their specific structure and function confer a wide range of applications. R-phycoerythrin is one of these soluble proteins and is widely used as a fluorescent probe or food colorant depending on its purity. Numerous studies have been conducted to extract and purify phycoerythrins using various methods.

11.1 INTRODUCTION

Marine macroalgae have a long history of use as foods and colorants in Asia but are much less common in Europe and North America. Mainly red and brown seaweeds are harvested and cultivated for use in the food, agricultural or medicine sectors (Dawczynski, Schubert, & Jahreis, 2007). Generally, marine algae are used in human or animal foods for their mineral content or for the functional properties of their polysaccharides. They are rarely promoted for the nutritional value of their proteins. However, some red seaweeds, such as *Porphyra tenera* and *Palmaria palmata*, can have protein contents higher than those recorded in pulses such as soybean (Fleurence, 1999b). Other algae, such as *Ulva*, *Undaria* or *Enteromorpha*, have protein levels comparable to those reported for common vegetables (Amano & Noda, 1990). Conditioned by their marine environment, macroalgae produce unique and interesting active compounds (Fitzgerald, Gallagher, Tasdemir, & Hayes, 2011).

Phycobiliproteins, which are the main light-harvesting pigments of photosystem II in Cyanobacteria and Rhodophyta, are the only water-soluble algal pigments. In most of these species, the phycobiliprotein content is relatively high and can reach around 50% of water-soluble proteins (Gantt & Lipschultz, 1974; Ke, 2001) and up to 60% in a filamentous blue-green cyanobacteria (Tandeau de Marsac, Castets, & Cohen-Bazire, 1980). Moreover, phycobiliproteins can represent up to 20% of dry algal material (Bogorad, 1975). They are constituted of one proteinic part linked by a covalent bond to a prosthetic group: the *chromophore* (also named bilin) (Glazer et al., 1976). These molecules are oligomeric proteins composed of two subunits, α and β, both coming probably from a common ancestor (Glazer, 1989). Contrary to other biliproteins, R-phycoerythrin is the only one to possess an extra γ subunit, increasing the stability of the pigment structure (Glazer & Hixson, 1977; Hilditch, Balding, Jenkins, Smith, & Rogers, 1991). Phycoerythrin (PE), like other phycobiliproteins, is highly water soluble and its proteinic nature has led to much interest and many applications (Sekar & Chandramohan, 2008).

Recently, methods have been improved to recover highly purified PEs using mild technologies as outlined in the final part of this chapter.

11.2 RED SEAWEED: BIOCHEMICAL COMPOSITION AND NUTRITIONAL VALUE

Most of the red seaweeds are consumed as sea vegetables. One example is the species *P. tenera* or *P. yezoensis*, well known under the name of nori (Nisizawa, Noda, Kikuchi, & Watanabe, 1987). The world production of this genus was established as 22,060,504 tons expressed in fresh weight (FAO, 2011). The red algae belonging to the genus *Gracilaria* are also cultivated for food, being used as sea vegetables (ogo-nori), for abalone breeding and mainly for agar production. In France, some red seaweed species are authorized for human consumption (Mabeau & Fleurence, 1993); *P. palmata* or dulse, *Gracilaria gracilis* or ogo-nori, and *Porphyra umbilicalis* or nori.

As described below, the biochemical composition and nutritional value of this resource justify the traditional consumption of red seaweeds in human or animal nutrition.

11.2.1 Biochemical Composition

11.2.1.1 Protein and Amino Acid Composition

Some red seaweeds are characterized by a high protein content (Table 11.1). For example, Japanese authors reported a protein content of 47% DW for the species *P. tenera* (Fujiwara-Arasaki, Mino, & Kuroda, 1984), which

Table 11.1 Seasonal Variation in Protein Content of Some Edible Red Seaweeds (Fujiware-Arasaki et al., 1984; Fleurence, 2004; Denis et al., 2010)

Species	*Porphyra tenera*	*Palmaria palmata*	*Grateloupia turuturu*	*Gracilaria gracilis*
Protein content (% DW)	33.0–47.0	8.0–35.0	14.0–27.5	7.0–25.0

is similar to that of soybean oilcake (50% protein). The red seaweed *P. palmata* (dulse) was described as containing up to 35% protein DW (Morgan, Wright, & Simpson, 1980). However, this amount was obtained from seaweeds cultivated in tanks supplied with nitrogen nutrients. In natural environments, such as on the coasts of Brittany (France), the maximal level of proteins recorded for this species was 25% DW (Galland-Irmouli et al., 1999).

In addition, protein levels in red seaweeds are subject to significant seasonal variation. In *P. palmata*, the protein content ranges from 8% to 35% DW (Table 11.1). A seasonal variation in protein level has also been reported for *Grateloupia turuturu*, an invasive red seaweed on the French coast, consumed as a sea vegetable in Japan (Denis et al., 2010). For this species, the lowest amounts of proteins were observed in summer (14% protein DW) and the highest at the end of winter (27.5% protein DW) (Table 11.1).

The amino acid content of red seaweeds is also subject to variations (Table 11.2). However, a species such as *P. palmata* is characterized by a high level of methionine (up to 4.5 g/100 g protein) in comparison with other red algae (Fleurence, 2004). This feature distinguishes this species both from other seaweeds and from plants such as legumes.

11.2.1.2 Carbohydrate Composition

Like the protein level, the carbohydrate content of red algae varies according to the species and the season. Some authors consider the approximate carbohydrate composition of red seaweeds to fall between 30% and 60% DW (Jensen, 1993). For *P. palmata*, it lies between 38% and 74% DW (Morgan et al., 1980). Carbohydrates mainly consist of acidic xylans and cellulose (Deniaud, Quemener, Fleurence, & Lahaye, 2003). The carbohydrate level can also vary according to the cultivation conditions. The cultivation of *G. gracilis*, previously called *Gracilaria verrucosa*, a seaweed producing agar, is a well-known example. In this species, the carbohydrate content is 35.9%

Table 11.2 Methionine Content of Some Red Seaweeds Compared to Other Food Sources (Egg, Soybean) (Fujiwara-Arasaki et al., 1984; Fleurence, 2004; Fleurence et al., 2012)

Food source	*Porphyra tenera*	*Palmaria palmata*	*Grateloupia turuturu*	Soybean	Ovalbumin
Methionine content	1.1	4.5	2.0	1.4	3.1

g amino acid/100 g protein

DW when the nitrogen input in the cultivation medium is minimal and 14.7% in conditions where the nitrogen input is maximal (Bird, 1984).

11.2.1.3 Lipid and Fatty Acid Composition

Generally, seaweeds are poor in lipids (<3% DW). For red species such as *P. tenera*, a level of 0.8% DW was recorded (Anonymous, 1969). Regarding the species consumed in Europe, such as *Porphyra dioica*, previously called *P. umbilicalis*, *Chondrus crispus*, *P. palmata* and *G. gracilis*, the amounts reported are 3.4%, 0.6%, 1.6% and 1.3% DW, respectively (Fleurence, Gutbier, Mabeau, & Leray, 1994). There are few studies on the seasonal variations in lipid content in red seaweeds. Recent research on *G. turuturu* showed an absence of variation in lipid content in relation to the season (2.31–2.78% DW) (Denis et al., 2010). However, small variations in lipid content (3.3–4.1% DW) were reported on *G. turuturu* collected in the same places but in another year (2009) (Kendel et al., 2013). In this last study, no marked variations were observed for glycolipids (42.3–46.8% DW), whereas phospholipids and neutral lipids varied significantly between summer and winter.

Red algae are known to contain large amounts of C20 polyunsaturated fatty acids. A high level of eicosapentaenoic acid (EPA) (C20:5 n-3) was reported for *Porphyra* sp. from Japan and Korea (Dawczynski et al., 2007). Similar data were observed for other red seaweeds collected on the coasts of Brittany. The levels of EPA in *P. umbilicalis*, *P. palmata* and *G. gracilis* were 48%, 46% and 26%, respectively, of the total fatty acid fraction (Fleurence et al., 1994).

11.2.2 Nutritional Value

Their high protein content is the main nutritional interest for the use of red seaweeds in human and animal foods (Fleurence et al., 2012). Among the species consumed in Europe, *P. palmata* has a high nutritional value. First, the protein content in spring (25% DW) is similar to that reported for the soybean plant, which is a widely used food protein source. Secondly, essential amino acids constitute almost 46% of the total amino acid fraction (Fleurence, 2004). This proportion is close to that reported for ovalbumin, which is the reference for determining the nutritional value of food protein.

However, the nutritional interest of the seaweed protein fraction is not limited to the protein content and composition. The digestibility of proteins is an important criterion to assess their true nutritional value. Unfortunately, the *in vitro* digestibility of *P. palmata* proteins is limited by the presence of water-soluble xylans, which are antinutritional factors (Galland-Irmouli

Table 11.3 *In vitro* Digestibility after 6 h of Casein, *Gracilaria gracilis* and *Palmaria palmata* Proteins (Marrion et al., 2005)

	Casein	*G. gracilis* proteins	*P. palmata* proteins
In vitro digestibility (%)	38.7	30.4	10.3

et al., 1999). However, their digestibility can be significantly improved by carrying out previous fermentation treatments on the thalli (Marrion, Schwertz, Fleurence, Guéant, & Villaume, 2003).

Although enzymatic or fermentation treatment is useful for *P. palmata*, it is not necessary for other species, such as the red seaweed *G. gracilis* which can be readily eaten as vegetable under the name ogo-nori. This species has a protein level similar (25% DW) (Table 11.1) to that reported for *P. palmata*. A comparative study on the digestibility of protein fractions from both these species showed that *G. gracilis* proteins were more digestible than those of *P. palmata*. Effectively, without any previous enzymatic digestion or fermentation, *G. gracilis* protein digestibility was 30% after 6 h of digestion compared to 10% for *P. palmata* proteins in the same conditions (Table 11.3) (Marrion et al., 2005).

11.3 R-PHYCOERYTHRIN AND PHYCOBILIPROTEINS: FUNCTION, STRUCTURE AND UTILIZATION

11.3.1 Location and Function of Phycobilisomes

The phycobiliproteins are subnumerary pigments organized *in vivo* in supramolecular structures called *phycobilisomes* (Gantt, 1980).

The main function of this remarkable photon collector is the survival of living organisms at low light intensities. In fact, as yellow and green radiations are quickly vanished according to depth, the phycobilisomes are able to absorb light in spectral zones where chlorophyll a cannot, and to transmit the energy to the photosystem II reactive center (Glazer, 1989; Ke, 2001).

Phycobilisomes are located on the external structure of thylakoid membranes, in the stroma (Figure 11.1). They have a supramolecular structure with a diameter ranging from 30 to 60 nm. They are constituted by an association of phycobiliproteins which allows the transfer of light energy according to the pathway described in Figure 11.2 (Bald, Kruip, & Rögner, 1996; Glazer & Hixson, 1977; Sun, Wang, Gong, & Chen, 2004).

Thus, these structures can transmit the light energy from chromophores absorbing at green wavelengths (such as PEs) to chromophores absorbing at red ones (such as allophycocyanins (APCs)) (Glazer & Hixson, 1977).

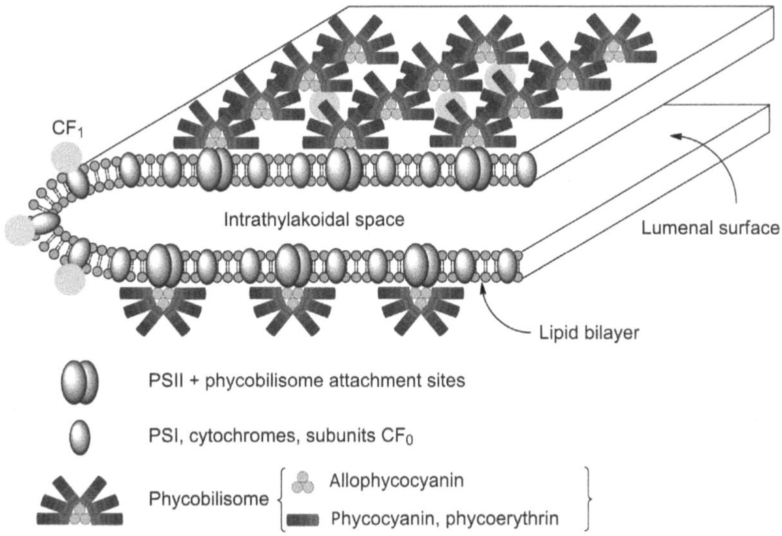

Figure 11.1 Schematic representation of the thylakoid membrane and phycobilisomes. CF, coupling factor; PS, photosystem. (See the colour plate.)

Figure 11.2 Light energy transfer pathway through phycobilisomes.

Four types of phycobilisome structure have been highlighted (bundle-shaped, hemi-discoidal, hemi-ellipsoidal and block-shaped), but only two are predominant:

- The hemi-ellipsoidal type which is found in red algae. Phycobilisome of this type are 40 nm long, 19 nm wide and 28 nm high.
- The globular type, which is more complex, is found in only some Rhodophyta.

The design of the phycobilisome (Figure 11.3) is such that excitation energy is delivered from any one of these many chromophores to a reaction center in the photosynthetic membrane with an efficiency approaching 100% (Glazer, 1984).

As described above, phycobilisomes have two morphologically distinct substructures: the rods and the core. The core is essentially composed of APCs and is linked to polypeptides responsible for the thylakoid connection. The core is composed of three cylinders, each constituted of four disks, which consist of APC trimer. The other major biliproteins are confined to

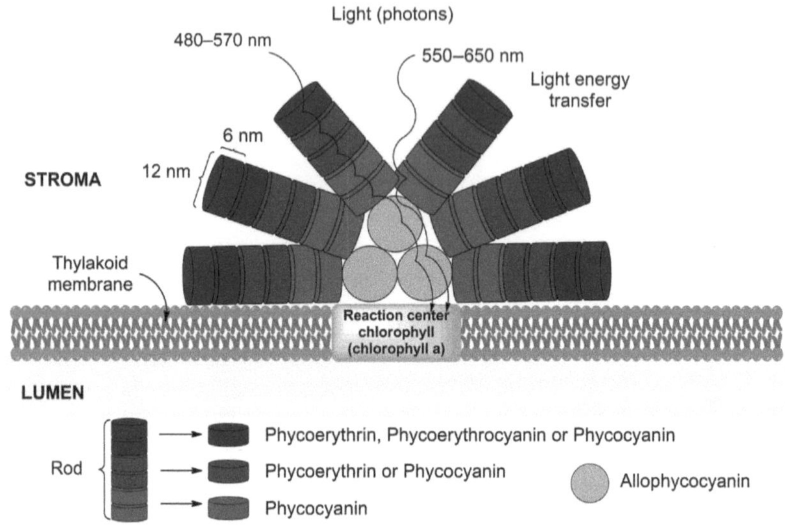

Figure 11.3 Typical phycobilisome structure in red algae. (See the colour plate.)

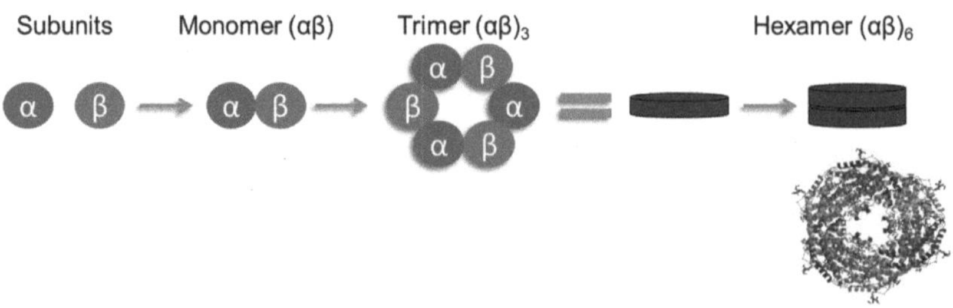

Figure 11.4 Diagram of phycobiliprotein subunit assemblage.

the rod substructure, shaped like a stick. Each stick is composed of disks made up of hexamers of distinct phycobiliproteins and linked with a polypeptide. A diagram of disk formation is given in Figure 11.4.

The linkage polypeptides represent around 15% (according to the species) of the proteinic part of the phycobilisome (Tandeau de Marsac & Cohen-bazire, 1977). Their main functions are first to fix the phycobilisome to the thylakoid membrane and second to stabilize the phycobilisome structure. Moreover, they ensure phycobiliprotein cohesion in the phycobilisome. In fact, while phycobiliproteins are acidic and hydrophilic, linkage polypeptides are basic and hydrophobic, thus allowing some interactions between these constituents (Ke, 2001). The polypeptides can be divided into four groups according to their molecular weight (Ducret et al., 1998; Glazer, 1984):

- Group I or L_{CM} (70–120 kDa): phycobilisome fixation to thylakoid membranes. They act as final emitters of the collecting antenna and contribute to the light energy transfer to photosystem II.
- Group II or L_R, L_{RC} (25–35 kDa): stabilization of phycobiliprotein in the stick structure and fixation of the rod to the core.
- Group III (12–22 kDa): this group is constituted of phycobiliprotein subunits, an 18.3 kDa polypeptide and the α subunit of APC B.
- Group IV or L_C (9–12 kDa): assembly and fixation of the core.

Hence, the linkage polypeptides determine the aggregation state of the phycobiliprotein and influence both different subunit positions and chromophore conformation (Glazer, 1984).

11.3.2 Chromophore Structure

Bilin chromophores are tetrapyrrolic structures, closely related to the mammalian blue pigments biliverdin and bilirubin (Brown, Holroyd, & Vernon, 1984). Unlike chlorophylls, these chromophores are not cyclized. Depending on their origin, phycobilisomes contain between 300 and 800 tetrapyrrol chromophores (Glazer, 1984), which confer spectral properties on the phycobiliproteins. Four types of chromophore are found in Cyanobacteria and Rhodophyta. The first two described were phycocyanobilin (PCB) and phycoerythrobilin (PEB). Thereafter, two other chromophores were identified: phycourobilin (PUB) and phycobiliviolin (PXB). All four structures are represented in Figure 11.5.

Chromophores differ according to their absorption wavelength maxima as depicted in Figure 11.5 (Apt, Collier, & Grossman, 1995). PXB is a minor chromophore only located on the α subunit of phycoerythrocyanin, so it has been little studied to date.

11.3.3 Structure of Phycobiliproteins

All the major biliproteins are oligomers of an αβ monomer, where α and β are dissimilar polypeptide chains of approximately 160–180 amino acid residues (Glazer, 1984). Glazer summarized several studies showing that all the polypeptides that make up the biliproteins are encoded by a family of closely related genes (Glazer, 1984). Originally, they were divided into two categories according to their colour: PEs with a soft pink colour and orange fluorescence and phycocyanins (PCs) with a blue colour and red fluorescence. They were then classified according to the chemical nature of their chromophores and the spectral properties of the native proteins (Bogorad, 1975; Glazer & Hixson, 1975). Hence, prefixes were introduced

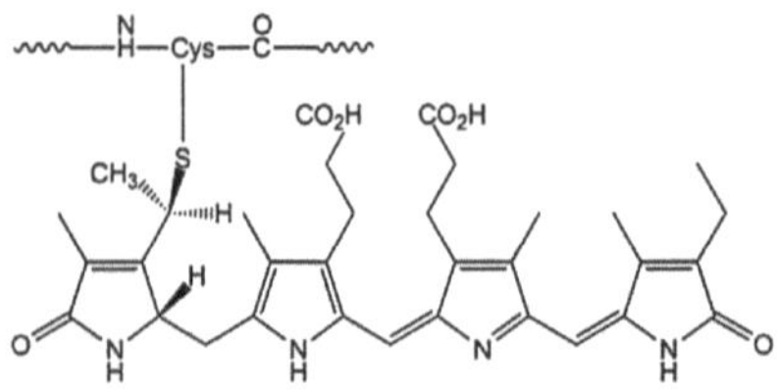

Peptide-linked Phycobiliviolin (purple)
PXB

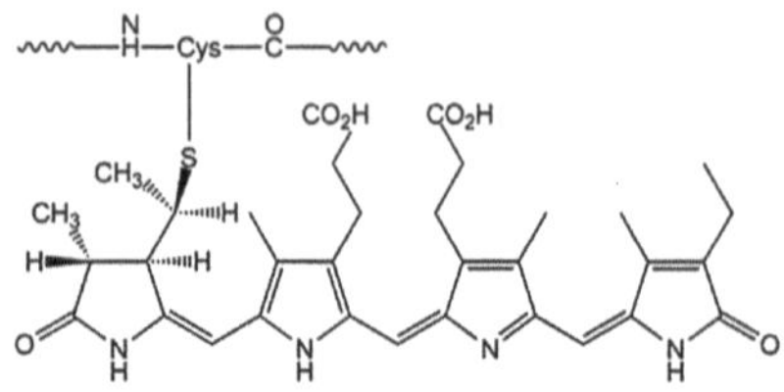

Peptide-linked Phycocyanobilin (blue)
PCB

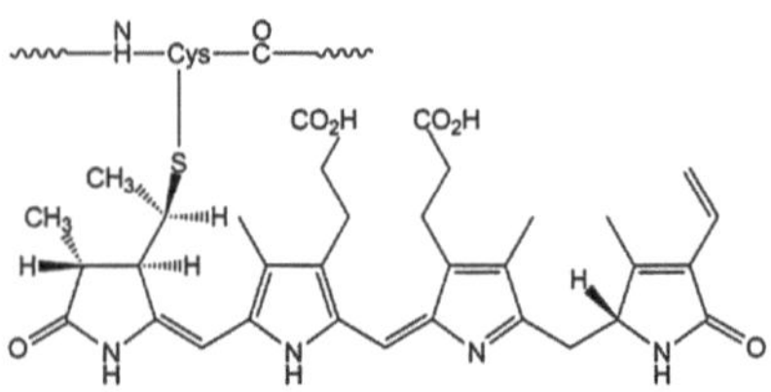

Peptide-linked Phycoerythrobilin (red)
PEB

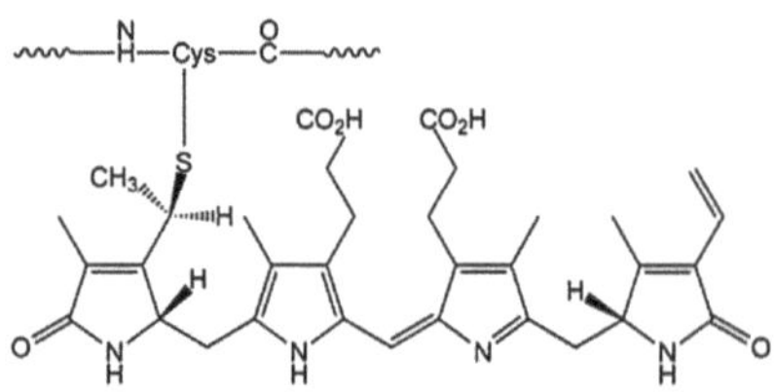

Peptide-linked Phycourobilin (turquoise-blue)
PUB

Figure 11.5 Chemical structures of bilin chromophores.

Table 11.4 Spectral Properties of Phycobiliproteins According to Bryant (1982)

Phycobiliprotein	Colour	Prefix	Max. Absorption (nm)	Shoulder Absorption (nm)	Max. Fluorescence Emission (nm)
Phycoerythrocyanin	Orange	–	568/570	595	625
Allophycocyanin	Blue	–	618/671	–	675
Phycocyanin	Blue	C	620	–	640
		R	555/617	–	636
Phycoerythrin	Pink	R	565/540	498	575
		B	565/545	499	576
		b	545	565	570
		C	560	–	577

to distinguish the phycobiliprotein subclasses: C for Cyanobacteria species, R for Rhodophyta and B for Bangiales (Kornprobst, 2011). A third class has been highlighted: the phycoerythrocyanins, which possess similar spectral properties to PE (Bryant, Glazer, & Eiserling, 1976). Table 11.4 summarizes the main differences between the different phycobiliproteins.

11.3.3.1 Phycoerythrocyanin

Phycoerythrocyanin was first described in 1976 by Bryant et al. This phycobiliprotein is found in a few Cyanobacteria unable to synthesize PE. Thus,

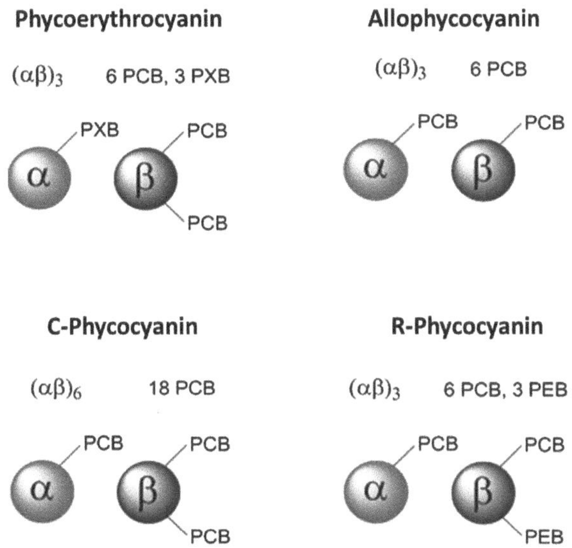

Figure 11.6 Phycobiliprotein structures.

phycoerythrocyanin is an efficient alternative for light absorption at wavelengths close to those absorbed by PE (568 nm instead of 565 nm). It can be found in trimeric $(\alpha\beta)_3$ or hexameric $(\alpha\beta)_6$ form. The PXB chromophore, specific to phycoerythrocyanin, is located on the α subunit while the β subunit possesses two PCB chromophores (Figure 11.6).

11.3.3.2 Allophycocyanin

APC is mostly defined by a 110 kDa molecular weight trimer $(\alpha\beta)_3$. Both α and β subunits possess one PCB chromophore (Figure 11.6). Its maximum absorption wavelength is 650 nm, corresponding approximately to the second maximum of chlorophyll b, leading to its nickname: blue-green and red algae chlorophyll b (Bogorad, 1975). APC is a minor pigment representing only 5% of water-soluble proteins in *Porphyridium cruentum* (Gantt & Lipschultz, 1974). It possesses an intermediary function in the energy transfer mechanism through the chlorophyll a reaction centers. Hence, APC is one of the core constituents near the start of the rod (PEs).

11.3.3.3 Phycocyanin

PC is found in Cyanobacteria, Rhodophyta and Cryptophyta. Hence, according to the classification mentioned above, two types have been described: C-PC and R-PC. C-PC is the most widespread. Like in APC, both α and β subunits possess a PCB chromophore but the β subunits have two chromophores (Figure 11.6). The conformation of APC can change according to pH; it is hexameric $(\alpha\beta)_6$ at pH 5–6 and trimeric $(\alpha\beta)_3$ at pH 7.

11.3.3.4 Phycoerythrin

PE is the main pigment found in Rhodophyta but it can also be recovered from Cyanobacteria (Glazer, 1994). Like other phycobiliproteins, different forms occur according to the algal species: R-PE (for Rhodophyta), B-PE (for Bangiales), b-PE and C-PE (for Cyanobacteria). B-PE and R-PE are the most abundant PEs in red algae. Both are composed of $(\alpha\beta)_6\gamma$ complexes and have a molecular weight between 240 and 260 kDa (Sun et al., 2009) while C-PE and b-PE are (αβ) complexes without any γ subunit (Figure 11.7).

11.3.3.4.1 R-Phycoerythrin

R-PE is the major phycobiliprotein in red algae (Glazer, 1984) and can also be found in Cyanobacteria (Bryant, 1982). It is a hexameric $(\alpha\beta)_6\gamma$ 240 kDa molecular weight protein (Galland-Irmouli et al., 2000; Glazer, 1989). The γ subunit is located in the center of the molecule and links the $(\alpha\beta)_3$ trimers conferring a high stability on the pigment (Wang, Zhou, & Zeng, 1998).

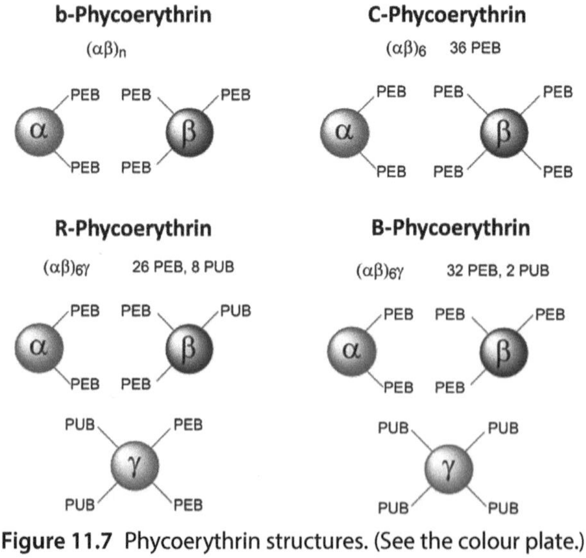

Figure 11.7 Phycoerythrin structures. (See the colour plate.)

11.3.3.4.2 B-Phycoerythrin

B-PE is characteristic of the Bangiales algae (Gantt & Lipschultz, 1974). It is a 245 kDa molecular weight protein and represents 2–6% DW of the algae (Rebolloso Fuentes, Acién Fernández, Sánchez Pérez, & Guil Guerrero, 2000). It also has a hexameric $(\alpha\beta)_6\gamma$ structure and possesses the same chromophores as R-PE but its absorption spectrum is slightly different as depicted in Table 11.4 (Glazer, 1982; Glazer & Hixson, 1977; Tchernov, Minkova, Georgiev, & Houbavenska, 1993).

11.3.3.4.3 b-Phycoerythrin

b-PE is synthesized by the same organisms as B-PE but is composed of an $(\alpha\beta)_n$ complex (Figure 11.7) with n ranging from 1 to 6. It is a smaller molecule than B-PE but shows strong structural similarities. The unique feature of this PE is due to its chromophore composition where only PEB is represented.

11.3.3.4.4 C-Phycoerythrin

C-PE is synthesized by Cyanobacteria and isolated as hexameric or trimeric complexes. Unusually, the β subunit is linked to four chromophores (all PEB) while in the other PEs, the β subunit carries only three chromophores.

11.3.4 Utilization of Phycoerythrins

Phycobiliproteins have a variety of uses and offer interesting industrial perspectives.

11.3.4.1 Colorant

PC from *Arthrospira platensis*, previously called *Spirulina*, is currently commercialized as a blue colorant in the food industry under the name of Lina Blue (Dainippon Ink & Chemicals, Sakura). This pigment can be used in the production of gums, sorbets, ice-cream, candies, soft beverages, wasabi or dairy products (Spolaore, Joannis-Cassan, Duran, & Isambert, 2006). Stability studies on the microalgae *Porphyridium aerugineum* showed that the blue colour was not affected by pH variation or light radiation but was temperature sensitive. However, its stability was significantly increased in dry preparations (Sekar & Chandramohan, 2008).

B-PE and b-PE extracted from *P. cruentum* are used as red colorants in the food industry, especially for jellified desserts and dairy products. R-PE from the seaweed *Porphyra* sp. can also be used as a food colorant (Dufossé et al., 2005). There are many patents regarding the utilization of PEs as food colorants (Sekar & Chandramohan, 2008), mainly intended for Asiatic countries.

11.3.4.2 Fluorescent Probe

PE possesses a high molar extinction coefficient ($\varepsilon = 2 \times 10^6 M^{-1} cm^{-1}$ for 240 kDa) and a high fluorescence yield, at least 10 times more than those usually observed (Kronick, 1986; Yu, Glazer, Spencer, & West, 1981). Moreover, the background noise occurring in orange wavelength emission is lower than for other spectrum areas. These fluorescent properties are widely used in various applications. R-PE and phycobiliproteins can also be exploited as a moiety in fluorescent energy transfer, fluorescent labels, tags, tracers and markers, which are useful in flow cytometry, fluorescent immunoassays, immunophenotyping and other such fluorescent studies (Glazer, 1994; Isailovic, Sultana, Phillips, & Yeung, 2006; Sekar & Chandramohan, 2008).

11.3.4.3 Bioactivity

Lastly, R-PE also possesses some biological activities, such as antitumoral ones in which R-PE subunits have been indicated as an attractive option for improving the selectivity of photodynamic therapy in mouse tumor cells and human liver carcinoma cells (Sekar & Chandramohan, 2008). Other effects reported include antioxidant (Fitzgerald et al., 2011; Pangestuti & Kim, 2011), antidiabetic, antitumoral (Fitzgerald et al., 2011), immunosuppressive and antihypertensive (Cian, Martínez-Augustin, & Drago, 2012). R-PE subunits could constitute a good alternative to increase the selectivity of photochemotherapy in cancer treatments (Bei, Guang-Ce, Chen-Kui, & Zhen-Gang, 2002).

11.4 PHYCOERYTHRIN EXTRACTION AND PURIFICATION PROCEDURES

11.4.1 Extraction Processes

11.4.1.1 Classic Processes

PEs can be extracted by soaking seaweeds in water for one or several days (Niu, Wang, & Tseng, 2006; Siegelman & Kycia, 1978; Wang, 2002). This method extracts the proteins by osmotic shock (Fleurence & Guyader, 1995). However, this type of extraction takes a long time and its main disadvantage is the partial degradation of PEs by proteases. Generally, PEs are extracted in diluted phosphate buffer (5–50 mM, pH 7) from fresh algae (D'Agnolo,

Rizzo, Paoletti, & Murano, 1994; Hilditch et al., 1991; Senthilkumar et al., 2013; Sun et al., 2009), frozen algae (Bermejo et al., 2003; Bermejo, Ruiz, & Acien, 2007; Chuner et al., 2012; Liu, Chen, Zhang, Zhang, & Zhou, 2005) or freeze-dried algae (Galland-Irmouli et al., 2000; Kaixian, Franklin, & Borowitzka, 1993; Munier et al., in press; Rossano, Ungano, D'Ambrosio, Liuzzi, & Riccio, 2003).

Currently, most procedures used are based on cell wall breakage. The macroalgae are ground in liquid nitrogen and the resulting powder is homogenized in sodium phosphate buffer (5–50 mM, pH 7) (D'Agnolo et al., 1994; Galland-Irmouli et al., 2000; Hilditch et al., 1991; Munier, Dumay, Morançais, Jaouen, & Fleurence, 2013). Grinding in liquid nitrogen is used to facilitate the destruction of the cell wall, which is the main obstacle to accessing and extracting algal proteins. However, this approach is not totally efficient for cell wall degradation and is also costly on an industrial scale.

Alternative extraction methods for PEs and seaweed proteins were investigated and enzymatic hydrolysis of the cell wall was suggested as another way of accessing algal proteins (Amano & Noda, 1990; Fleurence, 1999a).

11.4.1.2 Enzymatic Processes

Nowadays, as described earlier, PE is mainly extracted by grinding the algal sample in liquid nitrogen, then homogenizing it in phosphate buffer before centrifugation (Denis, Massé, Fleurence, & Jaouen, 2009; Sun et al., 2009). The presence of various polysaccharides in large quantities in the cell wall of many rhodophyceae reduces the extraction efficiency of classic extraction methods. In fact, seaweed proteins are not as bioavailable as expected due to their polysaccharide content (xylans, agar, carrageenan or alginates), which limits the accessibility of protein fractions (Horie, Sugase, & Horie, 1995), and also due to the strong covalent bonding between mix-linked xylan and glycoprotein complexes, as reported for *P. palmata* (Deniaud, Fleurence, & Lahaye, 2003). However, new kinds of extraction techniques have recently appeared, such as enzymolysis and microwave-assisted extraction (Wijesinghe & Jeon, 2012). Preliminary studies have demonstrated the benefits of using enzymes to recover high added-value seaweed compounds (Wijesinghe & Jeon, 2012), R-PE (Joubert & Fleurence, 2008) and seaweed proteins (Fleurence et al., 2012; Lahaye & Vigouroux, 1992). It is well known that enzymatic tissue disruption may be an efficient alternative for releasing many kinds of compounds, usually confined and inaccessible. Response surface methodology is a useful technique for investigating complex processes such as enzymatic hydrolysis. Some papers have reported the optimization of processing conditions, such as pH, temperature and enzyme concentration, for protein or peptide recovery (Diniz & Martin, 1996; Guerard, Sumaya-Martinez, Laroque, Chabeaud, & Dufossé, 2007). A recent paper describes the potential of using enzymes, especially xylanases, for *P. palmata* cell wall disruption with a focus on the extraction yield of R-PE. In this study, the R-PE extraction yield was 62 times greater than without enzyme treatment and 16 times greater than without optimization. Enzymatic optimization enhanced the Purity Index up to 16 times (Dumay, Clément, Morançais, & Fleurence, 2013). In this way, hydrolysis extracts could also be purified according to the techniques described below.

11.4.2 Purification Processes of Phycoerythrins

The traditional methods used in the purification of PEs involve ammonium sulfate precipitation, ion-exchange chromatography, gel filtration and hydrophobic interaction, hydroxyapatite and electrophoresis.

Ammonium sulfate precipitation is usually employed as a preliminary purification step with different saturation percentages (25–85%) depending on the species (Bermejo, Talavera, & Alvarez-Pez, 2001; Hilditch et al., 1991; Kaixian et al., 1993; Liu et al., 2005; Sun et al., 2009). The solubility of proteins varies according to the ionic strength of the solution, and hence according to the salt concentration. Since proteins differ markedly in their solubility at high ionic strength, salting-out is a very useful procedure to facilitate the purification of a given protein. The precipitated protein is then removed by centrifugation and the ammonium sulfate concentration is increased to a value that will precipitate most of the proteins of interest while leaving the maximum amount of protein contaminants still in solution. The precipitated protein of interest is recovered by centrifugation and dissolved in fresh buffer for the next stage of purification. Fractional precipitation with $(NH_4)_2SO_4$ eliminates the high-molecular-weight fraction. With this technique, the purity index of R-PE has been increased by around 2.7, 3.5 and 4.53 for *Polysiphonia urceolata* (Liu et al., 2005), *Portieria hornemannii* (Senthilkumar et al., 2013) and *Heterosiphonia japonica* (Sun et al., 2009), respectively.

The second step of purification can be performed using chromatographic methods.

The most popular for proteins is ion-exchange chromatography, a process that separates ions and polar molecules based on their net charge. Ion-exchange chromatography retains molecules on the column based on coulombic (ionic) interactions. The elution is carried out either with a pH gradient (Liu et al., 2005) or with a step or linearly increasing gradient of salt concentration (Kaixian et al., 1993; Senthilkumar et al., 2013). Through ammonium sulfate precipitation followed by DEAE column chromatography, Kaixan et al. (1993) and Liu et al. (2005) extracted the R-PE from *Ceramium isogonum* and *P. urceolata* with a purity index of 2.1 and 5.6, respectively. Likewise, with 55% salting-out and Q-Sepharose column chromatography, Senthilkumar et al. (2013) achieved a maximum purity index of 5.2 in *P. hornemannii*.

Protocols with gel filtration chromatography have also been developed to purify proteins. In this case, the separation is based on size. D'Agnolo et al. (1994) purified the R-PE of *Gracilaria longa* using a Superdex 200 column and obtained a purity index of 4.5.

The combination of these two methods has also been carried out with *Corallina officinalis* (Hilditch et al., 1991) and *H. japonica* (Sun et al., 2009).

Affinity chromatography is another powerful and generally applicable means of purifying proteins. This technique takes advantage of the high affinity of many proteins for specific chemical groups.

Hydrophobic interaction chromatography (HIC) is a separation technique that uses the properties of hydrophobicity to separate proteins from one another. Separation is based on the reversible interaction between a protein and the hydrophobic ligand bound to the chromatography matrix.

Hydroxyapatite chromatography is considered to be a 'pseudo-affinity' chromatography or 'mixed-mode' ion exchange. It has proved to be an effective purification mechanism in a variety of processes, providing biomolecule selectivity complementary to more traditional ion-exchange or hydrophobic interaction techniques.

HIC has been carried out in combination either with ion-exchange chromatography for the purification of the B-PE of *P. cruentum* (Bermejo et al., 2003; Bermejo et al., 2007) and the R-PE of *P. palmata* (Wang, 2002) and *P. urceolata* (Niu et al., 2006) or with hydroxyapatite chromatography for the purification of the R-PE of *P. urceolata* (Niu et al., 2006).

Hydroxyapatite chromatography has been used for the purification of R-PE and R-PC of *P. yezoensis* (Chuner et al., 2012) and in association with gel filtration chromatography for *Corallina elongata* (Rossano et al., 2003).

Preparative electrophoresis may also be used to purify proteins. The separation is based on ionic charge. This method has been described to purify the R-PE from *P. palmata* (Galland-Irmouli et al., 2000).

All these procedures are resumed in Table 11.5.

Table 11.5 Procedures Used for Phycoerythrin Purification

Protocol	Species	Extraction Yield (mg/g DW)	PI	References
$(NH_4)_2SO_4$ + GF + IE + UF + GF	*Corallina officinalis* (R-PE)	0.117	4.7	Hilditch et al. (1991)
$(NH_4)_2SO_4$ + IE + UF	*Ceramium isogonum* (R-PE)	0.038	2.1	Kaixan et al. (1993)
$(NH_4)_2SO_4$ + GF	*Gracilaria longa* (R-PE)		4.5	D'Agnolo et al. (1994)
Preparative electrophoresis	*Palmaria palmata* (R-PE)	0.562	3.2	Galland-Irmouli et al. (2000)
HIC + IE	*Palmaria palmata* (R-PE)	0.152	3.6	Wang et al. (2002)
Hydroxyapatite + GF	*Corallina elongata* (R-PE)	0.600	6.67	Rossano et al. (2003)
HIC + IE	*Porphyridium cruentum* (B-PE)	0.825	4.6	Bermejo et al. (2003), Bermejo, Ruiz, and Acien, (2007)
$(NH_4)_2SO_4$ + IE	*Polysiphonia urceolata* (R-PE)	0.841	5.59	Liu et al. (2005)
HIC + hydroxyapatite	*Polysiphonia urceolata* (R-PE)	0.425	3.90	Niu et al. (2006)
HIC + IE	*Polysiphonia urceolata* (R-PE)	0.500	3.26	Niu et al. (2006)
2GF + IE	*Heterosiphonia japonica* (R-PE)	0.940	4.89	Sun et al. (2009)
$(NH_4)_2SO_4$ + hydroxyapatite	*Porphyra yezoensis* (R-PE et R-PC)	0.540	3.2	Chuner et al. (2012)
$(NH_4)_2SO_4$ + IE	*Portiera hornemannii* R-PE	0.810	5.2	Senthilkumar et al. (2013)

HIC, hydrophobic interaction chromatography; IE, ion exchange; GF, gel filtration.

11.5 CONCLUDING REMARKS AND FUTURE PROSPECTS

In conclusion, red seaweeds are a promising source of proteins for human and animal nutrition. Their protein digestibility can be improved by the application of food processes such as mild fermentations but this is not necessary for all species; it depends on the presence or absence of large quantities of soluble fibers.

The use of red seaweed proteins, instead of soybean proteins, appears to be especially relevant for the nutrition of fish and mollusks (e.g. abalone) (Fleurence et al., 2012).

PEs have been identified as abundant pigments in marine seaweeds and Cyanobacteria and could provide an efficient target for the valorization of these resources.

Currently, the extraction and purification of PEs are being widely investigated but the methodologies employed still remain expensive and time-consuming, as they require multiple steps. Further studies could be conducted to improve the efficiency of these methods and to investigate less-studied red seaweed species.

Finally, to facilitate the introduction of new applications of PE, such as novel food, additive, colorant or cosmetic applications, new legislation must be established, especially in European countries, where only 21 red seaweed species are considered edible.

REFERENCES

Amano, H., & Noda, H. (1990). Proteins of protoplasts from red alga *Porphyra yezoensis*. *Nippon Suisan Gakkaishi*, *56*(11), 1859–1864.

Anonymous. (1969). Annual report of Fishery Institute of Iwate prefecture.

Apt, K. E., Collier, J. L., & Grossman, A. R. (1995). Evolution of the Phycobiliproteins. *Journal of Molecular Biology*, *248*(1), 79–96.

Bald, D., Kruip, J., & Rögner, M. (1996). Supramolecular architecture of cyanobacterial thylakoid membranes: how is the phycobilisome connected with the photosystems? *Photosynthesis Research*, *49*(2), 103–118.

Bei, H., Guang-Ce, W., Chen-Kui, Z., & Zhen-Gang, L. (2002). The experimental research of R-phycoerythrin subunits on cancer treatment: a new photosensitizer in PDT. *Cancer Biotherapy & Radiopharmaceuticals*, *17*, 35–42.

Bermejo, R., Acien, F. G., Ibanez, M. J., Fernandez, J. M., Molina, E., & Alvarez-Pez, J. M. (2003). Preparative purification of B-phycoerythrin from the microalga *Porphyridium cruentum* by expanded-bed adsorption chromatography. *Journal of Chromatography B, Analytical Technologies in the Biomedical and Life Sciences*, *790*, 317–325.

Bermejo, R., Ruiz, E., & Acien, F. G. (2007). Recovery of B-phycoerythrin using expanded bed adsorption chromatography: scale-up of the process. *Enzyme and Microbial Technology*, *40*, 927–933.

Bermejo, R., Talavera, E. M., & Alvarez-Pez, J. M. (2001). Chromatographic purification and characterization of B-phycoerythrin from *Porphyridium cruentum* semipreparative high-performance liquid chromatographic separation and characterisation of its subunits. *Journal of Chromatography A, 917*, 135–145.

Bird, K.T. (1984). Seasonal variation in protein: carbohydrate ratios in a subtropical estuarine alga, *Gracilaria verrucosa*, and the determination of Nitrogen limitation status using these ratios. *Botanica Marina, 27*, 111–115.

Bogorad, L. (1975). Phycobiliproteins and complementary chromatic adaptation. *Annual Review of Plant Physiology, 26*(1), 369–401.

Brown, S. B., Holroyd, A., & Vernon, D. I. (1984). Biosynthesis of phycobiliproteins. Incorporation of biliverdin into phycocyanin of the red alga *Cyanidium caldarium*. *Journal of Biochemistry, 219*, 905–909.

Bryant, D., Glazer, A., & Eiserling, F. (1976). Characterization and structural properties of the major biliproteins of *Anabaena* sp. *Archives of Microbiology, 110*(1), 61–75.

Bryant, D. A. (1982). Phycoerythrocyanin and phycoerythrin: properties and occurrence in cyanobacteria. *Journal of General Microbiology, 128*(4), 835–844.

Chuner, C., Chunxia, L., Shuxian, W., Qing, W., Ziye, G., & Peimin, H. (2012). Large scale preparation of phycobiliproteins from *Porphyra yezoensis* using co-precipitation with ammonium sulfate. *Natural Science, 4*(8), 536–543.

Cian, R. E., Martínez-Augustin, O., & Drago, S. R. (2012). Bioactive properties of peptides obtained by enzymatic hydrolysis from protein byproducts of *Porphyra columbina*. *Food Research International, 49*(1), 364–372.

D'Agnolo, E., Rizzo, R., Paoletti, S., & Murano, E. (1994). R-phycoerythrin from the red alga *Gracilaria longa*. *Phytochemistry, 35*(3), 693–696.

Dawczynski, C., Schubert, R., & Jahreis, G. (2007). Amino acids, fatty acids, and dietary fibre in edible seaweed products. *Food Chemistry, 103*, 891–899.

Deniaud, E., Fleurence, J., & Lahaye, M. (2003). Interaction of the max-linked β-(1,3)/β-(1,4)-D-xylans in the cell walls of *Palmaria plamata* (Rhodophyta). *Journal of Phycology, 39*, 74–82.

Deniaud, E., Quemener, B., Fleurence, J., & Lahaye, M. (2003). Structural studies of the mix-linked β-(1→3)/β-(1→4)-D-xylans from the cell wall of *Palmaria palmata* (Rhodophyta). *International Journal of Biological Macromolecules, 33*(1–3), 9–18.

Denis, C., Massé, A., Fleurence, J., & Jaouen, P. (2009). Concentration and pre-purification with ultrafiltration of a R-phycoerythrin solution extracted from macro-algae *Grateloupia turuturu*: process definition and up-scaling. *Separation and Purification Technology, 69*(1), 37–42.

Denis, C, Morançais, M., Li, M., Deniaud, E., Gaudin, P., Wielgosz-Collin, G., et al. (2010). Study of the chemical composition of edible red macroalgae *Grateloupia turuturu* from Brittany (France). *Food Chemistry, 119*(3), 913–917.

Diniz, F. M., & Martin, A. M. (1996). Use of response surface methodology to describe the combined effects of pH, temperature and E/S ratio on the hydrolysis of dogfish (*Squalus acanthias*) muscle. *International Journal of Food Science & Technology, 31*(5), 419–426.

Ducret, A., Müller, S. A., Goldie, K. N., Hefti, A., Sidler, W. A., Zuber, H., et al. (1998). Reconstitution, characterisation and mass analysis of the pentacylindrical allophycocyanin core complex from the cyanobacterium *Anabaena* sp. PCC 7120. *Journal of Molecular Biology, 278*(2), 369–388.

Dufossé, L., Galaup, P., Yaron, A., Malis Arad, S., Blanc, P., Chidambara Murthy, K. N., et al. (2005). Microorganisms and microalgae as sources of pigments for food use: a scientific oddity of industrial reality? *Trends in Food Science & Technology, 16*, 389–406.

Dumay, J., Clément, N., Morançais, M., & Fleurence, J. (2013). Optimization of hydrolysis conditions of *Palmaria palmata* to enhance R-phycoerythrin extraction. *Bioresource Technology, 131*(0), 21–27.

FAO. (2011). *Fisheries and aquaculture*. Department. Fisheries Statistical Collection.
Fitzgerald, C., Gallagher, E., Tasdemir, D., & Hayes, M. (2011). Heart health peptides from macroalgae and their potential use in functional foods. *Journal of Agricultural and Food Chemistry*, *59*(13), 6829–6836.
Fleurence, J. (1999a). The enzymatic degradation of algal cell walls: a useful approach for improving protein accessibility? *Journal of Applied Phycology*, *11*(3), 313–314.
Fleurence, J. (1999b). Seaweed proteins: biochemical, nutritional aspects and potential uses. *Trends in Food Science & Technology*, *10*, 25–28.
Fleurence, J. (2004). Seaweed proteins. In R. Y. Yada (Ed.), *Proteins in food processing* (pp. 197–211). CRC press.
Fleurence, J., Gutbier, G., Mabeau, S., & Leray, C. (1994). Fatty acids from 11 marine macroalgae of the French Brittany coast. *Journal of Applied Phycology*, *6*, 527–532.
Fleurence, J., & Guyader, O. (1995). Contribution of electrophoresis of red seaweeds (*Gracilaria* sp.) used as food ingredients. *Sciences des aliments*, *15*, 43–48.
Fleurence, J., Morançais, M., Dumay, J., Decottignies, P., Turpin, V., Munier, M., et al. (2012). What are the prospects for using seaweed in human nutrition and for marine animals raised through aquaculture? *Trends in Food Science & Technology*, *27*(1), 57–61.
Fujiwara-Arasaki, T., Mino, N., & Kuroda, M. (1984). The protein value in human nutrition of edible marine algae in Japan. *Hydrobiologia*, *116–117*(1), 513–516.
Galland-Irmouli, A. V., Fleurence, J., Lamghari, R., Luçon, M., Rouxel, C., Barbaroux, O., et al. (1999). Nutritional value of proteins from edible seaweed *Palmaria palmata* (Dulse). *The Journal of Nutritional Biochemistry*, *10*(6), 353–359.
Galland-Irmouli, A. V., Pons, L., Luçon, M., Villaume, C., Mrabet, N. T., Guéant, J.-L., et al. (2000). One-step purification of R-phycoerythrin from the red macroalga *Palmaria palmata* using preparative polyacylamide gel electrophoresis. *Journal of Chromatography B*, *739*, 117–123.
Gantt, E. (1980). Structure and function of phycobilisomes: light harvesting pigment complexes in red and Blue-Green algae. In J. F. D.G.H. Bourne, & K. W. Jeon (Eds.), *International Review of Cytology* (Vol. 66(pp. 45–80). Academic Press.
Gantt, E., & Lipschultz, C. A. (1974). Phycobilisomes of *Porphyridium cruentum*. Pigment analysis. *Biochemistry*, *13*(14), 2960–2966.
Glazer, A. (1994). Phycobiliproteins — a family of valuable, widely used fluorophores. *Journal of Applied Phycology*, *6*(2), 105–112.
Glazer, A. N. (1982). Phycobilisomes: structure and dynamics. *Annual Review of Microbiology*, *36*(1), 173–198.
Glazer, A. N. (1984). Phycobilisome a macromolecular complex optimized for light energy transfer. *Biochimica et Biophysica Acta (BBA) - Reviews on Bioenergetics*, *768*(1), 29–51.
Glazer, A. N. (1989). Light guides. Directional energy transfer in a photosynthetic antenna. *Journal of Biological Chemistry*, *264*(1), 1–4.
Glazer, A. N., Apell, G. S., Hixson, C. S., Bryant, D. A., Rimon, S., & Brown, D. M. (1976). Biliproteins of cyanobacteria and Rhodophyta: homologous family of photosynthetic accessory pigments. *Proceedings of the National Academy of Sciences*, *73*(2), 428–431.
Glazer, A. N., & Hixson, C. S. (1975). Characterization of R-phycocyanin. Chromophore content of R-phycocyanin and C-phycoerythrin. *Journal of Biological Chemistry*, *250*(14), 5487–5495.
Glazer, A. N., & Hixson, C. S. (1977). Subunit structure and chromophore composition of rhodophytan phycoerythrins. *Porphyridium cruentum* B-phycoerythrin and b-phycoerythrin. *Journal of Biological Chemistry*, *252*(1), 32–42.
Guerard, F., Sumaya-Martinez, M. T., Laroque, D., Chabeaud, A., & Dufossé, L. (2007). Optimization of free radical scavenging activity by response surface methodology in the hydrolysis of shrimp processing discards. *Process Biochemistry*, *42*(11), 1486–1491.

Hilditch, C. M., Balding, P., Jenkins, R., Smith, A. J., & Rogers, L. J. (1991). R-phycoerythrin from the macroalga *Corallina officinalis* (Rhodophyceae) and application of a derived phycofluor probe for detecting sugar-binding sites on cell membranes. *Journal of Applied Phycology, 3*(4), 345–354.

Horie, Y., Sugase, K., & Horie, K. (1995). Physiological differences of soluble and insoluble dietary fibre fractions of brown algae and mushrooms in pepsin activity in vitro and protein digestibility. *Asian Pacific Journal of Clinical Nutrition, 4*, 251–255.

Isailovic, D., Sultana, I., Phillips, G. J., & Yeung, E. S. (2006). Formation of fluorescent proteins by the attachment of phycoerythrobilin to R-phycoerythrin alpha and beta apo-subunits. *Analytical Biochemistry, 358*(1), 38–50.

Jensen, A. (1993). Present and future needs for algae and algal products. *Hydrobiologia, 260–261*(1), 15–23.

Joubert, Y., & Fleurence, J. (2008). Simultaneous extraction of proteins and DNA by an enzymatic treatment of the cell wall of *Palmaria palmata* (Rhodophyta). *Journal of Applied Phycology, 20*, 55–61.

Kaixian, Q., Franklin, M., & Borowitzka, M. A. (1993). The study for isolation and purification of R-phycoerythrin from a red alga. *Applied Biochemistry and Biotechnology, 43*, 133–139.

Ke, B. (2001). *Photosynthesis photobiochemistry and photobiophysics* (Vol. 10). Kluwer Academic Publishers.

Kendel, M., Couzinet-Mossion, A., Viau, M., Fleurence, J., Barnathan, G., & Wielgosz-Collin, G. (2013). Seasonal composition of lipids, fatty acids, and sterols in the edible red alga *Grateloupia turuturu*. *Journal of Applied Phycology, 25*(2), 425–432.

Kornprobst, J. (2011). *Encyclopedia of marine natural products*. Wiley.

Kronick, M. N. (1986). The use of phycobiliproteins as fluorescent labels in immunoassay. *Journal of Immunological Methods, 92*(1), 1–13.

Lahaye, M., & Vigouroux, J. (1992). Liquefaction of dulse (*Palmaria palmata* (L.) Kuntze) by a commercial enzyme preparation and a purified endo-β-1,4-D-xylanase. *Journal of Applied Phycology, 4*, 329–337.

Liu, L.-N., Chen, X.-L., Zhang, X.-Y., Zhang, Y.-Z., & Zhou, B.-C. (2005). One-step chromatography method for efficient separation and purification of R-phycoerythrin from *Polysiphonia urceolata*. *Journal of Biotechnology, 116*, 91–100.

Mabeau, S., & Fleurence, J. (1993). Seaweed in food products: biochemical and nutritional aspects. *Trends in Food Science & Technology, 4*(4), 103–107.

Marrion, O., Fleurence, J., Schwertz, A., Guéant, J.-L., Mamelouk, L., Ksouri, J., et al. (2005). Evaluation of protein digestibility of *Palmaria palmata* and *Gracilaria verrucosa*. *Journal of Applied Phycology, 17*(2), 99–102.

Marrion, O., Schwertz, A., Fleurence, J., Guéant, J. L., & Villaume, C. (2003). Improvement of the digestibility of the proteins of the red alga *Palmaria palmata* by physical processes and fermentation. *Food/Nahrung, 47*(5), 339–344.

Morgan, K. C., Wright, J. L. C., & Simpson, F. J. (1980). Review of chemical constituents of the red alga *Palmaria palmata* (Dulse). *Economic Botany, 34*(1), 27–50.

Munier, M., Dumay, J., Morançais, M., Jaouen, P., & Fleurence, J. (2013). Variation in the biochemical composition of the edible seaweed *Grateloupia turuturu* Yamada harvested from two sampling sites on the Brittany Coast (France): the influence of storage method on the extraction of the seaweed pigment R-phycoerythrin. *Journal of Chemistry*.

Munier, M., Jubeau, S., Wijaya, A., Morançais, M., Dumay, J., Marchal, L. et al. (2014). Physicochemical factors affecting the stability of two pigments: R-Phycoerythrin of Grateloupia turuturu and B-Phycoerythrin of Porphyridium cruentum. Food Chemistry, *150*, 400–407.

Nisizawa, K., Noda, H., Kikuchi, R., & Watanabe, T. (1987). The main seaweed foods in Japan. *Hydrobiologia, 151–152*(1), 5–29.

Niu, J.-F., Wang, G.-C., & Tseng, C.-K. (2006). Method for large-scale isolation and purification of R-phycoerythrin from red alga *Polysiphonia urceolata* Grev. *Protein Expression and Purification*, *49*, 23–31.

Pangestuti, R., & Kim, S.-K. (2011). Biological activities and health benefit effects of natural pigments derived from marine algae. *Journal of Functional Foods*, *3*(4), 255–266.

Rebolloso Fuentes, M. M., Acién Fernández, G. G., Sánchez Pérez, J. A., & Guil Guerrero, J. L. (2000). Biomass nutrient profiles of the microalga *Porphyridium cruentum*. *Food Chemistry*, *70*(3), 345–353.

Rossano, R., Ungano, N., D'Ambrosio, A., Liuzzi, G. M., & Riccio, P. (2003). Extracting and purifying R-phycoreythrin from Mediterranean red algae *Corallina elongata* Ellis & Solander. *Journal of Biotechnology*, *101*, 289–293.

Sekar, S., & Chandramohan, M. (2008). Phycobiliproteins as a commodity: trends in applied research, patents and commercialization. *Journal of Applied Phycology*, *20*(2), 113–136.

Senthilkumar, N., Suresh, V., Thangam, R., Kurinjimalar, C., Kavitha, G., Murugan, P., et al. (2013). Isolation and characterization of macromolecular protein R-Phycoerythrin from *Portieria hornemannii*. *International Journal of Biological Macromolecules*, *55*(0), 150–160.

Siegelman, H. W., & Kycia, J. H. (1978). In J. Hellebust, & J. S. Craigie (Eds.), *Algal biliproteins* (pp. 71–79). Cambridge: Cambridge University Press.

Spolaore, P., Joannis-Cassan, C., Duran, E., & Isambert, A. (2006). Commercial applications of microalgae. *Journal of Bioscience and Bioengineering*, *101*(2), 87–96.

Sun, L., Wang, S., Gong, X., & Chen, L. (2004). A rod-linker-contained R-phycoerythrin complex from the intact phycobilisome of the marine red alga *Polysiphonia urceolata*. *Journal of Photochemistry and Photobiology B: Biology*, *76*(1–3), 1–11.

Sun, L., Wang, S., Gong, X., Zhao, M., Fu, X., & Wang, L. (2009). Isolation, purification and characteristics of R-phycoerythrin from a marine macroalga *Heterosiphonia japonica*. *Protein Expression and Purification*, *64*(2), 146–154.

Tandeau de Marsac, N., Castets, A. M., & Cohen-Bazire, G. (1980). Wavelength modulation of phycoerythrin synthesis in *Synechocystis* sp. 6701. *Journal of Bacteriology*, *142*(1), 310–314.

Tandeau de Marsac, N., & Cohen-bazire, G. (1977). Molecular composition of cyanobacterial phycobilisomes. *Proceedings of the National Academy of Sciences*, *74*(4), 1635–1639.

Tchernov, A. A., Minkova, K. M., Georgiev, D. I., & Houbavenska, N. B. (1993). Method for B-phycoerythrin purification from *Porphyridium cruentum*. *Biotechnology Techniques*, 7(12), 853–858.

Wang, G. (2002). Isolation and purification of phycoerythrin from red alga *Gracilaria verrucosa* by expanded-bed-adsorption and ion-exchange chromatography. *Chromatographia*, *56*, 509–513.

Wang, G., Zhou, B., & Zeng, C. (1998). Isolation, properties and spatial site analysis of γ subunits of B-phycoerythrin and R-phycoerythrin. *Science in China Series C: Life Sciences*, *41*(1), 9–17.

Wijesinghe, W. A., & Jeon, Y.-J. (2012). Enzyme-assistant extraction (EAE) of bioactive components: a useful approach for recovery of industrially important metabolites from seaweeds: a review. *Fitoterapia*, *83*(1), 6–12.

Yu, M. H., Glazer, A. N., Spencer, K. G., & West, J. A. (1981). Phycoerythrins of the red alga *Callithamnion*: Variation in phycoerythrobilin and phycoroubilin content. *Plant Physiology*, *68*(2), 482–488.

CHAPTER TWELVE

Bioactive Components from Seaweeds: Cosmetic Applications and Future Development

Gilles Bedoux[*,1], **Kevin Hardouin**[*,†], **Anne Sophie Burlot**[*] and **Nathalie Bourgougnon**[*]
[*]Université de Bretagne-Sud, Laboratoire de Biotechnologie et Chimie Marine, EA 3884, Institut Universitaire Européen de la Mer, Vannes, France
[†]OLMIX SA, Bréhan, France
[1]Corresponding author: e-mail address: gilles.bedoux@univ-ubs.fr

Contents

Advances in Botanical Research, Volume 71
ISSN 0065-2296
http://dx.doi.org/10.1016/B978-0-12-408062-1.00012-3

Abstract

Marine macroalgae biomass is often used for the production of ingredients in cosmetics. These ingredients can have one of the three main functions: (1) they are considered as additives which contribute to the organoleptic properties; (2) they are used for stabilisation and preservation of the product; (3) or finally, they are bioactive compounds which fulfil a real cosmetic function and activity. This chapter presents the bioactivities of molecules or extracts used for cosmetic applications, and discusses some perspectives for the development of new compounds and specific extraction methods. Metabolites derived from seaweeds have been shown to be active in antiaging skin care, anticellulite treatment and slimming, as well as having antioxidant, photoprotective, moisturising, and whitening properties. Among the various classes of seaweed components, sulphated polysaccharides, peptides, carotenoids, fatty acids, and phytohormones exhibit antiaging and antioxidant properties, while mycosporine-like amino acids, flavonoids have an antiphotoaging activity. Flavonoids (i.e. phlorotannins) are lipolytic agents which are isolated from macroalgae and which also inhibit melanogenesis. A better knowledge of algae genetics and the improvement of algae cultivation or co-cultivation will provide new opportunities for the development of bioactive compounds.

12.1 INTRODUCTION

The diversity of marine macroalgae (seaweeds) species and their widely ranging biochemical composition means that they represent a source of potential bioactive compounds for applications in the agri-food industry, cosmetics, pharmacology and, more recently, in the field of functional food and chemistry (Holdt & Kraan, 2011; Ioannou & Roussis, 2009, pp. 51; Mayer, Rodríguez, Berlinck, & Hamann, 2007). Macroalgae can be classified into three groups according to their pigmentation: Phaeophyceae (brown), Rhodophyceae (red) and Chlorophyceae (green). Seaweeds are the sources of mineral matter (11–55% dry weight, in the form of ash), polysaccharides (15–76% dry weight), proteins (1–50% dry weight), lipids (0.3–5% dry weight), phytohormones and pigments, as well as a wide variety of secondary metabolites (phenolic compounds, terpenoids and halogenated compounds, sulphur derivatives, and nitrogen derivatives). The biochemical composition varies greatly depending on the sampling site, the season of harvest and the environment (Cardozo, Guaratini, & Barros, 2007; Bourgougnon & Stiger-Pouvreau, 2012; Holdt & Kraan, 2011). Algae have been mainly exploited as a source of food in Asia where seaweed cultivation has become a major coastal economy, and as a source of technofunctional polysaccharides (agars, carrageenans and alginates).

Nowadays, there is a market trend in the cosmetic industry towards the development and manufacture of cosmetics with seaweed extracts. Seaweeds are rich in vitamins (A, B1, B2, B3 or PP, B5 or pantothenic acid, B6, B9 or folic acid, B12, C, D, E and K), essential amino acids, mineral macroelements and trace elements such as iodine (Ioannou & Roussis, 2009, pp. 51). Some examples of trade extracts are presented in Table 12.1. The main classes of actives in cosmetic care products are used for antiaging care, antiphotoaging, slimming care, photoprotection, moisturising and skin whitening. The presence of a particular compound not only means that these plants are of interest but also opens up the possibility of harvesting as well as cultivation and transformation under different conditions, leading to an enrichment of specific bioactive compounds. Cosmetic products not only have to comply with strict rules concerning the use of chemical substances (EU regulation No. 1223/2009 on cosmetic products) but must also satisfy consumer demand for products containing effective natural and nontoxic ingredients. Seaweed is used as a plant extract and there is no restriction for cosmetic use. In the first part of this article, we present an overview of macroalgae bioactive components or extracts derived from seaweeds for use in further cosmetic applications. We also discuss perspectives in marine biotechnology for the development of bioactive compounds and the use of some specific extraction methods is described. An improved knowledge of the genetics of algae and the cultivation of macroalgae will open up new strategies for the production of compounds of interest.

Table 12.1 Examples of Nomenclature INCI of Some Trade Seaweed Extracts (CosIng Database, 2014)

INCI Name	CAS No.	Description
Ecklonia laminaria extract	–	*Ecklonia laminaria* extract is the extract of the alga, *Ecklonia laminaria*. Skin conditioning.
Laminaria angustata extract	92128-82-0	*Laminaria angustata* extract is an extract of the alga, *Laminaria angustata*, Laminariaceae. Skin conditioning.
Laminaria digitata extract	90046-12-1	*Laminaria digitata* extract is an extract of the alga, *Laminaria digitata*, Laminariaceae. Skin protecting.
Algae extract	92128-82-0/ 68917-51-1	Algae extract is an extract of various species of algae; extract of the seaweed, *Fucus vesiculosus*, Furaceae. Emollient, humectant, masking, oral car, skin conditioning.

Continued

Table 12.1 Examples of Nomenclature INCI of Some Trade Seaweed Extracts (CosIng Database, 2014)—cont'd

INCI Name	CAS No.	Description
Fucus serratus extract	94167-02-9	*Fucus serratus* extract is an extract of the alga, *Fucus serratus* L., Fucaceae. Skin protecting.
Fucus vesiculosus extract	84696-13-9	*Fucus vesiculosus* extract is an extract of the dried thallus of the Bladderwrack, *Fucus vesiculosus* L., Fucaceae. Emollient, skin conditioning, smoothing, soothing.
Hydrolysed *Fucus vesiculosus* extract	84696-13-9/ 100085-36-7	Hydrolysed *Fucus vesiculosus* extract is a hydrolysed iodinated extract from the Bladderwrack, *Fucus vesiculosus* L., Furaceae. Skin conditioning, smoothing, soothing.
Hydrolysed *Fucus vesiculosus* protein	84696-13-9	Hydrolysed *Fucus vesiculosus* protein is the hydrolysate of the protein obtained from the Bladderwrack, *Fucus vesiculosus* L., Furaceae, derived by acid, enzyme or other method of hydrolysis. Skin conditioning.
Algae oligosaccharides	–	Algae oligosaccharides are oligosaccharides produced by the enzymatic degradation of agar that is obtained from *Gelidium* sp. and other algae. Skin conditioning.
Gelidium cartilagineum extract	94945-01-4	*Gelidium cartilagineum* extract is an extract of the alga, *Gelidium cartilagineum*, Gelidiaceae. Skin protecting.
Hydrolysed *Asparagopsis armata* extract	–	Hydrolysed *Asparagopsis armata* extract is the hydrolysate of *Asparagopsis armata* extract derived by acid, enzyme or other method of hydrolysis. Skin protecting.
Chondrus crispus extract	244023-79-8	*Chondrus crispus* extract is an extract of the carrageenan, *Chondrus crispus*, Gigartinaceae. Skin conditioning, viscosity controlling.
Chondrus crispus	–	*Chondrus crispus* is the material obtained from the whole alga, *Chondrus crispus*, Gigartinacea. Masking
Pelvetia canaliculata extract	223751-75-5	*Pelvetia canaliculata* extract is an extract of the alga, *Pelvetia canaliculata*, Fucaceae. Skin protecting.

Table 12.1 Examples of Nomenclature INCI of Some Trade Seaweed Extracts (CosIng Database, 2014)—cont'd

INCI Name	CAS No.	Description
Palmaria palmata extract	223751-74-4	*Palmaria palmata* extract is an extract of the alga, *Palmaria palmata*, Rhodymeniaceae. Skin protecting.
Sargassum muticum extract	–	*Sargassum muticum* extract is an extract of the alga, *Sargassum muticum*, Sargassaceae. Skin protecting.
Fucoxanthin	3351-86-8	Fucoxanthin is a carotenoid isolated from the alga, *Sargassum siliquastrum*. Skin conditioning.
Undaria pinnatifida extract	–	*Undaria pinnatifida* extract is an extract of the alga, *Undaria pinnatifida*, Alariaceae. Skin conditioning, skin protecting.
Codium tomentosum extract	223749-84-6	*Codium tomentosum* extract is an extract of the alga, *Codium tomentosum*, Codiaceae. Skin protecting.
Enteromorpha compressa extract	-	*Enteromorpha compressa* extract is an extract of the thallus of the alga, *Enteromorpha compressa*, Ulvaceae. Skin protecting.
Hydrolysed *Ulva lactuca* extract	97,281-59-9	Hydrolysed *Ulva lactuca* extract is the hydrolysate of *Ulva lactuca* extract derived by acid, enzyme or other method of hydrolysis. Skin conditioning.
Ulva ramulosa Callus culture extract	–	*Ulva ramulosa* Callus culture extract is the extract of a culture of the callus of *Ulva ramulosa*, Ulvaceae. Skin conditioning.
Kappaphycus alvarezii extract	–	Kappaphycus alvarezii extract is an extract of the alga, *Kappaphycus alvarezii*, Soleriaceae. Skin conditioning.
Jania rubens extract	–	Jania rubens extract is an extract of the seaweed, *Jania rubens*. Skin conditioning.
Halopteris scoparia extract	–	*Halopteris scoparia* extract is an extract of the alga, *Halopteris scoparia*, Stypocaulaceae. Skin conditioning.

INCI: international nomenclature of cosmetic ingredient.

12.2 METABOLITES OF INTEREST FOR COSMETICS

Several reviews have presented the detailed biochemical composition of marine macroalgae (Bourgougnon & Stiger-Pouvreau, 2012; Cardozo et al., 2007; Holdt & Kraan, 2011; Ioannou & Roussis, 2009, pp. 51; Mayer et al., 2007). Bioactive compounds are isolated by various methods depending on different factors: physicochemical properties, molecular size, and solubility. Phycocolloid applications have been described in previous reviews, as well as the benefits of different minerals and vitamins, so these aspects are not discussed further here.

12.2.1 Polysaccharides and Sulphated Polysaccharides

The different chemical structures of cell wall and storage polysaccharides have been previously described in a number of reviews (Bourgougnon & Stiger-Pouvreau, 2012; Holdt & Kraan, 2011; Kraan, 2012). Alginates, carrageenans and agars are used as stabilisers, thickeners and emulsifiers in various sticks, creams, lotions, soaps, shampoos, toothpastes, foams, and gels. Phaeophyceae contain alginates, laminarin, sargassan and fucans, the latter being classified into three groups: fucoidans, sulphated galactofucans and ascophyllans (Hemmingson, Falshaw, Furneaux, & Thompson, 2006; Logeart et al., 1997; Logeart, Prigent-Richard, Jozefonvicz, & Letourneur, 1997; Rocha et al., 2005). Fucoidan is a sulphated polysaccharide isolated from brown seaweed, which is characterised by a sulphated fucose backbone. Polysaccharides derived from Rhodophyceae are extremely variable, consisting of floridean starch (a glucan similar to amylopeptin), agars, carrageenans, xylans, galactans, sulphated galactans and sulphated rhamnans. Chlorophyceae contain sulphated galactans, xylans and ulvans. The chemical structures of cell wall and storage polysaccharides are still under investigation and these compounds make up a wide range of species-specific components with a high degree of seasonal and environmental variability.

12.2.2 Proteins and Amino Acids

One interesting feature of seaweeds is their richness in proteins and bioactive peptides. The composition and concentration of proteins varies with species, site and season of harvesting (Fitzgerald, Gallagher, Tasdemir, & Hayes, 2011). The protein fraction of brown algae is relatively low (1–24%

DW) compared with green (4–44% DW) and red macroalgae (5–50% DW) (Fitzgerald et al., 2011; Holdt & Kraan, 2011). These proteins include lectins and phycobiliproteins (Aneiros & Garateix, 2004). Most seaweeds contain all the essential amino acids and acidic amino acids, e.g. glutamic and aspartic acids (Fleurence, 2004). *Palmaria* and *Porphyra* species have been reported to enclose high amounts of glycine and arginine, the latter being a precursor of urea which is a component of the natural moisturising factor (NMF). Fleurence (2004) reports low levels of threonine, lysine, tryptophane, histidine and the sulphur-containing amino acids cysteine and methionine. However, threonine, valine, leucine, lysine, glycine and alanine contents are higher in brown seaweed. It is noteworthy that certain specific amino acids have much higher contents in some macroalgae: e.g. proline in *Ulva armoricana*, accounting for 5–10.5% of total amino acids, glycine in *Palmaria palmata*, arginine in *Chondrus crispus* and *Porphyra* sp., as well as alanine, glycine, arginine, leucine, valine, lysine and methionine in *Undaria pinnatifida* (Fleurence, 1999; Harnedy & FitzGerald, 2013; Holdt & Kraan, 2011). The carbohydrate-binding bioactive lectins found in *Ulva* sp., *Euchema* sp., and *Gracilaria* sp. give rise to biological activity by binding to specific cell glycoreceptors (Holdt & Kraan, 2011).

12.2.3 Phycobiliproteins

Phycobiliproteins are water-soluble and coloured components of the photosynthetic system in red macroalgae (Aneiros & Garateix, 2004; Chronakis, Galatanu, Nylander, & Lindman, 2000). Among this class of components, phycoerythrin makes up a large proportion of the red algal cell proteins (Chronakis et al., 2000; Wang, Sun, Fan, & Tseng, 2002). R-phycoerythrin, phycocyanin, and allophycocyanin are present in *Gracilaria gracilis* (Francavilla, Franchi, Monteleone, & Caroppo, 2013). R-phycoerythrin obtained through biotechnological processes is used as a natural dye, and represents an attractive alternative for reducing the use of synthetic pigments in the formulation of coloured creams and make up products (Harnedy & FitzGerald, 2011; Ruiz-Ruiz, Benavides, & Rito-Palomares, 2013).

12.2.4 Lipids

The main classes of lipids are present in seaweeds but at a low level compared to terrestrial plant. The lipids include liposoluble vitamins (A, D, E and K), fatty acids, triglycerides, phospholipids, glycolipids, sterols and carotenoids which have been previously reported (Holdt & Kraan, 2011). Each class is composed of various molecules and the major compounds have been characterised. Fatty acids are precursors in the biosynthesis of eicosanoids, eicosapentaenoic and docosahexaenoic polyunsaturated ω-3 fatty acids. The nature of the fatty acids is highly variable, composed of saturated, monounsaturated and polyunsturated carbon chain. Thirty one fatty acids were identified in the red seaweed *G. gracilis* (Francavilla et al., 2013). Besides fatty acids, the unsaponifiable fraction contains carotenoids in red and green algae and fucoxanthine in brown algae, as well as tocopherols and sterols (Holdt & Kraan, 2011). Among the group of steroids, phytosterols form an important group and may have particular biological activities including anti-inflammatory and antioxidative effects (Lee, Lee, Jung, Kang, & Shin, 2003). However, the study of the biological activity of individual phytosterols is difficult because of the fact that pure phytosterols are expensive and scarcely available (Fernandes & Cabral, 2007). Fucosterol and fucosterol derivatives are present in brown algae, while desmosterol, cholesterol and cholesterol derivatives, campesterol and 24-ethylenecholesterol, are found in red seaweeds. Ergosterol and 24-ethylcholesterol have been reported in Chlorophyceae (Francavilla et al., 2013; Sánchez-Machado, López-Hernández, Paseiro-Losada, & López-Cervantes, 2004).

12.2.5 Pigments

Green lipid-soluble chlorophylls are present in all organisms that carry out photosynthesis. Carotenoids are coloured compounds involved in the photosynthetic apparatus and in the process of photoprotection by allowing the dissipation of excess heat energy and serving as direct quenchers of singlet oxygen or other reactive oxygen species (ROS) (Rastogi, Richa, Sinha, Singh, & Häder, 2010). These compounds are present in all species of macroalgae with seasonal variation. Zeaxanthin, α- and β-carotene, and lutein are found in red algae (Mikami & Hosokawa, 2013; Okai, Higashi-Okai, Yano, & Otani, 1996; Schubert, Garcia-Mendoza, & Pacheco-Ruize, 2006). β-carotene, zeaxanthin, violaxanthin and antheraxanthin are found in brown algae, along with fucoxanthin as well (Shang, Kim, Lee, & Um, 2011). In green algae, carotenoids consist of zeaxanthin, neoxanthin, antheraxanthin, violaxanthin, lutein, α- and β-carotene, siphonein and siphonaxanthin (Mikami & Hosokawa, 2013; Sánchez et al., 2012).

12.2.6 Phytohormones

The phytohormones include a number of "hormone-like" compounds, such as cytokinins, auxins, giberellins, betaïnes, abscisic acid (ABA), jasmonic acid, polyamine, brassinosteroids and rhodomorphin, which have been detected by liquid chromatography in various seaweeds (Tarakhovskaya, Maslov, & Shishova, 2007). ABA, gibberellic acid, indole-3-acetic acid, indole-3-butyric acid (IBA), salicylic acid and kinetin riboside (KR) have been reported as being present in *Ulva* sp. and *Monostroma oxyspermum* (Gupta et al., 2011).

12.2.7 Terpenoids and Halogenated Compounds

Seaweeds produce a wide range of secondary metabolites, many of which exhibit a broad spectrum of activity (Da Gama, Pereira, Carvalho, Coutinho, & Yoneshigue-Valentin, 2002; Ioannou & Roussis, 2009, pp. 51). The ecological roles of these compounds have recently been investigated in studies that chiefly emphasise chemical mediation as a defence against herbivores (Gagnon, St-Hilaire-Gravel, Himmelman, & Johnson, 2006; Macaya, Rothäusler, Thiel, Molis, & Wahl, 2005). These compounds can also inhibit settlement or development of fouling organisms (including viruses, bacteria, fungi and other algae) (Bazes et al., 2006, 2008; De Nys & Steinberg, 2002; Hay, 1996; Hellio et al., 2001; Hellio, Bergé, Beaupoil, Le Gal, & Bourgougnon, 2002; Silkina, Bazes, Mouget, & Bourgougnon, 2012). Although a large body of evidence has been available for some time showing that algae are endowed with chemical defences, it has only recently emerged that many of these defences are induced following challenges by bioaggressors (Bourgougnon & Stiger-Pouvreau, 2012; Potin, Bouarab, Salaun, Pohnert, & Kloareg, 2002).

Macroalgae extracts show efficacious antifungal, antimicroalgal and antibacterial activities (Hirschfeld et al., 1973; König & Wright, 1997). Chlorophyceae contain cyclic and linear sesqui-, di-, and triterpenes while Rhodophyceae are characterised by a high structural diversity of halogenated secondary metabolites whose polyhalogenated monoterpenes exhibit a wide range of activities. As an example, red algae produced halogenated terpenoids such as halogenated monoterpene from *Plocanium costatum* (Figure 12.1).

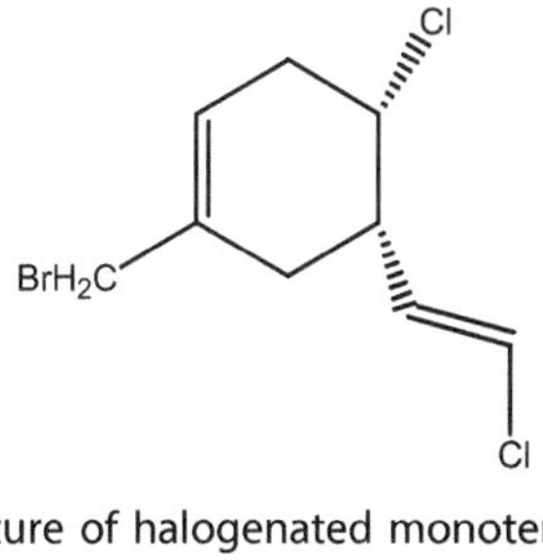

Figure 12.1 Chemical structure of halogenated monoterpene from *Plocanium costatum* (König, Wright, & de Nys, 1999).

Phaeophyceae differ from Rhodophyceae in having only minor halogenated hydrocarbons. Terpenes, phenolic compounds and meroterpenes make up the three major classes of secondary metabolites in brown seaweed (Kornprobst, 2014).

12.2.8 Phenolic Compounds, Polyphenols and Phlorotannins

In marine ecosystems, seaweeds of the upper intertidal zone are emerged for long periods during which they are exposed to strong, unfiltered UV radiation which induces the production of active oxygen species and free radicals. The production of phenolic compounds in marine algae has also been shown to be involved in various protection mechanisms such as against the oxidative or cytotoxic effects of UV damage. Phenolic compounds are secondary metabolites; hence, they are not directly involved in primary processes such as photosynthesis, cell division or reproduction of algae. Green and red seaweed have low contents of phenolic compounds compared to brown seaweeds (Hardouin et al., 2013; Mabeau & Fleurence, 1993) Phenolic compounds form a class of molecules divided into phloroglucinols (mono-, di-, tri-, tetra- and oligomeric) and phlorotannins, the latter being the most studied group in brown algae. Flavonoids are integral structural components of cell walls having an anti-UV function and acting as a chemical defence. Various compounds have been described, such as eckol, phlorofucofuroeckol A, dieckol (Figure 12.2), catechin and epigallocatechin (Heo et al., 2009; Rodríguez-Bernaldo de Quirós, Lage-Yusty, & López-Hernandez, 2010).

12.2.9 Mycosporine-like Amino Acids

The presence of mycosporine-like amino acids (MAAs) varies throughout the year due to the effect of insolation (Bedoux et al., in press; Bischof et al., 2006; Bourgougnon, Bedoux, Sangiardi, & Stiger-Pouvreau, 2011; Guinea, Franco, Araujo-Bazan, Rodriguez-Martin, & Gonzalez, 2012). MAAs are a family of secondary metabolites whose production is directly or indirectly related to the absorption of solar energy, and which protect marine organisms exposed to high UV radiation. MAAs have been detected in diverse organisms and especially in red macroalgae: *C. crispus*, *Palmaria* sp., *Gelidium* sp., *Porphyra* sp., *Gracilaria cornea*, *Asparagopsis* sp., *Desmarestia menziesii* (Karsten et Wiencke, 1999; Tsujino, Yade, & Sekikawa, 1980;Yuan, Westcott, Hu, & Kitts, 2009), *Grateloupia lanceola*, *Curdiea racovitzae* (Conde, Churio, & Previtali, 2000; Franklin, Kräbs, & Kuhlenkamp, 2001; Hoyer, Karsten, Sawall, & Wiencke, 2001; Huovinen, Matos,

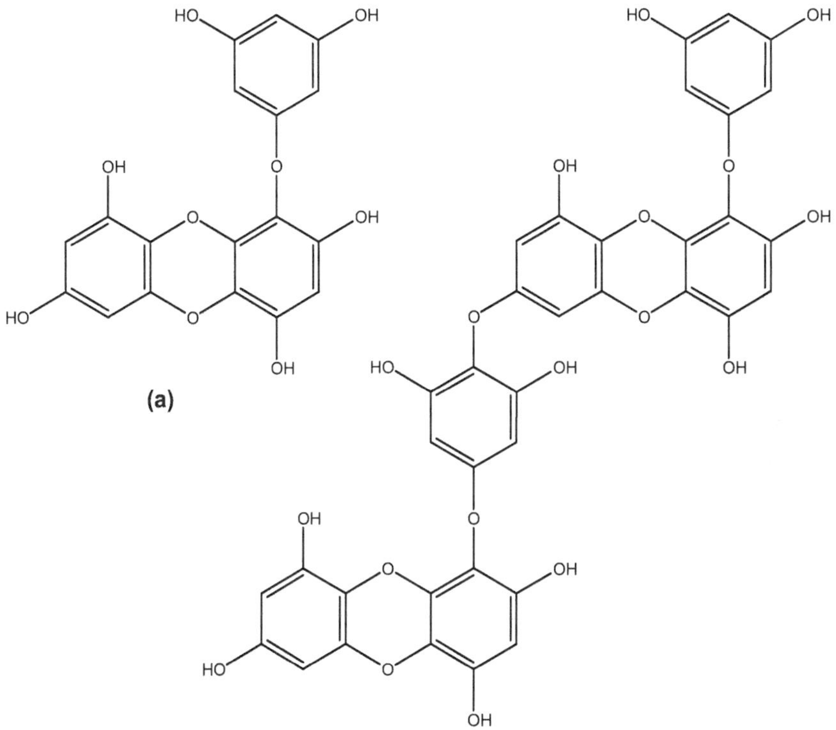

Figure 12.2 (a) Chemical structure of Eckol and Dieckol (Thomas & Kim, 2013).

Sousa Pinto, & Figueroa, 2006; Karentz, McEuen, Land, & Dunlap, 1991; Karsten, Sawall, & Wiencke, 1998; Peinado, Diaz, Figueroa, & Helbling, 2004; Sinha, Klisch, Gröniger, & Häder, 1998; Sinha, Klisch, Gröniger, & Häder, 2000), and more recently in *Solieria chordalis* (Bedoux et al., in press). Red macroalgae can contain up to 8 mg of MAAs per gram of dried matter (*C. racovitzae*). The content of MAAs is higher in summer and at a moderate depth (0–1 m) (Reef, Kaniewska, & Hoegh-Guldberg, 2009). The small intracellular compounds (<400 Da) of MAAs consist of a cyclohexenone or amino-cyclohexenimine ring linked to an amino acid, amino alcohol, or amino group, and are characterised by an absorption maximum between 320 and 360 nm. Generally, MAAs contain a glycine group on the C3 carbon and a second amino acid such as threonine, serine, taurine, and glutamate linked to the C1 carbon (Carreto & Carignan, 2011; Carreto, Carignan, & Montoya, 2001; Whitehead & Hedges, 2005; Yuan et al., 2009). MAAs may be potentially used in cosmetics and toiletries as UV protectors and activators of cell proliferation.

12.3 BIOLOGICAL ACTIVITIES AND SEAWEED COMPONENTS

Seaweeds have been described as having various activities as shown in Table 12.2. Molecules from brown macroalgae have antiaging, antiphotoaging, lipolytic and skin lightening (whitening) activities, while red seaweeds are characterized by antiaging, antiphotoaging, anti-UV and antioxidant activities.

12.3.1 Antiaging Care

The process of skin ageing is due to exogenous and endogenous factors. The genetic endogenous process is associated with a decrease in antioxidant status and cell proliferation capacity. Senescent cells express genes that produce degradative enzymes, growth factors and inflammatory cytokines. Exogenous factors include exposure to sunlight, pollution or nicotine, repetitive muscle movements, diet and medications (Farage, Miller, Elsner, & Maibach, 2008). Ageing of the skin occurs at several levels. The antioxidant defence system modulates the level of ROS that are produced during cellular respiration and energy metabolism. During skin ageing, cells lose their ability to regulate the generated ROS and suffer from oxidative stress. In the epidermis and dermis, a decrease in keratinocytes and melanocytes, as well as fibroblast proliferation and migration, leads to a decline of cellular activity and protein synthesis. Sulphated polysaccharides derived from macroalgae can have diverse biological and biochemical effects, including modulation of connective tissue proteolysis (Wijesinghe & Jeon, 2012b) and anti-inflammatory (Kang et al., 2011a; Senni et al., 2006), and can also act as free-radical scavengers and antioxidants for the prevention of oxidative damage in humans (Hu, Gen, Zhang, & Jiang, 2001; Rupérez, Ahrazem, & Leal, 2002; Zhang et al., 2004). Green algae, *Ulva lactuca* and *Codium tomentosum* (Chlorophyta) show antiaging, moisturising and lipolytic properties (Delaunay & Volle, 2011; Guglielmo & Montanari, 2008; Majmudar, 2012; Wang, Paul, & Luesch, 2013).

12.3.1.1 Sulphated Galactan and Sulfated Galactofucans

A fraction of the sulphated galactans was isolated from the red algae *Porphyra haitanensis* (Rhodophyta) by Zhang et al. (2004), using conventional hot-water extraction and ion-exchange cellulose chromatography. The 850 kDa-fraction was studied for *in vivo* antioxidant activity in ageing mice. Serum

Table 12.2 Examples of Seaweed used in Cosmetics, with Associated Metabolites and Activity

Macroalgae	Metabolites	Product Name	Activity	Manufacturer	Reference
Phaoephyceae					
Fucus vesiculosus	Fucoidan	–	Anticoagulant Antioxidant Skin fibroblast stimulation	–	Rupérez et al. (2002)
Fucus vesiculosus *Furcellaria lumbricalis*[a]	–	Topical cosmetic	Treating or preventing cellulite	Oriflame	Al-Bader, Davis, Laloeuf, and Rawlings (2013) (EP 2414047 B1)[b] Al-Bader et al. (2012)
Kjellmaniella crassifolia	Fucoidan		Antiaging Antiwrinkle	Takara Bio Inc.	Mizutani et al. (2010) (US 7678368 B2)
Padina pavonica	–	–	Keratinocytes differentiation Protein synthesis activation	Texinfine Patrinove	Gutierrez (1995) (EP 0655250 A1)
Laminaria digitata	Proteins (11%) Minerals (20%) Carbohydrates (69%)	Pheofiltrat	Lipolytic	Codif Int	Gedouin, Vallée, and Morvan (2006) (FR 2879098A1) Jang and Choung (2013)
Pelvetia canaliculata	Flanonoids Aminoacids Polyols Alginic acid Fucoidans	–	Antioxydant Collagen synthesis Stimulation Proteoglycans synthesis stimulation	–	Hupel, Lecointre, Meudec, Poupart, and Ar Gall (2011)
Ascophyllum nodosum	–	Algowhite	Tyrosinase inhibiting Anti-free-radical	Codif Int	http://www.codif-recherche-et-nature.com/
Hijikia fusiformis	Fucoxanthin	–	*In vivo* inducer of the Nrf2-ARE	–	Liu et al. (2011)

Continued

Table 12.2 Examples of Seaweed used in Cosmetics, with Associated Metabolites and Activity—cont'd

Macroalgae	Metabolites	Product Name	Activity	Manufacturer	Reference
Phaoephyceae					
Sargassum polycystum	Saponins, flavonoids, tannins, terpenoids, phenols, sugars, amino acids and amines	–	Antimelanogenesis or skin-whitening effect	–	Song et al. (2009) Chan et al. (2011)
Ascophyllum nodosum		Pheofiltrat	Skin conditioning regenerating and sebum regulating agent	Codif Int	http://www.codif-recherche-et-nature.com/
Saccharina longicruris	Galactofucan (1529 and 638 kDa)	–	Fibroblasts growth rate, synthesis of matrix metalloproteinase and collagen-I	–	Rioux et al. (2013)
Pelvetia wrightii	–	–	Anticellulite	–	Rozkin et al. (1991)
Ecklonia cava	Phlorotannins (dieckol)	–	Adipogenesis inhibitory effect	–	Ko et al. (2013)
Rhodophyceae					
Corallina pilulifera	–	–	Antiphotoaging	–	Ryu, Qian, Kim, Nam, and Kim (2009)
Corallina officinalis	Sulphated		Antioxidant		Yang, Liu, Wu, Chen, and Wang (2011)
Palmaria palmata	MAAs		Anti UV		Yuan et al. (2009)
Porphyra haitanensis	850 kDa sulphated galactans	–	*In vivo* antioxidant activity	–	Zhang et al. (2004)

Porphyra umbilicalis	MAAs		Anti-UVA Antioxydant		Zhang et al. (2004) Carreto and Carignan (2011)
Chlorophyceae					
Ulva lactuca	Clay and polysaccharide	Revertime	Antioxidant Antielastase Collagen synthesis Stimulation	Olmix	Demais et al. (2007) (EP 1786862 A1)
Ulva lactuca	Clay and lipopeptide	–	Antielastase Collagen synthesis Stimulation	Laboratoire Cosmétique de Lécousse	Delaunay and Volle (2011) (WO2011051591 A1)
Ulva lactuca	Fatty acids, keto-type C18 and keto-type C16	–	*In-vitro* and *in-vivo* Nrf2-ARE activation	–	Wang et al. (2013)
Ulva lactuca	Tripeptide composed of arginine–glycine–aspartic acid	–	Stimulation of collagen production via TGF-β, elastin, increase in the biosynthesis of collagen I	–	Guglielmo and Montanari (2008)
Codium tomentosum	–	–	Moisturizing	Mary Kay	Majmudar (2012) (US 8318178 B2)

[a]Red seaweed.
[b]Patent: skin care actives selected from the group consisting of glaucine, retinol, conjugated linoleic acid and caffeine.

malondialdehyde level was monitored as a marker of endogenous lipid peroxidation, along with the activities of the intracellular antioxidant enzymes superoxide dismutase (SOD), and glutathione peroxidase (GSH-Px), as well as total antioxidant capacity (TAOC). Treatment with the sulphated galactans fraction showed an effective inhibition of lipid peroxidation, inducing SOD and GSH-Px activities, and increasing the TAOC. Zhang et al. (2004) suggested that inhibition of lipid peroxidation might be attributed to the influence of sulphated galactans on the antioxidant enzymes and nonenzymatic defence system.

In the context of stimulating wound healing, galactofucan has been tested for its capacity to increase fibroblast cell migration and proliferation (Rioux, Moulin, Beaulieu, & Turgeon, 2013). Galactofucan (1529 and 638 kDa) was isolated from the brown seaweed *Saccharina longicruris*, subjected to depolymerisation and then added onto cultured dermal fibroblasts. Fibroblast growth rate and synthesis of matrix metalloproteinase and collagen-I were influenced according to the nature of the fractions (crude or depolymerised) and the molecular weight.

Furthermore, sulfated polysaccharides influence the migration and proliferation of fibroblasts, and studies have shown the capacity of these compounds for influencing growth factor modulation, stimulating oxidative phenomena, increasing skin thickness and elasticity, and inhibiting matrix metalloproteinases (Logeart, Letourneur, Jozefonvicz, & Kern, 1996; Senni et al., 2003; Senni et al., 2006).

12.3.1.2 Peptides

The role of proteins (elastin and collagen) and proteoglycans in the human skin network has been described by Gelse, Pöschl, and Aigner (2003). Several amino acids (serine, threonine and alanine) and amino acid derivatives (citrulline and pyrrolidone carboxylic acid) are components of the NMF in the *stratum corneum*. Over the last decade, the use of active peptides (less than 20 amino acids) in skin and hair care has attracted considerable attention in the industry. Up to the present, most active peptides have been synthesised using, for example, standard solid phase peptide synthesis methodology on Wang resin with a microwave synthesiser and Fmoc-protected amino acids as previously reported by Yang et al. (2013), or via classical chemical reaction using protected amino acids. Partial enzymatic hydrolysis of proteins (Hardouin et al., 2013) can also lead to oligopeptides. Although the use of L-amino acid ligase as a recombinant protein technology is able to produce peptides, this approach is not currently applied in the cosmetic field because it implies the use of genetic modified organisms (Arai, Arimura, Ishikura, & Kino, 2013; Gorouhi & Maibach, 2009). Macroalgae are an important source of amino acids, amino acid derivatives, and peptides. Those have shown the ability to stimulate collagen production in skin. Furthermore, macroalgal protein hydrolysates from various seaweeds (*U. pinnatifida*, *Sargassum* sp., *Porphyra yezoensis*, *P. palmata* and *Enteromorpha prolifera*) demonstrate antioxidant activities (Harnedy & FitzGerald, 2013). A polypeptide from *U. lactuca*, called *Aosa* biopeptide, is used in cosmetic products; it is composed of glycine, arginine, lysine, valine and aspartic acid. The tripeptide composed of arginine–glycine–aspartic acid stimulates the production mechanism of collagen, via tissue growth factor, elastine, and produces an increasing biosynthesis of collagen I in human fibroblasts (Guglielmo & Montanari, 2008). Progress in biotransformation design implies that biopeptides from macroalgae could be used as a potential alternative source to synthetic ingredients and may contribute to antiaging. Enzymatic hydrolysis of algal proteins would allow the preparation of peptides that might lead to the development of skin bioactive compounds (Hardouin et al., 2013).

12.3.1.3 Keto-type C18 Fatty Acid

Naturally occurring molecules have been identified that are inducers of the nuclear factor E2-related factor 2 (Nrf2) antioxidant-response element (ARE) pathways. This defence system is induced by the cell to counteract oxidative stress. Wang et al. (2013) tested 30 macroalgae using an ARE-luciferase reporter gene assay and observed that *U. lactuca* extracts activated the reporter. After purification, three fatty acids, two keto-type C18 and one keto-type C16, showed *in vitro* and *in vivo* ARE activation. The electrophilic Mickael acceptor group appeared to be essential for the ARE activities.

12.3.1.4 Fucoxanthine

The carotenoid isolated from *Hijikia fusiformis* has been found to induce the Nrf2-ARE in mouse liver cells. We should mention that the electrophilic Mickael acceptor group is present in fucoxanthin (Liu, Chiu, & Hu, 2011). Zheng, Piao, Keum, Kim, and Hyun (2013) have demonstrated the ability of fucoxanthine to reduce apoptosis, protecting keratinocytes against oxidative damage by scavenging ROS.

12.3.1.5 Phytohormone Abscisic Acid

Phytohormones have shown potential inhibition of skin proteases and ROS. Systemic sclerosis (SSc) is a chronic inflammatory disease resulting in skin fibrosis. The production of extracellular matrix (ECM) components, mainly type I collagen, is specifically upregulated in SSc fibroblasts causing thickening of the dermis. Bruzzone et al. (2012) have studied the effect of ABA on skin fibroblasts from SSc patients; ABA is reported to be an endogenous hormone in humans. ABA modifies some of the functions altered in SSc fibroblasts to a normal phenotype, suggesting that the skin of SSc patients could benefit from exposure to exogenous ABA.

12.3.2 Anticellulite and Slimming Care

Alterations of the skin of the thighs and buttocks, commonly referred to as cellulite, affect underlying adipose tissue of the hypodermis region. Adipose tissue contains collagen, reticulin and proteoglycan (Lotti, Ghersetich, Grappone, & Dini, 1990). Adipocytes produce triacylglycerols which represent the main component of adipose tissue in addition to cholesterol. Cellulite involves changes in the subcutaneous adipose layer biochemistry and in the structure of the supporting connective ECM together with the overlying dermal layer. These alterations include enhanced lipogenesis, decreased lipolysis and increased lipid storage within the adipocytes (Al-Bader et al., 2012; Hexsel & Soirefmann, 2011; Rosenbaum et al., 1998). Energy homeostasis is regulated by the chronobiology of adipocytes through the mechanisms of lipogenesis and lipolysis. Lipogenesis depends on α-adrenergic receptors on the adipocytes, cAMP production, antagonist cAMP phosphodiesterase, receptor adenosine A1, activation of the phosphodiesterase by insulin, and lipoprotein–lipase stimulation. The cAMP-dependent pathway is one of the mechanisms activating lipolysis in adipocytes (Carmen & Victo, 2006). Lipolysis is enhanced by β-receptors regulating cAMP production. Some topical treatments for cellulite include seaweed extracts, such as those used for increasing microcirculation flow, reducing lipogenesis and promoting lipolysis, which restore the normal structure of the dermis and subcutaneous tissue, and which scavenge free radicals or prevent their formation. The seaweeds most frequently used in anticellulite and slimming preparations are brown seaweeds, *Fucus vesiculosus*, *Furcellaria lumbricalis* (Al-Bader et al., 2012), *Laminaria digitata* (Jang & Choung, 2013) and *Pelvetia wrightii* (Rozkin, Levina, Efimov, & Usov, 1991). *Laminaria digitata* stimulates blood flow and aids the body in shedding excess fluids. In this way, seaweed can reduce the appearance of cellulite, but not the cellulite itself. Some constituents of macroalgae are lipolytic agents: flavonoids (Acheson et al., 2004; Kuppusamy & Das, 1992, Kuppusamy & Das, 1994; Ruckstuhl, Beretz, Anton, & Landry, 1979), quercetin (flavonoid belonging to the flavonols)

(Kuppusamy & Das, 1994) and phlorotannins (Ko et al., 2013). In the latter study, the potential inhibitory effect of extracts from five brown seaweed species was measured by adipogenesis assay according to the differentiation of 3T3-L1 preadipocytes into mature adipocytes by using the Oil-Red O staining protocol. The *Ecklonia cava* extract tested by Ko et al. (2013) showed evidence of an adipogenesis inhibitory effect on adipogenesis, compared to the results obtained from the other four brown seaweed extracts. Three phlorotannins were obtained by Ko et al. (2013) who observed their inhibitory effect on adipogenesis. Among the phlorotannins, dieckol exhibits potential adipogenesis inhibition and downregulates the expression of the peroxisome proliferator-activated receptor (PPAR) and other transcription factors implied in the expression of genes. PPAR is a protein which belongs to the family of nuclear receptors acting as a transcription factor of genes involved in lipid metabolism and adipogenesis. These results demonstrate the inhibitory effect of dieckol on adipogenesis through the activation of the AMP activated protein kinase signal pathway. Otherwise, Kang et al. (2011b) have shown that fucoxanthin isolated from *Petalonia binghamiae* has adverse effects.

12.3.3 Antioxidant—Photoprotection

Phenolic compounds are commonly characterized as being stress-induced compounds, and are involved in chemical protection mechanisms against biotic factors such as contamination due to bacteria and abiotic factors such as UV radiation and metal contamination (Llewellyn & Airs, 2010; Stengel, Connan, & Popper, 2011). Since phenolic compounds are produced in response to the generation of ROS, they exhibit anti-ROS properties (i.e. antioxidative) which ensure an efficient cytoprotective system. The antioxidant activity of seaweed phenols seems to depend on their structure and especially on the degree of polymerisation of phloroglucinol (Nakamura, Nagayama, Uchida, & Tanaka, 1996). The antioxidant and free-radical scavenging activities of phenols have essentially been identified in brown seaweeds (Ahn, Jeon, Kang, Shin, & Jung, 2004; Athukorala, Lee, Kim, & Jeon, 2007; Heo, Park, Lee, & Jeon, 2005; Park, Shahidi, & Jeon, 2004; Siriwardhana et al., 2008), and sometimes in red seaweeds (Cian, Martinez-Augustin, & Drago, 2012; Wang et al., 2010). In the short-wavelengths solar spectrum, the harmful effects of UV radiation (200–400 nm) are well documented (references in Pessoa, 2012; Rastogi et al., 2010; Talarico & Maranzana, 2000). All plant, animal and microbial organisms appear to be sensitive to UV radiation, but to a highly variable extent (Aiamla-or, Kaewsuksaeng,

Shigyo, & Yamauchi, 2010; Sinha et al., 2000; Srilaong, Aiamla-or, Soontornwat, Shigyo, & Yamauchi, 2011; Volkman & Gorbushina, 2006). Although sunlight-induced skin damage has been attributed mainly to the UV-B wavebands (285–320 nm), more recently, the important contribution of UV-A (320–400 nm) has been well demonstrated (Agar et al., 2004). UV radiation induces diverse effects such as the damaging of human skin (sunburn) and the formation of radical species, while causing the destruction of proteins, DNA and other biomolecules as well as leading to acute physiological stress (Torres et al., 2004). It can also lead to the proliferation of oncogenes which can mutate and cause cancer. Various natural substances extracted from plants could potentially have properties to protect against UV radiation. Macroalgae have developed mechanisms which counteract the damaging effects of UV-B and UV-A, by producing screen pigments such as carotenoids and phenolic compounds or UV-absorbing MAAs (Figueroa et al., 2003; Rosic, 2012). UV-absorbing compounds are bioactive compounds that can protect human fibroblast cells from UV-induced cell death and suppress UV-induced ageing in human skin. MAAs are UV-screen and antioxidant substances found in red algae, representing very efficient photoprotectors with high-energy dissipation and high photostability, but without generating oxidant photoproducts. Moreover, a product called Helioguard® 365 that contains MAAs has been commercialised (references in Bourgougnon et al., 2011; Bourgougnon & Stiger-Pouvreau, 2012; Richa et al., 2011).

12.3.4 Moisturising

The two external tissue layers of the skin surface are made up of dermis and epidermis. The epidermis is a multilamellar system whose outermost layer corresponds to the *stratum corneum* composed of corneocytes. Corneocytes contain keratin. Epidermis is mainly composed of keratinocytes, whereas dermis fibroblasts are embedded in an ECM composed of water, salts, proteins (fibrous collagen and elastin), adhesive glycoproteins and proteoglycans forming a colloidal gel. A core protein is linked to the polysaccharide fraction also called glycosaminoglycan (GAG) to form proteoglycan. Nonsulfated hyaluronic acid and sulphated GAGs represent more than 95% of the total mass of the molecule and are part of the skin's connective tissue giving elasticity and firmness. Proteoglycans act as a reservoir responsible for hosting most of the skin's water. The dermis ECM structure determines the physical characteristics of the tissue and the biological properties of its embedded cells (Esko, 1999). The moisturising properties of algae are due to the presence of several compounds belonging to the group of NMFs (Watabe et al., 2013). Polysaccharides and oligosaccharides provide a high capacity for water storage, and these macromolecules can be linked to keratin through hydrogen bonds (Choi et al., 2013; Du, Bian, & Xu, 2013). *Mastocarpus* extract containing floridoside and isethionic acid has been shown to increase water retention of the stratum corneum (Deslandes & Bodeau, 2007).

12.3.5 Whitening

Melanin biosynthesis involves tyrosinase that catalyses the hydroxylation of L-tyrosine to 3,4-dihydroxyphenylalanine (L-DOPA) (Parvez et al., 2006). The inhibition of tyrosinase is the most common approach used to achieve skin whiteness. *Sargassum polycystum* extracts have been studied for their antimelanogenesis or skin-whitening effect by means of the cell-free mushroom tyrosinase assay (Song et al., 2009). The inhibition of cellular tyrosinase derived from melanin and melanogenesis was also assessed using kojic acid as a positive control. Ethanolic and ethyl acetate extracts, as well as the non-cytotoxic hexane fraction were found to reduce cellular tyrosinase activity while the water fraction showed only a small inhibitory effect according to the cell-free mushroom tyrosinase assay. The phytochemical composition of the active fraction includes saponins, flavonoids, tannins, terpenoids, phenols, sugars, amino acids and amines (Chan, Kim, & Cheah, 2011). Phenolic compounds are well-known components from brown seaweeds and this finding strengthens the role of these compounds in the inhibition of melanogenesis (Sabudak, Demirkiran, Ozturk, & Topcu, 2013). In addition, it has been reported that antioxidants may reduce hyperpigmentation and favour skin health (Ma et al., 2001).

12.4 FUTURE DEVELOPMENT

12.4.1 Seaweed Cultivation or Cocultivation for the Production of Actives

Out of the 16.8 million tons of seaweed produced in 2008, 93.8% were cultivated mainly in Asia (FAO, 2010). The versatility of seaweeds, whether used whole or for extraction in the hydrocolloid, pharmaceutical and cosmetics industries, has created a demand that cannot be satisfied by harvesting of wild stocks alone. The development of active ingredients from seaweed depends on the domestication of the species concerned. Indeed, it is imperative to know the biology of algae to be able to ensure high-quality culture, or even to enhance the content of a compound of interest.

In addition, the cultivation of seaweeds is essential in providing a sustainable and reproducible good-quality product (Guillemin et al., 2008). As the harvesting of wild populations of *Palmaria* is a seasonal activity, open-sea cultivation using a "Kuralon" string is a new technique producing a reliable crop that can be harvested as and when required (Edwards & Dring, 2011).

12.4.2 Specific Extraction Methods

Isolating a compound from seaweed first involves harvesting the algal raw material, storing (freeze drying), and then performing the extraction itself. Pretreatment of seaweed is of growing importance to achieve good extraction efficiency and the required selectivity. It is possible to obtain extracts with different compositions and yields by changing the extraction design. Therefore, depending on the type of compounds of interest, optimum extraction conditions can be utilised for selective isolation of specific groups of compounds. It is noteworthy that the extraction of target substances is carried out following an ecodesign strategy according to the six basic principles proposed by Chemat (2011). The algal raw material forms a variable but abundant and renewable resource. Moreover, reproducible cultivation techniques should be established to obtain a more standardised raw material (Edwards & Dring, 2011). The use of ecofriendly and nontoxic solvents coupled to efficient and sustainable extraction techniques is greatly encouraged (Anastas & Warner, 1998). The extract is not purified to an excessive degree in order to minimise the production of waste and, ultimately to move toward full use of the biomass. Moreover, the number of successive operations is limited as much as possible to maintain a simple and low-cost process. Finally, the extract is designed with a view to being incorporated as a natural ingredient into cosmetic products. This approach is expected to enhance the efficiency of the product. Xiao, Yuan, and Li (2013) report the isolation of fucosterol and phytol from the brown seaweeds *U. pinnatifida* and *Sargassum fusiforme* by microwave-assisted extraction coupled with high-speed countercurrent chromatography. Supercritical fluid extraction and subcritical water extraction are used for the extraction of lipids and antioxidants compounds, whereas fucoidan is obtained by ultrasound- and microwave-assisted extraction techniques (Herrero, Cifuentes, & Ibañez, 2006). Fractions of fucoidan promote the proliferation of dermal fibroblasts, and protect the dermis elastic fibre network (Logeart et al., 1996; Senni et al., 2003; Senni et al., 2006). Different extraction procedures were compared by Hahn, Lang, Ulber, and Muffler (2012) as regards to the yield and chemical structure of the fucoidan, as well as the extraction time. The use of ultrasound- and microwave-assisted extraction techniques allows high yields and a low extraction time without modifying the sulphate ester linkage. Microwave and ultrasound-assisted extraction are promising tools for establishing a more sustainable and efficient purification of fucoidan, and these techniques are competing with the enzyme-assisted release of sulphated polysaccharides. The enzyme-based technique appears to be advantageous for the large-scale production of fucoidan, but the economic efficiency of the process is dependent on the supply of the required enzymes.

12.4.3 Enzymatic Hydrolysis and Preparation of Cosmetic Actives

The cell wall consists of complex embedded polymers and it is a real challenge to carry out isolation of active oligosaccharides or peptides of high purity using water or aqueous organic solvent extraction processes. Extraction assisted by enzymatic hydrolysis is a promising soft technique to generate lower molecular weight oligosaccharide or peptide fractions, while also improving the extraction efficiency (quality and yield) of bioactive compounds derived from seaweeds. To obtain active peptides, sequential enzymatic digestion is recommended. Promising results have been obtained by several workers (Hardouin et al., 2013; Samarakoon & Jeon, 2012; Wijesinghe & Jeon, 2012a). The enzyme or multienzyme mixes tested in most studies are commercial preparations which are already used in the food industry. Moreover, the experimental conditions can be designed to conserve the native structure and function. Up to present, enzymatic degradation is hampered by the limited availability of the appropriate enzymes and demand for specific enzymes is already on the increase for the future development of bioactive extracts. Thus, the aim is to achieve a successful optimisation of heterologous carbohydrase or protease production, and the subsequent use of immobilised enzymes to control the degradation process is a promising approach. The purpose of obtaining natural antioxidants would be to supply a large-scale process for the production of water-soluble extracts (Heo et al., 2005; Heo, Jeon, Lee, Kim, & Lee, 2003, Wang et al., 2010).

12.4.4 Potential Antimicrobial Agents

Several cosmetic products are used for counteracting microbial colonisation (deodorants and antiperspirants), dental biofilm formation (toothpaste) and skin infection or inflammation (antiacne). Those products often contain synthetic antimicrobial agents listed in the Annex V (positive list) of the cosmetic regulation (EU cosmetic regulation No. 1223/2009). The exploration of secondary metabolites from seaweeds could represent a new source of preservatives for cosmetics and has revealed important chemical relevant prototypes in the search for new agents (El Gamal, 2010; Ioannou & Roussis, 2009, pp. 51; Kornprobst, 2014; Mayer, 2002; Mayer & Hamann, 2004; Mayer et al., 2007). From 1960, up to the present, more than 15,000 novel compounds have been isolated from marine organisms (Cardozo et al., 2007).

12.5 CONCLUSION

Macroalgae represent a vegetative resource characterized by a high content of proteins, carbohydrates and minerals. Significant developments are taking place to introduce these marine plants in the field of cosmetic products, and there is growing public interest in the use of this natural resource for the production of industrially useful ingredients that may improve health and wellbeing. Seaweed extracts display a range of different activities in relation to antiaging, photoprotection, lipolysis, anti-inflammatory effects, moisturizing and whitening. Sulphated polysaccharide fractions and oligosaccharides, as well as peptides, keto-type fatty acids, fucoxanthin and phytohormones have been shown to have antiaging activities. Flavonoids and the particular class of phlorotannins, which are present in brown seaweeds, show lipolytic and anticellulite properties. Red seaweeds contain various MAAs, pigments and phenolic compounds which contribute to anti-photo-ageing and anti-free-radical activities. Green seaweeds, especially *U. lactuca* extracts, have been shown to counteract enzyme proteolysis of dermis components, while stimulating collagen biosynthesis. All these results have spurred great interest in the search for novel ingredients that might contribute to revealing new activities in the relevant applications. New strategies are being explored to monitor macroalgae growth and to study the biosynthesis and genetics of seaweeds, which will open up opportunities for developing new active compounds. However, it is important to study how the properties of the actives can be preserved in the formulations, and methods should be developed to enhance their bioavailability as functional cosmetics. Figure 12.3 presents a strengths–weaknesses–opportunities–threats (SWOT) diagram which shows that management of seaweed resources and the stability of actives are the main threats and weaknesses while the increased demand of consumer for natural product combined with the safety of seaweed extracts and the proved biological activities represent strengths and opportunities.

Figure 12.3 Strengths–weaknesses–opportunities–threats diagram related to the development of cosmetics containing seaweed actives.

ACKNOWLEDGEMENTS

The authors thank Dr Florent Yvergnaux for his critical and very beneficial reading of the manuscript and Michael Carpenter for rereading the manuscript.

REFERENCES

Acheson, K. J., Gremaud, G., Meirim, I., Montigon, F., Krebs, Y., Fay, L. B., et al. (2004). Metabolic effects of caffeine in humans: lipid oxidation or futile cycling? *American Journal of Clinical Nutrition*, *79*, 40–46.

Agar, N. S., Halliday, G. M., Barneston, R., Ananthaswamy, H. N., Wheeler, M., & Jones, A. M. (2004). The basal layer in human squamous tumors harbors more UVA than UVB fingerprint mutations: a role for UVA in human skin carcinogenesis. *Proceedings of the National Academy of Sciences*, *101*, 4954–4959.

Ahn, C. B., Jeon, Y. J., Kang, D. S., Shin, T. S., & Jung, B. M. (2004). Free radical scavenging activity of enzymatic extracts from a brown seaweed *Scytosiphon lomentaria* by electron spin resonance spectrometry. *Food Research International, 37*, 253–258.

Aiamla-or, S., Kaewsuksaeng, S., Shigyo, M., & Yamauchi, N. (2010). Impact of UV-B irradiation on chlorophyll degradation and chlorophyll-degrading enzyme activities in stored broccoli (*Brassica oleracea* L. Italica Group) florets. *Food Chemistry, 120*, 645–651.

Al-Bader, T., Byrne, A., Gillbro, J., Mitarotonda, A., Metois, A., Vial, F., et al. (2012). Effect of cosmetic ingredients as anticellulite agents: synergistic action of actives with in vitro and in vivo efficacy. *Journal of Cosmetic Dermatology, 11*, 17–26.

Al-Bader, T., Davis, M., Laloeuf, A., & Rawlings, A. V. (2013). *Topical cosmetic compositions for treating or preventing cellulite.* . EP 2414047 B1.

Anastas, P. T., & Warner, J. C. (1998). *Green chemistry: Theory and practice.* New York: Oxford University Press.

Aneiros, A., & Garateix, A. (2004). Bioactives peptides from marine sources: pharmacological properties and isolation procedures. *Journal of Chromatography B, 803*, 41–53.

Arai, T., Arimura, Y., Ishikura, S., & Kino, K. (2013). L-amino acid ligase from *Pseudomonas syringae* producing tabtoxin can be used for enzymatic synthesis of various functional peptides. *Applied Environmental Microbiology, 79*, 5023–5029.

Athukorala, Y., Lee, K. W., Kim, S. K., & Jeon, Y. J. (2007). Anticoagulant activity of marine green and brown algae collected from the Jeju Island in Korea. *Bioresource Technology, 98*, 1711–1716.

Bazes, A., Silkina, A., Defer, D., Quéméner, E., Braud, J. P., & Bourgougnon, N. (2006). Allelopathic substances from *Ceramium botryocarpum* used as antifouling products. *Aquaculture, 258*, 664–674.

Bazes, A., Silkina, A., Douzenel, P., Faÿ, F., Kervarec, N., Morin, D., et al. (2008). Investigation of the antifouling constituents from the brown alga *Sargassum muticum* (Yendo) Fensholt. *Journal of Applied Phycology, 21*, 395–403.

Bedoux, G., Hardouin, K., Marty, C., Taupin, L., Vandanjon, L., Bourgougnon, N. Chemical characterization and photoprotective activity measurement of extracts from the red macroalgae *Solieria chordalis. Botanica Marina*, in press.

Bischof, K., Gomes, I., Molis, M., Hanelt, D., Karsten, U., Luder, U., et al. (2006). Ultraviolet radiation shapes seaweed communities. *Review of Environmental Science and Biotechnology, 5*, 141–146.

Bourgougnon, N., Bedoux, G., Sangiardi, A., & Stiger-Pouvreau, V. (2011). Las algas: potencial nutritivo y aplicaciones cosméticas. In *Las algas como recurso, Valorización, Aplicaciones industriales y tendencias, Centro Tecnológico del* Mar - Fundación CETMAR (pp. 81–94). 978-84-615-3593-4.

Bourgougnon, N., & Stiger-Pouvreau, V. (2012). Chemodiversity and bioactivity within red and brown macroalgae along the French coasts, metropole and overseas departements and territories. In *Handbook of marine macroalgae, biotechnology and applied phycology* (pp. 58–105). Wiley-Blackwell.

Bruzzone, S., Battaglia, F., Mannino, E., Parodi, A., Fruscione, F., Basile, G., et al. (2012). Abscisic acid ameliorates the systemic sclerosis fibroblast phenotype *in vitro*. *Biochemical and Biophysical Research Communications, 422*, 70–74.

Cardozo, K. H. M., Guaratini, T., & Barros, M. P. (2007). Metabolites from algae with economical impact. *Comparative Biochemistry and Physiology C: Toxicology and Pharmacology, 146*, 60–78.

Carmen, G. Y., & Victo, S. M. (2006). Signalling mechanisms regulating lipolysis. *Cellular Signalling, 18*, 401–408.

Carreto, J. I., & Carignan, M. O. (2011). Mycosporine-like amino acids: relevant secondary metabolites. Chemical ecological aspects. *Marine Drugs, 9*, 387–446.

Carreto, J. I., Carignan, M. O., & Montoya, N. G. (2001). Comparative studies on mycosporine-like amino acids, paralytic shellfish toxins and pigment profiles of the toxic dinflagellates *Alexandrium tamarense, A. catenella* and *A. minutum. Marine Ecology Progress Series, 223*, 49–60.

Chan, Y. Y., Kim, K. H., & Cheah, S. H. (2011). *Inhibitory effects of Sargassum polycystum on tyrosinase activity and melanin formation in B16F10 murine melanoma cells.*

Chemat, F. (2011). In Dunod (Ed.), *Eco-extraction du végétal, Procédés innovants et solvants alternatifs.* 978-2-10-056543-6.

Choi, J. S., Moon, W. S., Choi, J. N., Do, K. H., Moon, S. H., Cho, K. K., et al. (2013). Effects of seaweed *Laminaria japonica* extracts on skin moisturizing activity *in vivo*. *Journal of Cosmetic Science, 64,* 193–205.

Chronakis, I. S., Galatanu, A. N., Nylander, T., & Lindman, B. (2000). The behavior of protein preparations from blue-green algae (*Spirulina platensis* strain Pacifica) at the air/water interface. *Colloids and Surfaces A: Physicochemical and Engineering Aspects, 173,* 181–192.

Cian, R. E., Martinez-Augustin, O., & Drago, S. R. (2012). Bioactive properties of peptides obtains by enzymatic hydrolysis from protein byproducts of *Porphyra columbina*. *Food Research International, 49,* 364–372.

Conde, F. R., Churio, M. S., & Previtali, C. M. (2000). The photoprotector mechanism of mycosporine-like amino acids, excited-state properties and photostability of porphira-334 in aqueous solution. *Journal of Photochemistry Photobiology B: Biology, 56,* 139–144.

CosIng Database. Website of the European Commission, Health and Consumers. Consulted the 31th of January 2014. http://ec.europa.eu/consumers/cosmetics/cosing.

Da Gama, B. A. P., Pereira, R. C., Carvalho, A. G. V., Coutinho, R., & Yoneshigue-Valentin, Y. (2002). The effects of seaweed secondary metabolites on biofouling. *Biofouling, 18,* 13–20.

Delaunay, D., & Volle, I. (2011). *Composition dermatologique et/ou cosmétique utilisée pour la régénération de la peau.* . WO 2011051591 A1.

De Nys, R., & Steinberg, P. D. (2002). Linking marine biology and biotechnology. *Current Opinion in Biotechnology, 13,* 244–248.

Demais, H., Brendle, J., Le Deit, H., Laza Anca, L., Lurton, L., & Brault, D. (2007). *Argiles intercalées.* EP 1786862 A1.

Deslandes, E., & Bodeau, C. (2007). *Composition cosmétique comprenant un extrait d'algue rouge comprenant une association de floridoside et d'acide iséthionique.* EP 1 743 628 A1.

Du, B., Bian, Z., & Xu, B. (2013). Skin health promotion effects of natural beta-glucan derived from cereals and microorganisms: a review. *Phytotherapy Research, 28* (2), 159–166. http://dx.doi.org/10.1002/ptr.4963.

Edwards, M. D., & Dring, M. J. (2011). Open-sea cultivation trial of the red alga, *Palmaria palmata* from seeded tetraspores in Strangford Lough, Northern Ireland. *Aquaculture, 317,* 203–209.

El Gamal, A. A. (2010). Biological importance of marine algae. *Saudi Pharmaceutical Journal, 18,* 1–25.

Esko, J. D. (1999). Proteoglycans and glycosaminoglycans. In A. Varki, R. Cummings, J. Esko, et al. (Eds.), *Consortium of glycobiology. Essentials of glycobiology* (2nd ed.). New York: Cold Spring Harbor Laboratory Press.

FAO. (2010). *The state of world fisheries and aquaculture.* Rome: FAO. 197 pp.

Farage, M. A., Miller, K. W., Elsner, P., & Maibach, H. I. (2008). Intrinsic and extrinsic factors in skin ageing: a review. *International Journal of Cosmetic Science, 30,* 87–95.

Fernandes, P., & Cabral, J. M. S. (2007). Phytosterols: applications and recovery methods. *Bioresource Technology, 98,* 2335–2350.

Figueroa, F. L., Escassi, L., Pérez-Rodriguez, E., Korbee, N., Delia Giles, A., & Johnsen, G. (2003). Effects of short-term irradiation on photoinhibition and accumulation of mycosporine-like amino acids in sun and shade species of the red algal genus Porphyra. *Journal of Photochemistry and Photobiology B: Biology, 69,* 21–30.

Fitzgerald, C., Gallagher, E., Tasdemir, D., & Hayes, M. (2011). Heart health peptides from macroalgae: isolation and potential use in functional foods. *Journal of Agriculture and Food Chemistry, 59,* 6829–6836.

Fleurence, J. (1999). Seaweed proteins: biochemical, nutritional aspects and potential uses. *Trends Food Science and Technology*, *10*, 25–28.

Fleurence, J. (2004). Seaweed proteins. In R.Y. Yada (Ed.), *Proteins in food processing* (pp. 197–213). Cambridge: Woodhead Publishing.

Francavilla, M., Franchi, M., Monteleone, M., & Caroppo, C. (2013). The red seaweed *Gracilaria gracilis* as a multi products source. *Marine Drugs*, *11*, 3754–3776.

Franklin, L. A., Kräbs, G., & Kuhlenkamp, R. (2001). Blue light and UV-A radiation control the synthesis of mycosporine-like amino acids in *Chondrus crispus* (Florideophycea). *Journal of Phycology*, *37*, 257–270.

Gagnon, P., St-Hilaire-Gravel, L.V., Himmelman, J. H., & Johnson, L. E. (2006). Organismal defenses versus environmentally mediated protection from herbivores: unraveling the puzzling case of *Desmarestia viridis* (Phaeophyta). *Journal of Experimental Marine Biology and Ecology*, *334*, 10–19.

Gedouin, A., Vallée, R., & Morvan, P.Y. (2006). *Use of algae extract to stimulate the oxygen uptake by the cells having lipolytic effect to produce ATP molecules.*. FR 2879098 A1.

Gelse, K., Pöschl, E., & Aigner, T. (2003). Collagens – structure, function, and biosynthesis. *Advanced Drug Delivery Reviews*, *55*, 1531–1546.

Gorouhi, F., & Maibach, H. I. (2009). Role of topical peptides in preventing or treating aged skin. *International Journal of Cosmetic Science*, *31*, 327–345.

Guglielmo, M., Montanari, D. (2008). Cosmetic composition with a lifting effect for sustaining relaxed tissues. Patent WO 2008146116 A2.

Guillemin, M. L., Faugeron, S., Destombe, C., Viard, F., Correa, J. A., & Valero, M. (2008). Genetic variation in wild and cultivated populations of the haploid-diploid red alga *Gracilaria chilensis*: how farming practices favor asexual reproduction and heterozygosity. *Evolution*, *62*, 1500–1519.

Guinea, M., Franco, V., Araujo-Bazan, L., Rodriguez-Martin, I., & Gonzalez, S. (2012). In vivo UVB-photoprotective activity of extracts from commercial marine macroalgae. *Food and Chemical Toxicology*, *50*, 1109–1117.

Gupta, V., Kumar, M., Brahmbhatt, H., Reddy, C. R. K., Seth, A., & Jha, B. (2011). Simultaneous determination of different endogenetic plant growth regulators in common green seaweeds using dispersive liquid liquid microextraction method. *Plant Physiology and Biochemistry*, *49*, 1259–1263.

Gutierrez, G. (1995). *Compositions of Padina algae or their extracts, and their pharmaceutical, food compositions, or use for the culture of molluscs or arthropods*. EP 0655250 A1.

Hahn, T., Lang, S., Ulber, R., & Muffler, K. (2012). Novel procedures for the extraction of fucoidans from brown algae. *Process Biochemistry*, *47*, 1691–1698.

Hardouin, K., Burlot, A. S., Umami, A., Tanniou, A., Stiger-Pouvreau, V., Widowati, I., et al. (2013). Biochemical and antiviral activities of enzymatic hydrolysates from different invasive French seaweeds. *Journal of Applied Phycology*. http://dx.doi.org/10.1007/s10811-013-0201-6.

Harnedy, P. A., & FitzGerald, R. J. (2011). Bioactive proteins, peptides, and amino acids from macroalgae. *Journal of Phycology*, *47*, 218–232.

Harnedy, P. A., & FitzGerald, R. J. (2013). Bioactive proteins and peptides from macroalgae, fish, shellfish and marine processing waste. In S.-K. Kim (Ed.), *Marine proteins and peptides, biological activities and applications* (pp. 5–39). John Wiley & Sons UK.

Hay, M. E. (1996). Marine chemical ecology: what's known and what's next? *Journal of Experimental Marine Biology and Ecology*, *200*, 103–134.

Hellio, C., Bergé, J. P., Beaupoil, C., Le Gal, Y., & Bourgougnon, N. (2002). Screening of marine algal extracts for anti-settlement activities against microalgae and spores of macroalgae. *Biofouling*, *18*, 205–215.

Hellio, C., Thomas-Guyon, H., Culioli, G., Piovetti, L., Bourgougnon, N., & Le Gal, Y. (2001). Marine antifoulants from *Bifurcaria bifurcata* (Phaeophyceae, Cystoseiraceae) and other brown macroalgae. *Biofouling*, *17*, 189–201.

Hemmingson, J. A., Falshaw, R., Furneaux, R. H., & Thompson, K. (2006). Structure and antiviral activity of the galactofucan sulfates extracted from *Undaria pinnatifida* (Phaeophyta). *Journal of Applied Phycology, 18*, 185–193.

Heo, S. J., Jeon, S. K., Lee, J., Kim, H. T., & Lee, K. W. (2003). Antioxidant effect of enzymatic hydrolysate from a kelp, *Ecklonia cava*. *Algae, 18*, 341–347.

Heo, S. J., Ko, S. C., Cha, S. H., Kang, D. H., Park, H. S., Choi, Y. U., et al. (2009). Effect of phlorotannins isolated from *Ecklonia cava* on melanogenesis and their protective effect against photo-oxidative stress induced by UV-B radiation. *Toxicology* in Vitro, *23*, 1123–1130.

Heo, S. J., Park, H. S., Lee, K. W., & Jeon, Y. J. (2005). Antioxidant activities of enzymatic extracts from brown seaweeds. *Bioresource Technology, 96*, 1613–1623.

Herrero, M., Cifuentes, A., & Ibañez, E. (2006). Sub- and supercritical fluid extraction of functional ingredients from different natural sources: plants, food-by-products, algae and microalgae: a review. *Food Chemistry, 98*, 136–148.

Hexsel, D., & Soirefmann, M. (2011). Cosmeceuticals for cellulite. *Seminary Cutaneous and Medecine Surgery, 30*, 167–170.

Hirschfeld, D. R., Fenical, W., Lin, G. H. Y., Wing, R. M., Radlick, P., & Sims, J. J. (1973). Marine natural products. VIII. Pachydictyol A, an exceptional diterpene alcohol from the brown alga, *Pachydictyon coriaceum*. *Journal of American Chemical Society, 95*, 4049–4050.

Holdt, S. L., & Kraan, S. (2011). Bioactive compounds in seaweed: functional food applications and legislation. *Journal of Applied Phycology, 23*, 543–597.

Hoyer, K., Karsten, U., Sawall, T., & Wiencke, C. (2001). Photoprotective substances in Antarctic macroalgae and their variation with respect to depth distribution, different tissues and developmental stages. *Marine Ecology Progress Series, 211*, 117–129.

Hu, J. F., Gen, M. Y., Zhang, J. T., & Jiang, H. D. (2001). An in vitro study of the structure-activity relationships of sulfated polysaccharide from brown algae to its antioxidant effect. *Journal of Asian Natural Products Research, 3*, 353–358.

Huovinen, P., Matos, J., Sousa Pinto, I., & Figueroa, F. L. (2006). The role of ammonium in photoprotection against high irradiance in red alga *Grateloupia lanceola*. *Aquatic Botany, 84*, 308–316.

Hupel, M., Lecointre, C., Meudec, A., Poupart, N., & Ar Gall, E. (2011). Comparison of photoprotective responses to UV radiation in the brown seaweed *Pelvetia canaliculata* and the marine angiosperm *Salicornia ramosissima*. *Journal of Experimental Marine Biology and Ecology, 401*, 36–47.

Ioannou, E., & Roussis, V. (2009). *Natural products from seaweeds*. Springer Science.

Jang, W. S., & Choung, S. Y. (2013). Antiobesity effects of the ethanol extract of *Laminaria japonica* areshoung in high-fat-diet-induced obese rat. *Evidence-based Complementary and Alternative Medicine*. 17 pp. http://dx.doi.org/10.1155/2013/492807.

Kang, S. M., Kim, K. N., Lee, S. H., Ahn, G., Cha, S. H., Kim, A. D., et al. (2011a). Anti-inflammatory activity of polysaccharide purified from AMG-assistant extract of *Ecklonia cava* in LPS-stimulated RAW264.7 macrophages. *Carbohydrate Polymers, 85*, 80–85.

Kang, S. Il, Ko, H. C., Shin, H. S., Kim, H. M., Hong, Y. S., Lee, N. H., et al. (2011b). Fucoxanthin exerts differing effects on 3T3-L1 cells according to differentiation stage and inhibits glucose uptake in mature adipocytes. *Biochemical and Biophysical Research Communications, 409*, 769–774.

Karentz, D., McEuen, F. S., Land, M. C., & Dunlap, W. C. (1991). Survey of mycosporine-like amino acid compounds in Antarctic organisms: potential protection from ultraviolet exposure. *Marine Biology, 108*, 157–166.

Karsten, U., Sawall, T., & Wiencke, C. (1998). A survey of the distribution of UV-absorbing substances in tropical macroalgae. *Phycology Research, 46*, 271–279.

Karsten, U., & Wiencke, C. (1999). Factors controlling the formation of UV-absorbing mycosporine-like amino acids in the marine red alga *Palmaria palmata* from Spitsbergen (Norway). *Journal of Plant Physiology, 155*, 407–415.

Ko, S. C., Myoungsook, L., Ji-Hyeok, L., Seung-Hong, L., Yunsook, L., & Jeon, Y. J. (2013). Dieckol, a phlorotannin isolated from a brown seaweed, *Ecklonia cava*, inhibits adipogenesis through AMP-activated protein kinase (AMPK) activation in 3T3-L1 preadipocytes. *Environmental Toxicology and Pharmacology, 36*, 1253–1260.

König, G. M., & Wright, A. D. (1997). *Laurencia rigida*: chemical investigations of its antifouling dichloromethane extract. *Journal of Natural Products, 60*, 967–970.

König, G. M., Wright, A. D., & de Nys, R. (1999). Halogenated monoterpenes from *Plocamium costatum* and their biological activity. *Journal of Natural Products, 62*, 383–385.

Kornprobst, J. M. (2014). *Encyclopedia of marine natural products. 1. General aspects, microorganisms, algae and fungi* (2nd ed.). Wiley-Blackwell.

Kraan, S. (2012). Algal polysaccharides, novel applications and outlook. In *Carbohydrates, comprehensive studies on glycobiology and glycotechnology*. (chapter 22), 44 pp.

Kuppusamy, U. R., & Das, N. P. (1992). Effects of flavonoids on cyclic AMP phosphodiesterase and lipid mobilization in rat adipocytes. *Biochemical Pharmacology, 44*, 1307–1315.

Kuppusamy, U. R., & Das, N. P. (1994). Potentiation of β-adrenoceptor agonist-mediated lipolysis by quercetin and fisetin in isolated rat adipocytes. *Biochemical Pharmacology, 47*, 521–529.

Lee, S., Lee, Y. S., Jung, S. H., Kang, S. S., & Shin, K. H. (2003). Anti-oxidant activities of fucosterol from the marine algae *Pelvetia siliquosa*. *Archives of Pharmacal Research, 26*, 719–722.

Liu, C. L., Chiu, Y. T., & Hu, M. L. (2011). Fucoxanthin enhances HO-1 and NQO1 expression in murine hepatic BNL CL2 cells through activation of the Nrf2/ARE system partially by its pro-oxidant activity. *Journal of Agriculture and Food Chemistry, 59*, 11344–11351.

Llewellyn, C. A., & Airs, R. L. (2010). Distribution and abundance of MAAs in 33 species of microalgae across 13 classes. *Marine Drugs, 8*, 1273–1291.

Logeart, D., Letourneur, D., Jozefonvicz, J., & Kern, P. (1996). Collagen synthesis by vascular smooth muscle cells in the presence of antiproliferative polysaccharides. *Journal of Biomedical Biomaterials Research, 30*, 501–508.

Logeart, D., Prigent-Richard, S., Boisson-Vidal, C., Chaubet, F., Durand, P., Jozefonvicz, J., et al. (1997a). Fucans, sulfated polysaccharides extracted from brown seaweeds, inhibit vascular smooth muscle cell proliferation. II. Degradation and molecular weight effect. *European Journal of Cell Biology, 74*, 385–508.

Logeart, D., Prigent-Richard, S., Jozefonvicz, J., & Letourneur, D. (1997b). Fucans, sulfated polysaccharides extracted from brown seaweeds, inhibit vascular smooth muscle cell proliferation. I. Comparison with heparin for antiproliferative activity, binding and internalization. *European Journal of Cell Biology, 74*, 376–384.

Lotti, T., Ghersetich, I., Grappone, C., & Dini, G. (1990). Proteoglycans in so-called cellulite. *International Journal of Dermatology, 29*, 272–274.

Mabeau, S., & Fleurence, J. (1993). Seaweed in food products: biochemical and nutritional aspects. *Trends in Food Science and Technology, 4*, 103–107.

Macaya, E. C., Rothäusler, E., Thiel, M., Molis, M., & Wahl, M. (2005). Induction of defenses and within-alga variation of palatability in two brown algae from the northern-central coast of Chile: effects of mesograzers and UV radiation. *Journal of Experimental Marine Biology and Ecology, 325*, 214–227.

Majmudar, G. (2012). *Compositions of marine botanicals to provide nutrition to aging and environmentally damaged skin*. US 8318178 B2.

Ma, W., Wlaschek, M., Tantcheva-Poor, I., Schbeider, L. A., Naderi, L., Razi-Wolf, Z., et al. (2001). Chronological aging and photoaging of the fibroblast and the dermal connective tissue. *Clinical and Experimental Dermatology, 26*, 592–599.

Mayer, A. M. S. (2002). Current marine pharmacology contributions to new drug development in the biopharmaceutical industry. *Pharmaceutical News, 9*, 479–482.

Mayer, A. M. S., & Hamann, M. T. (2004). Marine pharmacology in 2000: marine compounds with antibacterial, anticoagulant, antifungal, antiinflammatory, antimalarial, antiplatelet, antituberculosis, and antiviral activities; affecting the cardiovascular, immune, and nervous systems and other miscellaneous mechanisms of action. *Marine Biotechnology, 6*, 37–52.

Mayer, A. M. S., Rodríguez, A. D., Berlinck, R. G. S., & Hamann, M. T. (2007). Marine pharmacology in 2003–4: marine compounds with anthelmintic antibacterial, anticoagulant, antifungal, antiinflammatory, antimalarial, antiplatelet, antiprotozoal, antituberculosis, and antiviral activities; affecting the cardiovascular, immune and nervous systems, and other miscellaneous mechanisms of action. *Comparative Biochemistry and Physiology C: Toxicology and Pharmacology, 145*, 553–581.

Mikami, K., & Hosokawa, M. (2013). Biosynthetic pathway and health benefits of fucoxanthin, an algae-specific xanthophyll in brown seaweeds. *International Journal of Molecular Sciences, 14*, 13763–13781.

Mizutani, S., Deguchi, S., Kobayashi, E., Nishiyama, E., Sagawa, H., & Kato, I. (2010). *Fucoidan containing cosmetics*. US 7678368 B2.

Nakamura, T., Nagayama, K., Uchida, K., & Tanaka, R. (1996). Antioxydant activity of phlorotannins isolated from the brown alga *Eisenia bicyclis*. *Fisheries Science, 62*, 923–926.

Okai, Y., Higashi-Okai, K., Yano, Y., & Otani, S. (1996). Identification of antimutagenic substances in extracts of edible red alga, *Porphyra tenera* (Asadusa-nori). *Cancer Letters, 100*, 235–240.

Park, P. J., Shahidi, F., & Jeon, Y. J. (2004). Antioxidant activities of enzymatic extracts from an edible seaweed *Sargassum horneri* using ESR spectrometry. *Journal of Food Lipids, 11*, 15–27.

Parvez, S., Kang, M., Chung, H. S., Cho, C., Hong, M. C., Shin, M. K., et al. (2006). Survey and mechanism of skin depigmenting and lightening agents. *Phytotherapy Research, 20*, 921–934.

Peinado, N. K., Diaz, R. T. A., Figueroa, F. L., & Helbling, E. W. (2004). Ammonium and UV radiation stimulate the accumulation of mycosporine-like amino acids in *Porphyra columbina* (Rhodophyta) from Patagonia, Argentina. *Journal of Phycology, 40*, 248–259.

Pessoa, M. F. (2012). Algae and aquatic macrophytes responses to cope to ultraviolet radiation – a review. *Emirates Journal of Food and Agriculture, 24*, 527–545.

Potin, P., Bouarab, K., Salaun, J. P., Pohnert, G., & Kloareg, B. (2002). Biotic interactions of marine algae. *Current Opinion in Plant Biology, 5*, 308–317.

Rastogi, R. P., Richa, Sinha, R. P., Singh, S. P., & Häder, D. P. (2010). Photoprotective compounds from marine organisms. *Journal of Industrial Microbiology and Biotechnology, 37*, 537–558.

Reef, R., Kaniewska, P., & Hoegh-Guldberg, O. (2009). Coral skeletons defend against ultraviolet radiation. *PLoS One, 4*, e7995.

Richa Rastogi, R. P., Kumari, S., Singh, K. L., Kannaujiya, V. K., Singh, G., Kesheri, M., et al. (2011). Biotechnological potential of mycosporine-like amino acids and phycobiliproteins of Cyanobacterial origin. *Biotechnology Bioinformatics and Bioengineering, 1*, 159–171.

Rioux, L. E., Moulin, V., Beaulieu, M., & Turgeon, S. L. (2013). Human skin fibroblast response is differentially regulated by galactofucan and low molecular weight galactofucan. *Bioactive Carbohydrates and Dietary Fibre, 1*, 105–110.

Rocha, H. A. O., Moraes, F. A., Trindade, E. S., Franco, C. R. C., Torquato, R. J. S., Veiga, S. S., et al. (2005). Structural and haemostatic activities of a sulfated galactofucan from the brown alga *Spatoglossum schroederi*. An ideal antithrombotic agent? *Journal of Biological Chemistry, 280*, 41278–41288.

Rodríguez-Bernaldo de Quirós, A., Lage-Yusty, M. A., & López-Hernandez, J. (2010). Determination of phenolic compounds in macroalgae for human consumption. *Food Chemistry, 21*, 634–638.

Rosenbaum, M., Prieto, V., Hellmer, J., Boschmann, M., Krueger, J., Leibel, R. L., et al. (1998). An exploratory investigation of the morphology and biochemistry of cellulite. *Plastic Reconstructive Surgery, 101*, 1934–1939.

Rosic, N. (2012). Phylogenic analysis of genes involved in mycosporine-like amino acid biosynthesis in symbiotic dinoflagellates. *Applied Microbiology and Biotechnology, 94*, 29–37.

Rozkin, Mla, Levina, M. N., Efimov, V. S., & Usov, A. I. (1991). The anticoagulant and lipolysis-stimulating activity of polysaccharides from marine brown algae. *Farmakology Toksicology, 54*, 40–42.

Ruckstuhl, M., Beretz, A., Anton, R., & Landry, Y. (1979). Flavonoids are selective cyclic GMP phosphodiesterase inhibitors. *Biochemical Pharmacology, 28*, 535–538.

Ruiz-Ruiz, F., Benavides, J., & Rito-Palomares, M. (2013). Scaling-up of a B-phycoerythrin production and purification bioprocess involving aqueous two-phase systems: practical experiences. *Process Biochemistry, 48*, 738–745.

Rupérez, P., Ahrazem, O., & Leal, J. A. (2002). Potential antioxydant capacity of sulfated polysaccharides from the edible marine brown seaweed *Fucus vesiculosus*. *Journal of Agriculture and Food Chemistry, 50*, 840–845.

Ryu, B., Qian, Z. J., Kim, M. M., Nam, K. W., & Kim, S. K. (2009). Anti-photoaging activity and inhibition of matrix metalloproteinase (MMP) by marine red alga, *Corallina pilulifera* methanol extract. *Radiation Physics and Chemistry, 78*, 98–105.

Sabudak, T., Demirkiran, O., Ozturk, M., & Topcu, G. (2013). Phenolic compounds from *Trifolium echinatum Bieb.* and investigation of their tyrosinase inhibitory and antioxidant activities. *Phytochemistry, 96*, 305–311.

Samarakoon, K., & Jeon, Y. J. (2012). Bio-functionalities of proteins derived from marine algae – a review. *Food Research International, 48*, 948–960.

Sánchez-Machado, D. I., López-Hernández, J., Paseiro-Losada, P., & López-Cervantes, J. (2004). An HPLC method for the quantification of sterols in edible seaweeds. *Biomedical Chromatography, 18*, 183–190.

Sánchez, M. G., Korbee, N., Pérez-Rufaza, I. M., Marcos, C. B., Dominguez, F. L., Figueroa, A., et al. (2012). Physiological response and photoacclimatation capacity of *Caulerpa prolifera* (Forsskål) J.V. Lamouroux and *Cymodocea nodosa* (Ucria) Ascherson meadows in the Mar Menor lagoon (SE Spain). *Marine Environmental Research, 79*, 37–47.

Schubert, N., Garcia-Mendoza, E., & Pacheco-Ruize, I. (2006). Carotenoid composition of marine red algae. *Journal of Phycology, 42*, 1208–1216.

Senni, K., Gueniche, F., Foucault-Bertaud, A., Igondjo-Tchen, S., Fioretti, F., Colliec-Jouault, S., et al. (2006). Fucoidan a sulfated polysaccharide from brown algae is a potent modulator of connective tissue proteolysis. *Archives of Biochemistry and Biophysics, 445*, 56–64.

Senni, K., Pellat, B., Gogly, B., Blondin, C., Jozefonvicz, J., Letourneur, D., et al. (2003). European Patent 1039916 (WO9932099).

Shang, Y. F., Kim, S. M., Lee, W. J., & Um, B. H. (2011). Pressurized liquid method for fucoxanthin extraction from *Eisenia bicyclis* (Kjellman) Setchell. *Journal of Bioscience and Bioengineering, 111*, 237–241.

Silkina, A., Bazes, A., Mouget, J. L., & Bourgougnon, N. (2012). Comparative efficiency of macroalgal extracts and booster biocides as antifouling agents to control growth of three Diatom species. *Marine Pollution Bulletin, 64*, 2039–2046.

Sinha, R. P., Klisch, M., Gröniger, A., & Häder, D. P. (1998). Ultraviolet-absorbing/screening substances in cyanobacteria, phytoplankton and macroalgae. *Journal of Photochemistry and Photobiology B: Biology, 47*, 83–94.

Sinha, R. P., Klisch, M., Gröniger, A., & Häder, D. P. (2000). Mycosporine-like amino acids in the marine red alga *Gracilaria cornea* – effects of UV and heat. *Environmental and Experimental Botany, 43*, 33–43.

Siriwardhana, N., Kim, K. N., Lee, K. W., Kim, S. H., Ha, J. H., Song, C. B., et al. (2008). Optimisation of hydrophilic antioxidant extraction from *Hizika fusiformis* by intergrating treatments of enzymes, heat and pH control. *International Journal of Food Science and Technology*, 587–596.

Song, T.Y., Chen, C. H., Yang, N. C., Fu, C. S., Chang, Y.T., & Chen, C. L. (2009). The correlation of in vitro mushroom tyrosinase activity with cellular tyrosinase activity and melanin formation in melanoma cells A2058. *Journal of Food and Drug Analysis*, *17*, 156–162.

Srilaong, V., Aiamla-or, S., Soontornwat, A., Shigyo, M., & Yamauchi, N. (2011). UV-B irradiation retards chlorophyll degradation in lime (*Citrus latifolia Tan.*) fruit. *Postharvest Biology and Technology*, *59*, 110–112.

Stengel, D. B., Connan, S., & Popper, Z. A. (2011). Algal chemodiversity and biodiversity: source of natural variability and implications for commercial application. *Biotechnology Advances*, *29*, 483–501.

Talarico, L., & Maranzana, G. (2000). Light and adaptive responses in red macroalgae: an overview. *Journal of Photochemistry and Photobiology B: Biology*, *56*, 1–11.

Tarakhovskaya, E. R., Maslov, Y. I., & Shishova, M. F. (2007). Phytohormones in algae. *Russian Journal of Plant Physiology*, *54*, 163–170.

Thomas, N.V., & Kim, S. K. (2013). Beneficial effects of marine algal compounds in cosmeceuticals. *Marine Drugs*, *11*, 146–164.

Torres, A., Hochberg, M., Pergament, I., Smoum, R., Niddam, V., Dembitsky, V. M., et al. (2004). A new UV absorbing mycosporine with photo protective activity from the lichenized ascomycete *Collema cristatum*. *European Journal of Biochemistry*, *271*, 780–784.

Tsujino, I., Yade, K., & Sekikawa, I. (1980). Isolation and structure of a new amino acid, Shinorine, from the red alga *Chondrus yendoi*, Yamada et Mikami. *Botanica Marina*, *23*, 65–68.

Volkman, M., & Gorbushina, A. A. (2006). A broadly applicable method for extraction and characterisation of mycosporines and mycosporine-like amino acids of terrestrial, marine and freshwater origin. *Federation of European Microbiological Societies Letters*, *255*, 286–295.

Wang, T., Jonsdottir, R., Kristinsson, H. G., Hreggvidsson, G. O., Jonsson, J. O., Thorkelsson, G., et al. (2010). Enzyme-enhanced extraction of antioxidant ingredients from red algae *Palmaria palmata*. *Food Science and Technology*, *43*, 1387–1393.

Wang, R., Paul, V. J., & Luesch, H. (2013). Seaweed extracts and unsaturated fatty acid constituents from the green alga *Ulva lactuca* as activators of the cytoprotective Nrf2-ARE pathway. *Free Radical Biology and Medecine*, *57*, 141–153.

Wang, G. C., Sun, H. B., Fan, X., & Tseng, C. K. (2002). Large-scale isolation and purification of R-phycoerythrin from red alga *Palmaria palmata* using the expanded bed adsorption method. *Acta Botanica Sinica*, *44*, 541–546.

Watabe, A., Sugawara, T., Kikuchi, K., Yamasaki, K., Sakai, S., & Aiba, S. (2013). Sweat constitutes several natural moisturizing factors, lactate, urea, sodium, and potassium. *Journal of Dermatological Science*, *72*, 177–182.

Whitehead, K., & Hedges, J. I. (2005). Photodegradation and photosensitization of mycosporine-like amino acids. *Journal of Photochemistry and Photobiology*, *80*, 115–121.

Wijesinghe, W. A. J.P., & Jeon, Y. J. (2012a). Enzyme-assistant extraction (EAE) of bioactive components: a useful approach for recovery of industrially important metabolites from seaweeds: a review. *Fitoterapia*, *83*, 6–12.

Wijesinghe, W. A. J.P., & Jeon, Y. J. (2012b). Biological activities and potential industrial applications of fucose rich sulfated polysaccharides and fucoidans isolated from brown seaweeds: a review. *Carbohydrate Polymers*, *88*, 13–20.

Xiao, X. H., Yuan, Z. Q., & Li, G. K. (2013). Preparation of phytosterols and phytol from edible marine algae by microwave-assisted extraction and high-speed counter-current chromatography. *Separation and Purification Technology*, *104*, 284–289.

Yang, B., Chen, W., Jin, J., Mei, Y., Wang, X., Chen, H., et al. (2013). Evaluation of hypoglycemic and oxidative effects of synthesized peptide MC 62. *Chemical Biology and Drug Design*, *82*, 99–105.

Yang, Y., Liu, D., Wu, J., Chen, Y., & Wang, S. (2011). In vitro antioxidant activities of sulfated polysaccharide fractions extracted from *Corallina officinalis*. *International Journal of Biological Macromolecules*, *49*, 1031–1037.

Yuan, Y. V., Westcott, N. D., Hu, C., & Kitts, D. D. (2009). Mycosporine-like amino acid composition of edible red alga *Palmaria palmata* (dulse) harvested from the west and east coasts of Grand Manan Island, New Brunswick. *Food Chemistry*, *112*, 321–328.

Zhang, Q., Li, N., Liu, X., Zhao, Z., Li, Z., & Xu, Z. (2004). The structure of a sulfated galactan from *Porphyra haitanensis* and its in vivo antioxidant activity. *Carbohydrate Research*, *339*, 105–111.

Zheng, J., Piao, M. J., Keum, Y. S., Kim, H. S., & Hyun, J. W. (2013). Fucoxanthin protects cultured human keratinocytes against oxidative stress by blocking free radicals and inhibiting apoptosis. *Biomolecules and Therapeutics*, *21*, 270–276.

CHAPTER THIRTEEN

Phlorotannins in Sargassaceae Species from Brittany (France): Interesting Molecules for Ecophysiological and Valorisation Purposes

Valérie Stiger-Pouvreau*,[1], Camille Jégou†, Stéphane Cérantola‡, Fabienne Guérard* and Klervi Le Lann*

*UMR 6539 CNRS UBO IRD IFREMER, LEMAR-IUEM-UBO, UEB, Technopôle Brest-Iroise, Plouzane, France

†EA 3882 LUBEM-UBO, Quimper, France

‡Service RMN-RPE, Université de Bretagne Occidentale, UFR Sciences et Techniques, Brest, France

[1]Corresponding author: e-mail address: valerie.stiger@univ-brest.fr

Contents

Advances in Botanical Research, Volume 71
ISSN 0065-2296
http://dx.doi.org/10.1016/B978-0-12-408062-1.00013-5

Abstract

Phlorotannins are metabolites synthesised by brown algae to protect against environmental stresses. Only a few studies presented structural elucidation of native molecules extracted from macroalgae, contrary to previous published studies, which characterised acetylated phlorotannins. The present work introduces quantitative and qualitative studies to characterise phlorotannins from the eight species of Sargassaceae settled in Brittany, using a colorimetric method used for quantification of total phenolic content (TPC), and one- and two-dimensional nuclear magnetic resonance (NMR) (^{1}H, heteronuclear multiple bond correlation) together with *in vivo* NMR (high-resolution magic-angle spinning (HR-MAS) NMR) analyses, with the aim to elucidate structural and fingerprint of phlorotannins signals. *Halidrys siliquosa*, which settles in low tide rock pools, exhibited high TPC (17.77 ± 1.68% dry weight (DW)) while the other species showed lower TPC (ranged from 0.85 ± 0.11% DW for *Cystoseira baccata* growing in low tide rock pools to 5.53 ± 0.63% DW in *Cystoseira humilis* growing in high tide rock pools). No relation between TPC and the position of the algae along the shore could be highlighted. Fingerprints using NMR HR-MAS were useful to compare the eight species in terms of phenolic content and in terms of phenolic signals as the species from Brittany produced different phlorotannins. Our study demonstrated that *Cystoseira tamariscifolia* produced the monomer (phloroglucinol), *C. humilis*, phloroglucinol and a phlorethol, *C. baccata* and *Cystoseira nodicaulis* produced many compounds, i.e. traces of phloroglucinol together with fucols, phlorethols, and fucophlorethols and finally fuhalols were identified in *Bifurcaria bifurcata* and *H. siliquosa*. The putative ecophysiological roles of these phlorotannins from Sargassaceae are discussed, together with their potential bioactivities.

Abbreviations

CPE	Centrifugal partition extraction
EA	Ethyl acetate
HMBC	Heteronuclear multiple bond correlation
HR-MAS	High-resolution magic-angle spinning
NMR	Nuclear magnetic resonance
PLE	Pressurised liquid extraction
SFE	Supercritical fluid extraction
SPE	Solid phase extraction
TPC	Total phenolic content

13.1 INTRODUCTION

13.1.1 Generalities

Among brown algae, the family Sargassaceae Kützing belongs to the class Phaeophyceae and the order Fucales, as the genera *Fucus, Ascophyllum, Pelvetia* well-known in Brittany, France. Only the most speciose and generic-rich family Sargassaceae shows a worldwide distribution including the tropics, but it is absent in Antarctica and southern South America (Draisma, Ballesteros, Rousseau, & Thibaut, 2010).

This family was described first by Kützing (1843) and knew many taxonomic rearrangements, as reviewed by Nizzamuddin (1962). Its delineation has been extensively modified since the 1990s with the emergence of polymerase chain reaction and the ability to isolate molecular markers from genomic DNA. Based on molecular markers, Rousseau, Leclerc, and de Reviers (1997) together with Horiguchi and Yoshida (1998) were the first authors to propose to merge Cystoseiraceae species within the family Sargassaceae, followed by Rousseau and de Reviers (1999). More recent works, based on morphological and molecular tools, confirmed the integration of the Cystoseiraceae taxonomical unit within the family Sargassaceae (Cho, Rousseau, de Reviers, & Boo, 2006; Phillips, Burrowes, Rousseau, de Reviers, & Saunders, 2008; Rousseau, Burrowes, Peters, Kuhlenkamp, & de Reviers, 2001). The few morphological characters with which to make delineations and their inconsistency among morphologically defined taxa have led to taxonomic confusions in the genera *Sargassum* and *Cystoseira* in many areas (see Bourgougnon & Stiger-Pouvreau, 2011 for a review).

Currently, 481 species are recognised within the family Sargassaceae, which is divided into more than 30 genera (Draisma et al., 2010; Guiry & Guiry, 2013). Among these numerous genera, the genus *Sargassum* is the most diversified, with more than 500 described species and 335 recognised species. *Sargassum* represents, as of 2014, the most species-rich brown algal genus of the marine macrophytes (Mattio & Payri, 2011), followed by the genera *Cystoseira, Cystophora* and *Turbinaria* with, respectively, 38, 28 and 22 recognised species; the other genera are represented by a small number of species varying from one to eight (Guiry & Guiry, 2013).

Species of Sargassaceae are distributed worldwide and some genera are especially well represented in tropical and intertropical regions, as the genera *Sargassum* and *Turbinaria*. In many coastal waters, some species dominate, forming dense submarine forests, an essential habitat for numerous marine species, in tropical (*Sargassum*: Mattio, Dirberg, Payri, & Andrefouet, 2008; Mattio, Payri, & Stiger-Pouvreau, 2008) and temperate (*Cystoseira* or kelp forests: Nizamuddin, 1962; Phillips, 1995; Steneck et al., 2002; Thibaut, Pinedo, Torras, & Ballesteros, 2005) waters.

Within this family, some species are economically important especially in Asian countries, as the edible *Sargassum fusiforme* (previously known as *Hizikia fusiformis*, Stiger, Horiguchi, Yoshida, Coleman, & Masuda, 2003) and many species are exploited by agro-food, textile, cosmetic and pharmaceutical industries. In some countries, some species of Sargassaceae are invasive, as *Sargassum muticum* in Europe (Critchley, 1983; Critchley, Farnham, Yoshida, & Norton, 1990; Incera, Olabarria, Troncoso, & Lopez, 2009; Le Lann & Stiger-Pouvreau, 2009; Le Lann, Connan, & Stiger-Pouvreau, 2012; Plouguerné et al., 2006; Tanniou, Vandanjon, et al., 2013), or abundant in some parts of tropical areas as *Sargassum* and *Turbinaria* species in French Polynesia (Le Lann, Connan, et al., 2012; Mattio, Dirberg, et al., 2008; Mattio, Payri, et al., 2008; Stiger & Payri, 1999a,b; Stiger & Payri, 2005). Natural products are particularly searched for in Sargassaceae, notably for potential application in cosmetic and pharmaceutical industries. However, all taxa have not been equally investigated from this point of view as the chemical properties vary from one species to another (Kornprobst, 2010, chap. 14; Prud'homme van Reine, 2002; Smit, 2004).

13.1.2 Sargassaceae in Brittany

In Brittany, eight species of Sargassaceae are present along the rocky shore: *Bifurcaria bifurcata, Cystoseira baccata, Cystoseira foeniculacea, Cystoseira humilis, Cystoseira nodicaulis, Cystoseira tamariscifolia, Halidrys siliquosa* and *S. muticum*. In the intertidal zone, these species are present in rock pools but *B. bifurcata, C. baccata, C. tamariscifolia, S. muticum* and *H. siliquosa* can form dense forests in the subtidal zone. In the intertidal zone, a pattern of distribution of these species, depending on the location of the pool along the shore has been introduced by Cabioc'h et al. (2014) and a bathymetric distribution inside some pools, analogous to the algal belts observed along the sea shore, has been evidenced at different level belts as Fucales along the intertidal zone (Jégou, 2011, Figure 13.1).

Among the eight species, the upper one is *C. humilis* present in upper rock pools of exposed and sheltered sites. Nevertheless, some other Sargassaceae species could be observed in high tide rock pools (Figure 13.1), as *S. muticum, C. foeniculacea* and *C. nodicaulis*. In midtide pools, the three last cited can co-occur with *C. baccata*. The diversity of Sargassaceae species is high at low tide levels, where *S. muticum, H. siliquosa, B. bifurcata, C. baccata* and *C. tamariscifolia* can be observed in a same area (Jégou, 2011, Figure 13.1).

Figure 13.1 Illustration of the eight species of Sargassaceae from Brittany (France); up first line, from left to right and following lines: *Bifurcaria bifurcata, Cystoseira baccata, C. foeniculacea, C. humilis, C. nodicaulis, C. tamariscifolia, Halidrys siliquosa* and *Sargassum muticum. Photo credits: Klervi.Le-Lann©Lemar-iuem-ubo for pictures 1 and 8; Camille. Jegou©Lubem-ubo for pictures 2,3,4,5,6 and Erwan.Amice©CNRS-iuem-ubo for picture 7.* (See the colour plate.)

13.1.3 Description of Species under Study

The different morphologies of the eight species, i.e. shape of ultimate axes of the species, have been drawn on Figure 13.1. Morphological character differences traditionally separating *Bifurcaria/Cystoseira/Halidrys* and *Sargassum* species are as follows: radial or alternate branching with thallus differentiation into apical and basal portions (*Bifurcaria/Cystoseira/Halidrys*) versus radial branching only (*Sargassum*) with negligible thallus differentiation. *B. bifurcata* presents a clonal growth and is characterised by a creeping, much branched rhizomatous base attached by many small suckerlike disks to the substratum, sending up numerous bifurcating cylindrical fronds (Le Lann, Connan, et al., 2012). Concerning the *Cystoseira* species studied in the present work, *C. foeniculacea* and *C. humilis* are highly morphologically related; both are cespitose, and moreover, the first one, *C. foeniculacea*, differing by its kind of ramification (in a plan), is not easy to distinguish from *C. humilis* when its axes are beginning to regenerate (Jégou, 2011). *C. nodicaulis* and *C. tamariscifolia* have in common particular cell structures, called iridescent bodies, as described in Jégou, Culioli, Kervarec, Simon, and Stiger-Pouvreau (2010). These structures are responsible for the purple-to-green colour of the thalli when immersed, often used as an identification criterion in taxonomy. *C. nodicaulis* is a non-cespitous species with characteristic tophules in adult plants, and ramifications in several plants bearing small epinous leaves, contrary to *C. tamariscifolia* which is an obvious epinous species (Cabioc'h et al., 2014). *C. baccata* is a non-cespitous species, constituted by a main axis with distichous and alternate ramifications which present characteristic large and visible aerocysts but the presence of these vesicles depends on the geographical location (Cabioc'h et al., 2014). *S. muticum* is characterised by a thallus composed of a fastening holdfast, one to several perennial short main axes ramified into branches called 'laterals' of several orders, which differentiate into fine leaves, spherical and short pedunculated vesicles (aerocysts) and in summertime into cylindrical receptacles (reproductive organs). Some confusion can be made between young stages of *S. muticum* and *C. baccata* or *C. foeniculacea* (Cabioc'h et al., 2014). *H. siliquosa* is characterised by a strong and flattened holdfast bearing pinnate ramifications that give a distinctly zigzag appearance to the thallus. Two types of ramification appear on the thallus: some permanent composed of long (silicate) aerocysts and some ephemerous in summertime, bearing receptacles. All Sargassaceae species encountered in Brittany are monoecious;

only one individual is present in the population, producing both female and male gametes.

13.1.4 Phylogenetic Relationships within Studied Sargassaceae Species

In the light of recent phylogenetic studies (Jégou et al., 2010; Rousseau et al., 2001; Rousseau et al., 1997; Rousseau & de Reviers, 1999), the genera *Bifurcaria*, *Cystoseira*, *Halidrys*, and *Sargassum* appeared polyphyletic and each should be split into two or more genera (Draisma et al., 2010). Other genera as *Turbinaria, Myagropsis*, or *Cystophora*, which belong to the family Sargassaceae are monophyletic genera. Regarding recent molecular works, Sargassaceae from Brittany are classified in several taxonomic clades as shown on Table 13.1, together with their geographical range distribution.

B. bifurcata belongs to the *Bifurcaria*-2, Group III following the classification proposed by Draisma et al. (2010). *C. foeniculacea* (Linnaeus) Greville and *C. humilis* belongs to the *Cystoseira*-4 (Amico et al., 1985; Draisma et al., 2010). *C. baccata* and *C. nodicaulis* are regrouped within a same clade named *Cystoseira*-6, as proposed by Draisma et al. (2010) who proposed the name of *Baccifer* for the genus. The last *Cystoseira* species in our study is *C. tamariscifolia*, which belongs to the clade *Cystoseira*-5. Finally, *H. siliquosa*, occurring along the coast of Europe from Portugal to northern Norway, is of the type of the genus *Halidrys*.

Table 13.1 Classification of Sargassaceae Settled in Brittany

Species	Clade	Geographical Distribution
Bifurcaria bifurcata	Bifurcaria-2	Atlantic Europe
Cystoseira baccata	Cystoseira-6	Mediterranean and northeast Atlantic
Cystoseira foeniculacea	Cystoseira-4	Mediterranean, northeast Atlantic, and Bermuda
Cystoseira humilis	Cystoseira-4	Mediterranean, northeast Atlantic, and Bermuda
Cystoseira nodicaulis	Cystoseira-6	Mediterranean and northeast Atlantic
Cystoseira tamariscifolia	Cystoseira-5	Mediterranean and northeast Atlantic
Halidrys siliquosa	Halydris	Atlantic and Arctic Europe
Sargassum muticum	Sargassum-1	Temperate waters

From Draisma et al., 2010.

13.1.5 Interests of this Family

This family is interesting at several levels, especially in terms of spatial distribution in Brittany. Moreover, the speciose genus *Cystoseira* has the potential to become a model for the *in situ* study of speciation in seaweeds as it is thought to be in the process of active speciation (Draisma et al., 2010). It has its centre of diversity in the Mediterranean Sea, where most of the occurring species are endemic (20 of 28 species), while the species of the remaining third extend their distribution into the northeast Atlantic Ocean.

For the present study, we collected at the Pointe de Penmarc'h (Brittany, France), the eight Sargassaceae species at different tide levels: *C. humilis* in high tide pools, *C. foeniculacea, C. nodicaulis* in midtide pools and finally the five other species, *S. muticum, H. siliquosa, B. bifurcata, C. baccata* and *C. tamariscifolia* in same low tide pools.

13.2 PHLOROTANNINS IN SARGASSACEAE

13.2.1 Structure and Size

Phlorotannins is a class of phenolic compounds only occurring in brown algae. They consist of oligomeric or polymeric compounds using phloroglucinol (1,3,5-trihydroxybenzene) as a basic unit (structure shown on Figure 13.2), linked to each other by various ways. Therefore, the size variation of phlorotannins is large, from 126 Da (free phloroglucinol) to 650 kDa (polymer of 4800 units), although the most commonly found are in the range of 10–100 kDa (Bourgougnon & Stiger-Pouvreau, 2011). According to their chemical structure, phlorotannins are classified in four groups, based on the type of linkage present within the molecule. The different units of phloroglucinol could be linked by only carbon–carbon, i.e. phenyl linkage as in fucols, by a diaryl ether linkage as in fuhalols and phlorethols, by both phenyl and ether linkages as in fucophlorethols and finally, a dibenzodioxin linkage can associate two phloroglucinol units as it is observed for eckols and carmalols (Figure 13.2). Most of these four types of compounds have halogenated or sulphated representatives in brown algae (Kornprobst, 2010, chap. 14).

Each phlorotannin class has been identified in Sargassaceae, as peracetylated compounds, like illustrated in Figure 13.2: trifuhalol in *H. siliquosa*, triphlorethol A in *Sargassum spinuligerum*, difucol in *Cystophora retroflexa*, fucophlorethol in *C. baccata*, eckol in *Myagropsis myagroides* and finally triphlorethohydroxycarmalol in *Carpophyllum maschalocarpum*. Nevertheless,

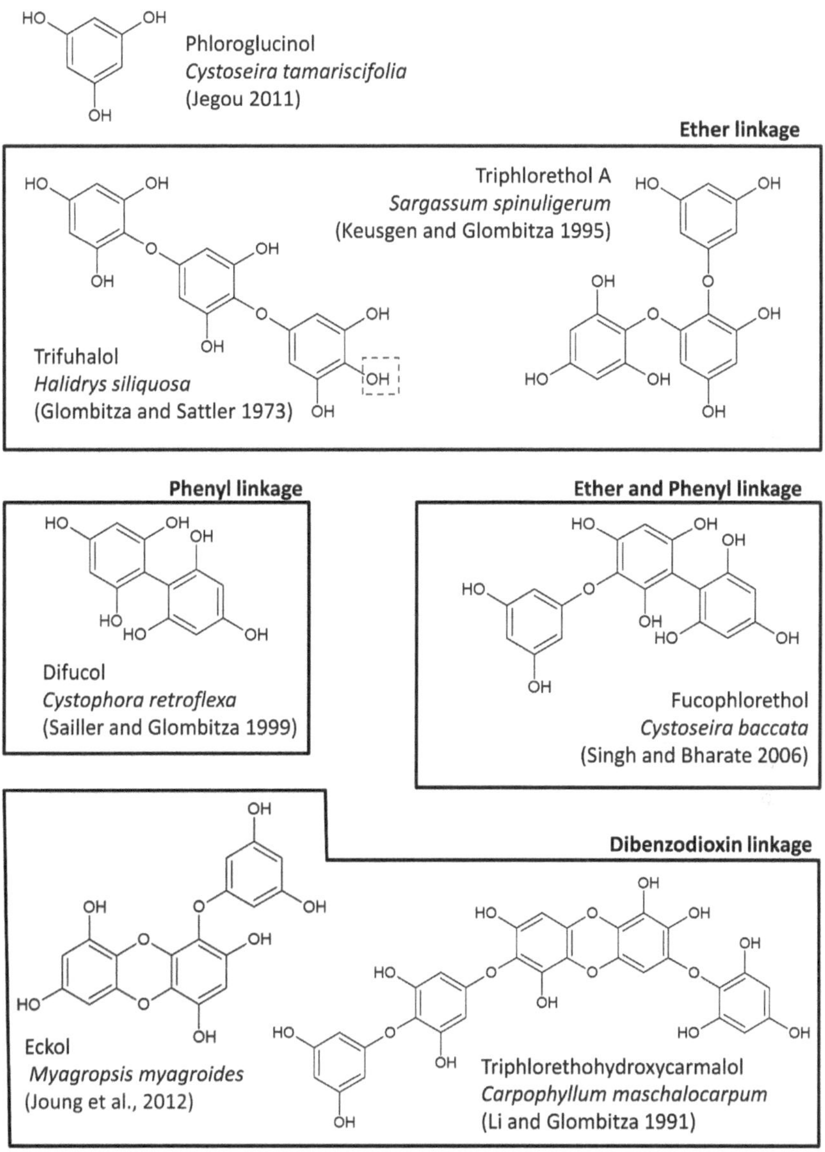

Figure 13.2 Examples of phlorotannins identified in Sargassaceae according to the type of linkage, i.e. phenyl linkage (fucols), diaryl ether linkage (fuhalols and phlorethols), both phenyl and ether linkages (fucophlorethols) and dibenzodioxin linkage (eckols and carmalols). Additional hydroxyl group in fuhalol is surrounded by dotted line.

neither eckol nor carmalol were actually identified in French Sargassaceae. Molecules can have a linear or polycyclic structure as shown on Figure 13.2. Because of some variation in the numbers of phloroglucinol units, the nature of the structural linkages between these units and the numbers of hydroxyl groups present, phlorotannins constitute an extremely heterogeneous group of compounds, which show various original chemical structures not found in terrestrial plants (Freile-Pelegrín & Robledo, 2013; Ragan & Glombitza, 1986; Singh & Bharate, 2006). Phlorotannins may amount up to nearly 20% of the dry weight (DW) in Fucales, although phenolic content in French Sargassaceae shows spatiotemporal variations according to both intrinsic and extrinsic factors (e.g. Bourgougnon & Stiger-Pouvreau, 2011; Connan, Delisle, Deslandes, & Ar Gall, 2006; Le Lann, Connan, et al., 2012; Le Lann, Ferret, et al., 2012; Plouguerné et al., 2006). They are both primary and secondary metabolites. Indeed, some of them are constituent of cell walls in Fucales, whereas the majority is found in specific components, named physodes (Amsler & Fairhead, 2005). As reviewed by these authors, phlorotannins have several putative ecological roles in brown algae usually associated with a chemical defence: protection against grazing, pathogen attack, epiphytism, microfouling and ultraviolet (UV) damages. In view of these ecological functions, various studies have explored the high potential of valorisation of these compounds (Bourgougnon & Stiger-Pouvreau, 2011). Indeed, as reviewed by Freile-Pelegrín and Robledo (2013) and Li, Wijesekara, Li, and Kim (2011), phlorotannins show various bioactivities such as antioxidant, antimicrobial and antitumoral. The interest of phlorotannins within Sargassaceae is discussed in Section 4.

13.2.2 Variations in Total Phenolic Content

Total phenolic contents (TPCs) were shown to vary also between and within species, and some species also show intrathallus variations (Connan, Deslandes, & Ar Gall, 2007; Connan, Goulard, Stiger, Deslandes, & Ar Gall, 2004; Ilvessalo & Tuomi, 1989; Le Lann, Connan, et al.,2012; Le Lann, Ferret, et al., 2012; Pavia & Aberg, 1996; Stiger, Deslandes, & Payri, 2004; Toth & Pavia, 2002).

Phenolic content in Phaeophyceae is governed by both intrinsic and extrinsic factors. The source of variability of TPC from brown macroalgae is then high and is visible at different scales of space (Le Lann, Ferret, et al., 2012; Steinberg, 1989; Targett, Coen, Boettcher, & Tanner, 1992; Van Alstyne, McCarthy, Hustead, & Kearns, 1999a) and time (Rönnberg & Ruokolahti, 1986; Steinberg, 1995; Stiger et al., 2004; Connan et al., 2004;

Connan et al., 2006; Plouguerné et al., 2006). TPC can vary with life history stage (Stiger et al., 2004; Van Alstyne, Whitman, & Ehlig, 2001), plant size (Denton, Chapman, & Markham, 1990; Plouguerné, 2006; Stiger et al., 2004) and tissue age (Pedersen, 1984), as well as maturity of the thallus (Le Lann & Stiger-Pouvreau, 2009; Plouguerné et al., 2006; Stiger et al., 2004; Van Alstyne, McCarthy, Hustead, & Kearns, 1999b). Toth and Pavia (2002) demonstrated that meristematic and cortical tissues of brown seaweeds usually contain more phenolic compounds than the cells of medulla; this is due to the differential abundance of physodes, i.e. cellular vacuoles accumulating mainly phenolic compounds (Ragan & Glombitza, 1986). Various environmental factors such as season (Connan et al., 2004; Le Lann, 2009; Plouguerné et al., 2006), light (Abdala-Díaz, Cabello-Pasini, Pérez-Rodríguez, CondeÁlvarez, & Figueroa, 2006; Plouguerné et al., 2006) and nutrient concentration (Yates & Peckol, 1993) can be responsible for variations in the TPC of brown macroalgae. In the present study we present interspecific variations in TPC for eight Sargassaceae species collected in June 2011 on a single site (Penmarc'h, France) but at different tidal levels.

13.2.3 Qualitative Characterisation Using *In vivo* ^{1}H High-resolution magic-angle spinning Nuclear magnetic resonance

In this study, *in vivo* ^{1}H high-resolution magic-angle spinning (HR-MAS) nuclear magnetic resonance (NMR) was used to observe the global chemical profile of phlorotannins in each taxon. This technique was already used to discriminate not only species of *Turbinaria* (Le Lann, Kervarec, Payri, Deslandes, & Stiger-Pouvreau, 2008) and *Cystoseira* (Jégou et al., 2010), but also diverse types of organisms such as bacteria (Salaün et al., 2010) or higher plants (e.g. Choze et al., 2013; Mucci, Parenti, Righi, & Schenetti, 2013).

This procedure is very simple; an algal fragment (around 5 mg) is placed in a 4 mm zirconium oxide MAS rotor. Approximately 30 µl of D_2O was added into the rotor with the algal sample for ^{2}H field locking. The analysis is made on a DRX 500 spectrometer (Bruker BioSpin, Wissembourg, France) equipped with an indirect HR-MAS ^{1}H/^{31}P probe head with gradient Z at 25 °C. A typical proton (^{1}H) HR-MAS NMR spectrum consisting of 64 scans was performed with presaturation of the water peak. Each spectrum was phased and baseline corrected using a polynomial function.

The sample is placed in a rotor spinning around an axis, which is oriented at the so-called magic angle of 54.7° with respect to the magnetic field B_0. Best homogenisation was obtained at a spinning rate of 5000 Hz. This resulted in a high-resolution NMR spectrum approaching the ones obtained with liquid samples, making spectra analysis possible. Furthermore, this technique is robust in the face to spatiotemporal variations (Le Lann, 2009). Indeed, the inclusion of samples coming from different sites, collected at different periods of the year, was carried out on *Cystoseira* and *Turbinaria* species and no intraspecific variability of the fingerprint was noted (Jégou, 2011; Le Lann, 2009).

13.2.4 Extraction of Phlorotannins

First researches on the extraction and isolation of phlorotannins were based on the isolation of non-native compounds, i.e. after modification of molecules in crude extract. Indeed, polyphenolic compounds are not stable and the structural and conformational analyses are difficult due to the rapid oxidation and the nucleophile reactivity of these compounds. The pioneering work of Ragan and Jensen (1978) and Ragan and Glombitza (1986) remained unquestioned for a long time. As phlorotannins could be potentially oxidised when exposed to air during a sufficient time, researchers protected interesting functions in the molecule from air oxidation using acetic anhydride-pyridine in order to acetylate phlorotannins functions (Ragan & Glombitza, 1986). In the present study, we gave priority to the extraction and purification of native compounds from the eight Sargassaceae species from Brittany and did not protect these compounds from oxidation.

Procedures to extract phlorotannins are diversified. Conventional solid/liquid extraction is frequently used with several solvents, most commonly aqueous mixtures of ethanol or methanol (Ragan & Glombitza, 1986). Generally, phenolic compounds are extracted with pure polar solvents or with hydroalcoholic mixtures put at 40 °C for 3 h under rotary agitation and in reduced light. The samples are centrifuged and the organic solvent is evaporated at 40 °C under vacuum with a Rotavapor system (Breton, Cérantola, & Ar Gall, 2011; Connan et al., 2004; Le Lann, Connan, et al.,2012; Le Lann, Ferret, et al., 2012; Le Lann et al., 2008; Plouguerné et al., 2006; Stiger et al., 2004; Tanniou, Vandanjon, et al., 2013). This classical procedure is time-consuming, and often uses large volumes of toxic organic solvents. To remedy these disadvantages, several innovative processes exist as reviewed by several authors (Ibañez, Herrero, Mendiola, & Castro-Puyana, 2012; Plaza & Rodríguez-Meizoso, 2013; Plaza et al., 2010). Tanniou, Serrano Leon, et al., 2013 compared classical solid/liquid extraction with these innovative methods such as (1) supercritical fluid extraction (SFE), (2) centrifugal partition extraction (CPE) and (3) pressurised liquid extraction (PLE) in order to identify the most efficient process to extract phlorotannins from *S. muticum*.

1. SFE is based on the use of solvent (as ethanol) mixed at high pressure with supercritical CO_2 to obtain a compressed fluid mixture. The fluid is heated up before entering in the extraction cell, where was loaded the algal powder mixed with sea sand (Tanniou, Serrano Leon, et al., 2013). Nevertheless, as demonstrated by these authors, the fluid mixture, which presented a polarity close to the hexane:ethanol (88:12) mixture, was too apolar to allow an efficient extraction of phlorotannins from *S. muticum*.
2. CPE uses a biphasic system. The solvent (organic phase) is kept stationary inside a column by a centrifugal acceleration while the aqueous phase (algal powder suspension) is pumped through the stationary phase with the pistons of a pump. CPE allowed obtaining very good yields in concentrated phlorotannins (Tanniou, Serrano Leon, et al., 2013).
3. Finally, PLE using an accelerated solvent extraction system permits to reduce time processing and to control precisely each extraction condition (Ibañez et al., 2012; Tanniou, Serrano Leon, et al., 2013; Zubia, et al., 2009). Generally, algal powder is loaded into stainless steel extraction cell. During the static extraction time, the extraction cell is filled with the solvent and the pressure is kept at the desired level. After this stage, the cell and the tubing are rinsed using fresh extraction solvent. The system is purged using N_2 gas and afterwards, a crude extract is obtained (Tanniou, Serrano Leon, et al., 2013; Zubia et al., 2009). This technique uses high pressure to maintain the solvent in liquid state at temperatures higher than their normal boiling point (Ibañez et al., 2012). PLE afforded to extract phlorotannins from *S. muticum* with an optimal efficiency (Tanniou, Serrano Leon, et al., 2013). These innovative and alternative extraction procedures allow to gain in phlorotannin productivity and interestingly to use nontoxic solvents (water and/or ethanol) in a sustainable management of the environment.

13.2.5 Purification of Phlorotannins

Usually, classical extracts of phlorotannins contains also other compounds like lipidic compounds or carbohydrates. In order to obtain more concentrated extracts, to provide information on the chemical composition of phlorotannin content and/or to characterise the chemical structure of these compounds, purification steps are needed. Several methods could be used such as dialysis, liquid/liquid purification, solid phase extraction (SPE) and several chromatographic techniques.

A way to fractionate the phlorotannins is to use their difference of size as described by Le Lann, Connan, et al. (2012), Le Lann, Ferret, et al. (2012). Briefly, three successive dialysis steps (corresponding to three size classes of cuts) on molecular weight cutoff 2000 Da, 5000 Da and 12,000–14,000 Da cellulose membranes were carried out in the dark. This process is long and water- and time-consuming. Nevertheless, it permitted the separation of compounds in temperate (Le Lann, Ferret, et al., 2012) and tropical (Le Lann, Connan, et al., 2012) Sargassaceae. Ultrafiltration could be a good alternative to classical dialysis. Nevertheless, time to filtrate towards cartridge is long, especially for the separation of small compounds. A comparative study about both these techniques is in progress in our lab (Tanniou, 2014).

In the 1970s, the phlorotannin crude extracts were already purified by liquid/liquid procedure. Chloroform/water or hexane/water mixture allowed removing lipophilic compounds and ethyl acetate/water mixture permitted to concentrate phlorotannins in the ethyl acetate phase (Ragan & Glombitza, 1986). After these works, analytical improvements were developed by Cérantola, Breton, Gall, and Deslandes (2006), Koivikko, Loponen, Pihlaja, and Jormalainen (2007) and Koivikko, Eränen, Loponen and Jormalainen (2008), which made possible the isolation of pure or semipure phenolic compounds.

High pressurised liquid chromatography (HPLC) is another helpful tool to quantify phlorotannins. The procedure was described by Koivikko et al. (2007). Briefly, a normal phase column (LiChrospher Si 60, 250 × 4 mmi.d.; 5 μm; Merck, Germany) and a binary mobile phase system, composed of a dichloromethane:methanol:water:acetic acid (82:14:2:2, v/v) mixture and a methanol:water:acetic acid (96: 2: 2, v/v) mixture, were used to separate efficiently phlorotannins of analysed fraction or extract (see Koivikko et al., 2007 for detail).

Recently, in the framework of the research Phlorotan-ING project (GIS Europôle mer) an interesting liquid/liquid purification process was carried out as described in Figure 13.3, and adapted for diverse genera in brown algae in order to increase the efficiency of the procedure in several genera, i.e. *Cystoseira* by Jégou (2011), *Pelvetia* (Ar Gall, Lelchat, Hupel, Jégou, & Stiger-Pouvreau, 2014), together with *Ascophyllum, Pylaeilla* and *Sargassum* (Tanniou, 2014; this study).

Briefly, the purification procedure is divided into three steps (Figure 13.3). The first step was three dichloromethane rinse cycles to eliminate a majority of lipidic compounds and chlorophylls from the crude

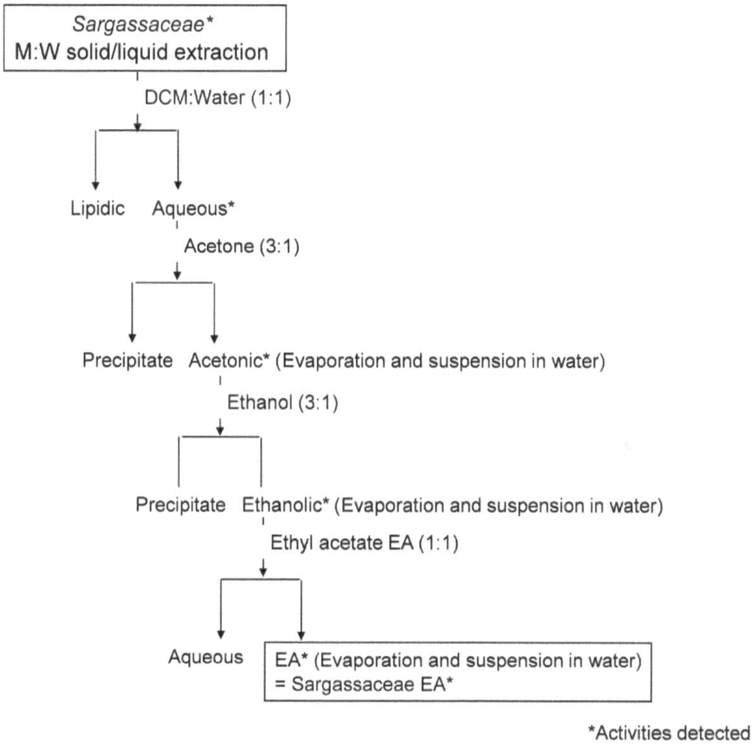

Figure 13.3 Procedure of liquid: liquid purification carried out on the crude extracts of each Sargassaceae species. M:W: methanol:water; DCM: dichloromethane.

extract. The second step consisted of rinsing the aqueous phase with acetone then ethanol to precipitate successively a majority of proteinic compounds and carbohydrates. The third step was three ethyl acetate rinses that permitted obtaining two semipurified fractions: the aqueous fraction and the ethyl acetate fraction (EAF). Before analysis, the EAF is evaporated at 40 °C under vacuum and dissolved with a small quantity of ethanol (not more than 10 ml) and the volume is brought to 100 ml with water. Then, this EAF could be washed with dichloromethane to eliminate the possible remaining of lipidic compounds.

SPE allowed fractionating crude extract according to a polarity gradient. Briefly, after conditioning with methanol and distilled water, the SPE cartridge, generally Strata C18-E 1000 mg/6 ml (Phenomenex, France), was loaded with the crude extract mixed beforehand with C18-silica. Fractionation was performed by stepwise elution with each of the following solvents: distilled water, 50% methanol, 100% methanol, dichloromethane:methanol

50:50% and 100% dichloromethane (Tanniou, Vandanjon, et al., 2013; Zubia et al., 2009). Other liquid chromatographic techniques such as low pressurised liquid chromatography or HPLC allowed fractionation of phlorotannins by their polarity (Jégou, 2011; Koivikko et al., 2007).

13.2.6 Structural Identification Using Two-dimensional NMR Analysis (Heteronuclear Multiple Bond Correlation)

First characterisations of Sargassaceae phlorotannins were carried out on non-native compounds in the 1970s (see Ragan & Glombitza, 1986 for a review). Among the useful techniques for elucidating structures of phlorotannins, researches used ^{1}H and ^{13}C NMR spectroscopy. It is always a favourite tool to determine chemical structure of compounds. Indeed, the global composition of crude extracts and fractions could be easily checked by ^{1}H NMR analyses. The aromatic protons of phlorotannins show characteristic chemical shift resonances between 5.5 and 6.5 ppm. The nature of the phlorotannins present in fractions was established thanks to heteronuclear multiple quantum coherence and heteronuclear multiple bond correlation (HMBC) experiments and comparison of the chemical shifts of the ^{1}H and ^{13}C resonances with literature data (Cérantola et al., 2006).

In the procedure presented here, we work on native compounds without any chemical modification and information obtained by two-dimensional (2D) NMR analysis (HMBC) allowed to furnish the class of phlorotannins (monomer, fuhalol, phlorethol, fucol, fucophlorethol, etc.) instead of giving information on the degree of heterogeneity of the total pool of the polyphenolics (Cérantola et al., 2006). As described in Cérantola et al. (2006), some specific areas, determined by their chemical shift (δ) are useful on HMBC analyses, which help determining the type of phlorotannins.

From HMBC analysis, main carbon atom resonances on spectra can be detailed:

1. δ = 95–105 ppm corresponding to methine groups, to quaternary carbons implicated in direct bonding between phloroglucinol units and to phenolic carbons for the two latter signals, respectively.
2. δ = 100 ppm signals for aryl–aryl carbons.
3. δ = 120–150 ppm signals for ether-linked phloroglucinol units and thus the occurrence of phlorethol-type units.
4. δ = 145–150 ppm signals for additional OH functions other than the 1,3,5 OH groups originally present in each phloroglucinol unit, thus the presence of fuhalol type units.

5. $\delta = 150$–165 ppm signals for phenolic carbons.
6. The relative intensity of signals at same δ (for example, signals at 160 or at 100 ppm) allows us to come up with a hypothesis concerning the form of the polymer (linear or branched).

The co-occurrence in the polymer of phloroglucinol units linked through aryl–aryl and diarylether bonds (fucol- and phlorethol-type units, respectively) combined with correlations between ^1H signals and ^{13}C signals for aryl–aryl and aryl-ether linkages, allowed one to conclude that this polymer belonged to the fucophlorethol class.

13.2.7 Quantification of Phlorotannins

Generally, TPC is determined by spectrophotometry with an adapted Folin–Ciocalteu assay known to be less affected by interfering compounds (see Sanoner, Guyot, Marnet, Molle, & Drilleau, 1999 for details). Interfering substances, however, are thought to account for less than 5% of the Folin–Ciocalteu-reactive compounds in brown seaweeds (Toth & Pavia, 2000). Briefly, for a procedure with microplates, the samples (20 µl) were mixed with 10 µl Folin–Ciocalteu reagent, 40 µl of 20% sodium carbonate solution and 130 µl of distilled water. Then, the mixture was allowed to stand at 70 °C in the dark for 10 min. After the production of a blue colour and shaking the microplates, the absorbance was read at 620 nm. Traditionally, TPCs are expressed in percentage phenolic compounds per gram dry weight (DW) of seaweed tissue (Le Lann, Connan, et al., 2012; Le Lann, Ferret, et al., 2012; Plouguerné et al., 2006; Tanniou, Serrano Leon, et al., 2013; Tanniou, Vandanjon, et al., 2013; Zubia et al., 2009; Zubia, Robledo, & Freile-Pelegrin, 2007).

13.3 QUANTITATIVE AND QUALITATIVE VARIABILITY OF PHLOROTANNINS IN SARGASSACEAE

13.3.1 TPC in French Sargassaceae

Figure 13.4 presents the variability of TPC within Sargassaceae from Brittany (France). TPC within Sargassaceae from Brittany presents significant interspecific variation ($p < 0.001$, Anova), with *H. siliquosa* producing high content of phenolic compounds (17.7 ± 1.68% DW) while *C. baccata* is a low producer of phenolic compounds (0.85 ± 0.11% DW).

The other species produced intermediate contents, *C. nodicaulis* has similar TPC to the one of *C. baccata* (0.85 ± 0.11% DW), the high tide rock pool species, *C. humilis* has an intermediate TPC (3.54 ± 0.399% DW), lower than *C. foeniculacea* (5.52 ± 0.63% DW) which is present in midtide

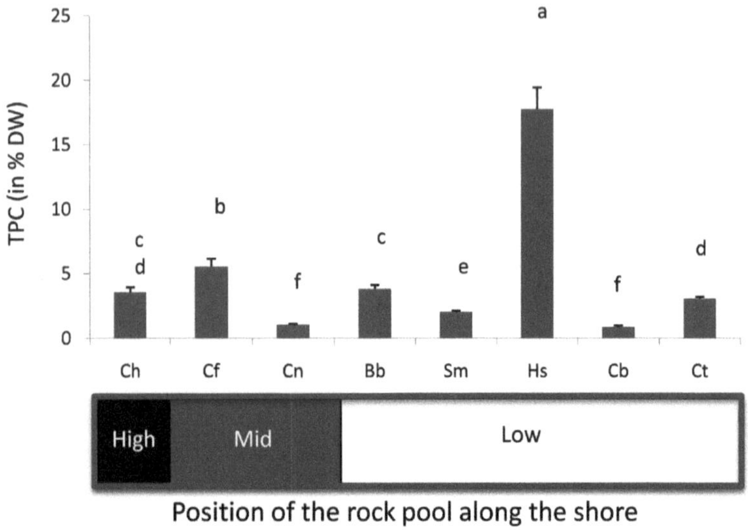

Figure 13.4 Total phenolic content (TPC) in the eight species of Sargassaceae from Brittany (France) in relation with their position on the rocky shore. Letters refer to statistical test (one-way Anova).

rock pools, and similar to the TPC from *C. tamariscifolia* which grows in lower parts of the rocky shore in the subtidal zone.

No relation between the position on the shore and TPC could be highlighted. Indeed, for species settling in a same pool (Cf/Cn and Bb/Cb/Ct/Hs/Sm) a great variability of TPC is noted (Figure 13.4). Others factors have to be taken into consideration to discuss our results, such as the state of maturity and length of individuals, for example (to appreciate the growth period). One should note interspecific differences of chemical defence within a same genus (*Cystoseira* in Figure 13.4) and more generally within Sargassaceae species within a same infralittoral rock pool (Figure 13.4). Our results demonstrated that at a same level within the seawater column, Sargassaceae species present different chemical strategies, probably in relation with their maturity period (allocation of energy).

13.3.2 Fingerprint of French Sargassaceae Using HR-MAS

HR-MAS NMR is useful in discriminating the Sargassaceae species in Brittany as shown in Figure 13.5. Focussing in the phlorotannins area, between 5.5 and 6.5 ppm, one should note the different profiles obtained for each Sargassaceae species from Brittany. The form of peaks associated with the aromatic area varies also in relation with the species. *C. tamariscifolia* presents an atypic

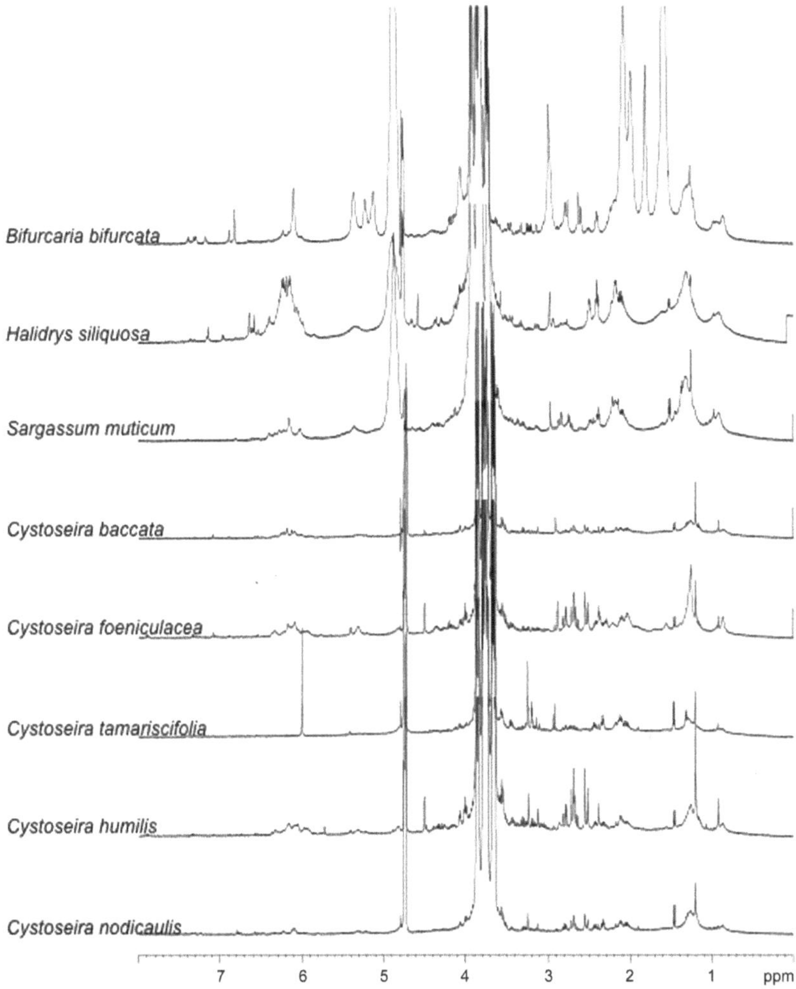

Figure 13.5 ^1H HRMAS NMR spectra obtained for the eight species of Sargassaceae from Brittany.

signal formed by a single peak while the other species present more massive peaks. *B. bifurcata* showed also a typical aromatic signal which is not seen in the others species. One should hypothesise that this signal represents meroditerpenoids compounds, as this kind of compounds, with a phenolic moiety, was described in *B. bifurcata* by Pellegrini, Piovetti, and Pellegrini (1997).

HR-MAS NMR is not a quantitative technique and permitted the obtaining of a chemical fingerprint (major compounds) of the eight Sargassaceae species; nevertheless, the spectral data can be compared between species. In this context, high signals of aromatic compounds observed for *H. siliquosa*, compared to the intense signals of other Sargassaceae species (Figure 13.5), showed that this species is probably the most active phlorotannin producer. This observation is confirmed by results previously presented on *H. siliquosa* TPC obtained by the colorimetric method of Folin–Ciocalteu (Figure 13.4).

Other parts of the spectrum are also useful to discriminate species, as the areas around 5 ppm characteristic to *B. bifurcata* (Figure 13.5), around 2 and 3 ppm which are different among the species, and peaks at 4.5 ppm which are present only in *C. humilis*, *C. foeniculacea* and *H. siliquosa* (Figure 13.5). From a general point of view, the lipidic compounds resonating at low frequencies are potential candidates to discriminate species of Sargassaceae in Brittany.

13.3.3 Identification of the Class of Phlorotannins Using HMBC

The diverse species of Sargassaceae from Brittany present, in their EAFs, different classes of phlorotannins (Figure 13.6), from the more simple (*C. tamariscifolia* on the left) to the most complex (*H. siliquosa* on the right). In this chapter, only HMBC analyses of *C. tamariscifolia, C. humilis, C. baccata* and *H. siliquosa* are presented.

In the EAF from *C. tamariscifolia*, only one major compound, identified as phloroglucinol, is characterised by a single signal at 5.78 ppm (^{1}H dimension) correlated to methine carbons (C-H from aromatic core at 95 ppm) and phenol carbons (C-OH) at 160 ppm in the ^{13}C dimension (Figure 13.6).

The EAF from *C. humilis* presents also the signal of phloroglucinol (singulet at 5.78 ppm on the ^{1}H dimension) and also other signals at 5.88 and 5.92 ppm. In observing information from both the ^{1}H and^{13}C dimensions, one should note a spot, which correlates the signal at 5.88 ppm and 95 ppm (which corresponds to methine carbons implicated in the aromatic core), and 160–163 ppm, with no aryl–aryl (absence of signal at 100 ppm) or diaryl-ether (absence of signal at 125–130 ppm). This compound looks like phloroglucinol but is different by the split of the spot C-OH near 160–163 ppm. The third compound was identified as a linear phlorethol. A diaryl-ether linkage between units was highlighted by the presence of spots at 125 ppm.

Phlorotannins in the EAF from *C. baccata* and *H. siliquosa* were shown to be more complex, with a broader distribution of the ^{1}H signals between 5.7 and 6.4 ppm, compared to the ones from *C. tamariscifolia* (one singulet) and *C. humilis* (Figure 13.6).

From observations of the HMBC obtained for the EAF of *C. baccata*, more compounds are present. Face to the signal at 5.98 ppm (^{1}H dimension) a large well-defined spot of correlation is present at 101 ppm (^{13}C dimension). Nevertheless, if no spot of correlation is present between 120 and 150 ppm then we can hypothesise the presence of a large linear

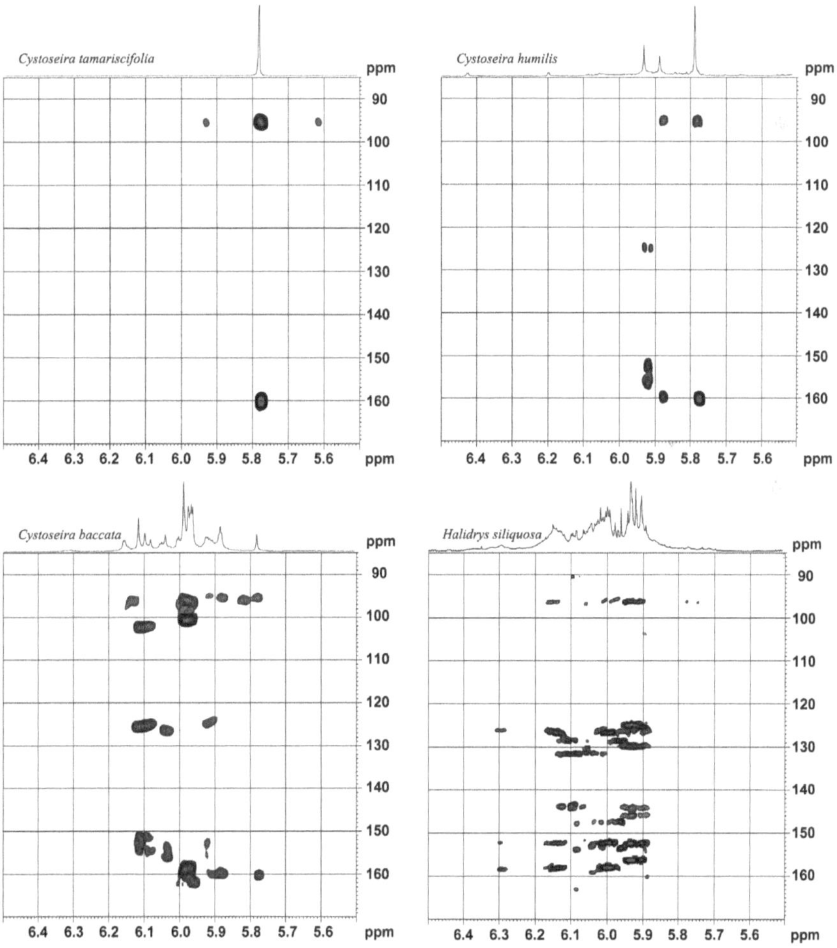

Figure 13.6 Class of the phlorotannins identified in the EA fractions from: up first line *Cystoseira tamariscifolia* (left), *C. humilis* (right), and second line *C. baccata* (left) and *Halidrys siliquosa* (right) (Fucales, Sargassaceae) from Brittany, established using heteronuclear multiple bond correlation (HMBC) analysis (^{1}H and ^{13}C NMR spectra).

fucol polymer in this fraction, constituted by phloroglucinol moieties linked via aryl bonds (Figure 13.6).

Following similar observations of presence/absence of spots at precise chemical shifts on both axis (^{1}H and ^{13}C), one should note the presence of phloroglucinol (δ = 5.78 ppm on the ^{1}H dimension) in small quantity compared to other compounds (low signal), a phlorethol (δ = 5.92 ppm), and a fucophlorethol (δ = 6.10 ppm).

Concerning the EAF of *H. siliquosa*, the complexity of the HMBC spectrum does not allow a detailed analysis, unlike for the other species. Nevertheless, the spectrum shows characteristic carbon atom resonances at (1) 96 ppm corresponding to methine groups, (2) between 123 and 132 ppm corresponding for ether-linked phloroglucinol units, (3) between 142 and 148 ppm for additional OH functions other than the 1,3,5 OH groups originally present in each phloroglucinol unit, and finally (4) between 151 and 159 ppm signals for phenolic carbons. Therefore, we can hypothesise the presence of fuhalol-type units. Following similar observations of presence/absence of spots at precise δ on both axis (^{1}H and ^{13}C), one should note the absence of phloroglucinol (δ = 5.78 ppm on the ^{1}H dimension).

Even if all spectra are not presented, we distinguished diverse compounds produced by the eight Sargassaceae species collected in June 2011 at Penmarc'h (Brittany, France):

1. *C. tamariscifolia* produced only the monomer phloroglucinol.
2. *C. humilis* produced both phloroglucinol and a linear phlorethol.
3. *C. baccata*, as *C. nodicaulis* (data not shown), produced some traces of phloroglucinol and more major compounds such as fucol, phlorethol and fucophlorethol.
4. *C. foeniculacea* produced phloroglucinol and more complex compounds (data not shown).
5. *S. muticum* produced only phlorethols (data not shown).
6. *B. bifurcata* produced mostly fuhalols (data not shown).
7. *H. siliquosa* produced mainly fuhalols.

The determination of the linkage within phlorotannin compounds was possible by the efficiency of the liquid:liquid purification procedure to remove contaminants (mannitol, fatty acids, proteins) and then to concentrate phlorotannins in the EAF. This purification procedure was used for others species belonging to red and green seaweeds and gave also the possibility to elucidate the structure of compounds using 2D NMR.

To complete the characterisation of phlorotannins, i.e. the number of phloroglucinol units, mass spectrometry could be useful as demonstrated in the literature (e.g. Nwosu et al., 2011; Steevensz et al., 2012; Tierney et al., 2013).

13.4 BIOLOGICAL ACTIVITIES OF PHLOROTANNINS

13.4.1 Ecophysiology

Brown macroalgae present TPC comprising between 1 and 10% DW as mentioned by Ragan and Glombitza (1986) but some species can be richer, especially some species from the orders Fucales and Dictyotales which contain high TPC, 20 and 30% DW observed, respectively, in some Fucales and Dictyotales species (Ragan & Glombitza, 1986; Targett, Boettcher, Targett, & Vrolijk, 1995).

Phlorotannins have various putative roles (see Amsler & Fairhead, 2005 for a review).

Phlorotannins play a primary role as integral structural components of cell walls in Fucales (Amsler & Fairhead, 2005; Schoenwaelder, 2002; Schoenwaelder & Clayton, 1998). The structural role of these compounds essentially distributed in peripheric cells, was demonstrated in several studies and indicates the contribution of apoplastic haloperoxidases (Potin & Leblanc, 2006, chap. 6) in numerous events, such as the cuticular protection, the reparation of lesions and mostly the bioadhesion of zygotes from brown macroalgae on the substrate. Phlorotannins liberated in the seawater (then soluble) are not polymerised and are normally secreted as halogenated monomers (Shibata, Hama, Miyasaki, Ito, & Nakamura, 2006). The roles and the variability of phlorotannins as primary and secondary metabolites, have been reviewed by Amsler and Fairhead (2005).

Moreover, some published reports show that the production of phenolic compounds on marine algae is usually associated with a chemical defence and is involved in various protection mechanisms, such as *e.g.*: against grazing (Fairhead, Amsler, McClintock, & Baker, 2006; Hemmi, Honkanen, & Jormalainen, 2004; Lüder & Clayton, 2004; Pavia & Toth, 2000; Stiger et al., 2004; Svensson, Pavia, & Toth, 2007), pathogen attack (Hay, 1996; Potin, Bouarab, Salaün, Pohnert, & Kloareg, 2002), epiphytism (Amsler & Fairhead, 2005; Brock, Aberg, & Pavia, 2001; Brock, Nylund, & Pavia, 2007), fouling (Fusetani, 2004; Maréchal & Hellio, 2009) and UV damages (Bjerke, Gwynn-Jones, & Callaghan, 2005; Henry & Van Alstyne, 2004; Fairhead et al., 2006; Swanson & Druehl, 2002). Moreover, these compounds exhibit antioxidative properties

(Le Lann, Jégou, & Stiger-Pouvreau, 2008; Nakai, Kageyama, Nakahara, & Miki, 2006) and are also involved in photoprotection mechanisms, particularly to counteract the cytotoxic effects of UV radiation (Swanson & Druehl, 2002).

13.4.2 Valorisation

In terms of valorisation, the potentials of phlorotannins are multiple (e.g. Freile-Pelegrín & Robledo, 2013; Li et al., 2011), even if their reactivity (instability, complexation with proteins, carbohydrates and others molecules) is problematic during extraction and purification steps. Nevertheless, recent works encouraged researchers in the exploration and isolation of bioactive compounds within phlorotannins. As an example, during the European project SeaHealth, (Zaragozá et al., 2008) demonstrated interesting activities (preventive and curative activities against atheroms of a hydroalcoholic extract from *Fucus*). Several antibacterial and antifungal (Hellio et al., 2001; Plouguerné, 2006; Sandsalen et al., 2003), antilarval and antialgal (Hellio et al., 2004; Ragan & Glombitza, 1986), and UV-protective (Connan, 2004; Le Lann et al., 2008) activities were highlighted for phlorotannins isolated from macroalgae. These compounds could also have some antidiabetic (Lee & Jeon, 2013), antimicrobial (Eom, Kim, & Kim, 2012), antioxidant, anticancer and radioprotective (Liu, Hansen, & Lin, 2011) together with anti-inflammatory effects (Kang, Eom, & Kim, 2013; Kim, Shin, et al., 2009). Moreover, studies have shown that phlorotannin derivatives increase alkaline phosphatase activity, mineralisation, total protein and collagen synthesis in human osteosarcoma cells (MG-63 cells) (Ali & Hasan, 2012; Ryu, Li, Qian, Kim, & Kim, 2009; Yeo, Jung, & Kim, 2012). These results suggest that phlorotannins could stimulate the osteoblast differentiation and regulate osteosarcoma differentiation, with an implication of phenolic compounds in the mitogen-activated protein kinase pathway (Ryu et al., 2009). Eckol-type compounds showed anti-HIV, antiallergenic, antiadipogenic and neuroprotective effects (see Li et al., 2011 for a review). In this way, they could have some therapeutic applications against Alzheimer disease (see, for example, Yoon, Chung, Kim, & Choi, 2008). From a pharmacological point of view, phlorotannin oligomers also present numerous properties. As an example, the phloroglucinol which is the base unit of phlorotannins is used in the drug Spasfon®, against human digestive troubles. Therefore, some reviews highlight the potential of phlorotannins in therapeutic application for human health (Li et al., 2011; Thomas & Kim, 2011).

Several studies showed the potential of phenolic extracts from Sargassaceae species (Table 13.2) and among the various bioactivities tested antioxidant seems to be the most frequently studied. Moreover, significant correlation was found between the total phenolic contents and the

Table 13.2 Review of Main Potential Biological Activities of Phlorotannins from Sargassaceae

Biological Activity	Species	References
Antioxidant	*Bifurcaria bifurcata*	Le Lann, Kervarec, et al., 2008; Jiménez-Escrig, Gómez-Ordóñez, & Rupérez, 2012; Zubia et al., 2009
	Cystoseira baccata	Jégou, 2011
	Cystoseira nodicaulis	
	Cystoseira tamariscifolia	Jégou, 2011; Zubia et al., 2009
	Halidrys siliquosa	Zubia et al., 2009
	Hizikia fusiformis	Siriwardhana et al., 2008; Siriwardhana, Lee, Jeon, Kim, & Haw, 2003
	Sargassum muticum	Le Lann, Kervarec, et al., 2008; Tanniou, Vandanjon, et al., 2013
	Sargassum ringgoldianum	Nakai et al., 2006
	Sargassum spinuligerum	Liu, Heinrich, Myers, & Dworjanyn, 2012
	Turbinaria ornata	Vijayabaskar & Shiyamala, 2012
	Sargassum kjellmanianum	Wei, Hu, & Xu, 2003
Antibacterial	*Sargassum hystrix*	Morales, Cantillo-Ciau, Sánchez-Molina, & Mena-Rejón, 2006
	Sargassum flipendula	
	Sargassum muticum	Tanniou, Vandanjon, et al., 2013
	Sargassum wightii	Vijayabaskar & Shiyamala, 2011
	Turbinaria ornata	
Antifungal	*Sargassum hystrix*	Morales et al., 2006
	Sargassum flipendula	
Anti-microalgae (anti-microfouling)	*Sargassum vulgare*	Plouguerné et al., 2010
Anti-macrofouling		
Antitumoral	*Bifurcaria bifurcata*	Zubia et al., 2009
	Cystoseira tamariscifolia	
	Halidrys siliquosa	

Continued

Table 13.2 Review of Main Potential Biological Activities of Phlorotannins from Sargassaceae—cont'd

Biological Activity	Species	References
Cytotoxic	*Sargassum* sp.	Mary, Vinotha, & Pradeep, 2012
	Sargassum thunbergii	Kim, Ham, et al., 2009
	Sargassum aquifolium	Le Lann, 2009
	Turbinaria conoides	
	Turbinaria decurrens	
	Turbinaria ornata	
Anti-inflammatory	*Sargassum aquifolium*	
	Sargassum polycystum	
	Turbinaria conoides	
	Turbinaria ornata	
	Myagropsis myagroides	Joung et al., 2012

antioxidant activities measured in phenolic extracts and/or fractions (e.g.: Jégou, 2011; Le Lann, Jégou, et al., 2008; Zubia et al., 2009).

In the study of Zubia et al. (2009), the finding of three species of the Sargassaceae family (i.e. *B. bifurcata*, *C. tamariscifolia* and *H. siliquosa*) among the four most active specimens regarding antioxidant and antitumoral activities, constituted more evidence of the bioactive potential of this group. Indeed, with a huge diversity of species (481 species, 26% of the Phaeophyceae species, Guiry & Guiry, 2013), Sargassaceae has elicited a high interest in a context of biomasses valorisation and study of phlorotannins of interest. But current direction of research on bioactivities of phlorotannins is clear. Now, it needs to identify which types of phlorotannins are responsible for the biological activity, as in many publications, only crude extracts are studied.

ACKNOWLEDGEMENTS

Authors are grateful to N. Kervarec and G. Simon for support in HR-MAS NMR analyses and for Figure 13.5. This work was supported by both the research project PHLOROTANING (GIS Europôle Mer) and the European Regional Development Fund (ERDF) – Atlantic Area Programme, MARMED project nr. 2011-1/164.

REFERENCES

Abdala-Díaz, R.T., Cabello-Pasini, A., Pérez-Rodríguez, E., Conde Álvarez, R.M., & Figueroa, F. L. (2006). Daily and seasonal variations of optimum quantum yield and phenolic compounds in *Cystoseira tamariscifolia* (Phaeophyta). *Journal of Phycology, 148*, 459–465.

Ali, T. F., & Hasan, T. (2012). Phlorotannin-incorporated mesenchymal stem cells and their promising role in osteogenesis imperfecta. *Journal of Medical Hypotheses and Ideas, 6*, 85–89.

Amico, V., Giaccone, G., Colombo, P., Colonna, P., Mannino, A. M., & Randazzo, R. (1985). Un nuovo approccio allo studio della sistematica del genere *Cystoseira* C. Agardh (Phaeophyta, Fucales). *Bulletin of the Sessions of the Academy of Natural Science in Catania, 18*, 887–986.

Amsler, C. D., & Fairhead, V. A. (2005). Defensive and sensory chemical ecology of brown algae. In J. A. Callow (Ed.), *Advances in botanical research* (pp. 1–91). Academic Press.

Ar Gall, E., Lelchat, F., Hupel, M., Jégou, C. Stiger-Pouvreau, V. Extraction and purification of phenols (phlorotannins) from brown algae. In D. Stengel, S. Connan (Eds.) *Natural products from marine algae: Methods and protocols.* (in press).

Bjerke, J. W., Gwynn-Jones, D., & Callaghan, T. V. (2005). Effects of enhanced UV-B radiation in the field on the concentration of phenolics and chlorophyll fluorescence in two boreal and arctic-alpine lichens. *Environmental and Experimental Botany, 53*, 139–149.

Bourgougnon, N., & Stiger-Pouvreau, V. (2011). Chemodiversity and bioactivity within red and brown macroalgae along the french coasts, metropole and overseas departements and territories. In S.-K. Kim (Ed.), *Handbook of marine macroalgae* (pp. 58–105). John Wiley & Sons, Ltd.

Breton, F., Cérantola, S., & Ar Gall, E. (2011). Distribution and radical scavenging activity of phenols in *Ascophyllum nodosum* (Phaeophyceae). *Journal of Experimental Marine Biology and Ecology, 399*, 167–172.

Brock, E., Aberg, P., & Pavia, H. (2001). Phlorotannins as chemical defense against macroalgal epiphytes on *Ascophyllum nodosum*. *Journal of Phycology, 37*, 8–9.

Brock, E., Nylund, G. M., & Pavia, H. (2007). Chemical inhibition of barnacle larval settlement by the brown alga *Fucus vesiculosus*. *Marine Ecology Progress Series, 337*, 165–174.

Cabioc'h, J., Floćh, J.-Y., Le Toquin, A., Boudouresque, C.-F., Meinesz, A., & Verlaque, M. (2014). *Algues des Mers d'Europe, Guide Delachaux*. Paris: Delachaux et Niestlé SA. 1992, 2006.

Cérantola, S., Breton, F., Gall, E. A., & Deslandes, E. (2006). Co-occurrence and antioxidant activities of fucol and fucophlorethol classes of polymeric phenols in *Fucus spiralis*. *Botanica Marina, 49*, 347–351.

Cho, G. Y., Rousseau, F., de Reviers, B., & Boo, S. M. (2006). Phylogenetic relationships within the Fucales (Phaeophyceae) assessed by the photosystem I coding psaA sequences. *Phycologia, 45*, 512–519.

Choze, R., Alcantara, G. B., Alves Filho, Ede G., e Silva, L. M. A., Faria, J. C., & Lião, L. M. (2013). Distinction between a transgenic and a conventional common bean genotype by ^{1}H HR-MAS NMR. *Food Chemistry, 141*, 2841–2847.

Connan, S., Delisle, F., Deslandes, E., & Ar Gall, E. (2006). Intra-thallus phlorotannin content and antioxidant activity in Phaeophyceae of temperate waters. *Botanica Marina, 49*, 39–46.

Connan, S., Deslandes, E., & Ar Gall, E. (2007). Influence of day-night and tidal cycles on phenol content and antioxidantc apacity in three temperate intertidal brown seaweeds. *Journal of Experimental Marine Biology and Ecology, 349*, 359–369.

Connan, S., Goulard, F., Stiger, V., Deslandes, E., & Ar Gall, E. (2004). Interspecific and temporal variation in phlorotannin levels in an assemblage of brown algae. *Botanica Marina, 47*, 410–416.

Critchley, A. T. (1983). The establishment and increase of *Sargassum muticum* (Yendo) Fensholt populations within the Solent area of southern Britain. I. An investigation of the increase in number of population individuals. *Botanica Marina, 26*, 539–545.

Critchley, A. T., Farnham, W. F., Yoshida, T., & Norton, T. A. (1990). A bibliography of the invasive alga *Sargassum muticum* (Yendo) Fensholt (Fucales, Sargassaceae). *Botanica Marina, 33*, 551–562.

Denton, A., Chapman, A. R. O., & Markham, J. (1990). Size-specific concentrations of phlorotannins (anti-herbivore compounds) in three species of *Fucus*. *Marine Ecology Progress Series, 65*, 103–104.

Draisma, S. G. A., Ballesteros, E., Rousseau, F., & Thibaut, T. (2010). DNA sequence data demonstrate the polyphyly of the genus *Cystoseira* and other Sargassaceae genera (Phaeophyceae). *Journal of Phycology, 46*, 1329–1345.
Eom, S.-H., Kim, Y.-M., & Kim, S.-K. (2012). Antimicrobial effect of phlorotannins from marine brown algae. *Food and Chemical Toxicology, 50*, 3251–3255.
Fairhead, V. A., Amsler, C. D., McClintock, J. B., & Baker, B. J. (2006). Lack of defense or phlorotannins induction by UV radiation or mesograzers in *Desmarestia anceps* and *D. meniesii* (Phaeophyceae). *Journal of Phycology, 42*, 1174–1183.
Freile-Pelegrín, Y., & Robledo, D. (2013). Bioactive phenolic compounds from algae. In B. Hernández-Ledesma, & M. Herrero (Eds.), *Bioactive compounds from marine foods* (pp. 113–129). John Wiley & Sons Ltd.
Fusetani, N. (2004). Biofouling and antifouling. *Natural Product Reports, 21*, 94–104.
Glombitza, K.-W., & Sattler, E. (1973). Trifuhalol, ein neuer triphenyldiäther aus *Halidrys siliquosa* (1). *Tetrahedron Letters, 14*, 4277–4280.
Guiry, M. D., & Guiry, G. M. (2013). *AlgaeBase*. Galway: World-wide Electronic Publication, National University of Ireland. searched on 20 November 2013 http://www.algaebase.org.
Hay, M. E. (1996). Marine chemical ecology, what's known and what's next? *Journal of Experimental Marine Biology and Ecology, 200*, 103–134.
Hellio, C., De La Broise, D., Dufosse, L., Le Gal, Y., & Bourgougnon, N. (2001). Inhibition of marine bacteria by extracts of macroalgae: potential use for environmentally friendly antifouling paints. *Marine Environnement Research* 52, 231–237.
Hellio, C., Maréchal, J.-P., Véron, B., Bremer, G., Clare, A.S., & Le Gal, Y. (2004). Seasonal Variation of Antifouling Activities of Marine Algae from the Brittany Coast (France). *Marine Biotechnology*, 6, 67–182.
Hemmi, A., Honkanen, T., & Jormalainen, V. (2004). Inducible resistance to herbivory in *Fucus vesiculosus*-duration, spreading and variation with nutrient availability. *Marine Ecological Progress Series, 273*, 109–120.
Henry, B. E., & Van Alstyne, K. L. (2004). Effects of UV radiations on growth and phlorotannins in *Fucus gardneri* (Phaeophyceae) juveniles and embryos. *Journal of Phycology, 40*, 527–533.
Horiguchi, T., & Yoshida, T. (1998). The phylogenetic affinities of *Myagropsis myagroides* (Fucales, Phaeophyceae) as determined from 18S rDNA sequences. *Phycologia, 37*, 237–245.
Ibañez, E., Herrero, M., Mendiola, J. A., & Castro-Puyana, M. (2012). Extraction and characterization of bioactive compounds with health benefits from marine resources: macro and micro algae, cyanobacteria, and invertebrates. In M. Hayes (Ed.), *Marine bioactive compounds* (pp. 55–98). Boston, MA: Springer US.
Ilvessalo, H., & Tuomi, J. (1989). Nutrient availability and accumulation of phenolic compounds in the brown alga *Fucus vesiculosus*. *Marine Biology, 101*, 115–119.
Incera, M., Olabarria, C., Troncoso, J. S., & Lopez, J. (2009). Response of the invader *Sargassum muticum* to variability in nutrient supply. *Marine Ecological Progress Series, 377*, 91–101.
Jégou, C. (2011). Etude du genre *Cystoseira* des côtes bretonnes: taxinomie, écologie et caractérisation de substances naturelles. Thèse de l'Université de Bretagne Occidentale.
Jégou, C., Culioli, G., Kervarec, N., Simon, G., & Stiger-Pouvreau, V. (2010). LC/ESI-MS^n and ^1H HR-MAS NMR analytical methods as useful taxonomical tools within the genus *Cystoseira* C. Agardh (Fucales; Phaeophyceae). *Talanta, 83*, 613–622.
Jiménez-Escrig, A., Gómez-Ordóñez, E., & Rupérez, P. (2012). Brown and red seaweeds as potential sources of antioxidant nutraceuticals. *Journal of Applied Phycology, 24*, 1123–1132.
Joung, E.-J., Lee, M.-S., Choi, J.-W., Kim, J.-S., Shin, T., Jung, B.-M., et al. (2012). Anti-inflammatory effects of phlorofucofuroeckol B-rich ethyl acetate fraction obtained from *Myagropsis myagroides* on lipopolysaccharide-stimulated RAW 264.7 cells and mouse edema. *International Immunopharmacology, 14*, 471–480.

Kang, Y.-M., Eom, S.-H., & Kim, Y.-M. (2013). Protective effect of phlorotannins from *Eisenia bicyclis* against lipopolysaccharide-stimulated inflammation in hepg2 cells. *Environmental Toxicology and Pharmacology, 35*, 395–401.

Keusgen, M., & Glombitza, K.-W. (1995). Phlorethols, fuhalols and their derivatives from the brown alga *Sargassum spinuligerum*. *Phytochemistry, 38*, 975–985.

Kim, A.-R., Shin, T.-S., Lee, M.-S., Park, J.-Y., Park, K.-E., Yoon, N.-Y., et al. (2009). Isolation and identification of phlorotannins from *Ecklonia stolonifera* with antioxidant and anti-inflammatory properties. *Journal of Agricultural and Food Chemistry, 57*, 3483–3489.

Kim, K. N., Ham, Y. M., Moon, J. Y., Kim, M. J., Kim, D. S., Lee, W. J., et al. (2009). In vitro cytotoxic activity of *Sargassum thunbergii* and *Dictyopteris divaricata* (Jeju Seaweeds) on the HL-60 tumour cell line. *International Journal of Pharmacology, 5*, 298–306.

Koivikko, R., Eränen, J., Loponen, J., & Jormalainen, V. (2008). Variation of phlorotannins among three populations of *Fucus vesiculosus* as revealed by HPLC and colorimetric quantification. *Journal of Chemical Ecology, 34*, 57–64.

Koivikko, R., Loponen, J., Pihlaja, K., & Jormalainen, V. (2007). High-performance liquid chromatographic analysis of phlorotannins from the brown alga *Fucus vesiculosus*. *Phytochemical Analysis, 18*, 326–332.

Kornprobst, J.-M. (2010). *Encyclopedia of marine natural products* . (Vol. 1)Wiley-Blackwell.

Kützing, F. T. (1843). *Phycologia Generalis: oder, Anatomie, physiologie und systemkunde der tange: Mit 80 farbig gedruckten Tafeln, gezeichnet und gravirt vom Verfasser*. Leipzig: F.A. Brockhaus.

Le Lann, K. (2009). Etude de la biodiversité des Sargassaceae (Fucales, Phaeophyceae) en milieux tempéré et tropical: écologie, chimiotaxonomie et source de composés bioactifs. Thèse de l'Université de Bretagne Occidentale.

Le Lann, K., Connan, S., & Stiger-Pouvreau, V. (2012). Phenology, TPC and size-fractioning phenolics variability in temperate Sargassaceae (Phaeophyceae, Fucales) from Western Brittany: native vs. Introduced species. *Marine Environmental Research, 80*, 1–11.

Le Lann, K., Ferret, C., vanmee, E., Spagnol, C., Lhuillery, M., Payri, C., et al. (2012). Total phenolic, size-fractionated phenolics and fucoxanthin content of tropical Sargassaceae (Fucales, Phaeophyceae) from the South Pacific Ocean: spatial and specific variability. *Phycological Research, 60*, 37–50.

Le Lann, K., Jégou, C., & Stiger-Pouvreau, V. (2008). Effect of different conditioning treatments on total phenolic content and antioxidant activities in two Sargassacean species: comparison of the frondose *Sargassum muticum* (Yendo) Fensholt and the cylindrical *Bifurcaria bifurcata* R. Ross. *Phycological Research, 56*, 238–245.

Le Lann, K., Kervarec, N., Payri, C. E., Deslandes, E., & Stiger-Pouvreau, V. (2008). Discrimination of allied species within the genus *Turbinaria* (Fucales, Phaeophyceae) using HRMAS NMR spectroscopy. *Talanta, 74*, 1079–1083.

Le Lann, K., & Stiger-Pouvreau, V. (2009). Spatio-temporal phenologies of temperate Sargassaceae: coexistence of invasive and native species. *Phycologia, 48*(4 Suppl.), 74.

Lee, S.-H., & Jeon, Y.-J. (2013). Anti-diabetic effects of brown algae derived phlorotannins, marine polyphenols through diverse mechanisms. *Fitoterapia, 86*, 129–136.

Li, S.-M., & Glombitza, K.-W. (1991). Carmalols and phlorethofuhalols from the brown alga *Carpophyllum maschalocarpum*. *Phytochemistry, 30*, 3417–3421.

Li, Y.-X., Wijesekara, I., Li, Y., & Kim, S.-K. (2011). Phlorotannins as bioactive agents from brown algae. *Process Biochemistry, 46*, 2219–2224.

Liu, M., Hansen, P. E., & Lin, X. (2011). Bromophenols in marine algae and their bioactivities. *Marine Drugs, 9*, 1273–1292.

Liu, L., Heinrich, M., Myers, S., & Dworjanyn, S. A. (2012). Towards a better understanding of medicinal uses of the brown seaweed *Sargassum* in traditional Chinese medicine: a phytochemical and pharmacological review. *Journal of Ethnopharmacology, 142*, 591–619.

Lüder, U. H., & Clayton, M. N. (2004). Induction of phlorotannins in the brown macroalga *Ecklonia radiata* (Laminariales, Phaeophyta) in response to simulated herbivory—the first microscopic study. *Planta, 218*, 928–937.

Maréchal, J. P., & Hellio, C. (2009). Challenges for the development of new non-toxic antifouling solutions. *International Journal of Molecular Sciences*, *10*, 4623–4637.

Mary, J. S., Vinotha, P., & Pradeep, A. M. (2012). Screening for in vitro cytotoxic activity of seaweed, *Sargassum* sp. against Hep-2 and MCF-7 cancer cell lines. *Asian Pacific Journal of Cancer Prevention*, *13*, 6073–6076.

Mattio, L., Dirberg, G., Payri, C. E., & Andrefouet, S. (2008). Diversity, biomass and distribution pattern of *Sargassum* beds in the South West lagoon of New Caledonia (South Pacific). *Journal of Applied Phycology*, *20*, 811–823.

Mattio, L., & Payri, C. (2011). 190 Years of *Sargassum* taxonomy, facing the advent of DNA phylogenies. *Botanical Review*, *77*, 31–70.

Mattio, L., Payri, C. E., & Stiger-Pouvreau, V. (2008). Taxonomic revision of *Sargassum* (Fucales, Phaeophyceae) from French Polynesia based on morphological and molecular analyses. *Journal of Phycology*, *44*, 1541–1555.

Morales, J. L., Cantillo-Ciau, Z. O., Sánchez-Molina, I., & Mena-Rejón, G. J. (2006). Screening of antibacterial and antifungal activities of six marine macroalgae from coasts of Yucatán Peninsula. *Pharmacien Biologiste*, *44*, 632–635.

Mucci, A., Parenti, F., Righi, V., & Schenetti, L. (2013). Citron and lemon under the lens of HR-MAS NMR spectroscopy. *Food Chemistry*, *141*, 3167–3176.

Nakai, M., Kageyama, N., Nakahara, K., & Miki, W. (2006). Phlorotannins as radical scavengers from the extract of *Sargassum ringgoldianum*. *Marine Biotechnology*, *8*, 409–414.

Nizamuddin, M. (1962). Classification and the distribution of the Fucales. *Botanica Marina*, *4*, 191–203.

Nwosu, F., Morris, J., Lund, V. A., Stewart, D., Ross, H. A., & McDougall, G. J. (2011). Anti-proliferative and potential anti-diabetic effects of phenolic-rich extracts from edible marine algae. *Food Chemistry*, *126*, 1006–1012.

Pavia, H., & Aberg, P. (1996). Spatial variation in polyphenolic content of *Ascophyllum nodosum* (Fucales, Phaeophyta). *Hydrobiologia*, *326/327*, 199–203.

Pavia, H., & Toth, G. (2000). Influence of light and nitrogen on the phlorotannin content of the brown seaweed *Ascophyllum nodosum* and *Fucus vesiculosus*. *Hydrobiologia*, *440*, 299–305.

Pedersen, A. (1984). Studies on phenol content and heavy metal uptake in fucoids. *Hydrobiologia*, *116/117*, 498–504.

Pellegrini, M., Piovetti, L., & Pellegrini, L. (1997). Chimiotaxonomie et marqueurs chimiques chez les algues brunes. *Lagascalia*, *19*, 145–154.

Phillips, N. (1995). Biogeography of *Sargassum* (Phaeophyta) in the Pacific basin. In I. A. Abbott (Ed.), *Taxonomy of economic seaweeds with reference to some Pacific species* (Vol. 5 (pp. 107–144). La Jolla, California: California Sea Grant College Program.

Phillips, N., Burrowes, R., Rousseau, F., de Reviers, B., & Saunders, G.W. (2008). Resolving evolutionary relationships among the brown algae using chloroplast and nuclear genes. *Journal of Phycology*, *44*, 394–405.

Plaza, M., & Rodríguez-Meizoso, I. (2013). Advanced extraction processes to obtain bioactives from marine foods. In B. Hernández-Ledesma, & M. Herrero (Eds.), *Bioactive compounds from marine foods* (pp. 343–371). John Wiley & Sons Ltd.

Plaza, M., Santoyo, S., Jaime, L., García-Blairsy Reina, G., Herrero, M., Señoráns, F. J., et al. (2010). Screening for bioactive compounds from algae. *Journal of Pharmaceutical and Biomedical Analysis*, *51*, 450–455.

Plouguerné, E., Hellio, C., Cesconetto, C., Thabard, M., Mason, K., Véron, B., et al. (2010). Antifouling activity as a function of population variation in *Sargassum vulgare* from the littoral of Rio de Janeiro (Brazil). *Journal of Applied Phycology*, *22*, 717–724.

Plouguerné, E., Le Lann, K., Connan, S., Jechoux, G., Deslandes, E., & Stiger-Pouvreau, V. (2006). Spatial and seasonal variation in density, reproductive status, length and phenolic content of the invasive brown macroalga *Sargassum muticum* (Tendo) Fensholt along the coast of Western Brittany (France). *Aquatic Botany*, *85*. 337–334.

Potin, P., Bouarab, K., Salaün, J.-P., Pohnert, G., & Kloareg, B. (2002). Biotic interactions of marine algae. *Current Opinion in Plant Biology*, *5*, 308–317.

Potin, P., & Leblanc, C. (2006). Phenolic-based adhesives of marine brown algae. In A. M. Smith, & J. A. Callow (Eds.), *Biological adhesives* (pp. 105–119). Berlin, Germany: Springer.

Prud'homme van Reine, W. F. (2002). *Sargassum* C. Agardh. In W. F. Prud'homme van Reine, & G. C. Trono (Eds.), *Plant resources of South-East Asia No. 15(1)* (pp. 240–246). (Cryptogams: Algae. Prosea Foundation, Bogor, Indonesia).

Ragan, M. A., & Glombitza, K.-W. (1986). Pholorotannins, brown algal polyphenols. *Progress in Phycological Research*, *4*, 129–241.

Ragan, M. A., & Jensen, A. (1978). Quantitative studies on brown algal phenols. II. Seasonal variation in polyphenol content of *Ascophyllum nodosum* (L.) Le Jol. and *Fucus vesiculosus* (L.). *Journal of Experimental Marine Biology and Ecology*, *34*, 245–258.

Rönnberg, O., & Ruokolahti, C. (1986). Seasonal variation of algal epiphytes and phenolic content of *Fucus vesiculosus* in a northern Baltic archipelago. *Annales Botanici Fennici*, *23*, 317–323.

Rousseau, F., Burrowes, R., Peters, A. F., Kuhlenkamp, R., & de Reviers, B. (2001). A comprehensive phylogeny of the Phaeophyceae based on nrDNA sequences resolves the earliest divergences. *Comptes Rendus de l'Académie des Sciences de Paris*, *324*, 305–319.

Rousseau, F., Leclerc, M. C., & de Reviers, B. (1997). Molecular phylogeny of European Fucales (Phaeophyceae) based on partial large-subunit rDNA sequence comparisons. *Phycologia*, *36*, 438–446.

Rousseau, F., & de Reviers, B. (1999). Phylogenetic relationships within the Fucales (Phaeophyceae) based on combined partial SSU+LSU rDNA sequence. *European Journal of Phycology*, *34*, 53–64.

Ryu, B., Li, Y., Qian, Z.-J., Kim, M.-M., & Kim, S.-K. (2009). Differentiation of human osteosarcoma cells by isolated phlorotannins is subtly linked to COX-2, INOS, MMPS, and MAPK signaling: implication for chronic articular disease. *Chemico-biological Interactions*, *179*, 192–201.

Sailler, B., & Glombitza, K.-W. (1999). Phlorethols and fucophlorethols from the brown alga *Cystophora retroflexa*. *Phytochemistry*, *50*, 869–881.

Salaün, S., Kervarec, N., Potin, P., Haras, D., Piotto, M., & La Barre, S. (2010). Whole-cell spectroscopy is a convenient tool to assist molecular identification of cultivatable marine bacteria and to investigate their adaptive metabolism. *Talanta*, *80*. 1758–1570.

Sandsalen, E., Haug, T., Stensvag, K., & Styrvold, O.B. (2003). The antibacterial effect of a polyhydroxylated fucophlorethol from the marine brown alga, *Fucus vesiculusus*. *World Journal of Microbiology & Biotechnology*, 19, 777–782.

Sanoner, P., Guyot, S., Marnet, N., Molle, D., & Drilleau, J.-F. (1999). Polyphenol profiles of french cider apple varieties (*Malus domestica* sp.). *Journal of Agricultural and Food Chemistry*, *47*, 4847–4853.

Schoenwaelder, M. E. A. (2002). The occurrence and cellular significance of physodes in brown algae. *Phycologia*, *41*, 125–139.

Schoenwaelder, M. E. A., & Clayton, M. N. (1998). The secretion of phenolic compound following fertilization in *Acrocarpia paniculata* (Fucales, Phaeophyta). *Phycologia*, *37*, 40–46.

Shibata, T., Hama, Y., Miyasaki, T., Ito, M., & Nakamura, T. (2006). Extracellular secretion of phenolic substances from living brown algae. *Journal of Applied Phycology*, *18*, 787–794.

Singh, I. P., & Bharate, S. B. (2006). Phloroglucinol compounds of natural origin. *Natural Product Reports*, *23*, 558–591.

Siriwardhana, N., Kim, K.-N., Lee, K.-W., Kim, S.-H., Ha, J.-H., Song, C. B., et al. (2008). Optimisation of hydrophilic antioxidant extraction from *Hizikia fusiformis* by integrating treatments of enzymes, heat and pH control. *International Journal of Food Science & Technology*, *43*, 587–596.

Siriwardhana, N., Lee, K.-W., Jeon, Y.-J., Kim, S.-H., & Haw, J.-W. (2003). Antioxidant activity of *Hizikia fusiformis* on reactive oxygen species scavenging and lipid peroxidation inhibition. *Food Science and Technology International, 9*, 339–3346.

Smit, A. J. (2004). Medicinal and pharmaceutical uses of seaweed natural products: a review. *Journal of Applied Phycology, 16*, 245–262.

Steevensz, A. J., MacKinnon, S. L., Hankinson, R., Craft, C., Connan, S., Stengel, D. B., et al. (2012). Profiling phlorotannins in brown macroalgae by liquid chromatography–high resolution mass spectrometry. *Phytochemical Analysis, 23*, 547–553.

Steinberg, P. D. (1989). Biogeographical variation in brown algal polyphenolics and other secondary metabolites: comparison between temperate Australasia and North America. *Oecologia, 78*, 373–382.

Steinberg, P. D. (1995). Seasonal variation in the relationship between growth rate and phlorotannin production in the kelp *Ecklonia radiata*. *Oecologia, 102*, 169–173.

Steneck, R. S., Graham, M. H., Bourque, B. J., Corbett, D., Erlandson, J. M., Estes, J. A., et al. (2002). Kelp forest ecosystems: biodiversity, stability, resilience and future. *Environmental Conservation, 29*, 436–459.

Stiger, V., Deslandes, E., & Payri, C. E. (2004). Phenolic contents of two brown algae, *Turbinaria ornata* and *Sargassum mangarevense* on Tahiti (French Polynesia): interspecific, ontogenic and spatio-temporal variations. *Botanica Marina, 47*, 402–409.

Stiger, V., Horiguchi, T., Yoshida, T., Coleman, A. W., & Masuda, M. (2003). Phylogenetic relationships inferred from ITS-2 nrDNA comparisons within the genus *Sargassum* (Fucales, Phaeophyceae) from the Pacific basin, with an emphasis on the taxonomic subdivision of the genus. *Phycological Research, 51*, 1–10.

Stiger, V., & Payri, C. E. (1999a). Spatial and seasonal variations in the biological characteristics of two invasive brown algae, *Turbinaria ornata* (Turner) J. Agardh and *Sargassum mangarevense* (Grunow) Setchell (Sargassaceae, Fucales) spreading on the reefs of Tahiti (French Polynesia). *Botanica Marina, 42*, 295–306.

Stiger, V., & Payri, C. E. (1999b). Spatial and temporal patterns of settlement of the brown macroalgae, *Turbinaria ornata* and *Sargassum mangarevense* in a coral reef on Tahiti. *Marine Ecology Progress Series, 191*, 91–100.

Stiger, V., & Payri, C. E. (2005). Natural settlement of a young population of *Turbinaria ornata* and phenological comparisons with older populations. *Aquatic Botany, 81*, 225–243.

Svensson, C. J., Pavia, H., & Toth, G. B. (2007). Do plant density, nutrient availability, and herbivore grazing interact to affect phlorotannin plasticity in the brown seaweed *Ascophyllum nodosum*. *Marine Biology, 151*, 2177–2181.

Swanson, A. K., & Druehl, L. D. (2002). Induction, exudation and the UV protective role of kelp phlorotannins. *Aquatic Botany, 73*, 241–253.

Tanniou, A. (2014). Etude de la production de biomolécules d'intérêt (phlorotannins, pigments, lipides) par l'algue brune modèle *Sargassum muticum* par des approches combinées de profilage métabolique et d'écophysiologie. Thèse de l'Université de Bretagne Occidentale.

Tanniou, A., Serrano Leon, E., Vandanjon, L., Ibanez, E., Mendiola, J. A., Cérantola, S., et al. (2013). Green improved processes to extract bioactive phenolic compounds from brown macroalgae using *Sargassum muticum* as model. *Talanta, 104*, 44–52.

Tanniou, A., Vandanjon, L., Incera, M., Serrano Leon, E., Husa, V., Grand, J., et al. (2013). Assessment of the spatial variability of phenolic contents and associated bioactivities in the invasive alga *Sargassum muticum* sampled along its European range from Norway to Portugal. *Journal of Applied Phycology*. (in press). http://dx.doi.org./10.1007/s10811-013-0198-x.

Targett, N. M., Boettcher, A. A., Targett, T. E., & Vrolijk, N. H. (1995). Tropical marine herbivore assimilation of phenolic-rich plants. *Oecologia, 103*, 170–179.

Targett, N. M., Coen, L. D., Boettcher, A. A., & Tanner, C. E. (1992). Biogeographic comparisons of marine algal polyphenolics: evidence against a latitudinal trend. *Oecologia*, *89*, 464–470.

Thibaut, T., Pinedo, S., Torras, X., & Ballesteros, E. (2005). Long-term decline of the populations of Fucales (*Cystoseira* spp. and *Sargassum* spp.) in the Albères coast (France, Northwestern Mediterranean). *Marine Pollution Bulletin*, *50*, 1472–1489.

Thomas, N. V., & Kim, S.-K. (2011). Health beneficial aspects of phloroglucinol derivatives from marine brown algae. In S.-K. Kim (Ed.), *Handbook of marine macroalgae* (pp. 378–386). John Wiley & Sons, Ltd.

Tierney, M. S., Smyth, T. J., Rai, D. K., Soler-Vila, A., Croft, A. K., & Brunton, N. (2013). Enrichment of polyphenol contents and antioxidant activities of Irish brown macroalgae using food-friendly techniques based on polarity and molecular size. *Food Chemistry*, *139*, 753–761.

Toth, G., & Pavia, H. (2000). Lack of phlorotannin induction in the brown seaweed *Ascophyllum nodosum* in response to increased copper concentrations. *Marine Ecology Progress Series*, *192*, 119–126.

Toth, G. B., & Pavia, H. (2002). Optimal defense theory: elasticity analysis as a tool to predict intraplant variation in defenses. *Ecology*, *83*, 891–897.

Van Alstyne, K. L., McCarthy, J. J., III, Hustead, C. L., & Kearns, L. J. (1999a). Geographic variation in polyphenolic levels of Northeastern Pacific kelps and rockweeds. *Marine Biology*, *133*, 371–379.

Van Alstyne, K. L., McCarthy, J. J., III, Hustead, C. L., & Kearns, L. J. (1999b). Phlorotannin allocation among tissues of Northeastern Pacific kelps and rockweeds. *Journal of Phycology*, *35*, 483–492.

Van Alstyne, K. L., Whitman, S. L., & Ehlig, J. M. (2001). Differences in herbivore preferences, phlorotannin production, and nutritional quality between juvenile and adult tissues from marine brown algae. *Marine Biology*, *139*, 201–210.

Vijayabaskar, P., & Shiyamala, V. (2011). Antibacterial activities of brown marine algae (*Sargassum wightii* and *Turbinaria ornata*) from the Gulf of Mannar biosphere reserve. *Advances in Biological Research*, *5*, 99–102.

Vijayabaskar, P., & Shiyamala, V. (2012). Antioxidant properties of seaweed polyphenol from *Turbinaria ornata* (Turner) J. Agardh, 1848. *Asian Pacific Journal of Tropical Biomedicine*, *2*, S90–S98.

Wei, Y., Hu, Y., & Xu, Z. (2003). Inhibition of mouse liver lipid peroxidation by high molecular weight phlorotannins from *Sargassum kjellmanianum*. *Journal of Applied Phycology*, *15*, 507–511.

Yates, J. L., & Peckol, P. (1993). Effects of nutrient availability and herbivory on polyphenolics in the seaweed *Fucus vesiculosus*. *Ecology*, *74*, 1757–1766.

Yeo, M., Jung, W.-K., & Kim, G. (2012). Fabrication, characterisation and biological activity of phlorotannin-conjugated PCL/beta-TCP composite scaffolds for bone tissue regeneration. *Journal of Materials Chemistry*, *22*, 3568–3577.

Yoon, N. Y., Chung, H. Y., Kim, H. R., & Choi, J. S. (2008). Acetyl- and butyrylcholinesterase inhibitory activities of sterols and phlorotannins from *Ecklonia stolonifera*. *Fisheries Science*, *74*, 200–207.

Zaragoza´, M. C., López, D., P. Sáiz, M., Poquet, M., Pérez, J., Puig-Parellada, P., et al. (2008). Toxicity and antioxidant activity *in vitro* and *in vivo* of two *Fucus vesiculosus* extracts. *Journal of Agricultural and Food Chemistry*, *56*, 7773–7780.

Zubia, M., Fabre, M. S., Kerjean, V., Le Lann, K., Stiger-Pouvreau, V., Fauchon, M., et al. (2009). Antioxidant and antitumoural activities of some Phaeophyta from Brittany coasts. *Food Chemistry*, *116*, 693–701.

Zubia, M., Robledo, D., & Freile-Pelegrin, Y. (2007). Antioxidant activities in tropical marine macroalgae from the Yucatan Peninsula, Mexico. *Journal of Applied Phycology*, *19*, 449–458.

CHAPTER FOURTEEN

The Antifouling Defence Mechanisms of Marine Macroalgae

Bernardo A.P. da Gama[1], Erwan Plouguerné and Renato C. Pereira
Marine Chemical Ecology Lab, Departament of Marine Biology, Biology Institute, Universidade Federal Fluminense, Niterói, Rio de Janeiro, Brazil
[1]Corresponding author: e-mail address: bapgama@pq.cnpq.br

Contents

Abstract

Marine macroalgae have long been recognised as important sources of natural products. However, the forces that have driven the selection for their production have been only scarcely understood, with the main focus on their role as defences against herbivores. As photosynthetic organisms, benthic algae are restricted to the euphotic zone of the world's oceans, where fouling growth and diversity of potential colonisers are higher. Epibionts

Advances in Botanical Research, Volume 71
ISSN 0065-2296
http://dx.doi.org/10.1016/B978-0-12-408062-1.00014-7

are known to have detrimental effects on basibiont algae, which was surely an important driving factor for the evolution of defences against epibiosis (i.e. antifouling mechanisms) among seaweeds. Physical antifouling defences such as epithallus sloughing, and chemical defences such as natural products and oxidative bursts, as well as interactive chemical signalling with surface bacterial communities (e.g. quorum sensing) are among the diverse strategies they employ to remain devoid of harmful epibionts. Chemical warfare seems to be the more widespread mechanism, and specialised structures (*viz., corps en cerise*, gland cells, physodes, etc.) were identified at the thallus cortex in a number of red algae. The present work nonexhaustibly describes what is known about algal defences against biofouling, and attempted to draw large-scale patterns and perspectives of study in this field. Latitudinal trends in the production of macroalgal antifouling compounds were not identified to date. Instead, a phylogenetic pattern, with red and brown algae usually as the most chemically prolific, emerged. Other antifouling mechanisms seem to be ubiquitous and part of the multifaceted algal defence system. Community-wide effects of algal antifoulants are still poorly known, and future research directions in this field are pointed out.

14.1 INTRODUCTION

Marine organisms, in particular benthic macroalgae, have long been the focus of interest of scientists, in particular as a source of interesting, novel natural products with a plethora of possible applications, predominantly as pharmacological drugs in the beginning of the marine natural products field (e.g. Fenical, 1982; Faulkner, 2002). However, very little was understood about the ecological roles played by marine natural products (Paul, 1992), and early attention focussed mainly on antipredator defence, i.e. algal defences against herbivores, a wide field from which much of the underlying theory in marine chemical ecology derives (Hay, 1996).

However, seaweeds are photosynthetic organisms, which means in practical terms that they are restricted to benthic habitats within the relatively shallow, euphotic zone of the world's oceans, where sunlight rapidly changes qualitatively and fades away with depth. As sessile organisms living attached to hard-bottom benthic substrata, where competition for space is known to be maximal (Jackson & Buss, 1975), macroalgae are exposed not only to consumers, but also to competitors for space, becoming themselves substrata for colonisation by other species, varying from bacteria to invertebrates, whose diversity is also known to peak at the euphotic zone. These competitors for space or epibiont organisms form a peculiar community collectively known as biofouling, and its process of development also affects all man-made structures in the sea, which led researchers to suggest the use of macroalgal antifouling (AF) defence mechanisms in new, bioinspired AF solutions for ships, an approach known as biomimicry (Chapman et al., 2014; Kirschner & Brennan, 2012; Salta et al., 2010).

It is not the purpose of this work to survey marine compounds exhibiting AF properties, since this information can be obtained in several reviews (e.g. Bhadury & Wright, 2004; Hellio, Maréchal, Da Gama, Pereira, & Clare, 2009; Qian, Xu, & Fusetani, 2010). Rather, our intention is to bring to light and attempt to provide an integrated view of aspects, recent advances and perspectives of the field of algal defence against biofouling.

14.2 WHY SEAWEEDS NEED ANTIFOULING DEFENCE?

14.2.1 The Detrimental Effects of Epibionts on Algae

Seaweed thalli are composed of relatively undifferentiated tissues, and all body exchanges occur through the thallus surface, including nutrient uptake and gas exchange. Moreover, algae rely on the sun irradiance reaching their surfaces in order to photosynthesise, and even reproductive stages such as gametes and spores are usually released from algal surfaces. This makes any surface clotting by epibionts undesirable per se, with potential to largely reduce algal fitness. In general, the type of epibiont will determine the degree of negative effects on the basibiont alga (e.g. Buschmann, Correa, Westermeier, Hernández-González, & Norambuena, 2001). The consequences of epibiosis for seaweeds, however, are still poorly understood, but appear to include direct effects, such as damage to tissues by mechanical anchoring and thallus penetration (Gonzalez & Goff, 1989), reduced growth and productivity through reductions in irradiance, depletion of nutrients and additional weight, reduced reproduction (Brawley, 1992; Buschmann & Gómez, 1993; Muñoz & Fotedar, 2009; Orth & van Montfrans, 1984; Williams & Seed, 1992), and increased drag with consequent tissue loss during storms (Dixon, Schroeter, & Kastendiek, 1981).

Known indirect effects include increased susceptibility to consumers allured to seaweeds bearing fouling organisms (Bernstein & Jung, 1979; Da Gama et al., 2008; Pereira, Da Gama, Teixeira, & Yoneshigue-Valentin, 2003), although epibionts are known to either increase or decrease the susceptibility of basibionts to consumption, depending on epibiont identity (e.g. Enderlein, Moorthi, Rohrscheidt, & Wahl, 2003; Karez, Engelbert, & Sommer, 2000). Epibiosis by the bryozoan *Membranipora membranacea* and haplosclerid sponges altered the susceptibility of the red seaweed *Cryptonemia seminervis* to sea urchin and amphipod grazing. Epibionts on seaweeds may act as lures to herbivores, i.e. attract consumers that otherwise would not feed significantly on the host plant. This could be due to an increased nutritional value of fouled compared to clean algae. Likewise, herbivores could be attracted to algal cues possibly resulting from fouling damage on seaweeds (Leonardi et al., 2006; Wahl & Hay, 1995). This negative effect of epibionts on basibionts was named 'shared doom' (Wahl & Hay, 1995) or 'co-consumption' (Karez et al., 2000). A possible positive effect of less palatable, or chemically defended epibionts would, however, provide a 'protective coating' to algae (*sensu* Karez et al., 2000), a treat also called 'associational resistance' (*sensu* Wahl & Hay, 1995).

Epibiosis on algae could also increase habitat invasibility. The encrusting bryozoan *M. membranacea* is an invasive species along the Atlantic coast of Nova Scotia, to which a disruption in sea urchin-kelp dynamics has been attributed (Scheibling, Hennigar, & Balch, 1999). By overgrowing kelp fronds, it causes thallus fragmentation and loss, decimating kelp beds. Through removal of the dominant canopy species, the bryozoan seems to have facilitated further invasion by the Asian green alga, *Codium fragile tomentosoides* (Scheibling et al., 1999). Thalli of the red alga *C. seminervis* overgrown by *M. membranacea* are frequently found adrift along the Brazilian coastline (southwestern Atlantic), although the bryozoan increase in *C. seminervis* fragmentation was not assessed (Da Gama, Santos, & Pereira, 2008). For a review of the effects of epibionts on basibionts, see Wahl (2008).

14.2.2 Do Epibionts Damage Algal Surfaces?

Some controversy has arisen on the relevance of testing whole-thallus extracts or surface extracts as chemical defences against macroalgal epibionts (Nylund, Gribben, de Nys, Steinberg, & Pavia, 2007). If epibionts do not damage or penetrate algal thalli, then they would only experience chemical defences present at the thallus surface. However, a growing body of evidence demonstrates that a wide diversity of epibionts may directly damage the thallus surface of algae (e.g. Gonzalez & Goff, 1989; Leonardi et al., 2006; Rawlence & Taylor, 1972) or at least alter or disorganise the thallus of algae at the attachment site (Muñoz & Fotedar, 2009). Therefore the question of whether or not epibionts would be exposed to within-thallus compounds becomes irrelevant. Moreover, AF defences may imply not only in settlement inhibition, but also in post-settlement mortality (Wikström & Pavia, 2004), an aspect rarely considered in AF studies. In general, the type of epibiont will determine the degree of the negative effects on the basibiont alga (e.g. Leonardi et al., 2006).

Gonzalez and Goff (1989) showed that the red alga *Microcladia californica*, found as an epibiont attached to the surface of the thalli of the brown alga *Egregia menziesii*, does not penetrate the thallus of the basibiont algae, while the congeneric *Microcladia coulteri*, usually an epibiont on a wide range of seaweeds, does penetrate the surface tissues of *E. menziesii*. From an anatomical perspective, the wide variety of algae that grows as epibionts on other algae represent a continuum between epiphytes and endophytes (Leonardi et al., 2006). These authors classified epibiont algae on farmed *Gracilaria chilensis* in five groups based on the invasiveness of the attaching structure, with the last three groups composed by epibiont algae that caused increased damage through penetration into *Gracilaria* thalli.

Australian *Gracilaria cliftonii* is fouled by a number of epibiont macroalgae, of which two red algae, *Polysiphonia forfex* and *Ceramium isogonum*, were shown to penetrate the cortex and outer medulla, while other species altered the cellular structure of the outer cell wall of *G. cliftonii* (Muñoz & Fotedar, 2009). Rawlence and Taylor (1972) suggested that *Polysiphonia lanosa* rhizoids use chemical, rather than mechanical penetration mechanisms into the brown alga *Ascophyllum nodosum*. The ability of different algal epibionts to digest the cortex and medullary cells of basibiont algae seems to be related to an enzymatic degradation of the algal cell wall. Muñoz and Fotedar (2009) observed the presence of diatoms on the penetration points, an indication that one epibiont species may actually facilitate the colonisation by others.

Epibionts may therefore not only interact at the surface level, but also with the interior of the thalli, despite the algal AF defence mechanisms. Some epibionts, on the other hand, seem to have evolved mechanisms to counteract the AF effects of its several basibiont algae (Gonzalez & Goff, 1989). It thus seems reasonable to hypothesise that macroalgae have been selected, over an evolutionary timescale, for the presence or induction of AF defences.

14.2.3 'Attach' and Defence

The negative effects of epibiosis are surely evolutionary pressures capable of driving the selection of the several traits seen today, such as chemical (natural products, oxidative bursts (OBs)), physical (epithallus sloughing, microtopography), or associational defences as effective AF mechanisms in modern macroalgae species. Epibionts are likely to have evolved equally effective mechanisms of settlement and attachment on algae regardless of their AF mechanisms. As the evolutionary arms race between epibionts – for space to settle and live, and basibionts – for a fully functional free surface, occurs at large timescales, difficult to be perceived by our own life span-limited science, epibiont countermeasures have rarely been investigated. Gonzalez and Goff (1989) suggested that the red alga *M. coulteri*, an epibiont that penetrates the thalli of several algae, evolved mechanisms to counteract the AF effects of its basibiont algae. This is particularly interesting considering that the sympatric *M. californica* became a specialist epibiont, growing only on the brown alga *E. menziesii* and not penetrating its thallus.

Some common traits in benthic organisms probably turned out to be very effective mechanisms to counteract AF defences of basibionts, such as coloniality (or modular growth): a number of marine macroalgae are known to be successfully overgrown by modular or colonial organisms despite their AF mechanisms, possibly due to the fact that colonial epibionts depend on the recruitment of a single larva to later overgrow large proportions of algal thalli (e.g. Da Gama, Santos, et al., 2008; Dixon et al., 1981; Scheibling et al., 1999).

It is noteworthy, however, that marine macroalgae, as benthic organisms, must have evolved both epibiont traits – for successful settlement even on other organisms, and basibiont traits – to remain clean of epibiosis growth, and many algal species are indeed found in the literature both as epibionts and basibionts of other algae under field conditions. AF defence studies are currently 'booming' due to the growing interest in new AF technologies (Hellio & Yebra, 2009). However, from a purely ecological point of view, there is no reason why studies on the mechanisms and adaptations to live as epibionts have not been similarly carried out. Notable exceptions, however, include the studies of *Ulva*'s (=*Enteromorpha*) 'first kiss' (Callow & Callow, 2002).

14.3 PHYSICAL ANTIFOULING DEFENCE IN SEAWEEDS

14.3.1 Epithallus Sloughing

The best known physical means of defence against biofouling in marine macroalgae is epithallus sloughing. Since its first discovery decades ago, periodical shedding of surface tissue has been reported in a number of species of crustose coralline algae (CCA – crustose red algae of the family Corallinaceae – Johnson & Mann, 1986; Keats, Groener, & Chamberlain, 1993; Keats, Knight, & Pueschel, 1997; Littler & Littler, 1999b; Masaki, Fujita, & Hagen, 1984; Pueschel & Miller, 1996), as well as in other red algae (Gonzalez & Goff, 1989; Nylund & Pavia, 2005; Sieburth & Tootle, 1981), in some brown algae, such as *A. nodosum* (Filion-Myklebust & Norton, 1981), *Halidrys siliquosa* (Moss, 1982), *E. menziesii* (Gonzalez & Goff, 1989), *Myagropsis myagroides*, and *Sargassum* spp. (Yamamoto, Endo, Yoshikawa, Ohki, & Kamiya, 2013) as well as in some green algae such as *Halimeda* sp. (Borowitzka & Larkum, 1977) and *Ulva intestinalis* (McArthur & Moss, 1977).

However, as most of the studies were performed under laboratory conditions, there is still scarce evidence of the ecological significance of surface peeling under natural conditions. Johnson and Mann (1986) showed that algae recruited significantly less on rocks with the CCA *Phymatolithon*, known to slough off the epithallus, while Keats et al. (1997) showed that other three coralline algae successfully removed epibionts by epithallial shedding in the field. Nylund and Pavia (2005) tested the effects of chemical defences and sloughing in the field, showing that the red alga *Dilsea carnosa* most likely prevents epibionts by peeling the fouled surfaces a few times per year. They postulated that even if shedding occurs rarely, it may be an effective mechanism for removing a significant part of fouling organisms in temperate regions, where a few properly timed sloughing events would suffice due to the strong seasonal distribution of fouling organisms (Nylund & Pavia, 2005).

14.3.2 Low Adhesion Mechanisms: Microtopography and Gelatinous Algal Surfaces?

Macroalgae are known to possess an outer cell wall usually with a mucilage and sheath consisting of an amorphous bedding matrix that could, theoretically, act as a low-adhesion, gelatinous covering. In brown algae this gelatinous matrix is usually alginic acid with traces of sulphated fucoidan. In red algae, xylans and several sulphated galactans, including agar and carrageenan, while in green algae, this consistency is the result of xylans and mannans (Davis, Volesky, & Mucci, 2003). Chapman et al. (2014) had to remove this cover in order to access, characterise and replicate the surface microtopography of the brown algae *Saccharina latissima* and *Fucus guiryi*, in an applied AF approach combining microtopography of these algae with a chemical defence, a brominated furanone isolated from the green alga *Ulva rigida* and known to cause inhibitory effects in the quorum sensing (QS) pathway of microorganisms. This led to a significant inhibition of biofouling compared to controls, in what is possibly the first report of an algal microtopography effect on biofouling. Despite the complexity of the study, it was not possible to separate the chemical from the physical effects of the bioinspired algal replicas produced. It is possible, if not likely, that microtopography is an additional AF defence trait in several macroalgae, although it is unlikely to work as a separate defence mechanism, as seen also for marine invertebrates (e.g. Bers et al., 2010; Magin, Cooper, & Brennan, 2010).

All physical AF defence traits of macroalgae (epithallial sloughing, and possibly gelatinous and microtopographically designed surfaces) are comparable to the foul-release coatings technology currently in use (e.g. siloxane paints) and still under development (Chambers, Stokes, Walsh, & Wood, 2006), and seem to work in synergy with other mechanisms, such as chemical defence, in what could be called a multiple AF system (Wahl, Kröger, & Lenz, 1998).

14.4 CHEMICAL ANTIFOULING DEFENCE

14.4.1 Natural Product-based Algal Antifoulants

Leal et al. (2013) estimated that 3129 marine natural products have been discovered from seaweeds between 1965 and 2012. The majority of those compounds were isolated from the division Rhodophyta (1658 natural products - NPs or 53% of the total). The division Ochrophyta appears in second position with 1213 NPs (39%), followed by the division Chlorophyta that produces only 8% of the total of natural products discovered (258 NPs). The same distribution between divisions regarding the numbers of discovered natural compounds has been observed in terms of algal extracts with AF activity. As reported by Da Gama, Carvalho, et al. (2008), among the algae collected along the coast of Brazil, extracts from red algae were the most active against attachment of juvenile mussels (*Perna perna*), followed by extracts from brown algae. Extracts from green algae were the less active. A previous study by Hellio, Berge, Beaupoil, Le Gal, and Bourgougnon (2002) already highlighted the same pattern between 30 seaweeds from the temperate coastal waters from Brittany (France). From the 18 extracts that demonstrated AF activity, 9 were obtained from red algae, 7 from brown algae and only 2 from Chlorophyta. However, it is noteworthy that sampling efforts are never equal among algal divisions. As there are many reports of AF extracts from algae, this section will cover only examples of AF compounds isolated from seaweeds.

14.4.1.1 Rhodophyta

AF compounds from red seaweeds are dominated by terpenoids and halogenated compounds. Among Rhodophyta and even among all macroalgae, the genus *Laurencia* emerged as the principal source of metabolites with more than 300 compounds discovered within its species (Hellio et al., 2009).

Elatol was first isolated from *Laurencia elata* in 1974, and remains to date one of the most promising AF terpenes (Fusetani & Clare, 2006). Konig and Wright (1997) isolated elatol and deschloroelatol from *Laurencia rigida* that inhibited settlement of *Balanus* (=*Amphibalanus*) *amphitrite* and *Bugula neritina*. However, both compounds were toxic to larvae. A highly halogenated monoterpene isolated from the Tasmanian red alga *Plocamium costatum* inhibited barnacle settlement at 1 $\mu g/cm^2$ (Konig, Wright, & de Nys, 1999). Vairappan, Daitoh, Suzuki, Abe, and Masuda (2001) highlighted the promising antibacterial activities of elatol and lembyne-A from *Laurencia* spp., and iso-obtusol from *Laurencia majuscula*. Another study by Vairappan, Suzuki, Abe, and Masuda (2001) isolated various halogenated compounds from the Okinawan *Laurencia* species and tested them against gram-negative marine bacteria. The most active metabolites were a C_{15} acetogenin, (12*E*)-lembyne-A, isolated from *Laurencia mariannensis*, the brominated sesquiterpene (6*R*,9*R*,10*S*)-10-bromo-9-hydroxy-chamigra-2,7(14)-diene contained in *L. majuscula*, and laurinterol and isolaurinterol produced by *Laurencia nidifica*. The alga *Laurencia caduciramulosa*, introduced into Brazilian coastal environments, was investigated for its potential AF activity against *P. perna* (Cassano, De-Paula, Fujii, Da Gama, & Teixeira, 2008). The dichloromethane extract of *L. caduciramulosa* inhibited mussel attachment at natural concentrations. Fractionation led to the isolation and identification of several sesquiterpenes.

Terpenes from *Sphaerococcus coronopifolius* inhibited the settlement of the barnacle *A. amphitrite* (Piazza et al., 2011). The most promising results were observed for bromosphaerol, which expressed an EC_{50} value of 0.23 mg/L and low toxicity levels ($LC_{50} > 100$ mg/L).

Brominated furanones from Australian *Delisea pulchra* were the first marine natural product inhibitors of QS reported (Dworjanyn, De Nys, and Steinberg, 2006).

Paul, de Nys, and Steinberg (2006) identified bromoform and dibromoacetic acid from *Asparagopsis armata* that were active against marine *Vibrio* spp. Furthermore, the increased densities of bacteria on the surface of metabolite-free algae indicated a probable role for these metabolites as AF agents.

Bioassay-guided fractionation of hexane surface extract from *Bonnemaisonia hamifera* revealed 1,1,3,3-tetrabromo-2-heptanone as the active compound that inhibits the growth of ecologically relevant fouling bacteria isolated from algal surfaces at natural concentrations (Nylund et al., 2008). Jah et al. (2013) suggested that 2-dodecanoyloxyethanesulfonate is responsible for the QS inhibition demonstrated by *Asparagopsis taxiformis*.

Lane et al. (2009) highlighted that bromophycolides and callophycoic acids from *Callophycus serratus* inhibit the growth of *Lindra thalassiae*, a marine fungal pathogen. A natural concentration of the bromophycolides was sufficient for fungal inhibition.

Floridoside, a D-galactopyranosyl-(1/2)glycerol, isolated from *Grateloupia turuturu* strongly inhibited larval settlement of *A. amphitrite* (EC_{50} < 1 mg/mL, Hellio, Simon-Colin, Clare, & Deslandes, 2004).

Kim, Kim, Seo, and Park (2007) reported that a mixture of floridoside, isethionic acid and betonicine obtained from *Ahnfeltiopsis flabelliformis* inhibited acyl homoserine lactone (AHL)-mediated QS.

Matsuo, Ishida, Matsumoto, Tatewaki, and Suzuki (1997) isolated yendolipin from *Neodilsea yendoana*. This lipobetaine is a potent morphogenetic substance that inhibits the morphological restoration of the green alga *Monostroma oxyspermum*.

14.4.1.2 Ochrophyta

The families Sargassaceae and Dictyotaceae produce two-thirds of all secondary metabolites discovered from brown algae (Blunt, Copp, Munro, Northcote, & Prinsep, 2011). It is therefore logical that most of the AF compounds described in the literature are synthesised by these groups. Concerning the class of compounds, the same pattern observed in Rhodophyta is repeated in Ochrophyta; terpenoids are the main class of compounds reported to prevent fouling. Other compounds such as glycolipids, glycerolipids, free fatty acids or chromanols also emerge as potential AF agents.

Taniguchi, Shiraishi, Kurata, and Suzuki (1989) reported a spatane diterpene from *Dilophus okamurai* that inhibits metamorphosis of abalone larvae at 5–10 ppb. Dictyol E, pachydictyol A, and dictyodial, antifeedant diterpenes of the brown algae *Dictyota* spp. inhibited the settlement of larvae from the bryozoan *B. neritina* (Schmitt, Hay, & Lindquist, 1995, Schmitt, Lindquist, & Hay, 1998). The diterpene dictyol C, isolated from a Mediterranean *Dictyota*, inhibited the growth of biofilm-forming bacteria at a low concentration (EC_{50} 30 mM) (Viano et al., 2009). Kim, Alamsjah, Hamada, Fujita, and Ishibashi (2006) isolated two cyclic diterpenes, dictyolactone and sanadaol, from *Dictyota dichotoma*, that exhibited activity against three phytoplankton species. A dolabellane diterpene from Brazilian *Dictyota pfaffii* inhibited the adhesion of mussel juveniles (*P. perna*) (Barbosa, Fleury, Da Gama, Teixeira, & Pereira, 2007). Hellio, De La Broise, et al. (2001), Hellio, Thomas-Guyon, et al. (2001) and Maréchal et al. (2004) isolated diterpenes from *Bifurcaria bifurcata* that inhibited the growth of marine bacteria and fungi. The same seaweed produces the AF agent bifurcane. Muñoz, Culioli, and Kock (2013) published a complete review of linear terpenes isolated from *B. bifurcata*.

Cho (2013) isolated chromanols from *Sargassum horneri* that inhibit settlement of mussel larvae (*Mytilus edulis*) with EC_{50} of 0.11–3.34 μg/mL, spore settlement of *Ulva pertusa* zoospores (EC_{50} of 0.01–0.43 μg/mL), and the diatom *Navicula annexa* (EC_{50} of 0.008–0.19 μg/mL). Morever, the biofouling bacteria *Pseudomonas aeruginosa* KNP-3 and *Alteromonas* sp. KNS-8 were sensitive to the tested compounds. The Mediterranean *H. siliquosa* produces meroditerpenes (i.e. terpenes of mixed biosynthetic origin) active against *A. amphitrite* cyprids (Culioli et al., 2008). Another class of meroditerpene and its terpenoid parts isolated from *Cystoseira baccata* inhibited the growth of two seaweeds and mussel phenoloxidase activity (Mokrini et al. 2008).

Katsuoka, Ogura, Etoh, Sakata, and Ina (1990) isolated galactosyl and sulphoquinovosyl-diacylglycerols from *Costaria costata* and *Undaria pinnatifida* that exhibit activity against *M. edulis*. Plouguerné et al. (2010) isolated galactoglycerolipids from *Sargassum muticum* that inhibited the growth of the bacteria *Shewanella putrefaciens*, *Polaribacter irgensii*, and the fungi *Halosphaeriopsis mediosetigera*, *Asteromyces cruciatus*, *Lulworthia uniseptata*, and *Monodictys pelagica* at 0.75 mg/L. Unsaturated fatty acids from this alga also showed AF properties. Ganti, Kim, Bhattarai, and Shin (2006) demonstrated the AF activity of fats and phthalic acid derivatives isolated from *Sargassum confusum*. Palmitic acid, isolated from *S. muticum*, was found to inhibit spore germination from the green alga *Ulva* at 3 mg/mL (Bazes et al., 2009). Glycerolipids from *Ishige sinicola* show algicidal properties (Hirao, Tara, Kuwano, Tanaka, & Ishibashi, 2012). Othmani et al. (2013) recently reinforced the importance of the glycerol moiety when isolating the antibacterial compounds 1-O-octadecenoyl-glycerol and sn-3-O-(geranyl geranyl) glycerol from *Dictyota* spp. Kubanek et al. (2003) isolated lobophorolide from *Lobophora variegata*, a metabolite active against marine fungi. Work by Lachnit, Fischer, Kunzel, Baines, and Harder (2013) identified fucoxantin from *Fucus vesiculosus* that may inhibit the attachment and settlement of Gammaproteobacteria (*Pseudoalteromonas carrageenovora*, *Pseudomonas* sp., *Zooshikella* sp.) and Firmicutes bacteria (*Bacillus aquimaris* and *Bacillus subtilis*). Another study by Saha et al. (2012) isolated dimethylsulphopropionate (DMSP) and proline from the surface of *F. vesiculosus*. These compounds inhibited bacterial attachment with $CC_{50} < 0.38$ ng/cm^2 for DMSP and <0.1 ng/cm^2 and <1.66 ng/cm^2 for proline.

14.4.1.3 Chlorophyta

In comparison to Rhodophyta and Ochrophyta, fewer AF compounds from Chlorophyta were described to date. Organic extracts from *Caulerpa prolifera* inhibited the growth of two gram-positive and one gram-negative unidentified marine bacteria (Smyrniotopoulos et al., 2003). The inhibition zone varied from 17 to 83% of that observed with the biocide [bis-(tributylin) oxide] (TBTO). The authors reported the extract to be as effective as the TBTO standard in terms of growth inhibition of the microalga *Phaedactylum tricornutum*, a representative of the primary surface colonisers. The extract was purified and acetylene sesquiterpenoid esters were isolated. The steroid yengaroside-A, and the clerosterol galactoside isolated from *Codium iyengarii* exhibited antibacterial activity (Ali, Saleem, Yamdagni, & Ali, 2002).

14.4.2 Oxidative Bursts

The formation of reactive oxygen species (ROS), commonly referred as an 'oxidative burst', has been recognised as a defensive system in green, brown and red macroalgae (Potin, 2008). Recent reviews detailed this new, unexplored, but exciting area of research (Bischof & Rautenberger, 2012; Dring, 2006; Lesser, 2006; Potin, 2008).

In general, this phenomenon comprises a complex biochemical process, which results in the production of oxygen radicals, such as superoxide ($O_2^{\cdot -}$), hydrogen peroxide (H_2O_2), hydroxyl ($^{\cdot}OH$), etc. The emission of ROS is most often triggered by cell-to-cell recognition, encompassing perception at the cell membrane, but the chemical signal can be an output from an invader organism or from the macroalgal surface (Potin, 2008). Thus, it is easy to imagine that this process can actually be a real, important chemical defence system to at least some macroalgae.

For example, in the red macroalga *Chondrus crispus*, the production of H_2O_2 inhibits the settlement of parasites (Dring, 2006; Weinberg, 2007; Weinberg et al., 2005). Other red, green, and brown algae also exhibit this type of reaction, although with different elicitors, sources and requirements for activation of defence (Potin, 2008). In fact, large amounts of ROS are produced and rapidly released in response to an attack by fouling organisms (Collén, Jiménez del Río, Garcia-Reina, & Pedérsen, 1995; Weinberger & Friedlander, 2000; Weinberger, Friedlander, & Hoppe, 1999).

In general, inducers of OB have been recognised as elicitors of defence responses in marine plant–microbe interactions (Potin, 2008; Potin, Bouarab, Küpper, & Kloareg, 1999). For example, lipopolysachcharides from *Salmonella arbutus equi* are the elicitors of OB in the brown macroalga *Laminaria digitata* (Küpper et al., 2006). The ability to recognise an elicitor is an initial and fundamental aspect in the generation of ROS and it is, unlike plants, very specific in macroalgae (Potin, 2008).

The OB may produce other effects that can also contribute to the chemical defence system against fouling organisms, as detailed in a review by Potin (2008). For example, they include the gene-regulated responses, production of phenolics, cell wall cross-linking, responses to wounding, lipid peroxidation, generation of oxylipins, and emission of volatile halogenated organic compounds. But independent of the stage of development of studies on OB-associated processes and its ecological roles for macroalgae, it is easy to imagine its ecological importance in the context of biofouling. For example, microorganisms are initial components of the biofouling (microfouling) and sometimes they are important for subsequent development of macrofouling communities. Therefore a better understanding of OB and related events is paramount for understanding the diverse aspects of algal AF mechanisms.

Production of ROS is also an important component of the stress produced in marine organisms, since they are exposed to a variety of stimuli arising from changes in environmental conditions such as thermal stress, exposure to ultraviolet radiation, pollution, etc. (Aguilera & Rautenberger, 2012; Lesser, 2006). ROS production may have a great impact on cellular health and integrity due to oxidation of proteins, pigments, DNA, and peroxidation of lipids in biological membranes (Aguilera & Rautenberger, 2012). Future studies not only need to clarify how seaweeds produce ROS as AF chemicals, but also need to produce high activity of antioxidant enzymes, large pool sizes of antioxidants, as well as their rapid regeneration.

14.4.3 Quorum Sensing Modulation: Is it an Associational Defence?

Bacteria are capable of complex assemblage behaviour through cell-to-cell communication using diffusible chemical signal molecules that accumulate to a threshold concentration that activates target genes (Swift et al., 2001). This communication, referred to as QS, is a density-dependent cell–cell signalling among bacteria that can regulate production of bioactive compounds, motility, and biofilm formation. Because bacterial biofilm formation appears to encourage surface fouling by invertebrate larvae and algal spores, there has been a strong emphasis on the discovery of compounds that inhibit or influence QS for marine AF technology (Choudhary & Schmidt-Dannert, 2010; Dobretsov, Dahms, & Qian, 2006; Dobretsov, Teplitski, & Paul, 2009; Qian et al., 2010; Xiong & Liu, 2010). In gram-negative bacteria, QS is affected by AHLs.

In macroalgae, the most complete study regarding QS system and AHLs has been carried out on the marine macroalga *D. pulchra*. This seaweed produces an unusual class of secondary metabolite called halogenated furanones or fimbrolides. This class of compound acts as a specific antagonist of the AHL regulatory system (QS) present in bacteria, thereby inhibiting bacterial colonisation through a nontoxic and nongrowth mechanism (Hentzer et al., 2002; Kjelleberg & Steinberg, 2001; Manefield et al., 2002; de Nys & Steinberg, 2002; Steinberg, de Nys, & Kjelleberg, 2001). Furanones inhibit AHL-regulated phenotypes in a wide range of gram-negative bacteria by shutting down the biofilm development process (Givskov et al., 1996; Manefield et al. 1999, 2002).

It is noteworthy that bacteria, although part of the biofouling process which has probably selected algae for defence traits, may also be valuable for basibiont macroalgae, as they often produce chemicals that can act as antifoulants to the benefit of basibiont algae (e.g. Burgess et al., 2003; Dobretsov et al., 2006), act as QS inhibitors preventing colonisation or biofilm development by other bacteria, and even provide a chemical camouflage to the basibiont alga (e.g. Wahl, 1989; Wahl, Goecke, Labes, Dobretsov, & Weinberger, 2012). From this perspective, however, a significant part of the literature assessing AF activity throughout solely the inhibition of marine bacteria lacks ecological significance, unless the bacteria tested were shown to be harmful (e.g. Hellio et al., 2009). The association between algae and beneficial epibiont bacteria could therefore constitute a type of associational defence against fouling.

14.5 METHODS FOR ASSESSING ALGAL ANTIFOULANTS

In general, fouling is caused by two basic types of benthic organisms, i.e. microfoulers (e.g. bacteria, microalgae – mainly diatoms, fungi, and protozoa), and macrofoulers (e.g. macroalgae and invertebrates such as barnacles, bryozoans, mussels, and tube worms). In general, studies of algal AF activity focused on microfouling organisms, with fewer tests against macrofoulers and even less field experiments (Briand, 2009; Da Gama et al., 2003).

A complex biological process involving numerous organisms with different life cycles and different processes of adhesion to the substrate is presumably difficult to be reproduced through laboratory bioassays. However, bioassays used to evaluate AF activity are varied and often involve testing against a single component of the fouling community. Unfortunately, very few field and more realistic assays have been used to this aim (e.g. Da Gama, Pereira, Carvalho, Coutinho, & Yoneshigue-Valentin, 2002, 2003; Nylund & Pavia, 2005). In addition, it is known that host secondary metabolites may control attachment of planktonic larvae either directly by toxic effects or deterrence, or indirectly by modification of the microbial community living on basibionts (Steinberg & de Nys, 2002; Wahl, 1997).

In general, the AF prospection includes tests with bacteria, diatoms, micro- and macroalgae, barnacles, mussels and few other species, encompassing evaluations of growth, settlement, toxicity and behaviour. Among the different tests used to verify AF properties, bacteria have been most frequently employed, followed by diatoms (Briand, 2009), verifying the growth of these organisms by using disc diffusion assays. Probably, this broad use of bacteria, mainly biofilm forming, is due to the presumable importance of the microbial community to promote or inhibit macrofouling colonisation. However, it is worth mentioning that many bacterial communities growing on macroalgae do benefit them, and may even protect them against further colonisation by other fouling organisms (Burgess et al., 2003; Dobretsov et al., 2006).

Among the fouling macroorganisms, most assays have used barnacle cypris larvae (e.g. Qiu, Hung, & Qian, 2008). Other tests include larvae of the bryozoan *Bugula nertitina* (Schmitt et al., 1998), the polychaete *Hydroides elegans*, or mussels (Da Gama et al., 2003).

Fouling communities are usually composed of dozens or even hundreds of species. Therefore, although time consuming and costly, field tests of seaweed antifoulants are essential for understanding their activity spectrum (Da Gama et al., 2003; Da Gama, Carvalho, et al., 2008), a fundamental step in order to assure they may have any practical appeal for the development of future AF technologies (Hellio et al., 2009).

14.6 ECOLOGY VS. APPLICABILITY

There is no doubt that macroalgae are able to inhibit the settlement and growth of fouling organisms (de Nys, Dworjanyn, & Steinberg, 1998; Qian et al., 2010). Many macroalgae harbour microbial surface communities differing from their surrounding environments (Lachnit, Wahl, & Harder, 2010), suggesting a self-control by these basibionts. Numerous reports of compounds obtained from macroalgae exhibit a wide range of AF properties, including action against microfouling and macrofouling (Hellio et al., 2009). There are, however, very few reports about antifoulants from macroalgae from a really ecological perspective (e.g. Dworjanyn, de Nys, & Steinberg, 2006; Lachnit et al. 2013; Paradas et al. 2010).

This interesting theme can be explored both in the context of ecological and applied research. The latter approach, however, has received much more attention than the former, probably due to the urgent need for the development of more environmentally friendly AF strategies (Hellio & Yebra, 2009; Silkina, Bazes, Mouget, & Bourgougnon, 2012). Therefore there is abundant literature gathering data on AF properties of crude extracts and/or pure substances obtained from macroalgae with an applied approach (e.g. Bhadury & Wright, 2004; Da Gama et al., 2008, Hellio et al., 2009; Qian et al., 2010; Fusetani, 2011), since the discovery and development of 'natural antifoulants' is currently among the most important areas of modern marine biotechnology.

Since several macroalgae have developed efficient defence mechanisms against microbial epibionts, there is an increasing interest in such organisms as a source of naturally released AF substances (Hellio, De La Broise, et al., 2001; Hellio, Thomas-Guyon, et al., 2001; Maréchal & Hellio, 2009). In addition, unveiling the mechanisms by which the macroalgal chemicals are stored and transported to the algal surface is an important step to confirm their AF properties, in addition to bioassays with fouling organisms (e.g. Paradas et al., 2010). Indeed, the ecological and applied approaches are related and sometimes complementary, but lie on distinct theoretical foundations. In an ecological context, it is very important to understand that seaweeds may employ distinct AF mechanisms to avoid fouling, and the relative importance of each mechanism (but see Nylund & Pavia, 2005). This is not, however, an important information for an AF application.

The concentration of substances on the algal surface is often less than that existing inside the thallus (Sudatti, Rodrigues, & Pereira, 2006), and sometimes insufficient to inhibit fouling (Sudatti et al., 2008). However, this concentration can be very variable since the transport of materials is a dynamic process induced by factors such as temperature (Paradas et al., 2010). Chemicals that come to the surface and can act as antifoulants are stored within three basic types of specialised structures (Figure 14.1): gland cells (or vesicle cells), in which a storage vesicle occupies almost the entire cellular space at the cortical layer (Dworjanyn, De Nys, & Steinberg, 1999); the *corps en cerise*, which are refractile cell inclusions found in the cortex of some species of *Laurencia* (Young, Howard, & Fenical, 1980; Salgado et al., 2008), and physodes from Ochrophyta – brown macroalgae (Schoenwaelder, 2002). A new structure was discovered in cortical cells of another

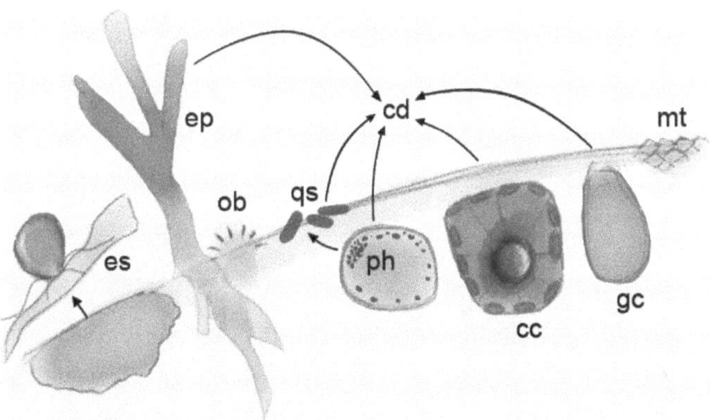

Figure 14.1 The multifaceted defence system of marine macroalgae against epibionts (**ep**). Physical defences such as epithallus sloughing (**es**) of colonised surfaces (a settling *Ulva* spore is represented) has been demonstrated in red, brown and green seaweeds, while thallus microtopography (**mt**) has only recently been investigated. Cellular structures related to AF chemical defence (**cd**), in particular halogenated natural products, were identified in thalli cortex from *Laurencia* spp. (*corps en cerise* – **cc**), *Delisea pulchra, Asparagopsis armata* (gland cells – **gc**), and *Plocamium brasiliense* and are possibly still unveiled in other red algae of the class Florideophyceae. Physodes (**ph**) are known intracellular structures related to polyphenolic chemical defences in brown algae, often suggested as antifoulants. Oxidative bursts (**ob**) and bacterial quorum sensing (**qs**) inhibition are AF mechanisms recently demonstrated in a number of algae. Despite the negative effects to thalli, defended macroscopic epibionts and specific bacteria may also produce compounds that chemically defend or camouflage the basibiont alga, therefore providing an associational defence. Mechanisms such as thallus abandonment/proliferation are not represented (see text for references). (See the colour plate.)

red alga, *Plocamium brasiliense* (Paradas, 2013). Several species of Ochrophyta also produce chemicals, the so-called phlorotannins, long presumed to play a role as AF compounds (Sieburth & Conover, 1965). However, AF activity may be negatively correlated with polyphenolic concentrations (Plouguerné, Cesconetto, Cruz, Pereira, & Da Gama, 2012).

14.7 GENERAL TRENDS AND PERSPECTIVES IN ALGAL ANTIFOULING DEFENCE

Biogeographical patterns of diversity such as latitudinal trends are well described for a number of biological groups, and chemical defence in particular has long been postulated to follow the latitudinal hypothesis, which predicts a decrease in diversity towards the poles (but see Pereira & Da Gama, 2008). However, latitudinal patterns of AF defence are difficult to draw due to the scarcity of systematic studies of seaweed AF defences along latitudinal gradients. To the authors' knowledge only one study assessed latitudinal trends of AF chemical defence on algae, using standardised methods, but found no traces of increased AF activity towards lower latitudes (Da Gama, Carvalho, et al., 2008), where fouling pressure is presumed to be higher (Railkin, 2003). Instead, a clear phylogenetic pattern emerged, with red macroalgae showing the highest proportion of AF activity (55% out of the 68 extracts or compounds from 42 algae), followed by brown macroalgae (14%), while green seaweeds never exhibited strong AF activity (i.e. ≥80% inhibition of mussel byssal attachment relative to controls). The majority (75%) of the laboratory results confirmed by field AF tests came from rhodophycean species (Da Gama, Carvalho, et al., 2008). This is further reinforced by the results from other studies (Hellio et al., 2002) and reviews (Hellio et al., 2009). However, Bhadury and Wright (2004) reported a more uniform distribution among algal groups.

Similarly, cellular structures possibly related to AF chemical defence include physodes (Lüder & Clayton, 2004), exclusive of brown algae, which have long been known to be related to polyphenolic (phlorotannin) defence, and, among red algae, gland cells, found in *Delisea* (Dworjanyn et al., 1999) and *Asparagopsis* (order Bonnemaisoniales – Paul, Cole, de Nys, & Steinberg, 2006), the *corps en cerise* (cherry bodies), usually found in cortical cells from many species of the *Laurencia* complex (order Ceramiales, Fuji et al., 2012), as well as similar spherical structures found in the cortex of *Plocamium brasiliense* (order Plocamiales, Paradas, 2013), all related to halogenated natural products. Interestingly, those structures were found only on the cortex of algae from these three orders from the class Florideophyceae. It seems very likely that similar structures, clearly related to surface defence, will be unveiled in other species from this group in the future.

Physical AF mechanisms such as epithallial sloughing (ES) were first described in a number of red algae, in particular from the order Corallinales (several genera), and then are also known in Gigartinales (e.g. *D. carnosa*, Nylund & Pavia, 2005). Among brown algae, ES was observed in Fucales (*Halidrys, Ascophyllum, Myagropsis* and *Sargassum* – Yamamoto et al., 2013) and Laminariales (*Egregia*), while in green algae this mechanism was described so far only in Bryopsidales (*Halimeda*) and Ulvales (*Ulva*), both from the class Ulvophyceae. Initially thought to be restricted to calcified seaweeds, ES does not seem to follow a phylogenetic pattern of occurrence. Similarly, recently discovered defence mechanisms such as OBs and QS seem to be ubiquitous mechanisms, and are likely to be part of the complex, multifaceted defence system of macroalgae (Figure 14.1). These results seem to suggest that research efforts should be focused on the more prolific red macroalgae in the quest for new natural product-based or bioinspired antifoulants from seaweeds. On the other hand, it seems clear that algal groups, such as green macroalgae, must have mechanisms of defence other than chemical against fouling. Littler and Littler (1999a) described blade abandonment/proliferation in the green alga *Avrainvillea longicaulis*, but this mechanism has never been described for the majority of green algae.

Modern marine benthic algae are formed by lineages that diverged c. one billion years ago. Recent results suggest that they feature natural immunity traits that are conserved, as well as others that appear to be phylum- or environment specific (Potin, Bouarab, Salatin, Pohnert, & Kloareg, 2002). Within this context, AF chemical defence in seaweeds appears to be phylum specific. On the other hand, from an analysis of the literature on AF activity from seaweed natural products, such as the review by Bhadury and Wright (2004), it may be concluded that brown macroalgae are the major producers of AF defences, in clear opposition to other authors, that described red macroalgae by far as more active against fouling. However, this could be a result of (1) a larger number of studies from temperate floras, in which brown macroalgae prevail, or (2) a large number of studies employing microbial assays in the laboratory as AF tests.

Variability in time or seasonality was seldom assessed in AF studies of macroalgae. Only two studies (Hellio et al., 2004; Maréchal et al., 2004) have addressed this important question, usually neglected in favour of single collection samplings, although the importance of understanding seasonal variation in AF defences increases with latitude. In lower latitudes, however, inducible defences make more sense than constitutive. Potin et al. (2002) suggested that the AF reactions of seaweeds can no longer be based on the search for constitutive defence chemicals alone. Da Gama, Carvalho, et al. (2008) performed assays to induce defence production through herbivore damage, and the majority of tested algae (88%) showed some degree of increased AF defence in comparison to control, undamaged algae. Weidner et al. (2004) found that of nine Brazilian algae tested for induced defence against herbivores, eight had at least a trend towards being less palatable after herbivore contact. The same trend seems to hold for AF chemical defences, although more relevant studies should be performed, perhaps simulating the detrimental effects of epibiosis to properly assess AF responses.

Chemical defence against epibionts is known in marine macroalgae since the early suspect that phlorotannins from *Sargassum* could retard fouling (Sieburth & Conover, 1965), while physical defence such as ES was unveiled much later (Keats et al., 1993). Although amazing progress in the knowledge concerning each of the known mechanisms of AF defence was achieved in recent years, the need to investigate deeply into the integration of those mechanisms within the algae is ever increasing. Similarly, from the original consumer-based ecology of algae, we need to further advance in order to understand the multiple ecological roles of algal traits shaped by evolution. As algae may be both basibionts and epibionts, we also need to further investigate the principles of algal settlement and recruitment on other organisms, since a trade-off between basibiont and epibiont traits on algae is possible due to the possible costs involved. Finally, community-wide effects of seaweed defences against epibionts are virtually unknown, but presumably exist, in particular in ecosystems largely dominated by macroalgae, such as kelp beds and isoyake or barren areas dominated by coralline algae (Masaki et al., 1984).

Much of the research in this field has been driven by the search for alternatives to the toxic AF paint biocides currently in use (Karlsson, Ytreberg, & Eklund, 2010). As science often imitates nature, different approaches have been suggested and attempted to develop new, bioinspired AF materials and coatings (e.g. Kirschner & Brennan, 2012; Salta et al., 2010; Scardino & de Nys, 2011). Parallels between natural algal AF mechanisms and commercial antifoulants include slow-release AF paints (e.g. self-polishing copolymers), microencapsulation (*viz. corps en cerise*, gland cells, physodes, etc.) and microtopography. Most likely, the best outcomes will be provided by approaches that combine more than one AF mechanism simultaneously, in a biomimicry of live marine macroalgae (Chapman et al., 2014).

Regardless of the virtual lack of commercial AF products from seaweeds, this may become reality soon, as White, Stewart, Grubbs, and Stoltz (2008) synthesised the algal antifoulant elatol, a necessary first step towards a commercial application of AF chemical defences from marine macroalgae, since the bottleneck is usually the production of natural compounds in large amounts. Other production alternatives have been extensively discussed (Bhadury & Wright, 2004).

Further experimental studies and collaborative research are needed to cast more light upon the dynamics of defence against epibionts in seaweeds, as well as the relative importance of epibiosis as a driving force for defence production.

ACKNOWLEDGEMENTS

BAPdG and RCP thank grant agencies CNPq and FAPERJ for Research Productivity and Young Scientist/Scientist of the State awards. BAPdG thanks MSc. R. Rogers for his fundamental help in Figure 14.1 improvement.

REFERENCES

Aguilera, J., & Rautenberger, J. (2012). Oxidative stress tolerance strategies of intertidal macroalgae. In D. Abele, J. P. Vázquez-Medina, & T. Tania Zenteno-Savın (Eds.), *Oxidative stress in aquatic ecosystems*. Blackwell.

Ali, M. S., Saleem, M., Yamdagni, R., & Ali, M. A. (2002). Steroid and antibacterial steroidal glycosides from marine green alga *Codium iyengarii* Borgesen. *Natural Product Letters*, *16*(6), 407–413.

Barbosa, J. P., Fleury, B. G., Da Gama, B. A. P., Teixeira, V. L., & Pereira, R. C. (2007). Natural products as antifoulants in the Brazilian brown alga *Dictyota pfaffii* (Phaeophyta, Dictyotales). *Biochemical Systematics and Ecology*, *35*, 549–553.

Bazes, A., Silkina, A., Douzenel, P., Fay, F., Kervarec, N., Morin, D., et al. (2009). Investigation of the antifouling constituents from the brown alga *Sargassum muticum* (Yendo) Fensholt. *Journal of Applied Phycology*, *21*, 395–403.

Bernstein, B. B., & Jung, N. (1979). Selective pressure and coevolution in a kelp canopy community in Southern California. *Ecological Monographs*, *493*, 335–355.

Bers, A. V., Díaz, E. R., Da Gama, B. A. P., Vieira-Silva, F., Dobretsov, S., Valdivia, N., et al. (2010). Relevance of mytilid shell microtopographies for fouling defence – a global comparison. *Biofouling*, *26*, 367–377.

Bhadury, P., & Wright, P. C. (2004). Exploitation of marine algae: biogenic compounds for potential antifouling applications. *Planta*, *219*, 561–578.

Bischof, K., & Rautenberger, R. (2012). Seaweed responses to environmental stress: reactive oxygen and antioxidative strategies. In C. Wiencke, & K. Bischof (Eds.), *Advances in seaweed biology. Novel insights into ecophysiology, ecology and utilization. Ecological studies* (Vol. 219). Springer.

Blunt, J. W., Copp, B. R., Munro, M. H. G., Northcote, P. T., & Prinsep, M. R. (2011). Marine natural products. *Natural Product Reports*, *28*, 196–268.

Borowitzka, M. A., & Larkum, A. W. D. (1977). Calcification in the green alga *Halimeda* I. An ultrastructure study of thallus development. *Journal of Phycology*, *13*, 6–16.

Brawley, S. H. (1992). Mesoherbivores. In D. H. John, S. J. Hawkins, & J. H. Price (Eds.), *Plant-animal interactions in the marine benthos. Systematics Association Special* (pp. 235–263). Oxford, UK: Clarendon Press.

Briand, J. F. (2009). Marine antifouling laboratory bioassays: an overview of their diversity. *Biofouling*, *25*, 297–311.

Burgess, J. G., Boyd, K. G., Armstrong, E., Jiang, Z., Yan, L. M., Berggren, M., et al. (2003). The development of a marine natural product-based antifouling paint. *Biofouling*, *19*, 197–205.

Buschmann, A. H., Correa, J. A., Westermeier, R., Hernández-González, M. C., & Norambuena, R. (2001). Red algal farming in Chile: a review. *Aquaculture*, *194*, 203–220.

Buschmann, A., & Gómez, P. (1993). Interaction mechanisms between *Gracilaria chilensis* (Rhodophyta) and epiphytes. *Hydrobiologia*, *260–261*, 345–351.

Callow, M. E., & Callow, J. A. (2002). Marine biofouling: a sticky problem. *The Biologist*, *49*, 10–14.

Cassano, V., De-Paula, J. C., Fujii, M. T., Da Gama, B. A. P., & Teixeira, V. L. (2008). Sesquiterpenes from the introduced red seaweed *Laurencia caduciramulosa* (Rhodomelaceae, Ceramiales). *Biochemical Systematics and Ecology*, *36*, 223–226.

Chambers, L. D., Stokes, K. R., Walsh, F. C., & Wood, R. J. K. (2006). Modern approaches to marine antifouling coatings. *Surface & Coatings Technology*, *202*, 412–413.

Chapman, J., Hellio, C., Sullivan, T., Brown, R., Russell, S., Kiterringham, E., et al. (2014). Bioinspired synthetic macroalgae: examples from nature for antifouling applications. *International Biodeterioration & Biodegradation*, *86*, 6–13.

Cho, J. Y. (2013). Antifouling chromanols isolated from brown alga *Sargassum horneri*. *Journal of Applied Phycology*, *25*, 299–309.

Choudhary, S., & Schmidt-Dannert, C. (2010). Applications of quorum sensing in biotechnology. *Applied Microbiology and Biotechnology*, *86*, 1267–1279.

Collén, J., Jiménez del Río, M., Garcia-Reina, G., & Pedérsen, M. (1995). Photosynthetic production of hydrogen peroxide by *Ulva rigida* C. Ag. (Chlorophyta). *Planta*, *196*, 225–230.

Culioli, G., Ortalo-Magne, A., Valls, R., Hellio, C., Clare, A. S., & Piovetti, L. (2008). Antifouling activity of meroditerpenoids from the marine brown alga *Halidrys siliquosa*. *Journal of Natural Products*, *71*, 1121–1126.

Da Gama, B. A. P., Carvalho, A. G. V., Weidner, K., Soares, A. R., Coutinho, R., Fleury, B. G., et al. (2008). Antifouling activity of natural products from Brazilian seaweeds. *Botanica Marina*, *51*, 191–201.

Da Gama, B. A. P., Pereira, R. C., Carvalho, M. G. V., Coutinho, R., & Yoneshigue-Valentin, Y. (2002). The effects of seaweed secondary metabolites on biofouling. *Biofouling*, *18*, 13–20.

Da Gama, B. A. P., Pereira, R. C., Soares, A. R., Teixeira, V. L., Yoneshigue-Valentin, Y. (2003). Is the mussel test a good indicator of antifouling activity? A comparison between laboratory and field assays. *Biofouling*, 19, 161–169.

Da Gama, B. A. P., Santos, R. P. A., & Pereira, R. C. (2008). The effects of epibionts on susceptibility to herbivory and fouling of the red seaweed *Cryptonemia seminervis*. *Biofouling*, *24*, 209–218.

Davis, T. A., Volesky, B., & Mucci, A. (2003). A review of the biochemistry of heavy metal biosorption by brown algae. *Water Research*, *37*, 4311–4330.

Dixon, J., Schroeter, S. C., & Kastendiek, J. (1981). Effects of the encrusting bryozoan, *Membranipora membranacea*, on the loss of blades and fronds by the giant kelp, *Macrocystis pyrifera* (Laminariales). *Journal of Phycology*, *17*, 341–345.

Dobretsov, S., Dahms, H. U., & Qian, P. Y. (2006). Inhibition of biofouling by marine microorganisms and their metabolites. *Biofouling*, *22*, 43–54.

Dobretsov, S., Teplitski, M., & Paul, V. (2009). Mini-review: quorum sensing in the marine environment and its relationship to biofouling. *Biofouling*, *25*, 413–427.

Dring, M. J. (2006). Stress resistance and disease resistance in seaweeds: the role of reactive oxygen metabolism. *Advances in Botanical Research*, *43*, 175–207.

Dworjanyn, S. A., De Nys, R., & Steinberg, P. D. (1999). Localisation and surface quantification of secondary metabolites in the red alga *Delisea pulchra*. *Marine Biology*, *133*, 727–736.

Dworjanyn, S. A., de Nys, R., & Steinberg, P. D. (2006). Chemically mediated antifouling in the red alga *Delisea pulchra*. *Marine Ecology Progress Series*, *318*, 153–163.

Enderlein, P., Moorthi, S., Rohrscheidt, H., & Wahl, M. (2003). Optimal foraging versus shared doom effects: interactive influence of mussel size and epibiosis on predator preference. *Journal of Experimental Marine Biology and Ecology*, *292*, 231–242.

Faulkner, D. J. (2002). Marine natural products. *Natural Products Reports*, *19*, 1–48.

Fenical, W. (1982). Natural products chemistry in the marine environment. *Science*, *215*, 923–928.

Filion-Myklebust, C., & Norton, T. A. (1981). Epidermis shedding in the brown seaweed *Ascophyllum nodosum* (L.) Le Jolis and its ecological significance. *Marine Biology Letters*, *2*, 45–51.

Fujii, M. T., Cassano, V., Senties, A., Diaz-Larrea, J., Machin-Sanchez, M., & Gil-Rodriguez, M. C. (2012). Comparative analysis of the corps en cerise in several species of *Laurencia* (Ceramiales, Rhodophyta) from the Atlantic Ocean. *Revista Brasileira de Farmacognosia-Brazilian Journal of Pharmacognosy*, *22*, 795–804.

Fusetani, N. (2011). Antifouling marine natural products. *Natural Products Report*, *28*, 400–410.

Fusetani, N., & Clare, A. S. (2006). *Antifouling compounds*. Berlin: Springer.

Ganti, V. S., Kim, K. H., Bhattarai, H. D., & Shin, H. W. (2006). Isolation and characterisation of some antifouling agents from the brown alga *Sargassum confusum*. *Journal of Asian Natural Products Research*, *8*, 309–315.

Givskov, M., DeNys, R., Manefield, M., Gram, L., Maximilien, R., Eberl, L., et al. (1996). Eukaryotic interference with homoserine lactone-mediated prokaryotic signaling. *Journal of Bacteriology, 178*, 6618–6622.

Gonzalez, M. A., & Goff, L. J. (1989). The red algal epiphytes *Microcladia coulteri* and *Microcladia californica* (Rhodophyceae, Ceramiaceae). II. Basiphyte specificity. *Journal of Phycology, 25*, 558–567.

Hay, M. E. (1996). Marine chemical ecology: what's known and what's next? *Journal of Experimental Marine Biology and Ecology, 200*, 103–134.

Hellio, C., Berge, J. P., Beaupoil, C., Le Gal, Y., & Bourgougnon, N. (2002). Screening of marine algal extracts for anti-settlement activities against microalgae and macroalgae. *Biofouling, 18*, 205–215.

Hellio, C., De La Broise, D., Dufosse, L., Le Gal, Y., & Bourgougnon, N. (2001). Inhibition of marine bacteria by extracts of macroalgae: potential use for environmentally friendly antifouling paints. *Marine Environmental Research, 52*, 231–247.

Hellio, C., Maréchal, J. P., Da Gama, B. A. P., Pereira, R. C., & Clare, A. (2009). Marine natural products with antifouling activities. In C. Hellio, & D. M. Yebra (Eds.), *Advances in marine antifouling coatings and technologies*. Cambridge: Woodhead Publishing.

Hellio, C., Simon-Colin, C., Clare, A. S., & Deslandes, E. (2004). Isethionic acid and floridoside isolated from the red alga, *Grateloupia turuturu*, inhibit settlement of *Balanus amphitrite* cyprid larvae. *Biofouling, 20*, 139–145.

Hellio, C., Thomas-Guyon, H., Culioli, G., Piovetti, L., Bourgougnon, N., & Le Gal, Y. (2001). Marine antifoulants from *Bifurcaria bifurcata* (Phaeophyceae, Cystoseiraceae) and other brown macroalgae. *Biofouling, 17*, 189–201.

Hellio, C., & Yebra, D. M. (2009). *Advances in marine antifouling coatings and technologies*. Cambridge: Woodhead Publishing.

Hentzer, M., Riedel, K., Rasmussen, T. B., Heydorn, A., Andersen, J. B., Parsek, M. R., et al. (2002). Inhibition of quorum sensing in *Pseudomonas aeruginosa* biofilm bacteria by a halogenated furanone compound. *Microbiology-Sgm, 148*, 87–102.

Hirao, S., Tara, K., Kuwano, K., Tanaka, J., & Ishibashi, F. (2012). Algicidal activity of glycerolipids from brown alga *Ishige sinicola* toward red tide microalgae. *Bioscience Biotechnology and Biochemistry, 76*, 372–374.

Jackson, J. B. C., & Buss, L. W. (1975). Allelopathy and spatial competition among coral reef invertebrates. *Proceedings of the National Academy of Sciences of the United States of America, 72*, 5160–5163.

Jha, B., Kavita, K., Westphal, J., Hartmann, A., & Schmitt-Kopplin, P. (2013). Quorum sensing inhibition by *Asparagopsis taxiformis*, a marine macro alga: separation of the compound that interrupts bacterial communication. *Marine Drugs, 11*, 253–265.

Johnson, C. R., & Mann, K. H. (1986). The crustose coralline alga, *Phymatolithon Foslie*, inhibits the overgrowth of seaweeds without relying on herbivores. *Journal of Experimental Marine Biology and Ecology, 96*, 127–146.

Karez, R., Engelbert, S., & Sommer, U. (2000). 'Coconsumption' and 'protective coating': two new proposed effects of epiphytes on their macroalgal hosts in mesograzer-epiphyte host interactions. *Marine Ecology Progress Series, 205*, 85–93.

Karlsson, J., Ytreberg, E., & Eklund, B. (2010). Toxicity of anti-fouling paints for use on ships and leisure boats to non-target organisms representing three trophic levels. *Environmental Pollution, 158*, 681–687.

Katsuoka, M., Ogura, C., Etoh, H., Sakata, K., & Ina, K. (1990). Galactosyldiacylglycerols and sulfoquinovosyldiacylglycerols isolated from the brown algae, *Undaria pinnatifida* and *Costaria costata* as repellents of the blue mussel, *Mytilus edulis*. *Agricultural and Biological Chemistry, 54*, 3043–3044.

Keats, D. W., Groener, A., & Chamberlain, Y. M. (1993). Cell sloughing in the littoral zone coralline alga, *Spongites yendoi* (Foslie) Chamberlain (Corallinales, Rhodophyta). *Phycologia, 32*, 143–150.

Keats, D.W., Knight, M. A., & Pueschel, C. M. (1997). Antifouling effects of epithallial shedding in three crustose coralline algae (Rhodopyhta, Corallinales) on a coral reef. *Journal of Experimental Marine Biology and Ecology*, *213*, 281–293.
Kim, J.Y., Alamsjah, M. A., Hamada, A., Fujita, Y., & Ishibashi, F. (2006). Algicidal diterpenes from the brown alga *Dictyota dichotoma*. *Bioscience Biotechnology and Biochemistry*, *70*, 2571–2574.
Kim, J. S., Kim, Y. H., Seo, Y. W., & Park, S. (2007). Quorum sensing inhibitors from the iced alga, *Ahnfeltiopsis flabelliformis*. *Biotechnology and Bioprocess Engineering*, *12*, 308–311.
Kirschner, C. M., & Brennan, A. B. (2012). Bio-inspired antifouling strategies. *Annual Review of Materials Research*, *42*, 211–229.
Kjelleberg, S., & Steinberg, P. (2001). Surface warfare in the sea. *Microbiology*, *28*.
Konig, G. M., & Wright, A. D. (1997). *Laurencia rigida*: chemical investigations of its antifouling dichloromethane extract. *Journal of Natural Products*, *60*, 967–970.
Konig, G. M., Wright, A. D., & de Nys, R. (1999). Halogenated monoterpenes from *Plocamium costatum* and their biological activity. *Journal of Natural Products*, *62*, 383–385.
Kubanek, J., Jensen, P. R., Keifer, P. A., Sullards, M. C., Collins, D. O., & Fenical, W. (2003). Seaweed resistance to microbial attack: a targeted chemical defense against marine fungi. *Proceedings of the National Academy of Sciences of the United States of America*, *100*, 6916–6921.
Küpper, F. C., Gaquerel, E., Boneberg, M., Morath, S., Salatin, J. O., & Potin, P. (2006). Early events in the perception of lipopolysaccharides in the brown alga *Laminaria digitata* include an oxidative burst and activation of fatty acid oxidation cascades. *Journal of Experimental Botany*, *57*, 1991–1999.
Lachnit, T., Fischer, M., Kunzel, S., Baines, J. F., & Harder, T. (2013). Compounds associated with algal surfaces mediate epiphytic colonization of the marine macroalga *Fucus vesiculosus*. *Fems Microbiology Ecology*, *84*, 411–420.
Lachnit, T., Wahl, M., & Harder, T. (2010). Isolated thallus-associated compounds from the macroalga *Fucus vesiculosus* mediate bacterial surface colonization in the field similar to that on the natural alga. *Biofouling*, *26*, 247–255.
Lane, A. L., Nyadong, L., Galhena, A. S., Shearer, T. L., Stout, E. P., Parry, R. M., et al. (2009). Desorption electrospray ionization mass spectrometry reveals surface-mediated antifungal chemical defense of a tropical seaweed. *Proceedings of the National Academy of Sciences of the United States of America*, *106*, 7314–7319.
Leal, M. C., Munro, M. H. G., Blunt, J. W., Puga, J., Jesus, B., Calado, R., et al. (2013). Biogeography and biodiscovery hot spots of macroalgal marine natural products. *Natural Product Reports*, *30*, 1380–1390.
Leonardi, P., Miravalles, A., Faugeron, S., Flores, V., Beltran, J., & Correa, J. (2006). Diversity, phenomenology and epidemiology of epiphytism in farmed *Gracilaria chilensis* (Rhodophyta) in northern Chile. *European Journal of Phycology*, *41*, 247–257.
Lesser, M. P. (2006). Oxidative stress in marine environments: biochemistry and physiological ecology. *Annual Review of Physiology*, *68*, 253–278.
Littler, M. M., & Littler, D. S. (1999a). Blade abandonment/proliferation: a novel mechanism for rapid epiphyte control in marine macrophytes. *Ecology*, *80*, 1736–1746.
Littler, M. M., & Littler, D. S. (1999b). Epithallus sloughing: a self-cleaning mechanism for coralline algae. *Coral Reefs*, *18*, 204.
Lüder, U. H., & Clayton, M. N. (2004). Induction of phlorotannins in the brown macroalga *Ecklonia radiata* (Laminariales, Phaeophyta) in response to simulated herbivory – the first microscopic study. *Planta*, *218*, 928–937.
Magin, C. M., Cooper, S. P., & Brennan, A. B. (2010). Non-toxic antifouling strategies. *Materials Today*, *13*, 36–44.
Manefield, M., de Nys, R., Kumar, N., Read, R., Givskov, M., Steinberg, P., et al. (1999). Evidence that halogenated furanones from *Delisea pulchra* inhibit acylated homoserine lactone (AHL)-mediated gene expression by displacing the AHL signal from its receptor protein. *Microbiology-UK*, *145*, 283–291.

Manefield, M., Rasmussen, T. B., Henzter, M., Andersen, J. B., Steinberg, P., Kjelleberg, S., et al. (2002). Halogenated furanones inhibit quorum sensing through accelerated LuxR turnover. *Microbiology-Sgm, 148*, 1119–1127.

Maréchal, J. P., Culioli, G., Hellio, C., Thomas-Guyon, H., Callow, M. E., Clare, A. S., et al. (2004). Seasonal variation in antifouling activity of crude extracts of the brown alga *Bifurcaria bifurcata* (Cystoseiraceae) against cyprids of *Balanus amphitrite* and the marine bacteria *Cobetia marina* and *Pseudoalteromonas haloplanktis*. *Journal of Experimental Marine Biology and Ecology, 313*, 47–62.

Maréchal, J. P., & Hellio, C. (2009). Challenges for the development of new non-toxic antifouling solutions. *International Journal of Molecular Sciences, 10*, 4623–4637.

Masaki, T., Fujita, D., & Hagen, N. (1984). The surface ultrastructure and epithallium shedding of crustose coralline algae in an 'isoyake' area of southwestern Hokkaido, Japan. *Hydrobiologia, 116/117*, 218–223.

Matsuo, Y., Ishida, R., Matsumoto, T., Tatewaki, M., & Suzuki, M. (1997). Yendolipin, a novel lipobetaine with an inhibitory activity toward morphogenesis in a foliaceous green alga *Monostroma oxyspermum*. *Tetrahedron, 53*, 869–876.

McArthur, D. M., & Moss, B. L. (1977). The ultrastructure of cell walls in *Enteromorpha intestinalis* (L.) Link. *British Phycological Journal, 12*, 359–368.

Mokrini, R., Ben Mesaoud, M., Daoudi, M., Hellio, C., Marechal, J. P., El Hattab, M., et al. (2008). Meroditerpenoids and derivatives from the brown alga *Cystoseira baccata* and their antifouling properties. *Journal of Natural Products, 71*, 1806–1811.

Moss, B. L. (1982). The control of epiphytes *Halidrys siliquosa* (L.) Lyngb. (Phaeophyta, Cystoseiraceae). *Phycologia, 21*, 185–188.

Muñoz, J., Culioli, G., & Kock, M. (2013). Linear diterpenes from the marine brown alga *Bifurcaria bifurcata*: a chemical perspective. *Phytochemistry Reviews, 12*, 407–424.

Muñoz, J., & Fotedar, R. (2009). Epiphytism of *Gracilaria cliftonii* (Withell, Millar & Kraft) from Western Australia. *Journal of Applied Phycology, 22*, 371–379.

Nylund, G. M., Cervin, G., Persson, F., Hermansson, M., Steinberg, P. D., & Pavia, H. (2008). Seaweed defence against bacteria: a poly-brominated 2-heptanone from the red alga *Bonnemaisonia hamifera* inhibits bacterial colonisation. *Marine Ecology Progress Series, 369*, 39–50.

Nylund, G. M., Gribben, P. E., de Nys, R., Steinberg, P. D., & Pavia, H. (2007). Surface chemistry versus whole-cell extracts: antifouling tests with seaweed metabolites. *Marine Ecology Progress Series, 329*, 73–84.

Nylund, G. M., & Pavia, H. (2005). Chemical versus mechanical inhibition of fouling in the red alga *Dilsea carnosa*. *Marine Ecology Progress Series, 299*, 111–121.

de Nys, R., Dworjanyn, S. A., & Steinberg, P. D. (1998). A new method for determining surface concentrations of marine natural products on seaweeds. *Marine Ecology Progress Series, 162*, 79–87.

de Nys, R., & Steinberg, P. D. (2002). Linking marine biology and biotechnology. *Current Opinion in Biotechnology, 13*, 244–248.

Orth, R. J., & van Montfrans, J. (1984). Epiphyte-seagrass relationships with an emphasis on the role of micrograzing: a review. *Aquatic Botany, 18*, 43–69.

Othmani, A., Bouzidi, N., Viano, Y., Alliche, Z., Seridi, H., Blache, Y., et al. (2013). Antimicrofouling properties of compounds isolated from several Mediterranean *Dictyota* spp. *Journal of Applied Phycology*, 1–12.

Paradas, W. C. (2013). *Mecanismos de armazenamento, biossíntese e liberação de metabolites secundários em macroalgas vermelhas (Rhodophyta). (PhD thesis), Pós-graduação em Biologia Marinha, Universidade Federal Fluminense, Niterói, Rio de Janeiro.*

Paradas, W. C., Salgado, L. T., Sudatti, D. B., Crapez, M. A. C., Fujii, M. T., Coutinho, R., et al. (2010). Induction of halogenated vesicle transport in cells of the red seaweed *Laurencia obtusa*. *Biofouling, 26*, 277–286.

Paul, V. J. (1992). *Ecological roles of marine natural products*. Ithaca, NY: Comstock.

Paul, N. A., Cole, L., de Nys, R., & Steinberg, P. D. (2006). Ultrastructure of the gland cells of the red alga *Asparagopsis armata* (Bonnemaisoniaceae). *Journal of Phycology*, *42*, 637–645.

Paul, N. A., de Nys, R., & Steinberg, P. D. (2006). Chemical defence against bacteria in the red alga *Asparagopsis armata*: linking structure with function. *Marine Ecology Progress Series*, *306*, 87–101.

Pereira, R. C., & Da Gama, B. A. P. (2008). Macroalgal chemical defenses and their roles in structuring tropical marine communities. In C. D. Amsler (Ed.), *Algal chemical ecology* (pp. 25–55). Germany: Springer-Verlag.

Pereira, R. C., Da Gama, B. A. P., Teixeira, V. L., & Yoneshigue-Valentin, Y. (2003). Ecologial roles of natural products from the Brazilian red seaweed *Laurencia obtusa*. *Brazilian Journal of Biology*, *63*, 665–672.

Piazza, V., Roussis, V., Garaventa, F., Greco, G., Smyrniotopoulos, V., Vagias, C., et al. (2011). Terpenes from the red alga *Sphaerococcus coronopifolius* inhibit the settlement of barnacles. *Marine Biotechnology*, *13*, 764–772.

Plouguerné, E., Cesconetto, C., Cruz, C. P., Pereira, R. C., & Da Gama, B. A. P. (2012). Within-thallus variation in polyphenolic content and antifouling activity in *Sargassum vulgare*. *Journal of Applied Phycology*, *24*, 1629–1635.

Plouguerné, E., Ioannou, E., Georgantea, P., Vagias, C., Roussis, V., Hellio, C., et al. (2010). Anti-microfouling activity of lipidic metabolites from the invasive brown alga *Sargassum muticum* (Yendo) Fensholt. *Marine Biotechnology*, *12*, 52–61.

Potin, P. (2008). Oxidative burst and related responses in biotic interactions of algae. In C. D. Amsler (Ed.), *Algal chemical ecology* (pp. 245–271). Germany: Springer-Verlag.

Potin, P., Bouarab, K., Küpper, F. C., & Kloareg, B. (1999). Oligosaccharides recognition signals and defence reactions in marine plant-microbe interactions. *Current Opinions in Microbiology*, *2*, 276–283.

Potin, P., Bouarab, K., Salatin, J. P., Pohnert, G., & Kloareg, B. (2002). Biotic interactions of marine algae. *Current Opinions in Plant Biology*, *5*, 308–317.

Pueschel, C. M., & Miller, T. J. (1996). Reconsidering prey specializations in an algal–limpet grazing mutualism: epithallial cell development in *Clathromorphum circumscriptum* (Rhodophyta, Corallinales). *Journal of Phycology*, *32*, 28–36.

Qian, P. Y., Xu, Y., & Fusetani, N. (2010). Natural products as antifouling compounds: recent progress and future perspectives. *Biofouling*, *26*, 223–234.

Qiu, J. W., Hung, O. S., & Qian, P. Y. (2008). An improved barnacle attachment inhibition assay. *Biofouling*, *24*, 259–266.

Railkin, A. I. (2003). *Marine biofouling: colonization processes and defenses*. Boca Raton, FL: CRC Press.

Rawlence, D. J., & Taylor, A. R. A. (1972). A light and electron microscopic study of rhizoid development in *Polysiphonia lanosa* (L.) Tandy. *Journal of Phycology*, *8*, 15–24.

Saha, M., Rempt, M., Gebser, B., Grueneberg, J., Pohnert, G., & Weinberger, F. (2012). Dimethylsulphopropionate (DMSP) and proline from the surface of the brown alga *Fucus vesiculosus* inhibit bacterial attachment. *Biofouling*, *28*, 593–604.

Salgado, L. T., Viana, N. B., Andrade, L. R., Leal, R. N., Da Gama, B. A. P., Attias, M., et al. (2008). Intra-cellular storage, transport and exocytosis of halogenated compounds in marine red alga *Laurencia obtusa*. *Journal of Structural Biology*, *162*, 345–355.

Salta, M., Wharton, J. A., Stoodley, P., Dennington, S. P., Goodes, L. R., Werwinski, S., et al. (2010). Designing biomimetic antifouling surfaces. *Philosophical Transactions of the Royal Society a - Mathematical Physical and Engineering Sciences*, *368*, 4729–4754.

Scardino, A. J., & de Nys, R. (2011). Mini review: biomimetic models and bioinspired surfaces for fouling control. *Biofouling*, *27*, 73–86.

Scheibling, R. E., Hennigar, A. W., & Balch, T. (1999). Destructive grazing, epiphytism, and disease: the dynamics of sea urchin – kelp interactions in Nova Scotia. *Canadian Journal of Fish Aquatic Science*, *56*, 2300–2314.

Schmitt, T. M., Hay, M. E., & Lindquist, N. (1995). Constraints on chemically mediated coevolution - multiple functions for seaweed secondary metabolites. *Ecology, 76*, 107–123.

Schmitt, T. M., Lindquist, N., & Hay, M. E. (1998). Seaweed secondary metabolites as antifoulants: effects of *Dictyota* spp. diterpenes on survivorship, settlement, and development of marine invertebrate larvae. *Chemoecology, 8*, 125–131.

Schoenwaelder, M. E. A. (2002). The occurrence and cellular significance of physodes in brown algae. *Phycologia, 41*, 125–139.

Sieburth, J. M., & Conover, J. T. (1965). *Sargassum* tannin, an antibiotic which retards fouling. *Nature, 208*, 52–53.

Sieburth, J. M., & Tootle, J. L. (1981). Seasonality of microbial fouling on *Ascophyllum nodosum* (L.) Lejol., *Fucus vesiculosus* L., *Polysiphonia lanosa* (L.) Tandy and *Chondrus crispus* Stackh. *Journal of Phycology, 17*, 57–64.

Silkina, A., Bazes, A., Mouget, J. L., & Bourgougnon, N. (2012). Comparative efficiency of macroalgal extracts and booster biocides as antifouling agents to control growth of three diatom species. *Marine Pollution Bulletin, 64*, 2039–2046.

Smyrniotopoulos, V., Abatis, D., Tziveleka, L. A., Tsitsimpikou, C., Roussis, V., Loukis, A., et al. (2003). Acetylene sesquiterpenoid esters from the green alga *Caulerpa prolifera*. *Journal of Natural Products, 66*, 21–24.

Steinberg, P. D., & de Nys, R. (2002). Chemical mediation of colonization of seaweed surfaces. *Journal of Phycology, 38*, 621–629.

Steinberg, P. D., de Nys, R., & Kjelleberg, S. (2001). Chemical mediation of surface colonization. *Marine Chemical Ecology*, 355–387 (unknown).

Sudatti, D. B., Rodrigues, S. V., Coutinho, R., Da Gama, B. A. P., Salgado, L. T., Amado-Filho, G. M., et al. (2008). Transport and defensive role of elatol at the surface of the red seaweed *Laurencia obtusa* (Ceramiales, Rhodophyta). *Journal of Phycology, 44*, 584–591.

Sudatti, D. B., Rodrigues, S. V., & Pereira, R. C. (2006). Quantitative GC-ECD analysis of halogenated metabolites: determination of elatol on surface and within-thallus of *Laurencia obtusa*. *Journal of Chemical Ecology, 32*, 835–843.

Swift, S., Allan Downie, J., Whitehead, N. A., Barnard, A. M. L., Salmond, G. P. C., & Williams, P. (2001). Quorum sensing as a population-density-dependent determinant of bacterial physiology. *Advances in Microbial Physiology, 45*, 199–270 (Academic Press).

Taniguchi, K., Shiraishi, K., Kurata, K., & Suzuki, M. (1989). Inhibitory effects of the settlement and the metamorphosis of the abalone Haliotis discus hannai veligers by the methanol extracts from the brown alga *Dilophus okamurai*. *Nippon Suisan Gakkaishi, 55*, 1133–1137.

Vairappan, C. S., Daitoh, M., Suzuki, M., Abe, T., & Masuda, M. (2001). Antibacterial halogenated metabolites from the Malaysian *Laurencia* species. *Phytochemistry, 58*, 291–297.

Vairappan, C. S., Suzuki, M., Abe, T., & Masuda, M. (2001). Halogenated metabolites with antibacterial activity from the Okinawan *Laurencia* species. *Phytochemistry, 58*, 517–523.

Viano, Y., Bonhomme, D., Camps, M., Briand, J. F., Ortalo-Magne, A., Blache, Y., et al. (2009). Diterpenoids from the mediterranean brown alga *Dictyota* sp. evaluated as antifouling substances against a marine bacterial biofilm. *Journal of Natural Products, 72*, 1299–1304.

Wahl, M. (1989). Marine epibiosis. I. Fouling and antifouling: some basic aspects. *Marine Ecology Progress Series, 58*, 175–189.

Wahl, M. (1997). Living attached: aufwuchs, fouling, epibiosis. In R. Nagabhushanam, & M. F. Thompson (Eds.), *Fouling organisms in the Indian Ocean: biology and control technology* (pp. 31–83). Rotterdam: A. A. Balkema.

Wahl, M. (2008). Ecological lever and interface ecology: epibiosis modulates the interactions between host and environment. *Biofouling, 24*, 427–438.

Wahl, M., Goecke, F., Labes, A., Dobretsov, S., & Weinberger, F. (2012). The second skin: ecological role of epibiotic biofilms on marine organisms. *Frontiers in Microbiology, 1*, 1–21.
Wahl, M., & Hay, M. E. (1995). Associational resistance and shared doom: effects of epibiosis on herbivory. *Oecologia, 102*, 329–340.
Wahl, M., Kröger, K., & Lenz, M. (1998). Non-toxic protection against epibiosis. *Biofouling, 12*, 205–226.
Weidner, K., Lages, B. G., Da Gama, B. A. P., Molis, M., Wahl, M., & Pereira, R. C. (2004). Effects of mesograzers and nutrient levels on the induction of defenses in several Brazilian macroalgae. *Marine Ecology Progress Series, 283*, 113–125.
Weinberger, F. (2007). Pathogen-induced defense and innate immunity in macroalgae. *Biological Bulletin, 213*, 290–302.
Weinberger, F., & Friedlander, M. (2000). Response of *Gracilaria conferta* (Rhodophyta) to oligoagars results in defense against agar-degrading epiphytes. *Journal of Phycology, 36*, 1079–1086.
Weinberger, F., Friedlander, M., & Hoppe, H. G. (1999). Oligoagars elicit a physiological response in *Gracilaria conferta*. *Journal of Phycology, 35*, 747–755.
Weinberger, F., Leonardi, P., Miravalles, A., Correa, J. A., Lion, U., Kloareg, B., et al. (2005). Dissecation of two distinct defense-related responses to agar oligosaccharides in *Gracilaria chilensis* (Rhodophyta) and *Gracilaria conferta* (Rhodophyta). *Journal of Phycology, 41*, 863–873.
White, D. E., Stewart, I. C., Grubbs, R. H., & Stoltz, B. M. (2008). The catalytic asymmetric total synthesis of elatol. *Journal of the American Chemical Society, 130*, 810–811.
Wikström, S. A., & Pavia, H. (2004). Chemical settlement inhibition versus post-settlement mortality as an explanation for differential fouling of two congeneric seaweeds. *Oecologia, 138*, 223–230.
Williams, G. A., & Seed, R. (1992). Interactions between macro- faunal epiphytes and their host algae. In D. H. John, S. J. Hawkins, & J. H. Price (Eds.), *Plant-animal interactions in the marine benthos. Systematics Association Special* (pp. 189–211). Oxford, UK: Clarendon.
Xiong, Y. H., & Liu, Y. (2010). Biological control of microbial attachment: a promising alternative for mitigating membrane biofouling. *Applied Microbiology and Biotechnology, 86*, 825–837.
Yamamoto, K., Endo, H., Yoshikawa, S., Ohki, K., & Kamiya, M. (2013). Various defense ability of four sargassacean algae against the red algal epiphyte *Neosiphonia harveyi* in Wakasa Bay, Japan. *Aquatic Botany, 105*, 11–17.
Young, D. N., Howard, B. M., & Fenical, W. (1980). Subcellular localization of brominated secondary metabolites in the red alga *Laurencia snyderae*. *Journal of Phycology, 16*, 182–192.

CHAPTER FIFTEEN

Haslea ostrearia-like Diatoms: Biodiversity out of the Blue

Romain Gastineau*, Nikolai Davidovich†, Gert Hansen‡, Jan Rines§, Angela Wulff¶, Irena Kaczmarska‖, James Ehrman‖, Dorothée Hermann*, Florian Maumus#, Yann Hardivillier*, Vincent Leignel*, Boris Jacquette*, Vona Méléder**, Gustaaf Hallegraeff††, Marian Yallop‡‡, Rupert Perkins§§, Jean-Paul Cadoret¶¶, Bruno Saint-Jean¶¶, Grégory Carrier¶¶ and Jean-Luc Mouget*,[1]

*LUNAM, FR 3473, MMS Mer, Molécules, Santé, Université du Maine, Le Mans, France
†Karadag Nature Reserve of the National Academy of Sciences, Feodosiya, Ukraine
‡Department of Biology, Marine Biological Section, University of Copenhagen, Copenhagen, Denmark
§Graduate School of Oceanography University of Rhode Island, Narragansett, RI, USA
¶Department of Biological and Environmental Sciences, University of Gothenburg, Göteborg, Sweden
‖Department of Biology, Mount Allison University, Sackville, NB, Canada
#Unité de Recherche en Génomique-Info, UR 1164, INRA Centre de Versailles-Grignon, Versailles, France
**LUNAM FR 3473, MMS Mer, Molécules, Santé, Université de Nantes, Nantes, France
††Institute for Marine and Antarctic Studies (IMAS), Hobart, Australia
‡‡School of Biological Sciences, University of Bristol, Bristol, UK
§§School of Earth and Ocean Sciences, Cardiff University, Cardiff, Wales, UK
¶¶Laboratoire de physiologie et biotechnologie des algues, IFREMER, Nantes, France
[1]Corresponding author: e-mail address: Jean-Luc.Mouget@univ-lemans.fr

Contents

Advances in Botanical Research, Volume 71
ISSN 0065-2296
http://dx.doi.org/10.1016/B978-0-12-408062-1.00015-9

Abstract

Diatoms are usually referred to as golden-brown microalgae, due to the colour of their plastids and to their pigment composition, mainly carotenoids (fucoxanthin, diadinoxanthin, diatoxanthin), which mask chlorophylls *a* and *c*. The species *Haslea ostrearia* Gaillon/Bory (Simonsen) appears unique because of its extraplastidial bluish colour, a consequence of the presence of a water-soluble blue pigment at cell apices, marennine. When released in seawater, marennine can be fixed on gills of oysters and other bivalves, which turn green. This greening phenomenon is economically exploited in Southwestern France, as it gives an added value to oysters. For decades, this singularity ascribed a worldwide distribution to *H. ostrearia*, first as *Vibrio ostrearius*, then *Navicula ostrearia*, last as *H. ostrearia*, when the genus *Haslea* was proposed by R. Simonsen (1974). Indeed, this 'birthmark' (presence of blue apices) made *H. ostrearia* easily recognisable without further scrutiny and identification of the microalga as well as its presence easily deduced from the greening of bivalves. Consequently, the widely admitted cosmopolitan character of *H. ostrearia* has only been questioned recently, following the discovery in 2008, of a new species of blue diatom in the Black Sea, *Haslea karadagensis*. The biodiversity of blue diatoms suddenly increased with the finding of other blue species in the Mediterranean Sea, the Canary Islands, etc., the taxonomic characterization of which is in progress. This review thus focuses on the unsuspected biodiversity of blue diatoms within the genus *Haslea*. Methods for species determination (morphometrics, chemotaxonomy, genomics), as well as a new species, are presented and discussed.

15.1 INTRODUCTION: THE DIATOMS

This chapter deals with peculiar diatoms, the ones with blue apices, hereafter called blue diatoms. Given the topic of this volume, one might ask the question: are diatoms 'sea plants'? At least for the ones living in seas and oceans, the answer to this trivial question is yes. Indeed, diatoms or Bacillariophyceae are microscopic algae, which possess main plant characters. Most of them are autotrophic organisms; they achieve oxygenic photosynthesis and have long been included among the 'flowerless plants', a polyphyletic group that encompassed organisms such as algae (both macro and microalgae), lichens, mosses, ferns, etc. Diatoms are unicellular, sometimes colonial organisms inhabiting almost every aquatic environment, from salt marshes to freshwaters, polar or tropical seas, ice or high-altitude lakes (Smol & Stoermer, 2010). Some diatoms can also be found on seabird feathers (Croll & Holmes, 1982), cetacean skin (Bennett, 1920) or turtle shells (Wetzel, Van de Vijver, & Ector, 2010). Some are endosymbionts of foraminifera (Lee, 2011), or can live inside lichens (Lakatos, Lange-Bertalot, & Büdel, 2004) or sponges (Bavestrello et al., 2000). Others are hosts of cyanobacteria (Stewart, Rowell, & Rai, 1983) or have their surface colonised by ciliates (Cupp, 1943).

Diatoms present unique features, as a silicified cell wall, the frustule, which consists of two valves, the epitheca and the hypotheca, that are held in place by silicified girdle bands, and which present a great variety of size and shapes (Round, Crawford, & Mann, 1990). Diatoms are also characterised by their mitotic cell division. When a diatom undergoes mitosis, each daughter cell receives one of the two valves of the frustule from the parent cell. The inherited valve is used as the outer valve, and the daughter cell synthesises a new inner valve. Hence, one of the daughter cells is identical in size to the parental cell, and the other is smaller, a phenomenon which usually leads to a reduction in the average cell size of the population and to its die-out (MacDonald, 1869; Pfitzer, 1869). In natural environments, this is usually counteracted by a phase of sexual reproduction, when zygotes turn into auxospores, which expand and form initial cells, which restore the maximal specific cell size. Diatoms have a diplontic life cycle. They are conveniently divided into two groups based on valve ornamentation and their mode of reproduction: centric forms and pennate forms (Sims, Mann, & Medlin, 2006). Centric diatoms display radial valve ornamentation and are oogamous (i.e. they produce small motile male gametes and large non-motile female gametes), whereas pennate diatoms are boat-shaped, and are aplanogamous (they do not release flagellate gametes) (Chepurnov, Mann, Sabbe, & Vyverman, 2004).

Diatoms are usually presented as 'golden-brown microalgae', in regard of their pigments, mainly chlorophylls *a* and *c*, masked by fucoxanthin (Round et al., 1990). In this respect, for decades, *Haslea ostrearia* Gaillon/Bory (Simonsen) (Figure 15.1) was considered the only species, worldwide in distribution (Figure 15.2), known for its capacity to synthesise a blue pigment, marennine, which accumulates at cell apices (for historical considerations, see Briée, 2010). Recently, new species of diatoms with blue apices, producing marennine-like pigments, have been discovered, which questions the supposed uniqueness of *H. ostrearia*, and unravels blue diatom biodiversity. This review presents recent advances in our knowledge of these blue diatoms.

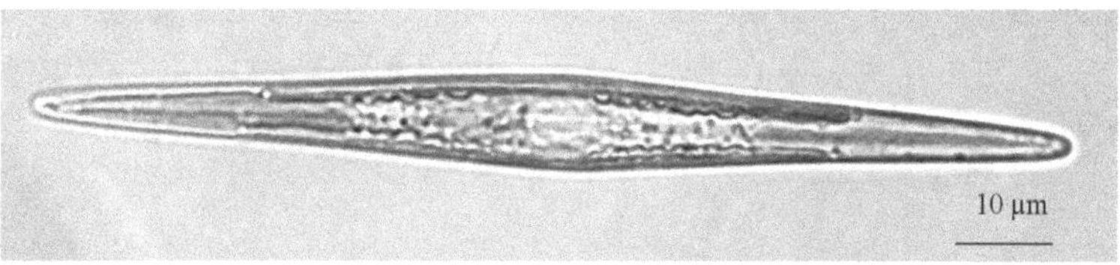

Figure 15.1 Live cell of *Haslea ostrearia* in light microscopy (LM). (See the colour plate.)

Figure 15.2 World localisation of *Haslea ostrearia* according to the literature. Each point indicates a site where the presence of *Haslea ostrearia* was assessed from observation of diatom with blue apices, or deduced from the occurrence of green oysters. *Map copyright free from* http://www.histgeo.ac-aix-marseille.fr.

15.2 THE GENUS *HASLEA*

The genus *Haslea* is one among many genera of pennate diatoms belonging to the family *Naviculaceae*. Simonsen (1974) created this genus by transferring to it several species formerly described in the genera *Navicula*, *Amphipleura*, *Pleurosigma*, *Schizonema* or *Stauroneis*. He did so after having observed significant common morphological features using scanning electron microscopy (SEM). Morphologically, species from the genus *Haslea* are characterised by their pennate shapes. The raphe, longitudinal groove on the frustule, is straight, but its endings may have different conformations (see below). The raphe is found on both valves. Living cells possess two parietal chloroplasts. Light microscopy cannot provide more information on frustule ornamentation, and striae are most of the time only barely visible. Using SEM however, it is obvious in external view that the frustules are made of longitudinal slits of silica, parallel to the raphe. Quadrate areolae can be seen in internal view. The name of the genus was given in honour of the Norwegian biologist Grethe Rytter Hasle (1920–2013). The type species for the genus *Haslea* is *H. ostrearia*, and the diagnosis by Simonsen (1974) is as follow:

Novum genus familiae Naviculacearum. Frustula membrana tenue, habitu fusiforme, apicibus acutis lateribus convexis, raro in media parte parallelis. Raphe recta, cum poris centralibus parvis, approximatis, poris apicalibus paulo evolutis. Areae axiales centralesque inconspicuae. Structura membranae valvae seriebus areolarum transapicalibus parallelibus apicalibusque rectis composita. Areolae in parte exteriore membrana tenuissima tectae, cum singula rima in unaquaque areola. Multae species planctonicae. Generotypus Haslea ostrearia (Gaillon) nov. comb.

Presently, the genus *Haslea* encompasses more than 20 species, which may present significant differences in morphology, ecology or physiology. For example, *Haslea wawrikae* is a planktonic diatom with very long and acute apices, and *Haslea gigantea* a mesoplanctonic species, with a cell length up to 500 μm (Simonsen, 1974). Most other *Haslea* species are known for being benthic organisms, and *Haslea crucigera* is a tube-dwelling species. In contrast with other *Haslea* species, *Haslea salstonica*, *H. crucigera*, and *Haslea spicula* display a thickening of their central ribs known as a pseudostauros, visible even using the light microscope (Massé et al., 2001; Round et al., 1990). The sigmoid shape of *Haslea nipkowii* is responsible for its previous classification as a *Gyrosigma* (Poulin et al., 2004). A first molecular phylogeny of some *Haslea* species constructed using chloroplast 16S rDNA (Pillet, de Vargas, & Pawlowski, 2011) supports the division of this genus into several sub-genera. This phylogeny groups the species with a pseudostauros (*H. crucigera* and *H. salstonica*) and discriminates between the sigmoid *H. nipkowii*, the planktonic *H. wawrikae*, the non-blue (and without pseudostauros) *Haslea pseudostrearia* and the blue *H. ostrearia*, which was the only blue species known at that time. With the recent discovery of new species of blue diatoms, the relative positions of the different clusters of species inside the genus *Haslea* and the status of the genus deserve further studies.

15.3 BIOLOGY OF *HASLEA OSTREARIA*

Haslea ostrearia is a marine pennate diatom with a specific diagnostic character, the water-soluble blue-pigment marennine, visible at the cell apices (Figure 15.1). Marennine is synthesised and released in seawater during growth, its production stimulated by high light, long photoperiod, blue light and possibly nutrient limitation (Mouget, Rosa, Vachoux, & Tremblin, 2005; Neuville & Daste, 1978; Robert, Morançais, Pradier, Mouget, & Tremblin, 2002). *H. ostrearia* is a unicellular tychoplanktonic microalga (Robert, 1983), mainly benthic, occasionally planktonic, but also epiphytic (Sauvageau, 1906). This species is euryhaline (Neuville & Daste, 1978; Wraige, Belt, Masse, Robert, & Rowland, 1998), and seems well adapted to shallow waters, such as oyster ponds. Indeed, it can acclimate to change in light quality (Mouget, Rosa, & Tremblin, 2004) and to high irradiance environments (Mouget, Tremblin, Morant-Manceau, Morançais, & Robert, 1999), and possesses defence mechanisms against UV stress, at doses similar to those experienced in oyster ponds (Fouqueray, 2007; Rech, 2004; Rech, Mouget, Morant-Manceau, Rosa, & Tremblin, 2005). It is an obligate photoautotroph, although it could exhibit photoheterotrophy or photomixotrophy, especially regarding some nitrogen and carbon organic substances (Neuville & Daste, 1978; Robert et al., 1982). Like all diatoms, *H. ostrearia* possesses two pathways for inorganic carbon assimilation, the Calvin–Benson cycle associated to the RuBPC and the PEPCK (Tremblin & Robert, 2001), and an anhydrase carbonic activity has been demonstrated (Rech, Morant-Manceau, & Tremblin, 2008). *H. ostrearia* mean biochemical composition in carbohydrate (50 mg 10^{-9} cells) and proteins (200 mg 10^{-9} cells) is stable during the ageing of the cultures, but as a function of the culture phase (exponential or stationary), lipid composition can vary between 60 and 200 mg 10^{-9} cells (Robert, 1983), with proportions of fatty acids, glycolipids and neutral lipids similar to those of other diatoms (Groth-Nard, 1994). The profile of galactolipids, the most important lipids in plastid membranes, is similar to those of other pennates, with more C_{18}/C_{16} forms of both mono- and digalactosyldiacylglycerols, but significantly different from those of centric diatoms, mainly composed of C_{20}/C_{16} (Dodson, Dahmen, Mouget, & Leblond, 2013). Key aspects of the *H. ostrearia* life cycle have been characterised, as the vegetative division (Neuville & Daste, 1972a), the range in cell size variation (c. 20–140 μm) and its dynamics (c. 4 μm decrease in apical length of cell per month) (Robert, 1978; Davidovich, Mouget, & Gaudin, 2009), and the maximum growth rates (c. 0.4–0.6/d) over a wide range of environmental conditions (Mouget et al., 1999; Robert, 1978, 1983; Robert et al., 2002).

Haslea ostrearia's sexual reproduction (auxosporulation) was first described by Neuville and Daste (1975, 1979) as homothallic, being intraclonal, and triggered by short photoperiod and low light conditions. The same conditions have been evidenced as inducing factors for heterothallic reproduction (Davidovich et al., 2009; Mouget, Gastineau, Davidovich, Gaudin, & Davidovich, 2009) between compatible pairs of clones of two different sexual types. The cardinal points, such as the size-threshold for auxosporulation or the maximal size of the initial cells, have been determined (Davidovich et al., 2009). It has been observed that the light quality (presence of red part of PAR spectra) has a role for auxosporulation induction (Mouget et al., 2009).

15.4 MARENNINE, A SPECIFIC BLUE PIGMENT

For decades, it has been known that marennine was the pigment produced by the single diatom species *H. ostrearia*, but its structure remains controversial. To obtain purified marennine is a prerequisite step before any characterisation could be achieved. Pouvreau, Morançais, Massé, et al. (2006) reviewed the different methods previously used to extract marennine (usually aqueous solvents), and to purify it, and presented a new protocol, which basically consists of an ultrafiltration and anion-exchange chromatography. Different hypotheses were proposed for its chemical nature, but none have been confirmed, such as metallic salt (Boyce & Herdman, 1896, pp. 663–669; Lankester, 1886; Mitchell & Barney, 1918; Sullivan, 1977; Valenciennes, 1841), carotenoid (Ranson, 1937), chlorophyll degradation product (Bachrach, 1935; Bachrach & Jouvent, 1933; Bachrach & Simonet, 1936; Genevès, Choussy, Barbier, Neuville, & Daste, 1976; Moreau, 1967; Robert, 1983), anthocyanin (Neuville & Daste, 1972b), or a phycobilin-like molecule (Bocat, 1907; Robert & Hallet, 1981; Ryder, 1884). Based on some biophysical and chemical characteristics, marennine is possibly a polyphenolic compound, existing as two forms, one intracellular and one extracellular, which differ by their spectral characteristics (UV–visible spectrophotometry, Raman spectroscopy), and molecular weight, 10.7 and 0.9 kDa, respectively (Pouvreau, Morançais, Fleury, et al., 2006). For both forms, marennine colour changes with pH, from violet-blue at acidic pH to green at basic pH (Pouvreau, Morançais, Fleury, et al., 2006). The existence of two different forms of marennine argues against passive release of the pigment at cell death (Bornet & Chatin, 1895), but reinforces the hypothesis of an active secretion (Robert, 1983; Robert et al., 2002) and chemical cleavage by the algal cells (Nassiri, Robert, Rincé, & Ginsburger-Vogel, 1998).

15.5 GREENING OYSTERS, SIGNATURE OF *HASLEA OSTREARIA*

Historically, *H. ostrearia* is intrinsically related to oysters as marennine is responsible for the greening of bivalves, whether it occurs in natural environments or in oyster ponds. The first mention of green oysters in the literature must be ascribed to Sprat (1667), who hypothesised how the combined action of sun and earth led to a green colouration of pond sediments in the Colchester area (England) before oysters turn green, thus postulating an abiotic origin. A major advance regarding a biological causal factor was made by Gaillon (1820) who observed small organisms collected at the surface of the shell of green oysters using a microscope device. Because these organisms were green and motile, Gaillon named them *Vibrio ostrearius*, in reference to the worm classification by Bruguière (1792). However, a few years later, Bory de Saint-Vincent proposed to group *V. ostrearius* into his 'psychodiaire' reign, which contained all organisms whose position between animal and plant was unclear, and to rename it *Navicula ostrearia* because of similarities to naviculoid diatoms (Bory de Saint-VincentBory de Saint-Vincent, 1824). As to the name of the specific blue pigment produced by *N. ostrearia*, marennine is the gallicised form of marennin, proposed by Lankester (1886) in reference to the Bay of Marenne, where green oysters have been produced for centuries.

Naturally occurring greening of oysters was observed in England and France, but also in Denmark (Petersen, 1916), in the United States of America (Virginia, Mitchell & Barney, 1918), in Canada (New-Brunswick, Medcof, 1945; Prince Edward Island, Kerswill, 1946) and in Australia (Moreton Bay, Great Oyster Bay). On the other hand, experimental greening of oysters or other invertebrates using cultures or supernatant of *H. ostrearia* was conducted by many authors, e.g. Dyer (1877), Puységur (1880) Sauvageau (1907), Gastineau et al. (2012a), demonstrating that the adsorption of marennine onto gills can be a common phenomenon.

15.6 MARENNINE BIOLOGICAL FUNCTION AND ACTIVITIES

Although occurring spontaneously in certain natural environments, it is doubtful whether greening bivalves is an issue per se for the diatom *H. ostrearia*. If some studies considered marennine as an autotoxine, or associated it with cell pathological states (Bachrach, 1935; Moreau, 1970; Neuville & Daste, 1978), the 'metabolic cost of production' of marennine in such quantity suggests it could play a significant ecological role, but its exact biological function remains unclear. As a pigment, a molecule that interacts with visible light, marennine is not directly involved in photosynthesis, neither used for light capture, nor for dissipation of excess energy excitation (Schubert, Tremblin, Robert, Sagert, & Rincé, 1995). However, marennine could play an indirect role in photosynthesis by modifying light quality and intensity as it passes through the water column (Mouget et al., 1999; Schubert et al., 1995; Tremblin, Cannuel, Mouget, Rech, & Robert, et al.,

2000). Marennine could help *H. ostrearia* outcompete other microalgae, as Pouvreau et al. (2007) demonstrated allelopathic activities on different diatom species encountered in oyster ponds, such as *Skeletonema costatum*. Moreover, it has been shown that purified marennine exhibits other biological activities, as antibacterial and antiviral (Gastineau et al., 2012a), or antioxidant (Pouvreau et al., 2008). This confirms former studies on *H. ostrearia* crude extracts, with demonstrated antiviral (Bergé et al., 1999) and antiproliferative activities (Carbonelle et al., 1999). A role in the protection against metals such as copper has also been suggested (Minier, Tutundjian, Galgani, & Robert, 1998).

It is worth noting that from a biotechnology and biorefinery perspective, marennine is not the only fraction of biomass possibly of interest. For instance, regarding lipids, *H. ostrearia* produces ω-3 poly-unsaturated fatty acids (Mimouni, Ulmann, Tremblin, & Robert, 2003), which are essential for health, and also peculiar C_{29} sterols, including up to 95% of 24-ethylcholesta-5-en-3β-ol (Barrett, Volkman, Dunstan, & LeRoi, 1995). It also produces haslenes, rare highly branched isoprenoids (HBIs) (Allard et al., 2001; Belt, Cooke, Robert, & Rowland, 1996; Johns et al., 2000; Wraige et al., 1997; Wraige et al., 1998; Wraige et al., 1999). Like squalene, haslenes have biological activities (Harivardhan Reddy & Couvreur, 2009), in particular cytostatic activities, as demonstrated using the NSCLC-N6 cell line derived from a human nonsmall-cell bronchopulmonary carcinoma (Rowland et al., 2001).

15.7 *HASLEA OSTREARIA* A PRESUMED COSMOPOLITAN SPECIES

For decades, blue diatoms have been observed in almost all seas and oceans in northern and southern hemispheres, from temperate to tropical waters, and identified as *H. ostrearia*. According to the literature (Figure 15.3), *H. ostrearia* is common on the French Atlantic coast, especially in anthroposised environments (oyster ponds) encountered in the Bay of Marennes-Oléron and Bay of Bourgneuf, but also in the North Sea, in Great Britain (Hustedt & Aleem, 1951), Belgium (M'harzi et al., 1998), Denmark (Petersen, 1916), Norway (Hendey, 1964), Sweden (Aleem, 1973; Cleve-Euler, 1955; Kuylenstierna, 1990; Wulff, Wängberg, Sundbäck, Nilsson, & Underwood, 2000) and in the Baltic Sea (Snoeijs & Kasperovičienė, 1996). Reports of blue diatoms were made elsewhere in the western Mediterranean Sea (e.g. Sauvageau, 1906), Sardinia and Sicily (Barone, 1979; Solazzi & Tolomio, 1976; Tolomio, 1978), the Adriatic Sea (Molisch, 1903; Munda, 2005), the

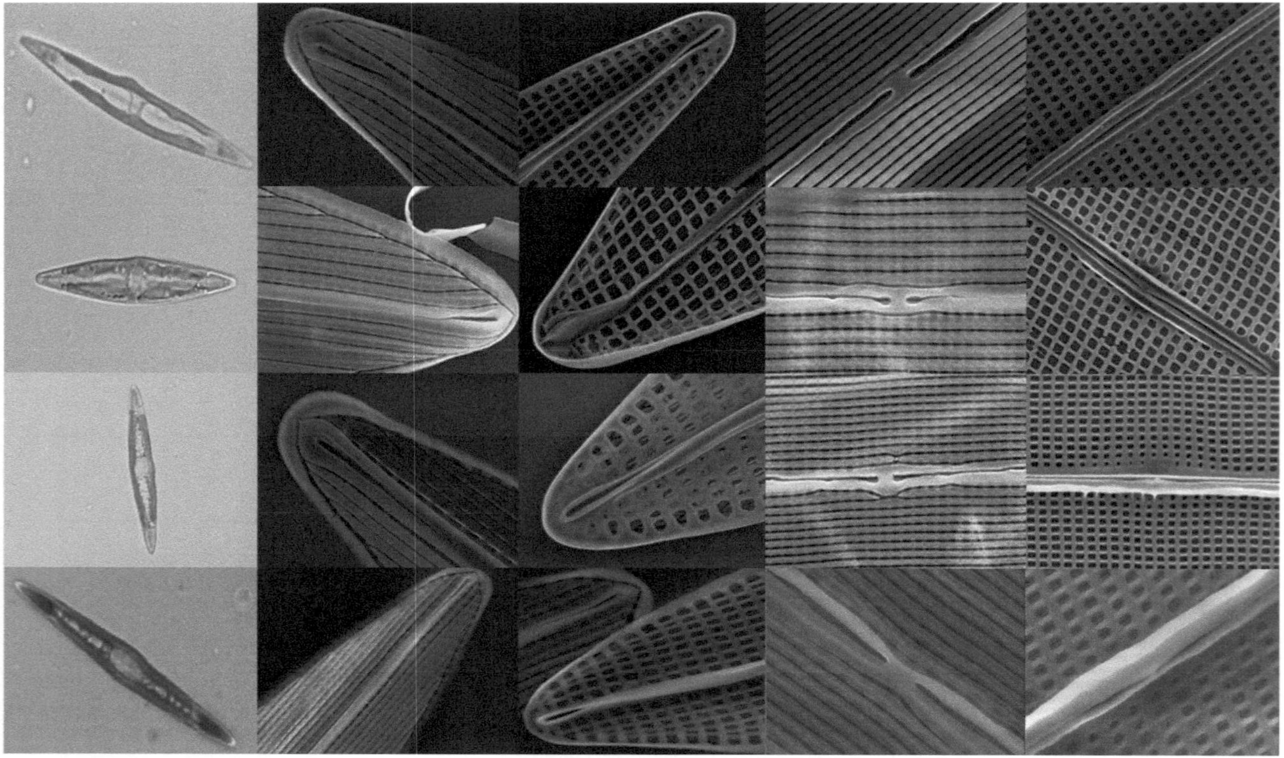

Figure 15.3 The four different species of blue diatoms. From up to down: *Haslea ostrearia*, species provisionally named *H. silbo* sp. inedit., species provisionally named *H. provincialis* sp. inedit., *H. karadagensis*. From left to right: live cells in LM showing apices filled with pigment, external view of apical ending in SEM, internal view of apical ending in SEM, external view of central raphe ending in SEM, internal view of central raphe ending in SEM.

Suez canal (Madkour, Dorgham, Hanafi, & Holligan, 2007) and in the Black Sea (Proshkina-Lavrenko, 1963). Along the eastern American coast, blue diatoms were observed in Rhode Island and North Carolina (Kenett & Hargraves, 1991), New Jersey (Sullivan, 1977), Virginia (Mitchell & Barney, 1918) and Florida (Julian Sprung, personal communication). The presence of blue diatoms has also been demonstrated in the Honduras (Grunow, 1877), in the San Juan Islands (Washington, USA) (Hardy, 1973), in the Indian Ocean (Simonsen, 1974), in Japan (Ranson, 1937), in Korea (Chung & Lee, 2008), in Tasmania (Hallegraeff et al., 2010) and in northern (Ricard, 1987), Western (Wasela Hosja, personal communication) and Eastern Australia (Volkman, Barrett, & Dunstan, 1994), thus presuming a worldwide distribution of *H. ostrearia*. In most cases, it is worth noting that usually a careful identification of the blue diatom did not rely on detailed studies of morphology and morphometrics, but was merely deduced from the presence of blue tips.

15.8 UNEXPECTED BIODIVERSITY OF BLUE DIATOMS

Despite a brief, controversial, and thus far uncorroborated hypothesis that marennine was produced by different species of diatoms during senescence (e.g. Bachrach, 1935; Bachrach & Jouvent, 1933; Bachrach & Simonet, 1936; Ranson, 1937), it was widely accepted that *H. ostrearia* was the only diatom species, worldwide in distribution, able to produce such blue pigment. However, recent discoveries have invalidated this dogma. In 2008, pennate diatoms with blue-grey apices were sampled from the shores of the Natural Reserve of Karadag, in Crimea (Ukraine) (Gastineau et al., 2012b). Cells appeared slightly different from *H. ostrearia*, and several methods of identification were used to study this intriguing organism. The classical morphological analysis, using SEM technique, demonstrated significant differences in the frustule ornamentation (longitudinal and transapical striae densities, see Table 15.1). Molecular biology corroborated the divergence between *H. ostrearia* and the Ukrainian blue diatom, by comparing genetic markers such as *rbc*L and the ITS1-5.8S-ITS2 sequences. The use of *in vivo* Raman spectrometry demonstrated that the pigments at cell apices in both diatoms belong to the same chemical family, but allowed discriminating between these molecules. The mechanisms of auxosporulation in the Ukrainian blue diatom have been described, together with the influence of some abiotic factors on the fecundity (Davidovich, Gastineau, Gaudin, Davidovich, & Mouget, 2012), and its sexual incompatibility with *H. ostrearia*. According to the concept of biological species (De Queiroz, 2007), the Ukrainian diatom

Table 15.1 Comparative Densities of Longitudinal and Transapical Striae for Several Species of Blue Diatoms

	Haslea ostrearia **(Bourgneuf)**	***Haslea*** **cf.** ***ostrearia*** **(Coffin Bay)**	***Haslea karadagensis***	***Haslea provincialis*** **sp. inedit.**	***Haslea silbo*** **sp. inedit.**
Transapical striae in 10 μm	34.8	34.0	31.4	34.9	37.7
Longitudinal striae in 10 μm	53.3	55.2	57.8	63.5	32.1

is thus a new species different from *H. ostrearia*. This second species of blue diatom was named *Haslea karadagensis*, after the original place of sampling. Although preliminary, studies on the blue-grey pigment produced by *H. karadagensis* documented biological activities, similar to those of marennine, including the capacity to colour oyster gills (Gastineau et al., 2013a).

Following this discovery, and thanks to international collaborations, new samples of blue diatoms were discovered and studied, becoming the cornerstone of an FP7 program dedicated to the study of the biodiversity of blue diatoms (BIOVADIA, 2011–2015). Diatoms with blue tips originating from La Gomera (Canary Islands), then from Boulouris (French Mediterranean Sea), were studied according to the multiple approach used for *H. karadagensis*, and revealed to be different species too, and received as provisional designation, the names of *Haslea silbo* sp. inedit. and *Haslea provincialis* sp. inedit., respectively (Gastineau, 2011). The characterisation as different species relies on significant morphological differences observed when comparing with both *H. ostrearia* and *H. karadagensis* (Gastineau, 2011). The frustule of the Canarian strain is blunter and displays square and large areolae with of lower density of striae, compared to the smaller rectangular areolae in both *H. ostrearia* and *H. karadagensis*. The frustule of the Mediterranean diatom more resembles those of *H. ostrearia* and *H. karadagensis*, with a similar lanceolate valve shape, and density of longitudinal striae being c. twice higher than density of transapical striae (Table 15.1) (Figure 15.3). Both are sexually incompatible with *H. ostrearia* and *H. karadagensis*. The molecular marker *cox*1 also confirms the latter blue diatoms are different species, with differences in base pairs (bp) up to 10%, whereas differences between subpopulations of *H. ostrearia* were about 0–2% (Table 15.2). A preliminary molecular phylogeny of the genus *Haslea* demonstrated that the 'blue species' form a single cluster inside the genus (Gastineau, 2011). It is worth

Table 15.2 Percentages of Identities between the Partial *cox*1 Markers for Several Species and Subspecies of Blue Diatoms

	Haslea ostrearia **(Narragansett)**	***Haslea ostrearia*** **(Kattegat)**	***Haslea*** **cf.** ***ostrearia*** **(Coffin Bay)**	***Haslea karadagensis***	***Haslea provincialis*** **sp. inedit.**	***Haslea silbo*** **sp. inedit.**
Haslea ostrearia (Bourgneuf)	100%	98%	96%	90%	91%	90%

Total DNA extraction was performed according to a modified Doyle and Doyle (1990). Primers for the cox1 fragment were COI1 (forward) and COI6 (reverse) (Gastineau et al., 2013b). Promega GoTaq® Flexi DNA Polymerase kit were used for PCR preparations. Chromas software was used to monitor Chromatogram quality (http://www.technelysium.com.au/chromas.html). Sequences identities were verified using BLAST (http://blast.ncbi.nlm.nih.gov/Blast.cgi), alignments were done using Clustal omega (Sievers et al., 2011) (http://www.ebi.ac.uk/Tools/msa/clustalo) and visualised using Genedoc (Nicholas & Nicholas, 1997) (http://www.nrbsc.org/gfx/genedoc/).

noting that using *in vitro* Raman spectrometry, in contrast with marennine, the extracellular forms of the purified pigments produced by *H. silbo* and *H. provincialis* do not seem to exhibit a different spectral signature compared to the intracellular ones. This may suggest that the ability to produce two distinct forms of pigments is not present among all species of the cluster of blue species, and that the mechanisms of pigment excretion differ. Data are currently lacking regarding this peculiarity for the species *H. karadagensis*.

More samples of blue diatoms have been collected from different geographical areas, and are currently being studied in our laboratory. These yet uncharacterised strains originate from France (Atlantic coast, Mediterranean Sea), Great Britain, the Canary Islands (Gran Canaria, Fuerteventura), USA (North Carolina), and Australia (see below).

15.9 INTO THE GENOMIC ERA

Molecular biology is the tool of choice for the study of blue diatoms, for instance those collected in the Kattegat strait (Sweden), which were compared to strains collected in the Bay of Bourgneuf and Bay of Marennes-Oléron (France). Strains derived from natural samples were identified as a population of *H. ostrearia*, using the multiple approach described above. Differences in a fragment of c. 720 pb of the *cox*1 marker allowed us to identify seven discriminating bases between the Swedish and the French clones. Since the Swedish populations were able to interbreed with clones from French populations, it has been possible to follow the distribution of the *cox*1 variants in their progenies (Gastineau et al., 2013b), demonstrating that the mitochondrial DNA inheritance is uniparental in this species. *H. ostrearia* is thus the first species of diatom for which mtDNA inheritance has been demonstrated, while inheritance of the plastid has been studied in *Pseudo-nitzschia delicatissima* (Levialdi-Ghiron, Amato, Montresor, & Kooistra, 2008).

With many biological specificities and a great potential of valorisation, *H. ostrearia* thus appears a promising model to be studied using new techniques of sequencing. A preliminary total DNA sequencing and an overview of the transposable elements (TEs) of its genome has been performed. TEs are DNA fragments that can be mobile in genomes, and may be a key factor in genome evolution (Hermann et al., in press; Shapiro & von Sternberg, 2005). In a previous study, *H. ostrearia* was investigated for the presence of mariner-like transposons, a family of TEs often used as mutagenesis tools but at that time, and based on nested PCR, none were found (Hermann, 2011). With the sequences gathered from the preliminary whole genome sequencing of *H. ostrearia*, an accurate investigation of the TEs has been performed using the REPET pipeline developed by Flutre, Duprat, Feuillet, and Quesneville (2011). This preliminary census of *H. ostrearia* TEs demonstrated the presence of both class I (retrotransposons) and class II (transposons) TEs, with many superfamilies being represented. The superfamilies Tc1-mariner, mutator, hAT and piggyBac have been found for the transposons, Ty1-Copia, Ty3-gypsy and LINE for the retrotransposons. Studying TEs in *H. ostrearia* is especially relevant for evolutionary and developmental biology since we have access to genetically distinct populations, we control their sexual reproduction, and therefore we could consider looking for their inheritance and distribution in F1 progeny clones.

15.10 EVIDENCE FOR *HASLEA OSTREARIA* TO BE A COSMOPOLITAN SPECIES

Molecular markers demonstrated that some of our strains isolated from geographically distant sites could belong to the same species, even if they were not genetically identical. Indeed, strains isolated from the Atlantic coast of France (Bourgneuf or Marennes-Oléron Bay), Sweden (Kattegat Bay), USA (Rhode Island, Narragansett Bay), and Australia (Coffin Bay) do differ genetically (see below), but they still interbreed and produce auxospores, thus revealing they represent a priori different populations of the same species, *H. ostrearia*. In particular, the strains established from the coast of Rhode Island display a complete genetic match with *H. ostrearia*, whatever the marker compared (see Table 15.2), which poses the question of how this mainly benthic organism could colonise both sides of the Atlantic Ocean (i.e. transportation with water ballast, translocation with oysters).

The case of the Australian strain is even more intriguing. One monoclonal culture of blue diatom was established from benthic samples collected in Coffin Bay, Southern Australia (34°37′S 135°28′E). Although no data are available yet regarding the pigment, the identification of this strain as *H. ostrearia* is currently based on morphological, reproductive and molecular data. The Australian strain exhibits all *Haslea* biometric generic features (Figures 15.4–15.9), with longitudinal slits parallel

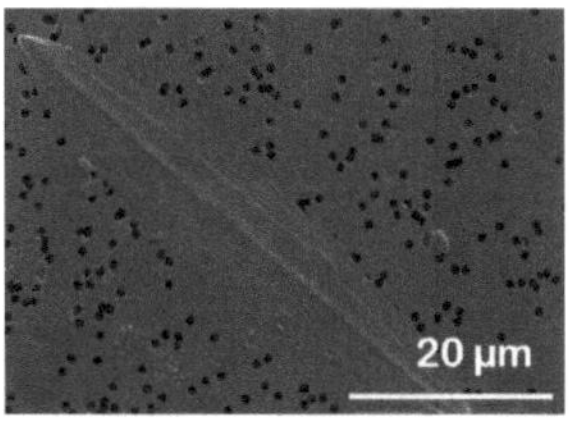

Figure 15.4 SEM pictures of the Australian blue *Haslea* strain. A cell in external view.

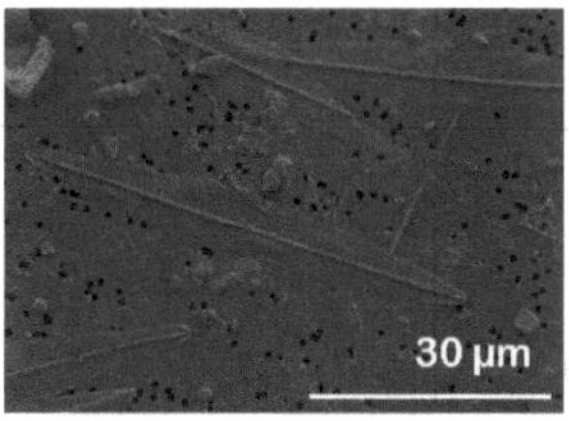

Figure 15.5 SEM pictures of the Australian blue *Haslea* strain. A cell in internal view.

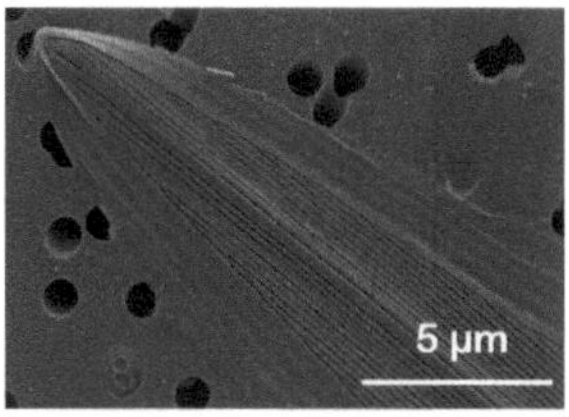

Figure 15.6 SEM pictures of the Australian blue *Haslea* strain. External view of apical ending.

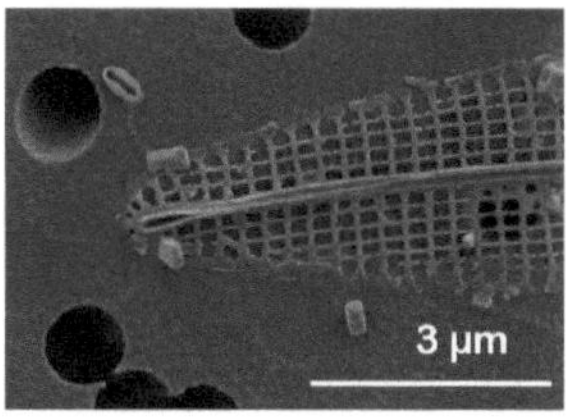

Figure 15.7 SEM pictures of the Australian blue *Haslea* strain. Internal view of apical ending.

to the raphe and quadrate areolae. The endings of its apices are straight in internal and external view, in both their central and apical positions. There is only one accessory rib on internal side. Cell measurements in SEM were 58.43 ± 0.08 in length and 7.30 ± 0.10 in width (mean ± SE, $n = 30$), with a length/width ratio of 8.00, as compared to 9.8 for *H. ostrearia* from Bay of Marennes–Oléron and Bay of Bourgneuf. Valves had 34.0 ± 0.2 transapical striae per 10 μm and 55.2 ± 0.1 longitudinal striae per 10 μm (mean ± SE, $n = 30$). These densities are very similar to those measured in French strains, respectively 34.8 ± 0.2 and 53.3 ± 0.2 for 10 μm (mean ± SE, $n = 30$).

For molecular characterisation in *Haslea* species, we favour using the *cox*1 marker, as it allowed for discrimination of geographically distant populations of diatoms (Hamsher et al., 2011) including *H. ostrearia* (Gastineau et al., 2013b). A 721 pb long fragment of the *cox*1 gene was retrieved from sequencing, identified and compared. Clustal omega alignment with referenced sequences of other *H. ostrearia* strains, followed by Genedoc analysis, showed that % difference between French or Swedish strains and the Australian was c. 4% for this gene, whereas the difference between French and Swedish

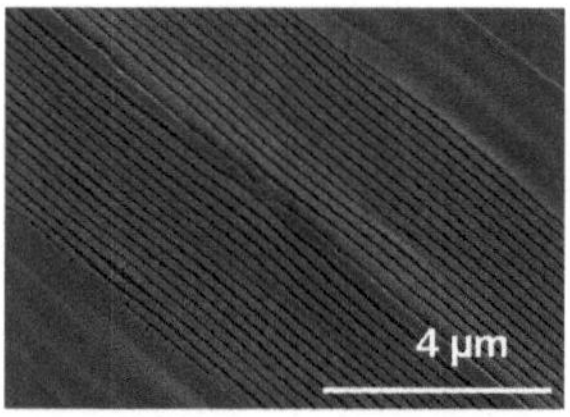

Figure 15.8 SEM pictures of the Australian blue *Haslea* strain. External view of central raphe ending.

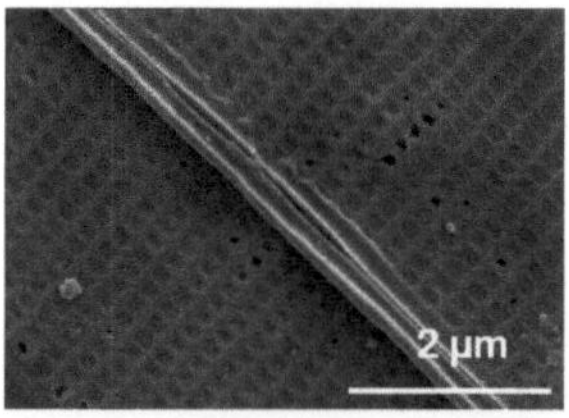

Figure 15.9 iSEM pictures of the Australian blue *Haslea* strain. Internal view of central raphe ending. Pictures have been taken with a JEOL 7600F scanning electron microscope operating at 15 kV, with frustules cleaned by digesting cells in a mixture of sulphuric and oxalic acids in potassium permanganate (Hendey, 1974) for pictures 5, 7 and 9, or left intact for pictures 4, 6 and 8, before being coated with platinum using a JUC5000 sputter coater.

strains was c. 2%. The number of diverging bases was higher if we compared Swedish to Australian (26 bp), than between French and Australian strains (22 bp). However, Swedish and Australian strains had three bases in common as compared to French strains (positions 189, 357 and 369). Protein translation and alignment evidenced the shifting from a Serine in French and Swedish strains to an Alanine in the Australian strain, at the 175th position of the 240-amino acid fragment. This is a structural polymorphism, which corresponds to a variation in the peptide sequence of the protein by one amino-acid residue.

A decisive criterion for the identification of the Australian strain as the member of *H. ostrearia* should be its ability to interbreed with strains of *H. ostrearia* from different geographical origin. Sexual compatibility with different populations of *H. ostrearia* was tested according to the standard protocol for crossing experiments (Davidovich et al., 2009; Mouget et al., 2009). Clonal cultures were mated by mixing in sterile Petri dishes, at a final concentration of 2000 cells/ml. Parental cultures were maintained in exponential phase of growth. Mixtures were placed under low irradiance (<30 μmol/m^2/s) and short photoperiod (8:16 h L:D cycle). Experiments were performed in triplicates, and signs of auxosporulation were monitored daily. The Australian strain was able to undergo auxosporulation with different strains of *H. ostrearia*. All stages were observed (pairing, gametogenesis, fertilisation, auxospore formation and elongation) and led to the production and release of initial cells (Figure 15.10). However, none of the initial cells, in three separate experimental matings, ever survived and divided, whenever it was picked up for isolation or left in the mixture of parental cells in mating dishes. This could result from a case of post-zygotic isolation. With an equivalent degree

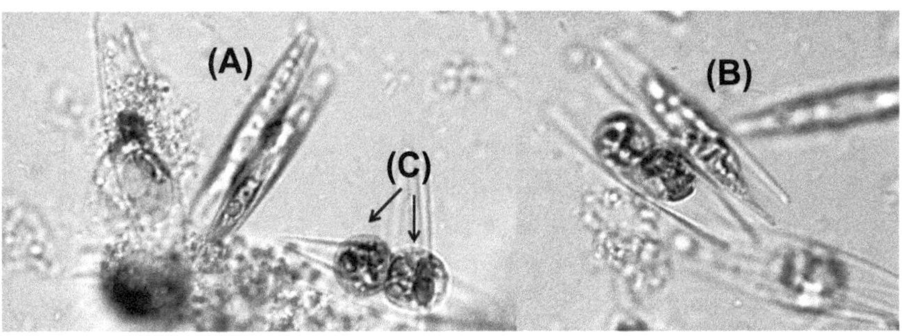

Figure 15.10 Different stages of auxosporulation occurring between *Haslea ostrearia* and the Australian strain with (A) pairing of parental cells, (B) gametogenesis, (C) formation of zygotes (arrows) after plasmogamy.

of divergence among some of their populations of *Sellaphora*, Evans, Wortley, and Mann (2007) succeeded in obtaining F1 clones, but no mention is made of their fate and long-term viability. Vanormelingen, Chepurnov, Mann, Sabbe, and Vyverman (2008) obtained sterile F1 clones from mating experiments between genetically different subpopulations of *Eunotia bilunaris*, but they used ITS as molecular marker instead of *cox*1. As only one strain of the Australian *H. ostrearia* was available, it has been impossible to test if another combination of clones or mating type would have given positive results. Empirically, it has been observed that different combination of parental clones, placed under identical conditions, could exhibit variable rate of fecundity. Thus, other combinations of parental clones, if available, may have exhibited different fecundity, or the fate of their progeny could have been different. Moreover, size reduction in Swedish strains of *H. ostrearia* prevented us for testing their compatibility with Australian strains, which call for further studies on the blue diatom biogeography.

15.11 CONCLUSIONS

To identify new species, especially in diatoms, one should rely on the decisive combination of traditional morphological traits, molecular barcoding using several genes, and information retrieved from reproductive biology. However, it is often difficult to perform a congruent diagnosis or identification of a new strain, when only one clone is available. The present review illustrates how this difficulty can be overcome. Morphological differences in the density of striae, although small, were founded to be statistically significant (Student's test) from those displayed by various populations of *H. ostrearia*. However, especially because only one Australian strain of *H. ostrearia* has been isolated and studied, this information alone is insufficient to conclude whether this is a new species, a subpopulation or just an atypical strain of *H. ostrearia*. Geographical separation usually calls for a higher degree of divergence, but this does not apply to the blue diatoms collected in Narragansett Bay (Rhode Island, U.S.A.), which were identical under all aspects, including molecular, to the *H. ostrearia* collected in the French Atlantic coast. This suggests a recent transfer of species, which could have occurred during the early twentieth century, when oysters were brought from France to the USA. Similar transportation could have previously happened in Australia, but the main mechanism by which alien species are introduced is more likely recent release of ship water ballasts, which is a real environmental issue (Gregg, Rigby, & Hallegraeff, 2009).

ACKNOWLEDGEMENTS

This study was supported by the EU through the FP7 program BIOVADIA. Partial support was also provided by Natural Sciences and Engineering Research Council of Canada Discovery Grant awarded to IK.

REFERENCES

Aleem, A. A. (1973). Contribution to the study of littoral diatoms of the West Coast of Sweden. *Botanica Marina, 16*, 193–200.

Allard, W. G., Belt, S. T., Massé, G., Naumann, R., Robert, J. M., & Rowland, S. (2001). Tetra-unsaturated sesterterpenoids (Haslenes) from *Haslea ostrearia* and related species. *Phytochemistry, 56*, 795–800.

Bachrach, E. (1935). Le bleuissement des Diatomées et le verdissement des huîtres. Revue critique. *Revue des travaux de l'Institut des pêches maritimes, 8*, 112–123.

Bachrach, E., & Jouvent, A. (1933). Sur la pigmentation bleue de certaines diatomées. *Bulletin mensuel de la Société linnéenne de Lyon, 2*, 36–37.

Bachrach, E., & Simonet, M. (1936). Le phénomène du bleuissement chez les diatomées. *Revue des travaux de l'Institut des pêches maritimes, 9*, 113–116.

Barone, R. (1979). La diatomee bentoniche del porto di Palermo. *Giornale Botanico Italiano, 3*, 187–295.

Barrett, S. M., Volkman, J. K., Dunstan, G. A., & LeRoi, J.-M. (1995). Sterols of 14 species of marine diatoms (*Bacillariophyta*). *Journal of Phycology, 31*, 360–369.

Bavestrello, G., Arillo, A., Calcinai, B., Cattaneo-Vietti, R., Cerrano, C., Gaino, E., et al. (2000). Parasitic diatoms inside antarctic sponges. *Biological Bulletin, 198*, 29–33.

Belt, S. T., Cooke, D. A., Robert, J. M., & Rowland, S. (1996). Structural characterisation of widespread polyunsaturated isoprenoid biomarkers: a C25 triene, tetraene and pentaene from the diatom *Haslea ostrearia* simonsen. *Tetrahedron Letters, 37*, 4755–4758.

Bennett, A. G. (1920). On the occurrence of diatoms on the skin of whales. *Proceedings of the Royal Society of London, 91*, 352–357.

Bergé, J. P., Bourgougnon, N., Alban, S., Pojer, F., Billaudel, S., Chermann, J. C., et al. (1999). Antiviral and anticoagulant activities of a water-soluble fraction of the marine diatom *Haslea ostrearia*. *Planta Medica, 65*, 604–609.

Bocat, L. (1907). Sur la marennine de la diatomée bleue; comparaison avec la phycocyanine. *Comptes rendus des séances de la Société de biologie et de ses filiales, 62*, 1073–1075.

Bornet, E., & Chatin, A. D. (1895). Le verdissement des huîtres. *Bulletin des séances de la Société royale et centrale d'agriculture, 55*, 429–439.

Bory de Saint-Vincent, J. B. (1824). Navicules. In *Encyclopédie méthodique* (pp. 562–565).

Boyce, R. W., & Herdman, W. A. (1896). *The possible infectivity of the oyster, and upon the green disease in oysters*. Communication to British Association for Advancement of Science.

Briée, C. (2010). *Le verdissement des huîtres : deux siècles de transformation d'un problème biologique*. (Unpublished PhD thesis). University of Nantes.

Bruguière, J. G. (1792). *Encyclopédie méthodique: histoire naturelle des vers*. (Vol. 1) Paris: Panckoucke.

Carbonnelle, D., Pondaven, P., Morançais, M., Massé, G., Bosch, S., Jacquot, C., et al. (1999). Antitumor and antiproliferative effects of an aqueous extract from the marine diatom *Haslea ostrearia* (Simonsen) against solid tumors: lung carcinoma (NSCLC-N6), kidney carcinoma (E39) and melanoma (M96) cell lines. *Anticancer Research, 19*, 621–624.

Chepurnov, V. A., Mann, D. G., Sabbe, K., & Vyverman, W. (2004). Experimental studies on sexual reproduction in diatoms. *International Review Cytology, 237*, 91–154.

Chung, M. H., & Lee, K. S. (2008). Species composition of the epiphytic diatoms on the leaf tissues of three *Zostera* species distributed on the southern coast of Korea. *Algae, 23*, 75–81.

Cleve-Euler, A. (1955). Die Diatomeen von Schweden und Finland. *Kungliga Svenska Vetenskaps Akademiens Handlinger Series, 4*, 1–161.

Croll, D. A., & Holmes, R. W. (1982). A note on the occurrence of diatoms on the feathers of diving seabirds. *The Auk, 99*, 765–766.

Cupp, E. E. (1943). *Marine plankton diatoms of the west coast of North America*. Berkeley: University of California Press.

Davidovich, N. A., Gastineau, R., Gaudin, P., Davidovich, O. I., & Mouget, J. L. (2012). Sexual reproduction in the second described blue diatom, *Haslea karadagensis*. *Fottea, 12*, 219–229.

Davidovich, N. A., Mouget, J.-L., & Gaudin, P. (2009). Heterothallism in the pennate diatom *Haslea ostrearia* (Bacillariophyta). *European Journal of Phycology, 44*, 251–261.

De Queiroz, K. (2007). Species concepts and species delimitation. *Systematic Biology, 56*, 879–886.

Dodson, V. J., Dahmen, J. L., Mouget, J.-L., & Leblond, J. D. (2013). Mono- and digalactosyldiacylglycerol composition of the marennine-producing diatom, *Haslea ostrearia*: comparison to a selection of pennate and centric diatoms. *Phycological Research, 61*, 199–207.

Doyle, J. J., & Doyle, J. L. (1990). Isolation of plant DNA from fresh tissue. *Focus, 12*, 13–15.

Dyer, W. T. T. (1877). Greening of oysters. *Nature (London), 16*, 397.

Evans, K. M., Wortley, A. H., & Mann, D. G. (2007). An assessment of potential diatom "barcode" genes (cox1, rbcL, 18 S and ITS rDNA) and their effectiveness in determining relationships in *Sellaphora* (Bacillariophyta). *Protist, 158*, 349–364.

Flutre, T., Duprat, E., Feuillet, C., & Quesneville, H. (2011). Considering transposable element diversification in de novo annotation approaches. *PLoS One, 6*(No. 1), e16526.

Fouqucray, M. (2007). *Dynamique de l'acclimatation des diatomées marines aux rayonnements ultraviolets : approches physiologiques, biochimiques et moléculaires*. (Unpublished PhD thesis). University of Le Mans.

Gaillon, B. (1820). Des huîtres vertes, et des causes de cette coloration. *Journal de physique, de chimie, d'histoire naturelle et des arts, 91*, 222–225.

Gastineau, R. (2011). *Biodiversité, reproduction et phylogénie des diatomées bleues du genre Haslea et valorisation de leurs pigments de type marennine*. (Unpublished PhD thesis). University of Le Mans.

Gastineau, R., Pouvreau, J. B., Hellio, C., Morançais, M., Fleurence, J., Gaudin, P., et al. (2012a). Biological activities of purified marennine, the blue pigment produced by the diatom *Haslea ostrearia* and responsible for the greening of oysters. *Journal of Agriculture and Food Chemistry, 60*, 3599–3605.

Gastineau, R., Davidovich, N. A., Bardeau, J. F., Davidovich, O. I., Rincé, Y., Gaudin, P., et al. (2012b). *Haslea karadagensis* sp. nov. (Bacillariophyta), a new diatom from the Black Sea producing a novel blue pigment. *European Journal of Phycology, 47*, 469–479.

Gastineau, R., Hardivillier, H., Leignel, V., Tekaya, N., Morançais, M., Fleurence, J., et al. (2013a). Greening effect on oysters and biological activities of the blue pigments produced by the diatom *Haslea karadagensis* (Naviculaceae). *Aquaculture, 368-369*, 61–67.

Gastineau, R., Leignel, V., Jacquette, B., Hardivillier, Y., Wulff, A., Gaudin, P., Bendahmane, D., Davidovich, N. A., Kaczmarska, I., & Mouget, J.-L. (2013b). Inheritance of Mitochondrial DNA in the Pennate Diatom Haslea ostrearia (Naviculaceae) during Auxosporulation Suggests a Uniparental Transmission. *Protist, 164*, 340–351.

Genevès, L., Choussy, M., Barbier, M., Neuville, D., & Daste, P. (1976). Ultrastructure et composition pigmentaire comparées des chromatophores de la Diatomée *Navicula ostrearia* (Gaillon) Bory normale et bleue. *Comptes Rendus de l'Académie des Sciences – Série D – Sciences Naturelles, 282*, 449–452.

Gregg, M., Rigby, G., & Hallegraeff, G. M. (2009). Review of two decades of progress in the development of management options for reducing or eradicating phytoplankton, zooplankton and bacteria in ship's ballast water. *Aquatic Invasions, 4*, 521–565.

Groth-Nard, C. (1994). *Les lipides des Diatomées: exemple d'Haslea ostrearia (Simonsen), responsable du verdissement des huîtres.* (Unpublished PhD thesis) University of Nantes.

Grünow, A. (1877). New diatoms from Honduras. *Monthly Microscopical Journal, 18*, 165–186.

Hallegraeff, G. M., Bolch, C. J. S., Hill, D. R. A., Jameson, I., LeRoi, J.-M., McMinn, A., et al. (2010). *Algae of Australia: Phytoplankton of temperature coastal waters.* Melbourne: CSIRO Publishing.

Hamsher, S. E., Evans, K. M., Mann, D. G., Poulíčková, A., & Saunders, C. W. (2011). Barcoding Diatoms: Exploring Alternatives to COI-5P. *Protist, 162*, 405–422.

Hardy, J. T. (1973). Ecology of phytoneuston in a temperate marine Lagoon. *Limnology and Oceanography, 18*, 525–533.

Harivardhan Reddy, L., & Couvreur, P. (2009). Squalene: a natural triterpene for use in disease management and therapy. *Advanced Drug Delivery Reviews, 61*, 1412–1426.

Hendey, N. I. (1964). Part V: bacillariophyceae (Diatoms). In *An introductory account of the smaller algae of british coastal waters.* London: Great Britain Ministry of Agriculture, Fisheries and Food.

Hendey, N. I. (1974). The permanganate method for cleaning freshly gathered diatoms. *Microscopy, 32*, 423–426.

Hermann, D. (2011). *Caractérisation d'éléments transposables de type mariner chez les microalgues marines.* (Unpublished PhD thesis). University of Le Mans.

Hermann, D., Egue, F., Tastard, E., Nguyen, D. H., Casse, N., Caruso, A., et al. (2014). Introduction to the vast world of transposable elements – what about the diatoms? *Diatom Research, 29*, 91–104.

Hustedt, F., & Aleem, A. A. (1951). Littoral diatoms from the salstone near Plymouth. *Journal of the Marine Biological Association of the United Kingdom, 30*, 177–196.

Johns, L., Belt, S. T., Lewis, C. A., Rowland, S., Massé, G., Robert, J.-M., et al. (2000). Configurations of polyunsaturated sesterterpenoids from the diatom *Haslea ostrearia*. *Phytochemistry, 53*, 607–611.

Kenett, D. M., & Hargraves, P. E. (1991). Benthic marine diatoms. In R. G. Sheath, & M. M. Harlin (Eds.), *Freshwater and marine plants of Rhode Island* (pp. 127–135). Dubuque: Kendall and Hunt Publishing Company.

Kerswill, C. J. (1946). Green-gilled clams and other bivalves on Prince Edward Island. *Acadian Naturalist, 2*, 102.

Kuylenstierna, M. (1990). *Benthic algal vegetation in the Nordre älv estuary (Swedish west coast).* (Unpublished PhD thesis). University of Gothenburg.

Lakatos, M., Lange-Bertalot, H., & Büdel, B. (2004). Diatoms living inside the thallus of the green algal lichen *Coenogonium linkii* in neotropical lowland rain forest. *Journal of Phycology, 40*, 70–73.

Lankester, E. R. (1886). On green oysters. *Quarterly Journal of Microscopical Science, 26*, 71–94.

Lee, J. J. (2011). Diatoms as endosymbionts. In J. Seckbach, & J. P. Kocioloek (Eds.), *The diatom world, cellular origin, life in extreme habitats and astrobiology, 19*, 437–464.

Levialdi-Ghiron, J. H., Amato, A., Montresor, M., & Kooistra, W. H. C.F. (2008). Plastid inheritance in the planktonic raphid pennate diatom *Pseudo-nitzschia delicatissima* (Bacillariophyceae). *Protist, 159*, 91–98.

MacDonald, J. D. (1869). On the structure of the diatomaceous frustule, and its genetic cycle. *Annals and Magazine of Natural History, 3*, 1–8.

Madkour, F. A., Dorgham, M. M., Hanafi, M. H., & Holligan, P. M. (2007). Comprehensive hydro-biological observations on the Suez Canal. *International Journal of Oceans and Oceanography, 2*, 125–137.

Massé, M., Rincé, Y., Cox, E. J., Allard, G., Belt, S. T., & Rowland, S. J. (2001). *Haslea salstonica* sp. nov. and *Haslea pseudostrearia* sp. nov. (Bacillariophyta), two new epibenthic diatoms from the Kingsbridge estuary, United Kingdom. *Comptes Rendus de l'Académie des Sciences de Paris, Sciences de la vie/Life Sciences, 324*, 617–626.

Medcof, J. C. (1945). Green oysters from New Brunswick. *Acadian Naturalist, 2*, 40.

Mimouni, V., Ulmann, L., Tremblin, G., & Robert, J.-M. (2003). Desaturation of linoleic acid in the marine diatom *Haslea ostrearia* Simonsen (Bacillariophyceae). *Cryptogamie Algologie, 24*, 269–276.

Minier, C., Tutundjian, R., Galgani, F., & Robert, J.-M. (1998). Copper tolerance in *Haslea ostrearia* assessed by measurements of in vivo esterase activity. *Marine Environmental Research, 46*, 579–582.

Mitchell, P. H., & Barney, R. L. (1918). The occurrence in Virginia of green-gilled oysters similar to those of Marennes. *Bulletin of the United States Bureau of Fisheries, 35*, 135–149.

Molisch, H. (1903). Notiz über eine blaue Diatomee. *Berichte der Deutschen Botanischen Gesellschaft, 21*, 23–26.

Moreau, J. (1967). Recherches préliminaires sur le verdissement en claires: l'évolution de leurs divers pigments liée au complexe pigmentaire de *Navicula ostrearia* Bory. *Revue des travaux de l'Institut des pêches maritimes, 31*, 372–382.

Moreau, J. (1970). Contribution aux recherches écologiques sur les claires à huîtres du bassin de Marennes-Oléron. *Revue des travaux de l'Institut des pêches maritimes, 34*, 381–462.

Mouget, J.-L., Gastineau, R., Davidovich, O., Gaudin, P., & Davidovich, N. A. (2009). Light is a key factor in triggering sexual reproduction in the pennate diatom *Haslea ostrearia*. *FEMS Microbiology and Ecology, 69*, 194–201.

Mouget, J.-L., Rosa, P., & Tremblin, G. (2004). Acclimation of *Haslea ostrearia* to light of different spectral qualities – confirmation of 'chromatic adaptation' in diatoms. *Journal of Photochemistry and Photobiology B: Biology, 75*, 1–11.

Mouget, J.-L., Rosa, P., Vachoux, C., & Tremblin, G. (2005). Enhancement of marennine production by blue light in the diatom *Haslea ostrearia*. *Journal of Applied Phycology, 17*, 437–445.

Mouget, J.-L., Tremblin, G., Morant-Manceau, A., Morançais, M., & Robert, J.-M. (1999). Long-term photoacclimation of *Haslea ostrearia* (Bacillariophyta): effect of irradiance on growth rates, pigment content and photosynthesis. *European Journal of Phycology, 34*, 109–115.

Munda, I. M. (2005). Seasonal fouling by diatoms on artificial substrata at different depths near Piran (Gulf of Trieste, Northern Adriatic). *Acta Adriatica, 46*, 137–157.

M'harzi, A., Tackx, M., Daro, M. H., Kesaulia, I., Caturao, R., & Podoor, N. (1998). Winter distribution of phytoplancton and zooplancton around some sandbanks of the belgian coastal zone. *Journal of Plankton Research, 20*, 2031–2052.

Nassiri, Y., Robert, J.-M., Rincé, Y., & Ginsburger-Vogel, T. (1998). The cytoplasmic fine structure of the diatom *Haslea ostrearia* (Bacillariophyceae) in relation to marennine production. *Phycologia, 37*, 84–91.

Neuville, D., & Daste, P. (1972a). Premières observations sur la reproduction, par division binaire, de la Diatomée *Navicula ostrearia* (Gaillon) Bory cultivée in vitro. *Comptes Rendus de l'Académie des Sciences de Paris Série D, 274*, 1083–1085.

Neuville, D., & Daste, P. (1972b). Production de pigment bleu par la Diatomée *Navicula ostrearia* (Gaillon) Bory, maintenue en culture unialgale sur un milieu synthétique carencé en azote nitrique. *Comptes Rendus de l'Académie des Sciences de Paris Série D, 274*, 2030–2033.

Neuville, D., & Daste, P. (1975). Observations préliminaires concernant l'auxosporulation chez la diatomée *Navicula ostrearia* (Gaillon) Bory en culture in vitro. *Comptes Rendus de l'Académie des Sciences de Paris Série D, 281*, 1753–1756.

Neuville, D., & Daste, P. (1978). Recherches sur le déterminisme de la production de marennine par la Diatomée marine *Navicula ostrearia* (Gaillon) Bory en culture in vitro. *Revue générale de botanique, 85*, 255–303.

Neuville, D., & Daste, P. (1979). Observations concernant les phases de l'auxosporulation chez la diatomée *Navicula ostrearia* (Gaillon) Bory en culture in vitro. *Comptes Rendus de l'Académie des Sciences de Paris Série D, 288*, 1496–1498.

Nicholas, K. B., & Nicholas, H. B. (1997). *GeneDoc: a tool for editing and annotating multiple sequence alignments.* www.psc.edu/biomed/genedoc.
Petersen, C. G. J. (1916). "Grüne Austern" in Dänemark in 1911–1912. *International Review of Hydrobiology and Hydrography, 7*, 39–41.
Pfitzer, E. (1869). Über Bau und Zellteilung der Diatomeen. *Botanische Zeitung, 27*, 774–776.
Pillet, L., de Vargas, C., & Pawlowski, J. (2011). Molecular identification of sequestered diatom chloroplasts and kleptoplastidy in foraminifera. *Protist, 162*, 394–404.
Poulin, M., Massé, G., Belt, S. T., Delavault, P., Rousseau, F., Robert, J. M., et al. (2004). Morphological, biochemical and molecular evidence for the transfer of *Gyrosigma nipkowii* Meister to the genus *Haslea* (Bacillariophyta). *European Journal of Phycology, 39*, 181–195.
Pouvreau, J. B., Housson, E., Le Tallec, L., Morançais, M., Rincé, Y., Fleurence, J., et al. (2007). Growth inhibition of several marine diatom species induced by the shading effect and allelopathic activity of marennine, a blue-green polyphenolic pigment of the diatom *Haslea ostrearia* (Gaillon/Bory) Simonsen. *Journal of Experimental Marine Biology and Ecology, 352*, 212–225.
Pouvreau, J. B., Morançais, M., Fleury, F., Rosa, P., Thion, L., Cahingt, B., et al. (2006). Preliminary characterisation of the blue-green pigment "marennine" from the tychopelagic diatom *Haslea ostrearia* (Gaillon/Bory) Simonsen. *Journal of Applied Phycology, 18*, 757–767.
Pouvreau, J. B., Morançais, M., Massé, G., Rosa, P., Robert, J. M., Fleurence, J., et al. (2006). Purification of the blue-green pigment "marennine" from the marine tychopelagic diatom *Haslea ostrearia* (Gaillon/Bory) Simonsen. *Journal of Applied Phycology, 18*, 769–781.
Pouvreau, J. B., Morançais, M., Taran, F., Rosa, P., Dufossé, L., Guérard, F., et al. (2008). Antioxidant and free radical scavenging properties of marennine, a blue-green polyphenolic pigment from the diatom *Haslea ostrearia* (Gaillon/Bory) Simonsen responsible for the natural greening of cultured oysters. *Journal of Agriculture and Food Chemistry, 56*, 6278–6286.
Proshkina-Lavrenko, A. I. (1963). *Diatomovye vodorosli bentosa Chernogo moria.* Leningrad: Akademii Nauk SSSR.
Puységur, M. (1880). Notice sur la cause du verdissement des huîtres. *Revue maritime et Coloniale, 64*, 248–256.
Ranson, G. (1937). Le verdissement des huîtres. *Sciences, 8*, 13–24.
Rech, M. (2004). *Effets de l'éclairement visible et ultraviolet sur la croissance et la photosynthèse de microalgues: incidence sur l'écophysiologie du phytoplancton de claires ostréicoles.* (Unpublished PhD thesis). University of Le Mans.
Rech, M., Morant-Manceau, A., & Tremblin, G. (2008). Carbon fixation and carbonic anhydrase activity in *Haslea ostrearia* (Bacillariophyceae) in relation to growth irradiance. *Photosynthetica, 46*, 52–62.
Rech, M., Mouget, J. L., Morant-Manceau, A., Rosa, P., & Tremblin, G. (2005). Long-term acclimation to UV radiation: effects on growth, photosynthesis and carbonic anhydrase activity in marine diatoms. *Botanica Marina, 48*, 407–420.
Ricard, M. (1987). Diatomophycées. In A. Sournia (Ed.), *Atlas du phytoplancton marin* (Vol. 2) (pp. 1–297). Paris: Editions du Centre National de la Recherche Scientifique.
Robert, J.-M. (1978). Variations biométriques de l'algue *Navicula ostrearia* Bory (Diatomée Pennée) en culture. *Bulletin – Société phycologique de France, 23*, 38–44.
Robert, J.-M., Maestrini, S. Y., Héral, M., & Zanette, Y. (1982). Production des micro-algues des claires ostréicoles en relation avec l'azote organique dissous excrété par les huîtres. *Oceanologica acta sp*, 389–395.
Robert, J. M. (1983). *Fertilité des eaux des claires ostréicoles et verdissement: utilisation de l'azote pas les diatomées dominantes.* (Unpublished PhD thesis). University of Nantes.

Robert, J. M., & Hallet, J. N. (1981). Absorption spectrum in vivo of the blue pigment "marennine" of the pennate diatom *Navicula ostrearia* Bory. *Journal of Experimental Botany, 32*, 341–345.
Robert, J.-M., Morançais, M., Pradier, E., Mouget, J.-L., & Tremblin, G. (2002). Extraction and quantitative analysis of the blue-green pigment "marennine" synthesized by the diatom *Haslea ostrearia*. *Journal of Applied Phycology, 14*, 299–305.
Round, F. E., Crawford, R. M., & Mann, D. G. (1990). *The diatoms. Biology and morphology of the genera*. Cambridge: Cambridge University Press.
Rowland, S. J., Belt, S. T., Wraige, E. G., Massé, G., Roussakis, C., & Robert, J.-M. (2001). Effects of temperature on of polyunsaturation in cytostatic lipids of *Haslea ostrearia*. *Phytochemistry, 56*, 597–602.
Ryder, J. A. (1884). On the cause of the greening of oysters. With a supplementary note on the coloration of the blood corpuscles of the oysters. In *Report of Commission of Fish and Fisheries for 1882* (Vol. X) (pp. 793–800). Washington: Government Printing Office.
Sauvageau, C. (1906). A propos de la présence de la diatomée bleue dans la Méditerranée. *Travaux des laboratoires (Arcachon), 6*, 46–59.
Sauvageau, C. (1907). Le verdissement des huîtres par la diatomée bleue. *Travaux des laboratoires (Arcachon), 2*, 1–128.
Schubert, H., Tremblin, G., Robert, J.-M., Sagert, S., & Rincé, Y. (1995). In-vivo fluorescence measurement of photosynthesis of *Haslea ostrearia* Simonsen in relation to marennine content. *Diatom Research, 10*, 341–349.
Shapiro, J. A., & von Sternberg, R. (2005). Why repetitive DNA is essential to genome function. *Biological Reviews, 80*, 1–24.
Sievers, F., Wilm, A., Dineen, D. G., Gibson, T. J., Karplus, K., Li, W., et al. (2011). Fast, scalable generation of high-quality protein multiple sequence alignments using Clustal Omega. *Molecular Systems Biology, 7*, 539.
Simonsen, R. (1974). The diatom plankton of the Indian Ocean expedition of RV "Meteor". *Herausgegeben von der Deutschen Forschungsgemeinschaft, 19*, 46–49.
Sims, P. A., Mann, D. G., & Medlin, L. K. (2006). Evolution of the diatoms: insights from fossil, biological and molecular data. *Phycologia, 45*, 361–402.
Smol, J. P., & Stoermer, E. F. (2010). *The diatoms: applications for the environmental and earth sciences*. Cambridge: Cambridge University Press.
Snoeijs, P., & Kasperovičienė, J. (1996). *Intercalibration and distribution of diatom species in the Baltic Sea* . (16). Baltic Marine Biologists Publications. 62.
Solazzi, A., & Tolomio, C. (1976). II fitoplancton primaverile ed estivo lungo le coste settentrionali della Sardegna. *Archo Oceanography Limnology, 18*, 389–409.
Sprat, T. (1667). *History of the Royal Society. London*.
Stewart, W. D. P., Rowell, P., & Rai, A. N. (1983). Cyanobacteria-Eukariotic plant symbioses. *Annales de l'Institut Pasteur Microbiology, 134*, 205–228.
Sullivan, M. J. (1977). Structural characteristics of a diatom community epiphytic on Ruppia maritime. *Hydrobiologia, 53*, 81–86.
Tolomio, C. (1978). Catalogo delle Diatomee e delle Peridinee pin signifactive segnalate nelle acque salmastre italiane. *Memorie di Biologia Marina e di Oceanografia, 8*, 129–150.
Tremblin, G., Cannuel, R., Mouget, J.-L., Rech, M., & Robert, J.-M. (2000). Change in light quality due to a blue-green pigment, marennine, released in oyster-ponds: effect on growth and photosynthesis in two diatoms, *Haslea ostrearia* and *Skeletonema costatum*. *Journal of Applied Phycology, 12*, 557–566.
Tremblin, G., & Robert, J.-M. (2001). Carbon fixation by the peculiar marine diatom *Haslea ostrearia*. *Photosynthetica, 39*, 215–220.
Valenciennes, A. (1841). Sur les causes de la coloration en vert de certaines huîtres. *Comptes rendus hebdomadaires des séances de l'Académie des sciences, 12*, 345–346.

Vanormelingen, P., Chepurnov, V. A., Mann, D. G., Sabbe, K., & Vyverman, V. (2008). Genetic divergence and reproductive barriers among morphologically heterogeneous sympatric clones of *Eunotia bilunaris* Sensu Lato (Bacillariophyta). *Protist*, *159*, 73–90.

Volkman, J. K., Barrett, S. M., & Dunstan, G. A. (1994). C_{25} and C_{30} highly branched isoprenoid alkenes in laboratory cultures of two marine diatoms. *Organic Geochemistry*, *21*, 407–414.

Wetzel, C. E., Van de Vijver, B., & Ector, L. (2010). *Luticola deniseae* sp. nov. A new epizoic diatom from the Rio Negro (Amazon hydrographic basin). *Vie et Milieu*, *60*, 177–184.

Wraige, E. J., Belt, S. T., Lewis, C. A., Cooke, D. A., Robert, J. M., Masse, G., et al. (1997). Variations in structures and distributions of C25 highly branched isoprenoid (HBI) alkenes in cultures of the diatom, *Haslea ostrearia* (Simonsen). *Organic Geochemistry*, *27*, 497–505.

Wraige, E. J., Belt, S. T., Masse, G., Robert, J. M., & Rowland, S. J. (1998). Variations in distributions of C25 highly branched isoprenoid (HBI) alkenes in the diatom, *Haslea ostrearia*: influence of salinity. *Organic Geochemistry*, *28*, 855–859.

Wraige, E. J., Johns, L., Belt, S. T., Masse, G., Robert, J. M., & Rowland, S. (1999). Highly branched C25 isoprenoids in axenic cultures of *Haslea ostrearia*. *Phytochemistry*, *51*, 69–73.

Wulff, A., Wängberg, S. Å., Sundbäck, K., Nilsson, C., & Underwood, G. J. C. (2000). Effects of UVB radiation on a marine microphytobentic community growing on a sand-substratum under different nutrient conditions. *Limnology and Oceanography*, *45*, 1144–1152.

CHAPTER SIXTEEN

Oxidative Stress Regulation in Lichens and Its Relevance for Survival in Coastal Habitats

Le Devehat Françoise*,[1], Thüs Holger†, Abasq Marie-Laurence*, Delmail David* and Boustie Joël*

*UMR 6226 ISCR, Equipe PNSCM, Faculté des Sciences Pharmaceutiques et Biologiques, Université de Rennes 1, Rennes Cedex, France

†Natural History Museum, Cromwell Road, London, UK

[1]Corresponding author: e-mail address: francoise.le-devehat@univ-rennes1.fr

Contents

Abstract

Lichens are symbiotic systems able to survive in harsh conditions like those on coastal rocks with rapid changes in the concentration of salt, immersion, desiccation–rehydration cycle, wind, waves and herbivorie. These stresses are at the origin of reactive oxygen species (ROS) production in lichens. Physical barriers, enzymatic systems and cellular contents of glutathione, ascorbic acid and tocopherols are involved in lichen defence systems against ROS. But which role have specific lichen compounds, e.g.: the depsides, depsidones, dibenzofuranes, diphenylethers as antioxidants? In order to answer this question, antioxidant studies were recorded and analysed for studies for lichens that occur on coastal rocks. Based on the methods used in the experiments (hydrogen atom transfer (HAT), electron transfer (ET), enzymatic assays, specific ROS scavenging, *in vivo* assays), lichen compounds seemed to be poor hydrogen acceptors and electron donors but appeared as potent ROS specific scavengers, they might act as antioxidants as proved by the positive results of many lichen extracts in antioxidant assays.

Advances in Botanical Research, Volume 71
ISSN 0065-2296
http://dx.doi.org/10.1016/B978-0-12-408062-1.00016-0

Abbreviations

CUPRAC	cupric ion reducing antioxidant capacity
DPPH	2,2-di(4-tert-octylphenyl)-1-picrylhydrazyl
ORAC	oxygen radical absorbance capacity
TBARS	thiobarbituric acid reactive substances
TEAC	trolox-equivalent antioxidant capacity
TRAP-TAR	total reactive antioxidant potential-total antioxidant reactivity

16.1 INTRODUCTION

Lichens could be qualified as born survivors in extreme environments (Purvis, Elix, Broomhead, & Jones, 1987). They can tolerate the most extreme conditions (hot dry places, arctic climate, salt spray and immersion in water) and are found on soil, woody debris, rocks, tree bark, tree leaves, other lichens, desert sand and animal bones. The association between two, and sometimes more, different organisms, which form a symbiotic relationship with each other helps their mutual survival. The coastal environment is one of these extreme environments and marine and maritime lichens have to respond to various stresses like extreme temperature variation, rapid change in water availability, nutrition depletion, presence and rapid changes in salt concentrations… All these conditions generate deleterious reactive oxygen species and without efficient antioxidant defence systems, lichens could not survive. The first barrier consists of the inactivation of the reactive oxygen species (ROS) by enzymatic systems and storage of lipophilic and hydrophilic constituents of cells. The concentration of secondary metabolites specific to lichens like depsides, depsidones and dibenzofuranes are known to increase under stress. Could they play a role by detoxification of ROS? Many researchers focused on the antioxidant activity of these compounds mostly motivated by discovering new antioxidants for food and chemicals. They evaluated them by numerous methods based on hydrogen atom transfer (HAT), electron transfer (ET), ROS-specific scavengering, *in cellulo, in vivo* assays. We proposed here to review all published antioxidant assays as far as they involve lichen taxa, which are either specific for coastal habitats or species that have a wider ecological amplitude but are also regularly found on coastal rocks in order to provide new insights into the supposed role of these compounds and their involvement in secondary defensive barriers against stress in coastal lichens.

16.2 COASTAL LICHENS: TAXONOMY, ECOLOGY AND ZONATION

Lichens are symbiotic associations of a fungus (mycobiont) and one or several photosynthetic partner(s) (photobiont). The mycobiont provides a physical outer framework (exhabitant) in which the alga or cyanobacterium is living as an inhabitant (Hawskworth & Honegger, 1994). The structural dominance of the fungus distinguishes lichens from mycophycobioses, fungal–algal associations where the fungus inhabits the thalli of seaweeds (Kohlmeyer, Hawksworth, & Volkmann-Kohlmeyer, 2004; Kohlmeyer & Volkmann-Kohlmeyer, 1998). Additional fungi, algae and heterotrophic bacteria can be present in or surrounding the lichen thallus and interact with the symbiotic system (Bates, Cropsey, Caporaso, Knight, & Fierer, 2011; Bjelland et al., 2011; Grube & Berg, 2009). Lichens are poikilohydric systems but in many species the rate of water uptake and the form of water entering the thallus as either liquid or gas can be influenced by the production of hydrophobic substances (Hamlett et al., 2011; Hauck, Jürgens, Brinkmann, & Herminghaus, 2008).

Coastal rocks in temperate areas are often covered by a lichen vegetation that can appear in vertically arranged belts. Based on characteristic species compositions on British shores, a classification of these belts has been proposed with six different categories: littoral, mesic-supralittoral, submesic-supralittoral, xeric-supralittoral, terrestrial halophilic and terrestrial halophobic (Fletcher, 1973a,b). Similar zonation patterns are common in temperate areas elsewhere. Only the littoral (intertidal) is regularly inundated on a daily bases and lichens in this zone are seen as truly marine. Those higher up the shore (aerohaline zone, exposed to saltwater spray and dry salt crystal deposition but lacking inundation by seawater and wave action) are distinguished as maritime lichens. In Britain alone, about 450 species grow in coastal habitats (Fletcher, 1980). On a worldwide scale, more than 700 species are regarded as coastal lichens (Hawksworth, 2000) but this figure does not include the large number of species in coastal mangrove forests in the tropics and subtropics (Benatti & Marcelli, 2007; Jagadeesh Ram, Aptroot, Sinha, & Singh, 2005; Stevens, 1979).

The focus on belt like zonations can mask smaller scaled mosaics of microhabitats. Apart from the often correlated gradients of inundation, salinity, wind and wave exposure further important factors are rock chemistry (e.g. surface pH and mineral composition) and texture, aspect, inclination, light, climate and the effects of freshwater seeping over the rocks. Exposure to both wind and waves translates into physical stress. High wind velocity causes rapid desiccation and this is linked with quick and large changes of salinity, water availability and surface temperatures. Biotic stress can result from grazing by invertebrates, which affects coastal lichens from the intertidal to the aerohaline zones and causes the exposure of algae to harmful light intensities. Strong fluctuations in alkalinity, and the concentration of nitrogen and phosphor compounds are associated with seal and bird colonies. During the winter, an increase in epiphyte cover by algae and cyanobacteria on intertidal lichens has been noticed. This cover reduces light intensity but also protects from rapid desiccation and thereby extends periods of physiological activity at a time of the year when temperatures and respiration rates are low but photosynthesis can yield positive carbon balances (Fletcher, 1980).

Species density decreases with inundation frequency (Fletcher, 1973a,b). Regular growth several metres below the low tide level is only documented for the Antarctic lichen *Verrucaria serpuloides* (Verrucariaceae) with its photobiont *Dilabifilum* (Ulvales, Chlorphyta) (Lamb, 1973). Some species of the genus *Collemopsidium* (Xanthopyreniaceae) and *Lichina pygmaea* (Lichinaceae) with cyanobacterial photobionts tolerate long periods of inundation but are usually not extending deeper than lowest tide level. Reports of other lichen-like associations in the marine environment refer to "lichenoids" where the contact between the symbionts is loose (Kohlmeyer & Volkmann-Kohlmeyer, 1988). Most lichens in the littoral are highly specialised species (e.g. do not occur in other habitats). All species of *Wahlenbergiella* (Verrucariaceae, mostly associated with *Dilabifilum,* Ulvales/Chlorophyta; one species with *Petroderma maculiformis*, Ralfsiales, Heterokontophyta) are intertidal (Gueidan et al., 2009; Gueidan, Thüs, & Pérez-Ortega, 2011), but in *Hydropunctaria* (Verrucariaceae; species associated with either *Dilabifilum* or *Heterococcus*, Tribonemataceae, Heterokontophyta) some taxa are restricted to either marine or freshwater habitats and only *H. orae* is found in the intertidal as well as in coastal freshwater seepages (Orange, 2012). Few species show a preference for brackish water with *Verrucaria ditmarsica* (Verrucariaceae) the most frequent representative in European estuaries. The number of marine species drops on a gradient from temperate areas towards the tropics. Temperature, either directly or indirectly via gas solubility, plays an important role for the submersion resistance of lichenised systems but the reduced extend of tidal height towards the equator may also play a role. A much higher diversity is found in areas that are only sprayed by sea water and salt crystals but lack inundation periods. This zone consists of a mix of more or less specialised species and taxa with a wider ecological amplitude but limited halotolerance (Grube & Blaha, 2005). Specialised lichen communities colonise sheltered overhangs, seepages of freshwater or the vicinity of bird colonies (Fletcher, 1973a, 1973b).

Coastal environments are home to a number of fungal–algal associations where the term lichen is inadequate, but which have stimulated the discussion of the concept of lichenisation and are ideal model systems to study the effects of increasingly complex interactions between the symbionts on the ecology and physiological performance of these systems (Pérez-Ortega, de los Rios, Crespo, & Sancho, 2010). There is no sharp delimitation between lichens and other fungal–algal associations. Even within a single species lichenised and saprophytic lifestyles can alternate (Wedin, Döring, & Gilenstam, 2004) and genera with both clearly lichenised and parasitic or more or less saprophytic species are not uncommon. The fungus *Mycophyscias ascophylii* lives as an inhabitant in the brown algae *Ascophyllum nodosum*. No lichen thallus is formed but the symbiotic system shows an increased tolerance to drought compared to the independently living thalli of this algal species (Garbary & London, 1995). In *Mastodia tesselata* (Verrucariaceae) the association is further developed with the fungus producing intraparietal haustoria, which are common in lichens although the fungus remains an inhabitant to the algal thallus (Pérez-Ortega et al., 2010). Together with its photobiont *Prasiola* sp. (Prasiolales/Chlorophyta), this lichen-like association occurs in drier habitats than the free-living *Prasiola* alga. Furthermore, the free-living alga relies on freshwater supply while the lichenised form is found in stronger salt-exposed sites and in the vicinity of nesting birds where high concentrations of ammonia and phosphates occur (Kovácik & Pereira, 2001). Closely related to *Mastodia* is the genus *Wahlenbergiella* (Gueidan et al., 2011), which contains exclusively lichenised taxa from coastal rocks and the photobiont is either the green alga *Dilabifilum* or the brown alga *Petroderma maculiforme* (Thüs et al., 2011). In both cases, the free-living algae are found in microsites that are less prone to desiccation compared to those of the lichenised systems on the same shores (Moe, 1997; Takeshita & Harada, 2001).

From the perspective of aquatic or sub-aquatic algae, lichenisation can be seen as a way to extend their range into habitats that are too dry or too much exposed to high salt and nutrient fluctuations for survival without a symbiotic partner. The association with a wide range of only distantly related algae from sub-aquatic or even fully submerged habitats instead of sub-aerophytic algae is a specific trait in freshwater and marine taxa of the family Verrucariaceae with *Hydropunctaria* and *Wahlenbergiella* the most important genera on coastal shores (Thüs et al., 2011). They dominate the intertidal together with the genera *Lichina* (Lichinaceae, associated with *Nostoc*, Nostocales/Cyanobacteria) and *Collemopsidium* (Xanthopyreniaceae, associated with *Hyella caespitosum* Chroococcales/Cyanobacteria). In the maritime zones, these taxa are replaced by other fungal families. The change in the fungal taxa is mirrored in the associated photobiont types. The closely related genera *Asterochloris*, *Trebouxia* and *Myrmecia* (Trebouxiophyceae) are among the most common photobionts of lichens from temperate areas, but only very few of them occur in the intertidal zone (Hawksworth, 2000). Surprisingly also species associated with the genus *Diplosphaera* (Prasiolales), which do occur in sub-aquatic lichens from freshwater habitats (Thüs et al., 2011) are absent from the intertidal but can be present in microhabitats where freshwater seepages occur. Towards the tropics the number of lichens with *Trentepohlia* as photobiont of lichens in the aerohaline zone increases, particularly on sheltered rocks (Chu, Seaward, & Hodgkiss, 2000) and as epiphyte on mangroves (Benatti & Marcelli, 2007; Jagadeesh Ram et al., 2005). In temperate areas with mild winters coastal lichens with this photobiont are still represented in shaded overhangs, but diminish towards the polar regions and more generally in exposed frost prone situations. While frost damage is a determining factor for *Trentepohlia*-associated lichens, extreme heat on sun-exposed coastal rocks in the tropics reduces the lichen cover in this habitat. Exceptions are microsites with the presence of additional freshwater from terrestrial sources, where specialists of the family Lichinaceae are found e.g. *Peltula* spp. with *Chroococcidiopsis* as cyanobacterial photobionts, (Aptroot & Seaward, 1999).

16.3 GENESIS OF OXIDATIVE COMPOUNDS AND PRIMARY ANTIOXIDANT BARRIER

Living near the seashore implies several physiological mechanisms to adapt and/or acclimate to natural disturbances and stress, especially as environmental conditions are extremely variable.

Light, heat and high salt concentrations may lead to many metabolism disruptions in all organisms and specifically in fixed ones like lichens. This symbiosis remains adnated or attached directly to their substrate and must cope all the time with autoecological parameters.

The sunshine can damage the main metabolic pathways in lichen photobionts through the degradation of important compounds, e.g. photosynthetic pigments. Such phenomena may involve the synthesis of free radicals and/or ROS, primarily reduced by antioxidant enzymatic system. However, heat can denaturate proteins, so other detoxification mechanisms must be activated to re-establish homeostasis efficiently. Moreover, salts decrease the hydric potential, which affects the water availability in the poikilohydric lichens. Indeed, their hydric potential tends to equilibrium with the ecosystem water status. When hydrated, the lichen metabolism is operational and active, but when their water content falls, they become dormant leading to a subsequent oxidative stress. In consequence, marine and maritime lichens have developed various biochemical and physiological mechanisms to respond to these stresses and thus acquire tolerance.

In coastal lichens, many acclimation processes are involved in their 'struggle for existence'. They are able to reverse physiological and metabolic disruptions and to maintain the homeostasis of organelles and cells. To limit oxidative stress, coastal lichens activate their antioxidant mechanisms. During light, heat and salt stress, ROS concentrate in symbiont cells, especially when light becomes excessive, photosynthesis is disrupted and declines with decreasing water content. The oxidative compounds lead to disturbances, e.g. they can alter protein skeleton and induce membrane-impermeability loss through liperoxidation (Delmail et al., 2013; Delmail & Labrousse, 2012; Green, Sancho, & Pintado, 2011; Kranner et al., 2005).

Coastal lichens use O_2 as the keystone of their photosynthetic metabolism via microalgae and/or cyanobacteria. This aerobic process may lead to the over-synthesis of ROS with ionic imbalance and photooxidation. In natural conditions, ROS are metabolism intermediaries and mediators which regulate homeostasis, signalling and apoptosis. However, under stress conditions, their levels can increase leading to disturbances and damages in whole organism (Delmail et al., 2013; Delmail & Labrousse, 2012; Kranner et al., 2005; Kranner et al., 2003).

A partial O_2 reduction among cytochromes from the respiratory chain implies oxidant synthesis, i.e. singlet oxygen ($^1O_2$) and superoxide radical ($O_2^{\cdot -}$) leading to the formation of hydrogen peroxide (H_2O_2) and two radicals, hydroxyl ($\cdot OH$) and hydroperoxyl ($\cdot O_2H$). The alkoxyl ($RO\cdot$) and peroxyl ($RO_2\cdot$) radicals are the consequence of the membrane phospholipid peroxidation by ROS (Delmail & Labrousse, 2012; Edreva, 2005; Lagadic, Caquet, & Amiard, 1997; Li, Trush, & Yager, 1994; Thompson, Legge, & Barber, 1987) (Figure 16.1).

In parallel, the photosynthetic electron transport chains can produce high concentrations of ROS. Electrons tetravalently reduce intracellular oxygen to water. But, some electrons can could leak from many sites along the electron transport chain, resulting in a univalent reduction of O_2 to form the extremely reactive $O_2^{\cdot -}$, which can dismutate to form H_2O_2 (Alscher, Erturk, & Heath, 2002) (Figure 16.1). This last reaction is

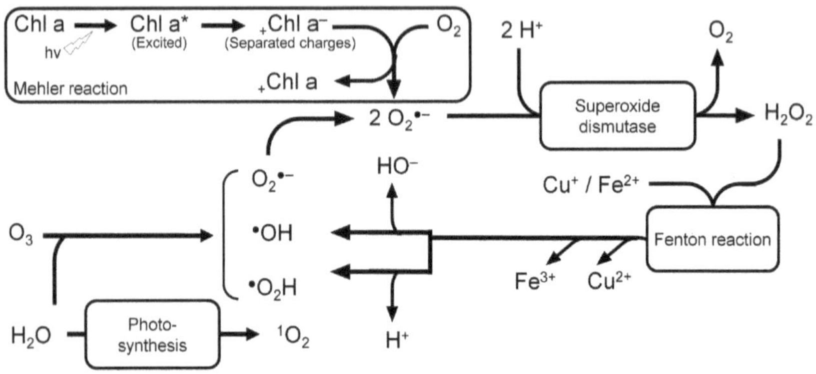

Figure 16.1 Main antioxidant pathways of ROS genesis. *Based on Delmail and Labrousse (2012).*

spontaneous or catalysed by one of the superoxide dismutases depending on the cell compartment where the reaction occurs: manganese-superoxide dismutase (microalga mitochondria and peroxisome), iron-superoxide dismutase (cyanobacteria cytosol, microalga chloroplast) or copper/zinc-superoxide dismutase (photobiont cytosol, microalga chloroplast) (Delmail et al., 2013; Delmail & Labrousse, 2012; Fornazier et al., 2002; Gill & Tuteja, 2010; Pereira, Molina, Lea, & Azevedo, 2002).

Hydrogen peroxide is not a free radical due to all its matched electrons. Nevertheless, it has a high toxic potential and it can diffuse far from its synthesis area. It can pass through biological membranes through aquaporins due to its chemical structure close to water (Bienert et al., 2007; Delmail & Labrousse, 2012; Parent, Capelli, & Dat, 2008). The levels of this compound are mainly mediated by proteins: ascorbate peroxidase, catalase and glutathione peroxidase. These enzymes use the nicotinamide adenine dinucleotide phosphate (NADPH) from photosynthesis for their proper functioning. However, ROS could shunt the photosynthetic electron transport pathway in thylakoid membranes leading to electron deflection. Without NADPH normal synthesis, photobionts use a cytosolic secondary catabolic pathway to produce it, the pentose phosphate pathway (Delmail & Labrousse, 2012; Kranner, 2002; Kruger & von Schaewen, 2003). This oxidative compound may be also produced through the bivalent reduction of oxygen by oxidases, e.g. peroxisomal glycolate oxidase or amine oxidase (Parent et al., 2008). Hydrogen-peroxide toxicity is also related to its involvement in the synthesis of hydroxyl and hydroperoxyl radicals through the Haber-Weiss and Fenton reactions. These short-lifespan radicals diffuse easily through membranes and affect all cell compartments. They are also mainly implied in lipoperoxidation (Delmail & Labrousse, 2012; Edreva, 2005; Lagadic et al., 1997). The resulting fatty-acid radical then reacts with molecular oxygen, thereby creating a peroxyl fatty acid radical that reacts with an another phospholipid, leading to a new radical and a lipid peroxide, or a cyclic peroxide if it reacts with itself. This cycle triggers a chain reaction mechanism (Schaich, 2005). The process ends up when two radicals react to produce a nonradical compound. It occurs when the radical concentration is sufficiently high. Living organisms have evolved different metabolites that improve the ROS reduction like scavengers mainly represented by phospholipophilic α-tocopherol (or vitamin E) and carotenoids (β-caroten, xantophylls) (Delmail et al., 2013; Delmail & Labrousse, 2012; Kranner & Lutzoni, 1999; Smirnoff, 1993). Hydrophilic antioxidants of low molecular weight also exist such as glutathione (γ-glutamyl-cysteinyl-glycine) and ascorbate. They act as scavengers resulting in the production of radicals reduced by NADPH-dependent enzymes; they protect cells from oxidative damages in liquid phases, e.g. in the cytosol (Beckett, Kranner, & Minibayeva, 2008; Delmail et al., 2013).

When primary antioxidant defences are overpassed, some secondary polyphenolic metabolites like flavonoids in plants could act (Fini, Brunetti, Di Ferdinando, Ferrini, & Tattini, 2011; Hernández, Alegre, Van Breusegem, & Munné-Bosch, 2009). Indeed, recent experiments confirmed the adaptative response of lichens to stress conditions, with an accumulation of such metabolites (Bialonska & Dayan, 2005) and it is natural to suppose that they have a role as active compounds in their exceptional resistance capacities. A great number of lichen extracts have been tested for their antioxidant activities after extraction either by apolar, middle polar or polar solvent and biological results varied along with the tested lichens and their content. Among the coastal lichens retrieved in literature, no extracts from marine (intertidal) taxa like *Collemopsidium, Hydropunctaria, Wahlenbergiella, Verrucaria* or *Lichina* have been tested for their antioxidant capacities. Nevertheless, activities from maritime species have been reported. They concerned the mostly coastal *Caloplaca regalis* (Bahattarai, Paudel, Hong, Lee, & Yim, 2008; Paudel et al., 2008), but also species that do occur frequently at inland localities, e.g. *Cladonia furcata* (Heng et al., 2010; Kosanic & Rankovic, 2011a,b; Kosanic, Rankovic, & Vukojevic, 2011), *Ochrolechia parella* (Rankovic, Rankovic, & Maric, 2010), *Parmelia saxatilis* (Gulluce et al., 2006; Kosanic, Rankovic, & Stanojkovic,

2012) and *Tephromela atra* (Kosanic & Rankovic, 2011a,b; Rankovic, Kosanic, & Stanojkovic, 2011). Except the extract of *Ochrolechia parella*, all extracts were active as antioxidant. Moreover, most of researchers have found high correlations between antioxidant activity and phenolic content determined by Folin-Ciocalteu method (Odabasoglu et al., 2005; Stojanovic, Stojanovic, Stankov-Jovanovic, Mitic, & Kostic, 2010). However, Aydin suggested that sugar alcohols such erythritol, sorbitol or mannitol could also protect cells against oxidative stress via scavenging oxyradical (Aydin & Turkez, 2011). Many secondary compounds are found across various unrelated taxa and although often characteristic and diagnostically valuable for certain species they are not specific. For this reason, our interest was to focus on the antioxidant properties of the substances rather than on their systematic occurrence in lichenised fungi.

16.4 LICHEN SECONDARY METABOLITES AND THEIR ANTIOXIDANT ACTIVITY

16.4.1 The Lichen Secondary Metabolites

The first reports on lichen compounds stem back in the end of the nineteenth century and the number of known substances from lichens is about 1000 while many more still need to be characterised. Most of these metabolites are phenols with unique structural features (depsides, depsidones, dibenzofurans, …) (Huneck & Yoshimura, 1996). Under stress of varied nature (sun, cold, desication-hydratation) and like for fungi (Wangemai, Luemail, Sunemail, & Zhu, 2011), concentrations of these metabolites in thalli are known to increase (Caviglia, Nicora, Giordani, Brunialti, & Modenesi, 2001; Kranner et al., 2005). These observations led the researchers to study the antioxidant properties of molecules produced by lichens that could participate in the oxidative stress regulation of the organism. As a result, an exponential number of publications relating to their role as antioxidants emerged and the number has more than doubled between 2007 and 2013. Twenty-nine research teams worked on this topic: ten are localised in Asia (four in India, one in Malaysia, two in Japan, one in Pakistan, two in South Korea), five in South America (one in Chile, one in Colombia, one in Mexico, one in Bolivia, one in Brazil) and fourteen in Europe (three in Turkey, one in Greece, two in Germany, three in Spain, one in France, two in Italia, two in Serbia). Out of all the known compounds, only sixty three molecules have been evaluated in antioxidant assays; fifty three molecules (Figure 16.2) have been selected to be analysed according the protocol described in Scheme 16.1. This research has been performed using the database SciFinder Scholar. They belong to several families (Tables 16.1–16.3, Figure 16.2) and the most described are depsides (21%), depsidones (31%) and monoaromatic phenols (17%). In these studies, a plethora of *in vitro* chemical methods have been used to evaluate the 'antioxidant capacity' of molecules or lichen extracts.

Depsides

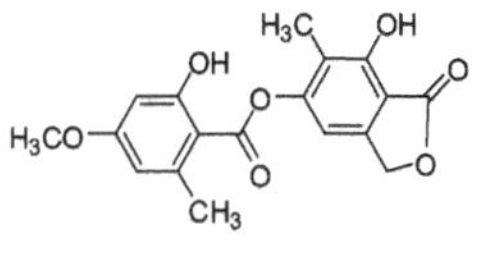

Hypotrachynic acid

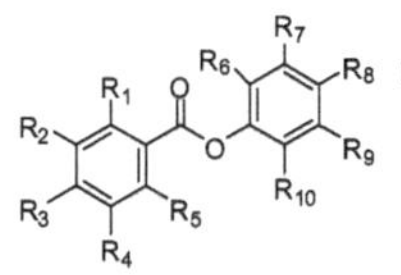

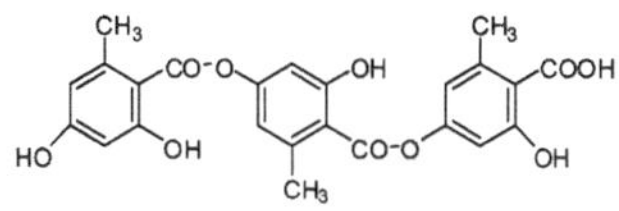

Gyrophoric acid

Atranorin $R_1 = CH_3$ $R_2 = H$ $R_3 = OH$ $R_4 = CHO$ $R_5 = H$ $R_6 = H$ $R_7 = CH_3$ $R_8 = COOCH_3$ $R_9 = OH$ $R_{10} = CH_3$
Decarboxythamnolic acid $R_1 = CH_3$ $R_2 = H$ $R_3 = OCH_3$ $R_4 = COOH$ $R_5 = H$ $R_6 = OH$ $R_7 = CHO$ $R_8 = OH$ $R_9 = H$ $R_{10} = CH_3$
Diffractaic acid $R_1 = CH_3$ $R_2 = H$ $R_3 = OCH_3$ $R_4 = CH_3$ $R_5 = CH_3$ $R_6 = H$ $R_7 = CH_3$ $R_8 = COOH$ $R_9 = OH$ $R_{10} = CH_3$
Divaricatic acid $R_1 = C_3H_7$ $R_2 = H$ $R_3 = OCH_3$ $R_4 = H$ $R_5 = H$ $R_6 = H$ $R_7 = OH$ $R_8 = COOH$ $R_9 = C_3H_7$ $R_{10} = H$
Erythrin $R_1 = OH$ $R_2 = H$ $R_3 = OH$ $R_4 = H$ $R_5 = CH_3$ $R_6 = H$ $R_7 = CH_3$ $R_8 =$ CO-erythritol $R_9 = OH$ $R_{10} = H$
Evernic acid $R_1 = OH$ $R_2 = H$ $R_3 = OCH_3$ $R_4 = H$ $R_5 = CH_3$ $R_6 = H$ $R_7 = OH$ $R_8 = COOH$ $R_9 = CH_3$ $R_{10} = H$
Lecanoric acid $R_1 = CH_3$ $R_2 = H$ $R_3 = OH$ $R_4 = H$ $R_5 = H$ $R_6 = H$ $R_7 = OH$ $R_8 = COOH$ $R_9 = CH_3$ $R_{10} = H$
Perlatolic acid $R_1 = C_5H_{11}$ $R_2 = H$ $R_3 = OCH_3$ $R_4 = H$ $R_5 = H$ $R_6 = H$ $R_7 = OH$ $R_8 = COOH$ $R_9 = C_5H_{11}$ $R_{10} = H$
Sekikaic acid $R_1 = C_3H_7$ $R_2 = H$ $R_3 = OCH_3$ $R_4 = H$ $R_5 = H$ $R_6 = H$ $R_7 = OH$ $R_8 = COOH$ $R_9 = C_5H_{11}$ $R_{10} = H$
Thamnolic acid $R_1 = CH_3$ $R_2 = H$ $R_3 = OCH_3$ $R_4 = COOH$ $R_5 = H$ $R_6 = OH$ $R_7 = CHO$ $R_8 = OH$ $R_9 = COOH$ $R_{10} = CH_3$

Depsidones

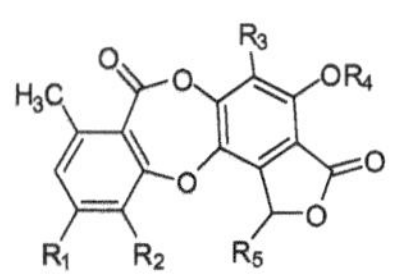

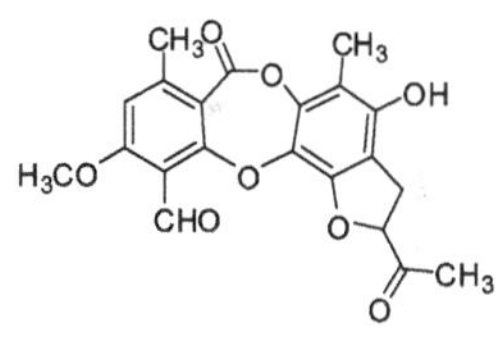

Pulmonarianin

Cryptostictinolide $R_1 = OCH_3$ $R_2 = CH_2OH$ $R_3 = CH_3$ $R_4 = H$ $R_5 = H$
Deoxy-stictic acid $R_1 = OCH_3$ $R_2 = CHO$ $R_3 = CH_3$ $R_4 = H$ $R_5 = H$
Isidiophorin $R_1 = OCH_3$ $R_2 = CHO$ $R_3 = CH_3$ $R_4 = H$ $R_5 = OH$
8'-menegazziaic acid $R_1 = OCH_3$ $R_2 = OH$ $R_3 = CH_3$ $R_4 = H$ $R_5 = OCH_3$
8'-methyl-stictic acid $R_1 = OCH_3$ $R_2 = CHO$ $R_3 = CH_3$ $R_4 = H$ $R_4 = OCH_3$
Norstictic acid $R_1 = OH$ $R_2 = CHO$ $R_3 = CH_3$ $R_4 = H$ $R_4 = OH$
Salazinic acid $R_1 = OH$ $R_2 = CHO$ $R_3 = CH_2OH$ $R_4 = H$ $R_4 = OH$
Stictic acid $R_1 = OCH_3$ $R_2 = CHO$ $R_3 = CH_3$ $R_4 = H$ $R_4 = OH$
Vesuvianic acid $R_1 = OCH_3$ $R_2 = CHO$ $R_3 = CH_3$ $R_4 = CH_2CH_3$ $R_5 = OH$

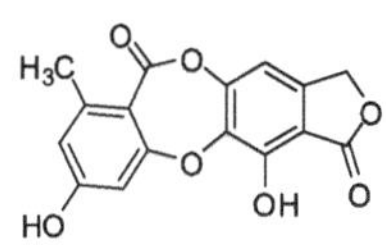

Variolaric acid

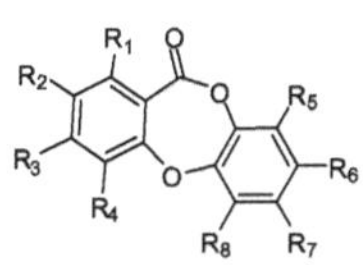

1'-chloro-pannarin $R_1 = CH_3$ $R_2 = Cl$ $R_3 = OH$ $R_4 = CHO$ $R_5 =$ CH3 $R_6 = OCH_3$ $R_7 = Cl$ $R_8 = CH_3$
Fumarprotocetraric acid $R_1 = CH_3$ $R_2 = H$ $R_3 = OH$ $R_4 = CHO$ $R_5 = CH_2O$-fumaric acid $R_6 = OH$ $R_7 = COOH$ $R_8 = CH_3$
Lobaric acid $R_1 = CO(CH_2)_3CH_3$ $R_2 = H$ $R_3 = OCH_3$ $R_4 = H$ $R_5 = H$ $R_6 = OH$ $R_7 = COOH$ $R_8 = (CH_2)_4CH_3$
Pannarin $R_1 = CH_3$ $R_2 = Cl$ $R_3 = OH$ $R_4 = CHO$ $R_5 =$ CH3 $R_6 = OCH_3$ $R_7 = H$ $R_8 = CH_3$
Physodic acid $R_1 = CH_2CO(CH_2)_4CH_3$ $R_2 = H$ $R_3 = OH$ $R_4 = H$ $R_5 = H$ $R_6 = OH$ $R_7 = COOH$ $R_8 = (CH_2)_4$
Protocetraric acid $R_1 = CH_3$ $R_2 = H$ $R_3 = OH$ $R_4 = CHO$ $R_5 = CH_2OH$ $R_6 = OH$ $R_7 = COOH$ $R_8 = CH_3$
Stictic acid in opened form $R_1 = CH_3$ $R_2 = H$ $R_3 = OCH_3$ $R_4 = CHO$ $R_5 = CH_3$ $R_6 = OH$ $R_7 = COOH$ $R_8 = CHO$
Vicanicin $R_1 = CH_3$ $R_2 = Cl$ $R_3 = OH$ $R_4 = CH_3$ $R_5 = CH_3$ $R_6 = OCH_3$ $R_7 = Cl$ $R_8 = CH_3$

Figure 16.2 Molecules described for their antioxidant activity.

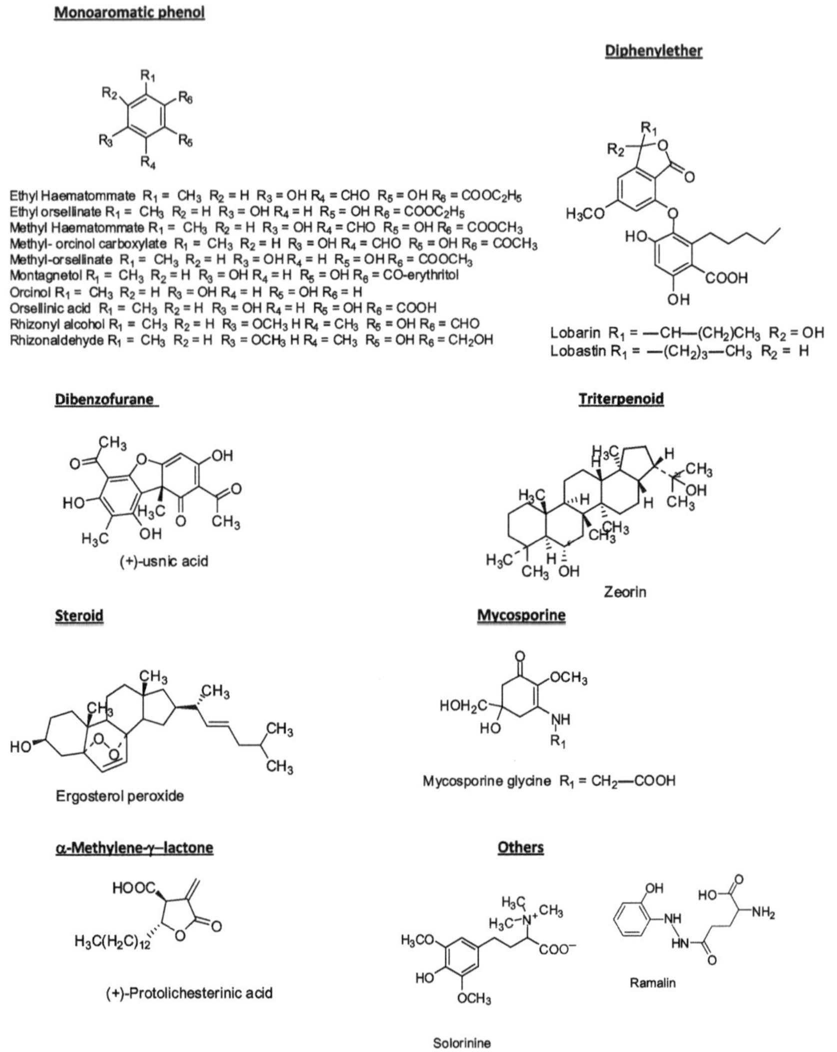

Figure 16.2 (*Continued*)

16.4.2 What Represents an Antioxidant Activity?

A biological antioxidant is defined as 'a molecule which, when present in small concentrations compared to the biomolecules, is supposed to protect, prevent or reduce the extent of oxidative destruction of biomolecules' (Halliwell, 1990). Antioxidants may act in organism by different processes: scavengers (radical or reducing processes), radical chain reaction inhibitors, ion metal chelators and modulation of gene expression by activation of

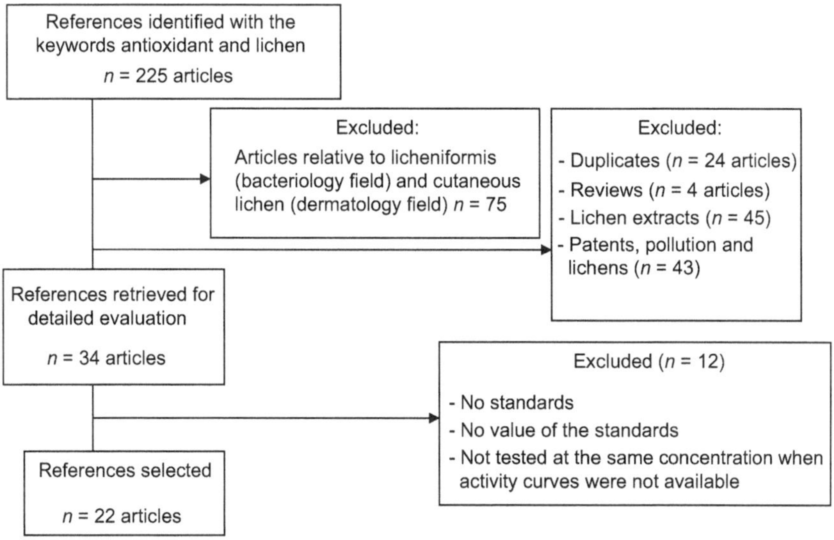

Scheme 16.1 Litterature search flow rate.

transcription factors, as it is described for polyphenol compounds (Dangles, 2012; Halliwell & Gutteridge, 2003; Lopez-Alarcon & Denicola, 2013). From these observations, several methods were developed to study the antioxidant properties of a molecule or an extract (Apak et al., 2013; Huang, Ou, & Prior, 2005; Lopez-Alarcon & Denicola, 2013). Few protocols use *in vivo* techniques, those are mainly *in vitro* assays, cellular or acellular. The results presented in this article are essentially those obtained with a chemical method, the most numerous.

The chemical *in vitro* methods are well described in reviews and were classified in two majors categories, following the main chemical mechanism involved (Apak et al., 2013; Huang et al., 2005). The methods are based on either ET or HAT measurements, involved in respectively reducing or radical scavenging processes. In addition, other assays exist to measure specifically the reaction between a substrate and a reactive oxygen/nitrogen species ROS/RNS such singlet oxygen, superoxide radical, NO radical, hydroxyl radical. In all these methods, measurements are essentially spectrophotometric or fluorometric, RPE and electrochemical techniques also exist but are sparse (Blasco, Crevillen, Gonzales, & Escarpa, 2007; Hillard et al., 2008; Le Bourvellec, Hauchard, Darchen, Burgot, & Abasq, 2008; Rene, Abasq, Hauchard, & Hapiot, 2010). Tables 16.1 and 16.2 list the *in vitro* chemical antioxidant assays used for studying lichenic compounds. HAT-based assays consider radical reactions, with or without chain propagation steps, respectively lipid peroxidation and autooxidation tests for the formers; oxygen radical absorbance capacity (ORAC), total reactive antioxidant potential (TRAP) tests for the latters. An HAT occurs, these tests represent the capacity of the molecule as H atom donor. ET-based assays consider electron transfer reactions and evaluate the capacity of the molecule as reducing agent.

Reviews summarise the chemistry behind the methods and their limitations involved (Apak et al., 2013; Huang et al., 2005). They warn against the inappropriate and misleading conclusions that can be drawn from the results. The 'antioxidant capacity' measured by a method reflects only the chemical reactivity in the specific conditions, applied in that test. Consequently, the values between different methods cannot be comparable. Inside a method, values are comparable if and only if the experimental conditions are strictly identical (concentrations, pH, solvents…). The question of the standard is also essential. Among the results we consulted, we refer in this paper only those that presented the values compared to a standard (expressed as equivalent of standard) or gave the value of the standard in the same experimental conditions. We expressed the results in the tables not by the direct value but by a multiplier number with regard to the standard (Tables 16.1–16.3).

Nine standards were used: ascorbic acid, BHA, BHT, propyl gallate, quercetin, anthralin, nordihydroguaieretic acid, trolox, α-tocopherol (Tables 16.2–16.4). BHA, BHT (Mulky, 1976), quercetin (Boots, Haenen, & Bast, 2008), nordihydroguairetic acid (Oliveto, 1972; Parke & Lewis, 1992), tocopherols and trolox (Evans & Reynhout, 1992; Kamal-Eldin, Kim, Tavadyan, & Min, 2008) are phenolics, which protect polyunsaturated fatty acids from oxidative degradation by donating their hydrogen of the hydroxyl to chain carrier lipid peroxyl radicals and preventing further chain radical reaction and also act as antiradical. Vitamin C (Sies & Stahl, 1995) is recognised as a good metal ion chelator. Anthralin (1,8-dihydroxy-9(10H)-anthracenone) is a drug used in the treatment of psoriasis and its antioxidant effect is moderate compared to nordihydroguaiaretic acid (Muller & Gurster, 1993) and it seems that possible antioxidant effects may be surpassed by the strong pro-oxidant actions mediated by active oxygen species. Ascorbic acid and trolox are water-soluble and exert their highest activity in aqueous solutions and *vice versa* for the others that are lipophilic standards. Half of the controls were of natural origin: quercetin (member of the flavonoids family and one of the most prominent dietary antioxidants as glycoside), nordihydroguaieretic acid (major constituent of the resinous exudates of *Larrea divaricata*), tocopherol (major lipid-soluble forms of vitamin E in

Table 16.1 Assays Based-on the Hydrogen Atom Transfer Mechanism

Class	Compound	Method/Standard (Reference)	Results
Depside	Atranorin	β-Carotene linoleate/BHA (Jayaprakasha & Rao, 2000)	↓ (3) for 500 μg/ml at 90 min
		Lipid peroxidation/nordihydroguairetic acid, anthralin (Kumar & Muller, 1999)	na
		TRAP-TAR/trolox (Melo et al., 2011)	↓ (3) at 0.3 mM
		TBARS/trolox (Melo et al., 2011)	Pro-oxidant at 0.3 mM
	Diffractaic acid	Lipid peroxidation/nordihydroguairetic acid, anthralin (Kumar & Muller, 1999)	na
		Lipid peroxidation – lecithine liposome test/BHT (Atalay et al., 2011)	↓ (29) after 7 days
		Lipid peroxidation – linoleic acide emulsion test/BHT (Atalay et al., 2011)	↓ (5.6) after 7 days
	Lecanoric acid	β-Carotene linoleate/BHA (Jayaprakasha & Rao, 2000)	↓ (2.7) for 500 μg/ml at 90 min
		Lipid peroxidation/ascorbic acid (Heng et al., 2010)	↓ (1.2) at 2 mg/ml
Depsidone	1′-Chloropannarin	Autoxidation/propyl gallate (Hidalgo, Fernandez, Quilhot, & Lissi, 1994)	↔
	Isidiophorin	Lipid peroxidation – lecithine liposome test/BHT (Atalay et al., 2011)	↓ (6.5) after 7 days
		Lipid peroxidation – linoleic acide emulsion test/BHT (Atalay et al., 2011)	↔ after 7 days
	Pannarin	Autoxidation/propyl gallate (Hidalgo et al., 1994)	↓ (1.94)
	Pulmonarianin	Lipid peroxidation – lecithine liposome test/BHT (Atalay et al., 2011)	↓ (6.5) after 7 days
		Lipid peroxidation – linoleic acide emulsion test/BHT (Atalay et al., 2011)	↓ (4) after 7 days
	Salazinic acid	ORAC/trolox (Amo de Paz et al., 2010)	↑ (2.7)
	Stictic acid	Lipid peroxidation – lecithine liposome test/BHT (Atalay et al., 2011)	↓ (15) after 7 days
		Lipid peroxidation – linoleic acide emulsion test/BHT (Atalay et al., 2011)	↓ (8.6) after 7 days
		ORAC/trolox (Amo de Paz et al., 2010)	↑ (2.3)

Continued

Table 16.1 Assays Based-on the Hydrogen Atom Transfer Mechanism—cont'd

Class	Compound	Method/Standard (Reference)	Results
Dibenzofurane	Usnic acid	Lipid peroxidation – lecithine liposome test/BHT (Atalay et al., 2011)	↓ (31) after 7 days
		Lipid peroxidation – linoleic acide emulsion test/BHT (Atalay et al., 2011)	↓ (1.3) after 7 days
		Lipid peroxidation/nordihydroguairetic acid, anthralin (Kumar & Muller, 1999)	na
		ORAC/trolox (Amo de Paz et al., 2010)	↑ (3)
Monoaromatic phenol	Ethyl haematommate	Lipid peroxidation/nordihydroguairetic acid, anthralin (Kumar & Muller, 1999)	na
	Methyl orcinol carboxylate	Lipid peroxidation/nordihydroguairetic acid, anthralin (Kumar & Muller, 1999)	na
	Methyl orsellinate	β-Carotene linoleate/BHA (Jayaprakasha & Rao, 2000)	↓ (2.4)
	Orsellinic acid	β-Carotene linoleate/BHA (Jayaprakasha & Rao, 2000)	↓ (1.9) for 500 μg/ml at 90 min
	Rhizonaldehyde	Lipid peroxidation – lecithine liposome test/BHT (Atalay et al., 2011)	↓ (15) after 7 days
		Lipid peroxidation – linoleic acide emulsion test/BHT (Atalay et al., 2011)	↓ (3) after 7 days
	Rhizonyl alcohol	Lipid peroxidation – lecithine liposome test/BHT (Atalay et al., 2011)	↔ after 7 days
		Lipid peroxidation – linoleic acide emulsion test/BHT (Atalay et al., 2011)	↓ (2) after 7 days

Mycosporine	Mycosporine glycine	β-Carotene bleaching method/ α-tocopherol (De la Coba, Aguilera, Figueroa, De Calvez, & Herrera, 2009)	↓ (5)
Steroid	Ergosterol peroxide	Lipid peroxidation – lecithine liposome test/BHT (Atalay et al., 2011)	↓ (12) after 7 days
		Lipid peroxidation – linoleic acide emulsion test/BHT (Atalay et al., 2011)	↓ (5) after 7 days
Tridepside	Gyrophoric acid	Lipid peroxidation/nordihydroguairetic acid, anthralin (Kumar & Muller, 1999)	na
Alpha-methylene-gamma-lactone	(+)-protolichesterinic acid	Lipid peroxidation/nordihydroguairetic acid, anthralin (Kumar & Muller, 1999)	na

TRAP-TAR, total reactive antioxidant potential-total antioxidant reactivity; ORAC, oxygen radical absorbance capacity; TBARS, thiobarbituric acid reactive substances; na, not active.

Table 16.2 Assays Based-on the ET Mechanism

Class	Compound	Method/Standard (Reference)	Results
Depside	Atranorin	DPPH/anthralin (Kumar & Muller, 1999)	↓ (2.1)
		TEAC/quercetin in trolox equivalent (Mollinedo, Vila, & Sterner, 2003)	↓ (8.5)
		DPPH/ascorbic acid (Ismed et al., 2012)	na
		DPPH/quercetin (Mollinedo et al., 2003)	na
		DPPH/BHA (Thadhani et al., 2011)	na
	Decarboxythamnolic acid	DPPH/quercetin (Mollinedo et al., 2003)	na
		TEAC/quercetin en equivalent trolox (Mollinedo et al., 2003)	↓ (5.1)
	Diffractaic acid	DPPH/anthralin (Kumar & Muller, 1999)	na
		DPPH/BHT (Atalay et al., 2011)	na
		DPPH/ascorbic acid (Brisdelli et al., 2013)	na
	Divaricatic acid	DPPH/BHA (Thadhani et al., 2011)	↓ (3.5)
	Erythrin	DPPH/BHA (Thadhani et al., 2011)	na
	Evernic acid	DPPH/ascorbic acid (Kosanic et al., 2013)	↓ (50)
		Reducing power/ascorbic acid (Kosanic et al., 2013)	↓ (2) at 250 μg/ml
	Lecanoric acid	DPPH/gallic acid (Lopes et al., 2008)	↓ (6147)
		DPPH/BHA (Heng et al., 2010)	↓ (2.5) at 2 mg/ml
		DPPH/BHA (Thadhani et al., 2011)	↓ (2.8)
	Perlatolic acid	DPPH/quercetin (Mollinedo et al., 2003)	na
		TEAC/quercetin in trolox equivalent (Mollinedo et al., 2003)	↓ (1.9)
	Sekikaic acid	DPPH/BHA (Thadhani et al., 2011)	↓ (2.9)
	Thamnolic acid	DPPH/quercetin (Mollinedo et al., 2003)	na
		TEAC/quercetin in trolox equivalent (Mollinedo et al., 2003)	↓ (3.6)

Depsidone	Cryptostictinolide	DPPH/quercetin (Lohézic-Le Dévéhat et al., 2007)	na
	Fumarprotocetraric acid	DPPH/quercetin (Lohézic-Le Dévéhat et al., 2007)	na
	Isidiophorin	DPPH/BHT (Atalay et al., 2011)	↓ (1.1)
		DPPH/BHA (Bahattarai, Kim, Oh, & Yim, 2013)	na
	Lobaric acid	DPPH/ascorbic acid (Ismed et al., 2012)	na
		DPPH/ascorbic acid (Brisdelli et al., 2013)	na
		DPPH/BHA (Thadhani et al., 2011)	na
	Norstictic acid	DPPH/ascorbic acid (Rankovic, Kosanic, Stanojkovic, Vasiljevic, & Manojlovic, 2012)	↓ (16)
		Reducing power (Rankovic et al., 2012)	↓ (1.1) at 250 μg/ml
	Open form of stictic acid	DPPH/quercetin (Lohézic-Le Dévéhat et al., 2007)	na
	Physodic acid	DPPH/ascorbic acid (Kosanic et al., 2013)	↓ (11)
		Reducing power/ascorbic acid (Kosanic et al., 2013)	↓ (1.1) at 125 μg·ml
	Protocetraric acid	DPPH/ascorbic acid (Manojlovic, Rankovic, Kosanic, Vasiljevic, & Stanojkovic, 2012)	↓ (23)
		Reducing power/ascorbic acid (Manojlovic et al., 2012)	↓ (13) at 250 μg/ml
	Pulmonarianin	DPPH/BHT (Atalay et al., 2011)	↓ (18) at 500 μg/ml
	Salazinic acid	DPPH/ascorbic acid (Manojlovic et al., 2012)	↓ (14.3)
		Reducing power/ascorbic acid (Manojlovic et al., 2012)	↓ (13) at 250 μg/ml
	Stictic acid	DPPH/BHT (Atalay et al., 2011)	↓ (5)
	Variolaric acid	DPPH/ascorbic acid (Brisdelli et al., 2013)	na
	Vicanicin	DPPH/ascorbic acid (Brisdelli et al., 2013)	na

Continued

Table 16.2 Assays Based-on the ET Mechanism—cont'd

Class	Compound	Method/Standard (Reference)	Results
Dibenzofurane	Usnic acid	DPPH/ascorbic acid (Brisdelli et al., 2013)	na
		DPPH/quercetin (Mollinedo et al., 2003)	na
		DPPH/BHT (Atalay et al., 2011)	na
		DPPH/ascorbic acid (Brisdelli et al., 2013)	na
		DPPH/BHA (Thadhani et al., 2011)	na
		DPPH/ascorbic acid (Manojlovic et al., 2012)	↓ (9.5)
		DPPH/anthralin (Kumar & Muller, 1999)	na
		DPPH/ascorbic acid (Rankovic et al., 2012)	↓ (20)
		TEAC/quercetin in trolox equivalent (Mollinedo et al., 2003)	↓ (51)
		Reducing power/ascorbic acid (Manojlovic et al., 2012)	↓ (3.5) at 250 μg/ml
		Reducing power (Rankovic et al., 2012)	↓ (1.4)
Diphenylether	Lobarin	DPPH/ascorbic acid (Ismed et al., 2012)	na
	Lobastin	DPPH/BHA (Bahattarai et al., 2013)	↓ (2.7)
Monoaromatic phenol	Ethyl haematommate	DPPH/anthralin (Kumar & Muller, 1999)	↓ (2)
	Ethyl orsellinate	DPPH/gallic acid (Lopes et al., 2008)	na
	Methyl haematommate	DPPH/BHA (Thadhani et al., 2011)	na
	Methyl orcinol carboxylate	DPPH/BHA (Thadhani et al., 2011)	na
		DPPH/ascorbic acid (Ismed et al., 2012)	na
		DPPH/anthralin (Kumar & Muller, 1999)	↓ (4)
	Methyl orsellinate	DPPH/BHA (Thadhani et al., 2011)	↓ (39)
		DPPH/gallic acid (Lopes et al., 2008)	na
	Montagnetol	DPPH/BHA (Thadhani et al., 2011)	na
	Orcinol	DPPH/BHA (Thadhani et al., 2011)	↓ (1.9)
		DPPH/gallic acid (Lopes et al., 2008)	na
	Orsellinic acid	DPPH/BHA (Thadhani et al., 2011)	↓ (2.4)
		DPPH/gallic acid (Lopes et al., 2008)	na
	Rhizonaldehyde	DPPH/BHT (Atalay et al., 2011)	↓ (4.5)
	Rhizonyl alcohol	DPPH/BHT (Atalay et al., 2011)	↓ (9)

Mycosporine	Mycosporine glycine	TEAC/ascorbic acid (De la Coba et al., 2009)	↔ at physiological pH
Steroid	Ergosterol peroxide	DPPH/BHT (Atalay et al., 2011)	↓ (2.4)
Tridepside	Gyrophoric acid	DPPH/anthralin (Kumar & Muller, 1999)	↓ (6.7)
Triterpenoid	Zeorin	DPPH/BHA (Thadhani et al., 2011)	na
Alpha-methylene-gamma-lactone	(+)-Protolichesterinic acid	DPPH/ascorbic acid (Brisdelli et al., 2013)	na
		DPPH/anthralin (Kumar & Muller, 1999)	na
Divers	Ramalin	DPPH/BHA (Paudel et al., 2011)	↑ (7.6)
		TEAC/trolox (Paudel et al., 2011)	↑ (27)
		Reducing power/BHT equivalent (Paudel et al., 2011)	↑ (2.5)
	Solorinine	DPPH/trolox (Hara, Endo, Kawakami, Komine, & Yamamoto, 2011)	↑ (1.25)

DPPH, 2,2-di(4-tert-octylphenyl)-1-picrylhydrazyl; TEAC, trolox-equivalent antioxidant capacity; TEAC, trolox-equivalent antioxidant capacity; ↓ (×), activity × times less than the control; ↑ (×), activity × times more than the control; ↔, activity equal to the control; na, not active.

Table 16.3 Assays Involving Reactive Oxygen Species Scavenging Activities

Class	Compound	Method/Standard/Reference	Results
Depside	Atranorin	Hydroxyl radical scavenging/trolox (Papadopoulou, Tzakou, Vagias, Kefalas, & Roussis, 2007)	↓ (12)
		Hydroxyl radical scavenging/nordihydroguairetic acid (Kumar & Muller, 1999)	↑ (23)
		Hydroxyl radical scavenging/trolox (Melo et al., 2011)	Pro-oxidant at 0.3 mM
		Superoxide anion scavenging/ascorbic acid (Ismed et al., 2012)	↓ (1.1)
		Superoxide anion scavenging/BHA, propylgallate (Thadhani et al., 2011)	↓ (62)
		NO radical scavenging/rutin (Thadhani et al., 2011)	↓ (1.8)
		NO radical scavenging/trolox (Melo et al., 2011)	Pro-oxidant at 0.3 mM
	Diffractaic acid	Hydroxyl radical scavenging/nordihydroguairetic acid (Kumar & Muller, 1999)	↑ (1.9)
	Divaricatic acid	Superoxide anion scavenging/BHA, propylgallate (Thadhani et al., 2011)	↓ (1.1)
		NO radical scavenging/rutin (Thadhani et al., 2011)	na
	Erythrin	Superoxide anion scavenging/BHA, propylgallate (Thadhani et al., 2011)	↓ (1.2)
		NO radical scavenging/rutin (Thadhani et al., 2011)	↓ (1.3)
	Evernic acid	Superoxide anion scavenging/ascorbic acid (Kosanic et al., 2013)	↓ (5)
	Hypotrachynic acid	Hydroxyl radical scavenging/trolox (Papadopoulou et al., 2007)	↓ (85)
	Lecanoric acid	Superoxide anion scavenging/BHA, propylgallate (Thadhani et al., 2011)	↔
		NO radical scavenging/rutin (Thadhani et al., 2011)	↓ (1.8)
	Sekikaic acid	Superoxide anion scavenging/BHA, propylgallate (Thadhani et al., 2011)	↔
		NO radical scavenging/rutin (Thadhani et al., 2011)	na

Depsidone	8′-Methylmenegazziaic acid	Hydroxyl radical scavenging/trolox (Papadopoulou et al., 2007)	↓ (8)
	8′-Methylstictic acid	Hydroxyl radical scavenging/trolox (Papadopoulou et al., 2007)	↓ (62)
	Cryptostictinolide	Hydroxyl radical scavenging/trolox (Papadopoulou et al., 2007)	na
	Deoxystictic acid	Hydroxyl radical scavenging/trolox (Papadopoulou et al., 2007)	↓ (14)
	Fumarprotocetraric acid	Superoxide anion scavenging/quercetin (Lohézic-Le Dévéhat et al., 2007)	↑ (1.3)
	Lobaric acid	Superoxide anion scavenging/ascorbic acid (Ismed et al., 2012)	↓ (2.3)
		Superoxide anion scavenging/BHA, propylgallate (Thadhani et al., 2011)	↔
		NO radicadical scavenging/rutin (Thadhani et al., 2011)	na
	Norstictic acid	Superoxide anion scavenging/quercetin (Lohézic-Le Dévéhat et al., 2007)	↑ (1.3)
		Superoxide anion scavenging/ascorbic acid (Rankovic et al., 2012)	↓ (1.1)
	Physodic acid	Superoxide anion scavenging/ascorbic acid (Kosanic et al., 2013)	↔
	Protocetraric acid	Superoxide anion scavenging/ascorbic acid (Manojlovic et al., 2012)	↓ (1.53)
	Salazinic acid	Superoxide anion scavenging/ascorbic acid (Manojlovic et al., 2012)	↓ (1.03)
	Stictic acid	Hydroxyl radical scavenging/trolox (Papadopoulou et al., 2007)	↓ (105)
Dibenzofurane	Usnic acid	Hydroxyl radical scavenging/nordihydroguairetic acid (Kumar & Muller, 1999)	↓ (6.4)
		Superoxide anion scavenging/BHA, propylgallate (Thadhani et al., 2011)	↓ (1.6)
		Superoxide anion scavenging/ascorbic acid (Manojlovic et al., 2012)	↓ (1.18)
		Superoxide anion scavenging/ascorbic acid (Rankovic et al., 2012)	↓ (1.7)
		NO radical scavenging/rutin (Thadhani et al., 2011)	↓ (1.7)
Diphenylether	Lobarin	Superoxide anion scavenging/ascorbic acid (Ismed et al., 2012)	↑ (3.9)

Continued

Table 16.3 Assays Involving Reactive Oxygen Species Scavenging Activities—cont'd

Class	Compound	Method/Standard/Reference	Results
Monoaromatic phenol	Ethyl haematommate	Hydroxyl radical scavenging/nordihydroguairetic acid (Kumar & Muller, 1999)	↓ (1.3)
	Methyl haematommate	Superoxide anion scavenging/BHA, propylgallate (Thadhani et al., 2011)	↓ (4.8)
		NO radical scavenging/rutin (Thadhani et al., 2011)	↓ (1.5)
	Methyl orcinol carboxylate	Hydroxyl radical scavenging/nordihydroguairetic acid (Kumar & Muller, 1999)	↓ (6.2)
		Superoxide anion scavenging/BHA, propylgallate (Thadhani et al., 2011)	↓ (4.7)
		Superoxide anion scavenging/ascorbic acid (Ismed et al., 2012)	na
		NO radical scavenging/rutin (Thadhani et al., 2011)	↔
	Methyl orsellinate	Superoxide anion scavenging/BHA, propylgallate (Thadhani et al., 2011)	↓ (4.7)
		NO radical scavenging/rutin (Thadhani et al., 2011)	↓ (1.5)
	Montagnetol	Superoxide anion scavenging/BHA, propylgallate (Thadhani et al., 2011)	↓ (10)
		NO radical scavenging/rutin (Thadhani et al., 2011)	↓ (1.5)
	Orcinol	Superoxide anion scavenging/BHA, propylgallate (Thadhani et al., 2011)	↓ (2.6)
		NO radical scavenging/rutin (Thadhani et al., 2011)	↓ (2)
	Orsellinic acid	Superoxide anion scavenging/BHA, propylgallate (Thadhani et al., 2011)	↓ (12.5)
		NO radical scavenging/rutin (Thadhani et al., 2011)	↓ (1.6)
Mycosporine	Mycosporine glycine	Superoxide anion scavenging/SOD (De la Coba et al., 2009)	na
Tridepside	Gyrophoric acid	Hydroxyl radical scavenging/nordihydroguairetic acid (Kumar & Muller, 1999)	↑ (1.9)
Triterpenoid	Zeorin	Superoxide anion scavenging/BHA, propylgallate (Thadhani et al., 2011)	↓ (62)
		NO radical scavenging/rutin (Thadhani et al., 2011)	na
Alpha-methylene-gamma-lactone	Protolichesterinic acid (+)	Hydroxyl radical scavenging/nordihydroguairetic acid (Kumar & Muller, 1999)	↑ (2.9)
Divers	Ramalin	Superoxide anion scavenging/ascorbic acid (Paudel et al., 2011)	↑ (1.8)

↓ (×), activity × times less than the control; ↑ (×), activity × times more than the control; ↔, activity equal to the control; na, not active.

Table 16.4 *In cellulo* and *In vivo* Antioxidant Assays

	Compounds	Protocol/Control	Assays	Results	References
In cellulo	Diffractaic acid (50 µM) Vicanicin (50 µM) Lobaric acid (50 µM) Variolaric acid (50 µM) Usnic acid (50 µM) Protolichesterinic acid (50 µM)	Evaluation of the protective effect against *t*-BHP (500 µM) on Hela cells	ROS production measured by DCF fluorescence	No antioxidant and pro-oxidant activity	Brisdelli et al. (2013)
	Ramalin (0.5; 1, 2 mg/ml)	Evaluation of the protective effects against *E. coli* lipopolysaccharide on murine macrophages raw/dexamethasone (4 µg/ml)	NO production measured by colorimetric assay	Better activity than dexamethasone at 2 µg/ml	Paudel et al. (2011)
			H_2O_2 production measured by flow cytometry	Better activity than ascorbic acid at all doses, in concentration dependant manner	Paudel et al. (2011)
In vivo	Diffractaic acid (25, 50, 100, 200 mg/kg)	6 rats/Evaluation of the protective activity against effects of indomethacin (25 mg/kg) vs ranitidine (50 mg/kg)	After $t = 6$ h, stomachs were removed and homogenates were used to evaluate enzyme activities: CAT, SOD, *Gpx*, GSH content, lipid peroxidation, MPx, iNOS, cNOS	↑SOD, GPx, cNOS, GSH ↓CAT, GR, iNOS, MPx, lipid peroxidation – marked effect at 100 mg/kg	Bayir et al. (2006)
	Usnic acid (25, 50, 100, 200 mg/kg)		After $t = 6$ h, stomachs were removed and homogenates were used to evaluate enzyme activities: CAT, *GR*, SOD, GSH content, lipid peroxidation, MPx, iNOS, cNOS	↑SOD, GPx, cNOS, GSH ↓CAT, GR, iNOS, MPx, lipid peroxidation – effects at all doses of usnic acid	Odabasoglu et al. (2006)

ROS, reactive oxygen species.

nature). L-ascorbic acid, the closed structures of gallates: BHA (E320), BHT (E321), propyl gallate, trolox (water-soluble derivative of vitamine E) are synthetic compounds.

The chemical *in vitro* methods give first information on the physico-chemical properties of a molecule but do not consider some very important *in vivo* parameters such as bioavailability, absorption and lipophilicity, metabolism and pharmacokinetics (Apak et al., 2013; Huang et al., 2005; Lopez-Alarcon & Denicola, 2013). These methods have to be considered as a screening process to evaluate the antioxidant properties of a molecule or an extract, before entering into a costly *in vivo* study. Several *in vitro* chemical methods must be considered to evaluate the antioxidant properties of a natural molecule: whole HAT, ET, specific ROS scavenging parameters are necessary to appreciate it.

A compound can also exert its antioxidant activity by taking part in some metal complexation stages. Some lichens tolerate high levels of pollutants including heavy metals and can live in extreme metalliferous environments (uranium, ferrous…). Considering that, studies were conducted to effect chemical weathering of mineral and rocks of lichens or to appreciate the role of lichen substances in metal homeostasis and pollution tolerance (Ascaso & Galvan, 1976; Ascaso, Galvan, & Ortega, 1976; Bialonska & Dayan, 2005; Hauk & Huneck, 2007; Iskandar & Syers, 1972; Jones & Wilson, 1985; Purvis, Bailey, McLean, Kasama, & Williamson, 2004; Sanita et al., 2005; Syers, 1969; Williams & Rudolph, 1974). Observations proved that the lichen thallus or lichen extracts contribute to detoxification processes, without however discriminating between the active substances, phenolic lichen substances and/or organic acids such as oxalic acid, carbonic acid, citric acid, which are very abundant in lichen thalli. So far, few experiments really studied the interaction of metal ion with phenolic lichen compounds to evaluate their chelation ability (Purvis et al., 1987; Takani, Yajima, Masuda, & Yamauchi, 2002). The results showed that complexation of these compounds occurs, a physicochemical property already described for flavonoid compounds (Dangles, 2012).

16.4.3 Are Lichen Compounds Antioxidant?

To retrieve the articles describing the antioxidant activities of lichen compounds, a protocol was applied (Scheme 16.1). After discarding the articles for which it was difficult to appreciate the antioxidant activity, 22 articles were selected. Each lichen compound was then classified according the mechanism of the bioassay involved (HAT, ET, ROS, enzyme activity). The ratio of its activity was compared to the standard, calculated and reported in Tables 16.1–16.3

16.4.3.1 Chemical In vitro Methods

Based on the HAT-based mechanism, nine experiments using β-carotene bleaching method, lipid peroxidation inhibition (in lecithine liposome and in linoleic acid emulsion tests), total reactive antioxidant potential-total antioxidant reactivity (TRAP-TAR), thiobarbituric acid reactive substances (TBARS), autoxidation and ORAC assays were conducted (Table 16.1). Twenty compounds belonging to several lichenic families (three depsides, six depsidones, six monoaromatic phenols, one dibenzofurane, one tridepside, one steroid, one gamma-lactone, one mycosporine) were tested. Unfortunately, most of them were tested once except atranorin, diffractaic acid, stictic acid and usnic acid. The depside atranorin exhibited a weak activity compared to the control and when it was active (Melo et al., 2011), the concentration at which it was tested revealed a pro-oxidant activity (Melo et al., 2011). Diffractaic acid showed also a weak activity. The depsidone stictic acid was not very active in two assays but exhibited a good activity in the ORAC assay, which is a test of reference (Apak et al., 2013). Activity of the well-known dibenzofurane usnic acid was ambiguous because it showed either a very good activity for ORAC assay in two experiments either a weak activity in three other assays. Among the 19 compounds six of them were evaluated twice (pulmonarianin, isidiophorin, lecanoric acid, rhizonaldehyde, rhizonyl alcohol, ergosterol peroxide). Lecanoric acid and rhizonyl alcohol were active in both assays while pulmonarianin, isidiophorin and rhizonaldehyde showed divergent results; ergosterol peroxide was less active. Among the compounds evaluated once, six compounds were active (1′-chloropannarin, pannarin, perlatolic acid methyl orsellinate, orsellinic acid) and salazinic acid exhibited an interesting activity in the ORAC assay. Thereby, compounds appeared as weak hydrogen donors compared to the well-known quercetin (Reiterer, Toborek, & Hennig, 2004).

A great number of compounds were tested using ET mechanism (2,2-di(4-tert-octylphenyl)-1-picrylhydrazyl (DPPH), ABTS, trolox-equivalent antioxidant capacity (TEAC), reducing power, Cu^{2+} reduction) because of the simple feasibility of their implementation (Table 16.2). Among the 43 molecules, seven were tested several times: atranorin, diffractaic acid, lecanoric acid, lobaric acid, methyl orcinol carboxylate, usnic acid and ramalin. Some divergent results were obtained for all except for ramalin. But the trend was to consider lecanoric acid and ramalin the most potent compared to the others. Several patents have been filled for ramalin, which showed no or very little cytotoxicity in human keratinocyte and fibroblast cells at its antioxidant concentration (Paudel et al., 2011). Thirty-five compounds were tested twice: 23 were weak ET donors and for five of them divergent results were obtained. Eight compounds were promising antioxidant compounds: divaricatic acid, ergosterol peroxide, ethyl haemmatommate, isidiophorin, lobastin, mycosporine glycine, sekikaic acid, rhizonaldehyde and solorinine but they were tested once. Even a great number of compounds have been evaluated, most of them showed a weak reducing activity in ET assays and so a structure activity relationship is difficult to be elicited.

For the specific assays (Table 16.3), 33 compounds were assayed to scavenge OH radical (15 assays), NO radical (15 assays) and superoxide anion radical (28 assays). Except usnic acid and atranorin, none of these compounds were tested more than twice as an ROS scavenger. Atranorin was a scavenger against hydroxyl radical but was also pro-oxidant at the tested concentration and usnic acid seems to be a good superoxide anion scavenger as potent as the controls in three independent experiments. Out of the 29 remaining compounds tested once, diffractaic acid, gyrophoric acid, ethyl haematommate and protolichesterinic acid were the most active against hydroxyl radical. Erythrin, lecanoric acid, methyl haematommate, methyl orsellinate, montagnetol and orsellinic acid were the most scavengers on NO radical. Out of the 28 compounds tested on superoxide anion, 50% of the molecules were active (divaricatic acid, erythrin, lecanoric acid, fumarprotocetraric acid, lobaric acid, norstictic acid, physodic acid, protocetraric acid, sekikaic acid, salazinic acid, methyl haematommate, methyl orsellinate, orcinol and ramalin). Compounds seemed to be more active against specific ROS and less divergent results were seen compared to HAT and ET assays.

16.4.3.2 Structure–Activity Relationship between the Lichen Compounds

Several arguments could explain the lack of reactivity of most compounds in HAT and ET experiments and were well explained by a study realised by Lopes (Lopes, Coelho,Yoshida, & Honda, 2008). He evaluated the structure–activity relationship of lecanoric acid and some synthesised derivatives of orsellinic acid and orsellinates (methyl, ethyl, *n*-propyl, *n*-butyl, *iso*-propyl, *sec*-butyl, *tert*-butyl orsellinates) using DPPH assay. All compounds were 500 times less active (IC_{50} = 1.84 mM) than gallic acid (IC_{50} = 6 μM). Nevertheless, he confirmed that:

- the carboxyl group could be an electron-withdrawing group in pH around 4.0 when the ionisation of the carboxyl group occurs and favours H-atom-transfer and electron-donating based on radical scavenging
- the establishment of hydrogen bonds between the oxygen of carbonyl and hydrogen of hydroxyl group in *o*-position also contribute to the lower activity of the orsellinic acid relative to the orcinol
- the esterification of the carboxyl group of the orsellinic acid produced a significant reduction in the radical scavenging activity (alkyl orsellinates showed decreased activities). The methyl group linked at *meta*-position relative to hydroxyl-groups of the orcinol (to make β-orcinol) has not significant effect on stabilisation of the radical by resonance
- the alkyl group in the *para*-position is favourable for electron delocalisation and stabilisation of the phenoxyl radical and the second hydroxyl group in the *ortho*- and *para*-position affects additional resonance stability because products of oxidation are so *ortho*- and *para*-quinones. For these lichen phenolics, the formation of quinone is not possible with these *meta*-dihydroxybenzene compounds. The chain elongation with methyl, ethyl, *n*-propyl, *n*-butyl esters increased activity and the ramified chain (*iso*-propyl, *tert*-butyl) were more active than *n*-propyl and tert-butyl groups. This observation is in favour of a better stabilisation of the phenoxyl radical compared with the corresponding linear chain. The alkyl chain connecting the phenolic ring increases its electron-donating activity and resulting in a well-stabilised phenoxy radical.

These arguments could also be extended to the observed results in HAT assays and could explain the lack of reactivity: hydrogen bond formation between –OH and COOH/CHO in *ortho*-position as for atranorin, lecanoric acid, divaricatic acid, rhizonylalcohol; methylation of electron attractor functions as for atranorin; position of aromatic hydroxyl groups in *meta*-position and not in *ortho*. However, even if these remarks could fit to some compounds as for isidiophorin, pulmonarianin, stictic acid, it cannot explain the good activity observed for lecanoric acid, 1′-chloropannarin and pannarin. In fact, the effects of substituents in the aromatic ring may cause different behaviour in the scavenging free radical activity.

16.4.3.3 *In cellulo* and *In vivo* Methods

Activities of lichen compounds were also tested on cell systems based on enzyme activities together with *in vivo* assays. In all these assays, stressed conditions were applied to cells or organs using an oxidant standard, and compounds were administrated to evaluate if they can prevent the production of ROS. Antioxidative activity of seven compounds (Table 16.4) were evaluated on cells but none was active except the hydrazide derivative ramalin, which can inhibit NO radical and H_2O_2 production in a better way than ascorbic acid, the positive control.

In vivo assays of two compounds the depside diffractaic acid and the dibenzofurane usnic acid were performed. On a previous study, indomethacin ingestion followed with an aqueous extract of *Usnea longissima* (50, 100, 200 mg/kg) vs ranitidine (150 mg/kg) was given orally to rats. Results showed a relative cytoprotective activity mediated by ROS (Halici et al., 2005). Then, the two major compounds of this lichen identified as diffractaic acid (Bayir et al., 2006) and usnic acid (Odabasoglu et al., 2006) were evaluated with the same protocol. Enzyme activities were modified in presence of these compounds in order to inactivate ROS generated by indomethacin. Diffractaic acid and usnic acid had an enhancing effect on antioxidant defence systems (superoxide dismutase, glutathione peroxidase, catalase, reduced glutathione, glutathione reductase) but an inhibiting one on neutrophil infiltration into gastric mucosal tissues. Compared with the results obtained with the *U. longissima* aqueous extract (Halici et al., 2005), effects are more pronounced with the compounds administrated alone. It may be explained by the fact that aqueous extract contain less (+)-usnic acid and diffractaic acid because hot water is not a suitable solvent for extracting them and that ranitidine dose was more appropriate in the second experiments (50 mg/kg instead of 150 mg/kg).

16.5 CONCLUSION

Coastal lichens and specifically marine lichens adapted their morphological and chemical organisation to ecological conditions. They are good models for studying the environmental impact on stress regulation. Indeed, most lichens in the littoral have a dense hyphal arrangement without air-filled spaces. Highly melanised cell walls are common in some of the intertidal species of Verrucariaceae but they lack the large amounts of extracellularly deposited substances (extrolites), which are common in many other lichens. The function of extrolites in the lichen symbiosis is not fully understood and at least some of them may have multiple roles, e.g. as antioxidant, herbivore repellent, antibiotic or for the modification of hydrophobicity in parts of the thallus. These functions may overlap or become important during different life stages of lichen. Most studies on antioxidant activities have focused on extrolites, which occur in maritime species. From the littoral, only the chemistry of *L. pygmaea* has been studied in greater detail (Roullier, Chollet-Krugler, Weghe, Devehat, & Boustie, 2010; Roullier et al., 2011) and its major compound mycosporine glycine exhibited a good activity in an assay based-on ET mechanism. Even for this lichen, it is still unknown if the mycosporine is synthesised by the fungal or cyanobacterial cells in the lichen system. Much more work remains to be done for the metabolic response of the algae from various different phyla associated with lichens of the families Verrucariaceae and Xanthopyreniaceae to lichenisation and their mechanisms of stress regulation in the lichenised and free-living state. Further groups that merit more detailed studies on their stress regulation are the specialised taxa in the surrounding of bird colonies, specialist taxa from brackish water or coastal freshwater seepages, the frost-sensitive *Trentepohlia*-associated lichens from sheltered rocks and the still insufficiently known diversity of lichens from tropical mangroves. Through literature, lichen compounds have been evaluated through a plethora of methods based on different chemical mechanisms (HAT, ET, specific assays, *in cellulo, in vivo* assays) and all articles showed that they could act as antioxidant molecules but less powerful than the flavonoids. However, in opposite to spermatophytes or other organisms, lichens contain two or three major compounds in large amounts *in situ* and under stress conditions the content of these secondary metabolites increased. So, it is reasonable to think that they play a role when primary antioxidant defences are overtaken by the ecologic conditions especially as their concentration becomes very important *in situ*. Their biological activity is not only a result of the different activities of individual components but may also be affected by antagonistic (Kosanic, Manojlovic, Jankovic, Stanojkovic, & Rankovic, 2013) or synergistic (Jayaprakasha & Rao, 2000; Luo et al., 2009; Luo et al., 2010) interactions among several components and fulfil different roles for the symbiotic partners in the lichen symbiosis.

REFERENCES

Alscher, R. G., Erturk, N., & Heath, L. S. (2002). Role of superoxide dismutases (SODs) in controlling oxidative stress in plants. *Journal of Experimental Botany*, *53*(372), 1331–1341.

Amo de Paz, G., Raggio, J., Gomez-Serranillos, M., Palomino, O., Gonzalez-Burgos, E., Carretero, M., et al. (2010). HPLC isolation of antioxidant constituents from *Xanthoparmelia spp. Journal of Pharmaceutical and Biomedical Analysis*, *53*, 165–171.

Apak, R., Gorinstein, S., Bohm, V., Schaich, K., Ozyürek, M., & Güçlü, K. (2013). Methods of measurement and evaluation of natural antioxidant capacity/activity (IUPAC technical report). *Pure and Applied Chemistry*, *85*, 957–998.

Aptroot, A., & Seaward, M. R. D. (1999). Annotated checklist of Hongkong lichens. *Tropical Bryology*, *17*, 57–101.

Ascaso, C., & Galvan, J. (1976). Studies on the pedogenic action of lichen acids. *Pedobiologia*, *16*, S321–S331.

Ascaso, C., Galvan, J., & Ortega, C. (1976). The pedogenic action of *Parmelia conspersa, Rhizocarpon geographicum* and *Umbilicaria pustulata*. *The Lichenologist*, *8*, 151–171.

Atalay, F., Halici, M., Mavi, A., Cakir, A., Odabasoglu, F., Kazaz, C., et al. (2011). Antioxidant phenolics from *Lobaria pulmonaria* (L.) Hoffm. and *Usnea longissima* Ach. lichen species. *Turkish Journal of Chemistry*, *35*, 647–661.

Aydin, E., & Turkez, H. (2011). Antioxidant and genotoxicity screening of aqueous extracts of four lichens collected from north east Anatolia. *Fresenius Environmental Bulletin*, *20*(8a), 2085–2091.

Bahattarai, H., Paudel, B., Hong, S., Lee, H., & Yim, J. (2008). Thin layer chromatography analysis of antioxidant constituents of lichens from Antarctica. *Journal of Natural Medicine*, *62*, 481–484.

Bahattarai, H., Kim, T., Oh, H., & Yim, J. (2013). A new pseudodepsidone from the antarctic lichen *Stereocaulon alpinum* and its antioxidant, antibacterial activity. *The Journal of Antibiotics*, 481–484.

Bates, S. T., Cropsey, G. W., Caporaso, J. G., Knight, R., & Fierer, N. (2011). Bacterial communities associated with the lichen symbiosis. *Applied and Environmental Microbiology*, *77*, 1309–1314.

Bayir, Y., Odabasoglu, F., Cakir, A., Aslan, A., Suleyman, H., Halici, M., et al. (2006). The inhibition of gastric mucosal lesion, oxidative stress and neutrophil-infiltration in rats by the lichen constituent diffractaic acid. *Phytomedicine*, *13*, 584–590.

Beckett, R. P., Kranner, I., & Minibayeva, F.V. (2008). *Stress physiology and the symbiosis.* Lichen biology. N. I. T.H. Cambridge. Cambridge University Press, 134–151.

Benatti, M. N., & Marcelli, M. P. (2007). Gêneros de fungos liquenizados dos manguezais do Sul-Sudeste do Brazil com enfoque no manguezal do Rio Itanhaém. *Acta Botanica Brasilica*, *21*, 863–878.

Bialonska, D., & Dayan, F. (2005). Chemistry of the lichen *Hypogymnia physodes* transplanted to an industrial region. *Journal of Chemical Ecology*, *31*, 2975–2991.

Bienert, G. P., Møller, A. L. B., Kristiansen, K. A., Schulz, A., Møller, I. M., Schjoerring, J. K., et al. (2007). Specific aquaporins facilitate the diffusion of hydrogen peroxide across membranes. *The Journal of Biological Chemistry*, *282*(2), 1183–1192.

Bjelland, T., Grube, M., Hoem, S., Jorgensen, S. L., Daae, F. L., Thorset, I. H., et al. (2011). Microbial metacommunities in the lichen-rock habitat. *Environmental Microbiology Reports*, *3*, 434–442.

Blasco, A., Crevillen, A., Gonzales, M., & Escarpa, A. (2007). Direct electrochemical sensing and detection of natural antioxidants and antioxidant capacity *in vitro* systems. *Electroanalysis*, *19*, 2275–2286.

Boots, A., Haenen, G., & Bast, A. (2008). Health effects of quercetin: from antioxidant to nutraceutical. *European Journal of Pharmacology*, *585*, 325–337.

Brisdelli, F., Perilli, M., Sellitri, D., Piovano, M., Garbarino, J., Nicoletti, M., et al. (2013). Cytotoxic activity and antioxidant capacity of purified lichen metabolites: an *in vitro* study. *Phytotherapy Research*, *27*, 431–437.

Caviglia, A. M., Nicora, P., Giordani, P., Brunialti, G., & Modenesi, P. (2001). Oxidative stress and usnic acid content in *Parmelia caperata* and *Parmelia soredians* (Lichenes). *Il Farmaco*, *56*, 379–382.

Chu, F. J., Seaward, M. R. D., & Hodgkiss, I. J. (2000). Effects of wave exposure and aspect on Hong Kong supralittoral lichens. *The Lichenologist*, *32*, 155–170.

Dangles, O. (2012). Antioxydant activity of plant phenols: chemical mechanisms and biological significance. *Current Organic Chemistry*, *16*, 692–714.

De la Coba, F., Aguilera, J., Figueroa, F., De Calvez, M., & Herrera, E. (2009). Antioxidant activity of mycosporine-like amino acids isolated from three red macroalgae and one marine lichen. *Journal of Applied Phycology*, *21*, 161–169.

Delmail, D., & Labrousse, P. (Eds.). (2012). *Plant ageing, a counteracting agent to xenobiotic stress. Senescence:* . InTech Publishers.

Delmail, D., Grube, M., Parrot, D., Cook-Moreau, J., Boustie, J., Labrousse, P., et al. (2013). *Halotolerance in lichens: Symbiotic coalition against salt stress*, 115–148.

Edreva, A. (2005). Generation and scavenging of reactive oxygen species in chloroplasts: a submolecular approach. *Agriculture, Ecosystems & Environment*, *106*, 119–133.

Evans, R., & Reynhout, G. (1992). Alternates to synthetic oxidants. *Developments in Food Science*, *29*, 27–42.

Fini, A., Brunetti, C., Di Ferdinando, M., Ferrini, F., & Tattini, M. (2011). Stress-induced flavonoid biosynthesis and the antioxidant machinery of plants. *Plant Signaling & Behavior*, *6*(5), 709–711.

Fletcher, A. (1973a). The ecology of marine (littoral) lichens on some rocky shores of Anglesey. *The Lichenologist*, *5*, 368–400.

Fletcher, A. (1973b). The ecology of maritime (surpalittoral) lichens on some rocky shores of Anglesey. *The Lichenologist*, *5*, 401–422.

Fletcher, A. (1980). Marine and maritime lichens of rocky shores: their ecology, physiology and biological interactions. In J. H. Price, D. E. G. Irvine & W. E. Farnham (Eds.), *The shore environment* 2 (pp. 789–842). London: Academic Press.

Fornazier, R. F., Ferreira, R. R., Vitória, A. P., Molina, S. M. G., Lea, P. J., & Azevedo, R. A. (2002). Effects of cadmium on antioxidant enzyme activities in sugar cane. *Biologia Plantarum*, *45*(1), 91–97.

Garbary, D. J., & London, J. (1995). The *Ascophyllum/Polysiphonia/Mycosphaerella* symbiosis. V. *Mycosphaerella* protects *A. nodosum* from desiccation. *Botanica Marina*, *38*, 529–533.

Gill, S. S., & Tuteja, N. (2010). Reactive oxygen species and antioxidant machinery in abiotic stress tolerance in crop plants. *Plant Physiology and Biochemistry*, *48*(12), 909–930.

Green, T. G. A., Sancho, L. G., & Pintado, A. (2011). *Ecophysiology of desiccation/rehydration cycles in mosses and lichens*, 10, 89–120. Springer.

Grube, M., & Berg, G. (2009). Microbial consortia of bacteria and fungi with focus on the lichen symbiosis. *Fungal Biology Reviews*, *23*, 72–85.

Grube, M., & Blaha, J. (2005). *Halotolerance and lichen symbioses*. Dordrecht: Springer.

Gueidan, C., Savi, S., Thüs, H., Roux, C., Keller, C., Tibell, L., et al. (2009). Generic classification of the Verrucariaceae (Ascomycota) based on molecular and morphological evidence: recent progress and remaining challenges. *Taxon*, *58*, 184–208.

Gueidan, C., Thüs, H., & Pérez-Ortega, S. (2011). Phylogenetic position of the brown algae-associated lichenized fungus *Verrucaria tavaresiae* (Verrucariaceae). *The Bryologist*, *114*, 563–569.

Gulluce, M., Aslan, A., Sokmen, F., Adiguzel, A., Agar, G., & Sokmen, A. (2006). Screening the antioxidant and antimicrobial properties of the lichens *Parmelia saxatilis, Platismatia glauca, Ramalina pollinaria, Ramalina polymorpha, Umbilicaria nylanderiana*. *Phytomedicine*, *13*, 515–521.

Halici, M., Odabasoglu, F., Suleyman, H., Cakir, A., Aslan, A., & Bayir, Y. (2005). Effects of water extract of *Usnea longissima* on antioxidant enzyme activity and mucosal damage caused by indomethacin in rats. *Phytomedicine*, *12*, 656–662.

Halliwell, B., & Gutteridge, J. (2003). *Free radicals in biology and medecine*. Oxford: Oxford Science Publications University Press.

Halliwell, B. (1990). How to characterize a biological antioxidant. *Free Radical Research Communications*, *9*, 1–32.

Hamlett, C. A. E., Shirtcliffe, N. J., Pyatt, F. B., Newton, M. I., McHale, G., & Koch, K. (2011). Passive water control at the surface of a superhydrophobic lichen. *Planta*, *234*, 1267–1274.

Hara, K., Endo, M., Kawakami, H., Komine, M., & Yamamoto, Y. (2011). Anti-oxidation activity of ethanol extracts from natural thalli of lichens. *Mycosystema, 30*(6), 950–954.

Hauck, M., Jürgens, S.-R., Brinkmann, M., & Herminghaus, S. (2008). Surface hydrophobicity causes SO_2 tolerance in lichens. *Annals of Botany, 101*, 531–539.

Hauk, M., & Huneck, S. (2007). Lichen substances affect metal adsorption in *Hypogymnia physodes*. *Journal of Chemical Ecology, 33*, 219–223.

Hawksworth, D. L. (2000). Freshwater and marine lichen-forming fungi. *Fungal Diversity, 5*, 1–7.

Hawskworth, D. L., & Honegger, R. (1994). *The lichen thallus: A symbiotic phenotype of nutritionally specialised fungi and its response to gall producers*. Fairlawn: Clarendon.

Heng, L., Yamamoto, Y., Liu, Y., Jung, J., Kahng, H.-Y., Koh, Y., et al. (2010). The *in vitro* anti-oxidant properties of chinese Highlands Lichens. *Journal of Microbiology and Biotechnology, 20*(11), 1524–1528.

Hernández, I., Alegre, L., Van Breusegem, F., & Munné-Bosch, S. (2009). How relevant are flavonoids as antioxidants in plants? *Trends in Plant Science, 14*(3), 125–132.

Hidalgo, M., Fernandez, E., Quilhot, W., & Lissi, E. (1994). Antioxidant activity of depsides and depsidones. *Phytochemistry, 37*(6), 1585–1587.

Hillard, E., de Abreu, F., Melo Ferreira, D., Jaouen, G., Fonseca Goulart, M., & Amatore, C. (2008). Electrochemical parameters and techniques in drug development, with an emphasis on quinones and related compounds. *Chemical Communications*, 2612–2628.

Huang, D., Ou, B., & Prior, R. (2005). The chemistry behind antioxidant capacity assays. *Journal of Agricultural and Food Chemistry, 53*, 1841–1856.

Huneck, S., & Yoshimura, I. (1996). *Identification of lichen substances*. Berlin: Springer-Verlag Berlin and Heidelberg GmbH & Co. K.

Iskandar, I., & Syers, J. (1972). Metal-complex formation by lichen compounds. *Journal of Soil Science, 23*(3), 255–265.

Ismed, F., Lohézic-Le Dévéhat, F., Delalande, O., Sinbandhit, S., Bakhtiar, A., & Boustie, J. (2012). Lobarin from the Sumatran lichen, *Stereocaulon halei*. *Fitoterapia, 83*(8), 1693–1698.

Jagadeesh Ram, T. A. M., . Aptroot, A., Sinha, G. P., & Singh, K. P. (2005). New species and new records of lichenized and non-lichenized pyrenocarpous ascomycetes from the Sundarbans Biosphere Reserve, India. *Mycotaxon, 91*, 455–459.

Jayaprakasha, G. K., & Rao, L. J. (2000). Phenolics constituents from the lichen *Parmotrema stuppeum* (Nyl.) Hale and their antioxidant activtity. *Zeitschrift für Naturforschung, 55c*, 1018–1022.

Jones, D., & Wilson, M. (1985). Chemical activity of lichens on mineral surfaces – a review. *International Biodeterioration, 21*(2), 99–104.

Kamal-Eldin, A., Kim, H. J., Tavadyan, L., & Min, D. (2008). Tocopherol concentrations and antioxidant efficacy. *Lipid Oxidation Pathways, 2*, 127–141.

Kohlmeyer, J., & Volkmann-Kohlmeyer, B. (1988). *Halographis* (Opegraphales), a new endolithic lichenoid from corals and snails. *Canadian Journal of Botany, 66*, 1138–1141.

Kohlmeyer, J., & Volkmann-Kohlmeyer, B. (1998). Mycophyscias, a new genus for the mycobionts of *Apophlaea, Ascophyllum and Pelvetia. Systema Ascomycetum, 16*, 1–7.

Kohlmeyer, J., Hawksworth, D. L., & Volkmann-Kohlmeyer, B. (2004). Observations on two borderline lichens: *Mastodia tesselata* and *Collemopsisdium pelvetiae*. *Mycological Progress, 3*, 51–56.

Kosanic, M., & Rankovic, B. (2011a). Antioxidant and antimicrobial properties of some lichen species and their constituents. *Journal of Medicinal Food, 14*(12), 1624–1630.

Kosanic, M., & Rankovic, B. (2011b). Lichens as possible sources of antioxidants. *Pakistan journal of Pharmaceutical Sciences, 24*(2), 165–170.

Kosanic, M., Rankovic, B., & Vukojevic, J. (2011). Antioxidant properties of some lichen species. *Journal of Food Science and Technology, 48*(5), 584–590.

Kosanic, M., Rankovic, B., & Stanojkovic, T. (2012). Antioxidant, antimicrobial and anticancer activity of 3 Parmelia species. *Journal of the Science of Food and Agriculture, 92*, 1909–1916.

Kosanic, M., Manojlovic, N., Jankovic, S., Stanojkovic, T., & Rankovic, B. (2013). *Evernia prunastri* and *Pseudevernia furfuraceae* lichens and their major metabolites as antioxidant, antimicrobial and anticancer agents. *Food and Chemical Toxicology, 53*, 112–118.

Kovácik, L., & Pereira, A. B. (2001). Green alga *Prasiola crispa* and its lichenized form *Mastodia tesselata* in Antarctic environment: general aspects. *Beihefte zur Nova Hedwigia, 123*, 465–478.

Kranner, I. (2002). Glutathione status correlates with different degrees of desiccation tolerance in three lichens. *New Phytologist, 154*(2), 451–460.

Kranner, I., Cram, W. J., Wornik, S., Yoshimura, I., Stabentheiner, E., & Pfeifhofer, H. (2005). Antioxidants and photoprotection in a lichen as compared with its isolated symbiotic partners. *Proceedings of the National Academy of Sciences, 102*(8), 3141–3146.

Kranner, I., & Lutzoni, F. (1999). Evolutionary consequences of transition to a lichen symbiotic state and physiological adaptation to oxidative damage associated with poikilohydry. In H. R. Lerner (Ed.), *Plant response to environmental stress: from phytohormones to genome reorganisation* (pp. 591–628). New York: Dekker.

Kranner, I., Zorn, M., Turk, B., Wornik, S., Beckett, R. P., & Batic, F. (2003). Biochemical traits of lichens differing in relative desiccation tolerance. *New Phytologist, 160*(1), 167–176.

Kruger, N. J., & von Schaewen, A. (2003). The oxidative pentose phosphate pathway: structure and organisation. *Current Opinion in Plant Biology, 6*(3), 236–246.

Kumar, S., & Muller, K. (1999). Lichen metabolites.I. Inhibitory action against leukotriene B4 biosynthesis by a non-redox mechanism. *Journal of Natural Product, 62*, 817–820.

Lagadic, L., Caquet, T., & Amiard, J. C. (1997). *Biomarqueurs en écotoxicologie: Principes et définitions*. (Paris).

Lamb, I. (1973). *Further observations on Verrucaria serpuloides M. Lamb, the only known permanently submerged marine lichen*. Occasional Papers of the Farlow Herbarium of Cryptogamic Botany. Harward University.

Le Bourvellec, C., Hauchard, D., Darchen, A., Burgot, J., & Abasq, M. (2008). Validation of a new method using the reactivity of electrogenerated superoxide radical in the antioxidant capacity determination of flavonoids. *Talanta, 75*, 1098–1103.

Li, Y., Trush, M. A., & Yager, J. D. (1994). DNA damage caused by reactive oxygen species originating from a copper-dependent oxidation of the 2-hydroxy catechol of estradiol. *Carcinogenesis, 15*(7), 1421–1427.

Lohézic-Le Dévéhat, F., Tomasi, S., Elix, J., Bernard, A., Rouaud, I., Uriac, P., et al. (2007). Stictic acid derivatives from the lichen usnea articulata and their antioxidant activities. *Journal of Natural Products, 70*, 1218–1220.

Lopes, T. I. B., Coelho, R. G., Yoshida, N. C., & Honda, N. K. (2008). Radical scavenging activity of orsellinates. *Chemical and Pharmaceutical Bulletin, 56*(11), 1551–1554.

Lopez-Alarcon, C., & Denicola, A. (2013). Evaluating the antioxidant capacity of natural products: a review on chemical and cellular-based assays. *Analytical Chimica Acta, 763*, 1–10.

Luo, H., Yamamoto, Y., Kim, J. A., Jung, J., Koh, Y., & Hur, J. (2009). Lecanoric acid, a secondary lichen substance with antioxidant properties from *Umbilicaria antarctica* in maritime Antarctica (King George Island). *Polar Biology, 32*, 1033–1040.

Luo, H., Yamamoto, Y., Liu, Y., Wang, L., Jung, J., Koh, Y., et al. (2010). Antioxidant activities of edible lichen *Ramalina conduplicans* and its free radical-scavenging constituents. *Mycosciences, 51*, 391–395.

Manojlovic, N., Rankovic, B., Kosanic, M., Vasiljevic, P., & Stanojkovic, T. (2012). Chemical composition of three Parmelia lichens and antioxidant, antimicrobial and cytotoxic activities of some major metabolites. *Phytomedicine, 19*, 1166–1172.

Melo, M., Dos Santos, J., Serafini, M., Caregnato, F., Pasquali, M., Rabelo, T., et al. (2011). Redox properties and cytoprotective actions of atranorin, a lichen secondary metabolite. *Toxicology In Vitro, 25*, 462–468.

Moe, R. (1997). *Verrucaria tavaresiae sp.* nov., a marine lichen with a brown algal photobiont. *Bulletin of the California Lichen Society, 4*, 7–11.

Mollinedo, P., Vila, J. L., & Sterner, O. (2003). Actividad antioxidante y fotoprotectora de metabolitos secundarios aislados de liquenes. *Revista Boliviana De Quimica, 20*(1), 21–27.

Mulky, M. (1976). BHA and BHT as foddstuff antioxidants. *Journal of the Oil Technologists Association of India, 8*(4), 134–139.

Muller, K., & Gurster, D. (1993). Hydroxyl radical damage to DNA sugar and model membranes induced by anthralin (dithranol). *Biochemical Pharmacology, 46*(10), 1695–1704.

Odabasoglu, F., Aslan, A., Cakir, A., Suleyman, H., Karagoz, Y., Bayir, Y., et al. (2005). Antioxidant activity, reducing power and total phenolic content of some lichen species. *Fitoterapia, 76*, 216–219.

Odabasoglu, F., Cakir, A., Suleyman, H., Aslan, A., Bayir, Y., Halici, M., et al. (2006). Gastroprotective and antioxidant effects of usnic acid on indomethacin-induced gastric ulcer in rats. *Journal of Ethnopharmacology, 103*, 59–65.

Oliveto, E. (1972). Nordihydroguaiaretic acid. A naturally occuring antioxidant. Chemistry and Industry. L. S. o. C. Industry. 17: 677–679.

Orange, A. (2012). Semi-cryptic marine species of Hydropunctaria (Verrucariaceae, lichenized Ascomycota) from north-west Europe. *The Lichenologist, 44*, 299–320.

Papadopoulou, P., Tzakou, O., Vagias, C., Kefalas, P., & Roussis, V. (2007). B-orcinol metabolites from the lichen *Hypotrachyna revoluta*. *Molecules, 12*, 997–1005.

Parent, C., Capelli, N., & Dat, J. (2008). Formes réactives de l'oxygène, stress et mort cellulaire chez les plantes. *Comptes Rendus Biologies, 331*(4), 255–261.

Parke, D., & Lewis, D. (1992). Safety aspects of food preservatives. *Food additives and contaminants. L. a. T. a. F. Ltd, 9*, 561–577.

Paudel, B., Bhattarai, H., Lee, J., Hong, S., Shin, H., & Yim, Y. (2008). Antioxidant activity of polar lichens from King George Island (Antarctica). *Polar Biology, 5*, 605–608.

Paudel, B., Bhattarai, H., Koh, Y., Lee, S., Han, S., Lee, H., et al. (2011). Ramalin, a novel nontoxic antioxidant compound from the Antarctic lichen *Ramalina terebrata*. *Phytomedicine, 18*, 1285–1290.

Pereira, G. J. G., Molina, S. M. G., Lea, P. J., & Azevedo, R. A. (2002). Activity of antioxidant enzymes in response to cadmium in *Crotalaria juncea*. *Plant and Soil, 239*(1), 123–132.

Pérez-Ortega, S., de los Rios, A., Crespo, A., & Sancho, L. (2010). Symbiotic lifestyle and phylogenetic relationships of the bionts of *Mastodia tessellata* (Ascomycota, Incertae Sedis). *American Journal of Botany, 97*, 738–752.

Purvis, O., Bailey, E., McLean, J., Kasama, T., & Williamson, B. (2004). Uranium biosorption by the lichen *Trapelia involuta* at a uranium mine. *Geomicrobiology Journal, 21*, 159–167.

Purvis, O., Elix, J., Broomhead, J., & Jones, G. (1987). The occurence of copper-norstictic acid in lichens from cupriferous substrata. *The Lichenologist, 19*(2), 193–203.

Rankovic, B., Rankovic, D., & Maric, D. (2010). Antioxidant and antimicrobial activity of some lichen species. *Microbiology, 79*(6), 809–815.

Rankovic, B., Kosanic, M., & Stanojkovic, T. (2011). Antioxidant, antimicrobial and anticancer activity of the lichens *Cladonia furcata, Lecanora atra* and *Lecanora muralis*. *BMC Complementary and Alternative Medicine, 11*, 97.

Rankovic, B., Kosanic, M., Stanojkovic, T., Vasiljevic, P., & Manojlovic, N. (2012). Biological activities of *Toninia candida* and *Usnea barbata* together with their norstictic acid and usnic acid constituents. *International Journal of Molecular Sciences, 13*, 14707–14722.

Reiterer, G., Toborek, M., & Hennig, B. (2004). Quercetin protects against linoleic acid-induced Porcine Endothelial cell Dysfunction. *Journal of Nutrition, 134*(4), 771–775.

Rene, A., Abasq, M., Hauchard, D., & Hapiot, P. (2010). How do phenolic compounds react toward superoxide ion? A simple electrochemical method for evaluating antioxidant capacity. *Analytical Chemistry, 82*, 8703–8710.

Roullier, C., Chollet-Krugler, M., Pferschy-Wenzig, E. M., Maillard, A., Rechberger, G. N., Legouin-Gargadennec, B., et al. (2011). Characterization and identification of mycosporines-like compounds in cyanolichens. Isolation of mycosporine hydroxyglutamicol from *Nephroma laevigatum* Ach. *Phytochemistry, 72*, 1348–1357.

Roullier, C., Chollet-Krugler, M., Weghe, P. V. D., Devehat, F. L. -L., & Boustie, J. (2010). A novel aryl-hydrazide from the marine lichen *Lichina pygmaea*: isolation, synthesis of derivatives, and cytotoxicity assays. *Bioorganic and Medicinal Chemistry Letters, 20*, 4582–4586.

Sanita di Toppi, L., Marabottini, R., Vattuone, Z., Musetti, R., Favali, M., Sorgona, A., et al. (2005). Cell wall immobilisation and antioxidant status of *Xanthoria parietina* thalli exposed to cadmium. *Functional Plant Biology, 32*, 611–618.

Schaich, K. M. (2005). Lipid oxidation: theoretical aspects. In F. Shahidi (Ed.), *Bailey's Industrial Oil and Fat products* (pp. 269–355). Hoboken: John Wiley & Sons.

Sies, H., & Stahl, W. (1995). Vitamins E and C, beta-carotene, and other carotenoids as antioxidants. *The American Journal of Clinical Nutrition, 62*(Suppl.), 1315S–1321S.

Smirnoff, N. (1993). The role of active oxygen in the response of plants to water deficit and desiccation. *New Phytologist, 125*(1), 27–58.

Stevens, G. N. (1979). Distribution and related ecology of macrolichens on mangroves on the east Australian coast. *The Lichenologist, 11*, 293–305.

Stojanovic, G., Stojanovic, I., Stankov-Jovanovic, V., Mitic, V., & Kostic, D. (2010). Reducing power and radical scavenging activity of four Parmeliaceae species. *Central European Journal of Biology, 5*(6), 808–813.

Syers, J. (1969). Chelating ability of fumarprotocetraric acid and *Parmelia conspersa*. *Plant and Soil, 21*(1), 205–208.

Takani, M., Yajima, T., Masuda, H., & Yamauchi, O. (2002). Spectroscopic and structural characterization of copper(II) and palladium(II) complexes of a lichen substance usnic acid and its derivatives. Possible forms of environmental metals retained in lichens. *Journal of Inorganic Biochemistry, 91*, 139–150.

Takeshita, S., & Harada, H. (2001). Maritime lichens and their photobionts (1). Distribution of photobionts at Choshi, Chiba Pref., central Japan. *Journal of Plant Research Supplement, 114*, 43–44.

Thadhani, V., Choudhary, M., Ali, S., Omar, I., Siddique, H., & Karunaratne, V. (2011). Antioxidant activity of some lichen metabolites. *Natural Product Research, 25*(19), 1827–1837.

Thompson, J. E., Legge, R. L., & Barber, R. F. (1987). The role of free radicals in senescence and wounding. *New Phytologist, 105*(3), 317–344.

Thüs, H., Muggia, L., Pérez-Ortega, S., Favero-Longo, S. E., Joneson, S., O'Brien, H., et al. (2011). Revisiting photobiont diversity in the lichen family Verrucariaceae (Ascomycota). *European Journal of Phycology, 46*, 399–415.

Wangemai, l., Luemail, Z., Sunemail, K., & Zhu, W. (2011). Effects of high salt stress on secondary metabolite production in the Marine-Derived fungus *Spicaria elegans*. *Marine drugs, 9*(4), 535–542.

Wedin, M., Döring, H., & Gilenstam, G. (2004). Saprotrophy and lichenization as options for the same fungal species on different substrata: environmental plasticity and fungal lifestyles in the *Stictis-Conotrema* complex. *New Phytologist, 164*, 459–464.

Williams, M., & Rudolph, E. (1974). The role of lichens and associated fungi in the chemical weathering of rock. *Mycologia, 66*, 648–660.

AUTHOR INDEX

Note: Page numbers with "*f*" denote figures; "*t*" tables.

A

B

C

E

F

G

H

I

J

K

L

N

O

P

S

T

U

V

W

X

Y

Z

SUBJECT INDEX

Note: Page numbers with "*f*" denote figures; "*t*" tables.

A

D

E

F

O

P

COLOUR PLATES

Jonas Collén *et al.*, Figure 3.2 The life cycle of *Chondrus crispus*. A nonmotile spermatium fertilises the carpogonium which is retained on the female gametophytic thallus (*syngamy*). Within the cystocarp, the zygote is mitotically amplified potentially liberating thousands of genetically identical diploid carpospores following a single fertilisation event. The diploid carpospores produce the diploid free-living tetrasporophyte. *Meiosis* occurs in the tetrasporophyte, releasing haploid tetraspores which form the free-living female and male gametophytes. *Images: S.A. Krueger-Hadfield.*

Alejandro H. Buschmann *et al.*, Figure 6.1 Kelp landings in the north of Chile: *Lessonia trabeculata* (A) and *Macrocystis pyrifera* (B).

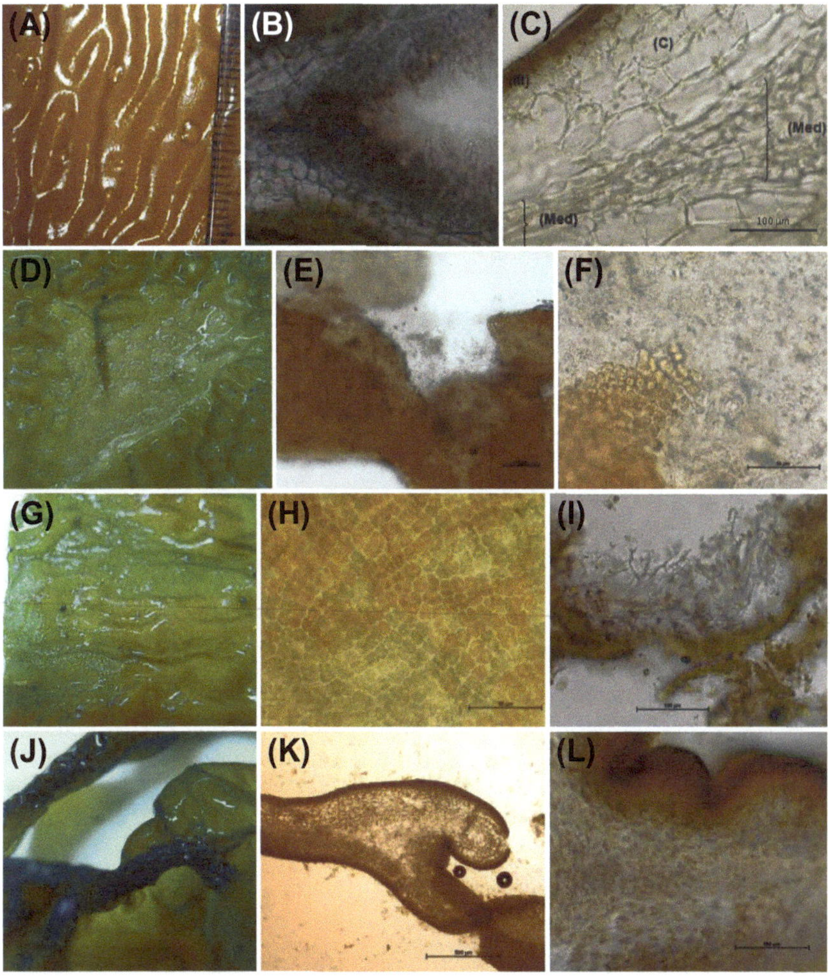

Alejandro H. Buschmann *et al.*, Figure 6.3 Kelp disease symptoms found in kelp cultivated in a 21 ha farm located in southern Chile. Protuberance of the lamina: (A) Superficial view of protuberance on the lamina; (B) Cross-section of the protuberance; (C) Opening of the medulla due to protuberance ((Med) = medulla and (C)= cortex). White spot on the lamina: (D) Superficial view of white spot; (E) and (F) Cross-section showing presence of bacteria. Green spot on the lamina: (G) and (H) Superficial view of green spot; (I) Cross-section of lamina showing decompression of the cells of the meristoderm, breakdown and loss of cortical cells. Plication of the lamina: (J) Superficial view of the frond; (K) and (L) Cross-section of twisted showing overgrowth of the meristodermatic area.

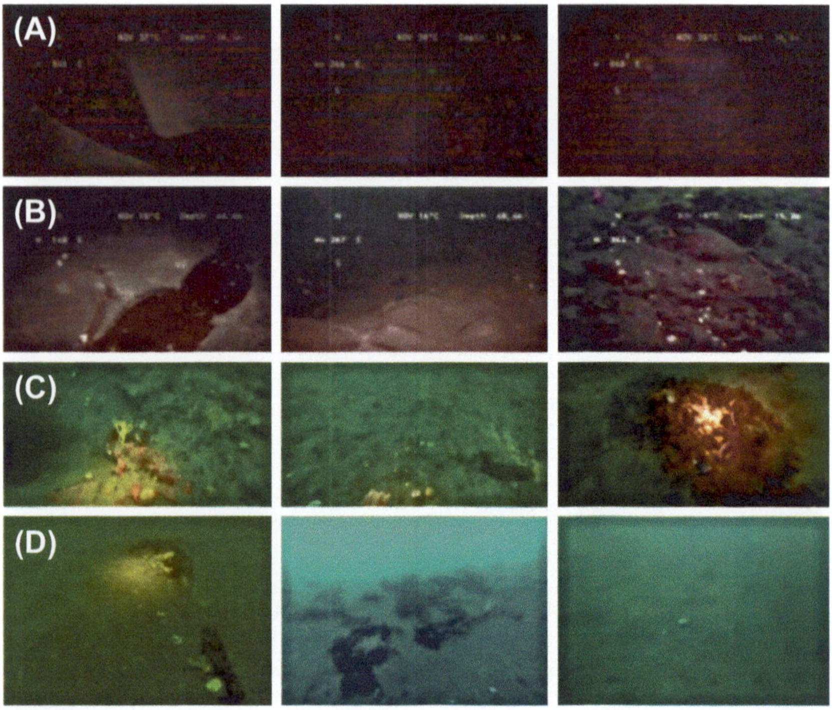

Alejandro H. Buschmann *et al.*, Figure 6.4 Underwater bottom images under the 21 ha kelp farm (42° 29′27″S; 73° 18′28″W). Images correspond to transects made at: (A) summer 2012, (B) winter 2012, (C) spring 2012, (D) summer 2013.

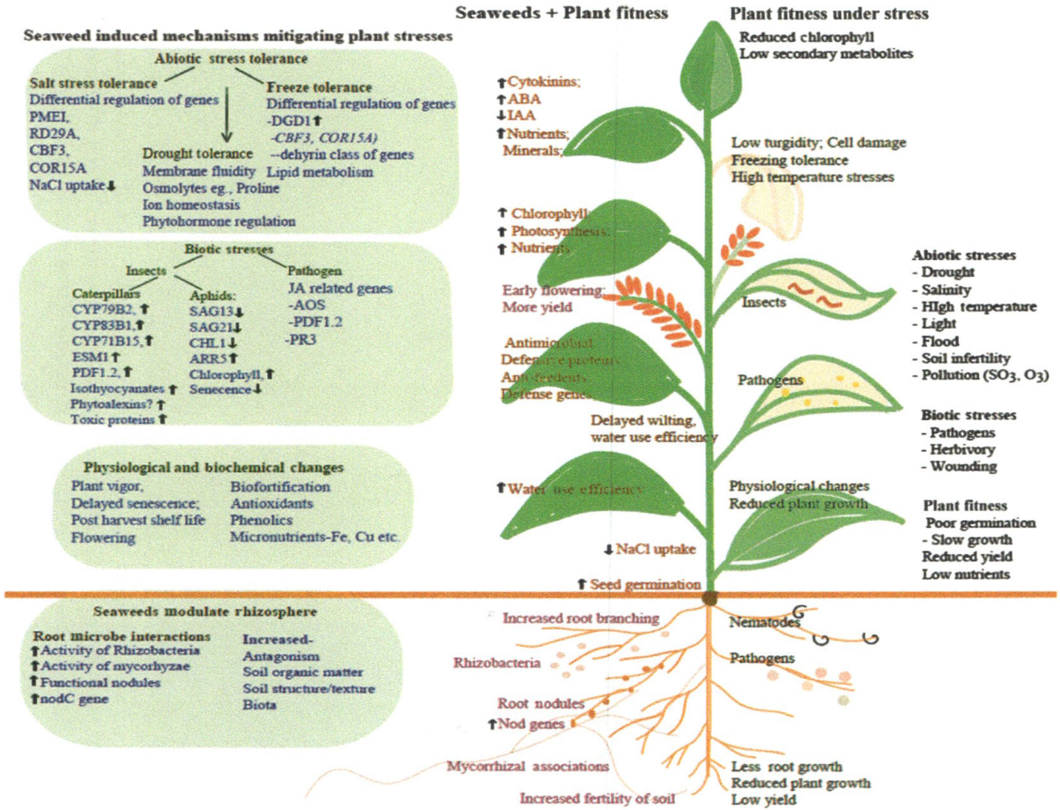

Balakrishnan Prithiviraj *et al.*, Figure 7.1 A schematic representation of effect and mechanism(s) of activity of seaweed extracts.

Marina Ciancia *et al.*, Figure 9.1 (A) General aspect of the tallus of *Codium vermilara*. (B) Detail of the tallus of *Codium decorticatum*. (C) General aspect of the dry thallus of *Codium fragile*. Scale bar = 1 cm. (D) Inner structure of the thallus of *C. fragile*. Scale bar = 250 μm (me: medullar region; co: photsynthetic cortex). (E) Detail of an utricle of *C. fragile*. Scale bar = 250 μm (mu: mucron). (F) Detail of an utricle of *C. decorticatum* (ga: gametangium). (G) Detail of an utricle of *C. vermilara*. *(C), (D), and (E) from Estevez et al., 2009; (G) from Fernández et al., 2010; (B) and (F) courtesy from María Paula Raffo (Cenpat-Conicet, Muerto Madryn, Argentina).*

Kevin Hardouin *et al.*, Figure 10.1 (A) *Kappaphyccus* sp. farming on ropes and (B) open air drying in Bali, Indonesia, April 2013. © *Balakrishnan Prithiviraj and Kevin Hardouin.*

Taxonomy:	
Empire:	Eukaryota
Kingdom:	Plantae
Phylum:	Chlorophyta
Class:	Ulvophyceae
Order:	Ulvales
Family:	Ulvaceae
Genus:	Ulva
Specie:	*Ulva sp.*

Kevin Hardouin *et al.*, Figure 10.2 Taxonomy and picture of *Ulva* sp. © *Kristian Peters.*

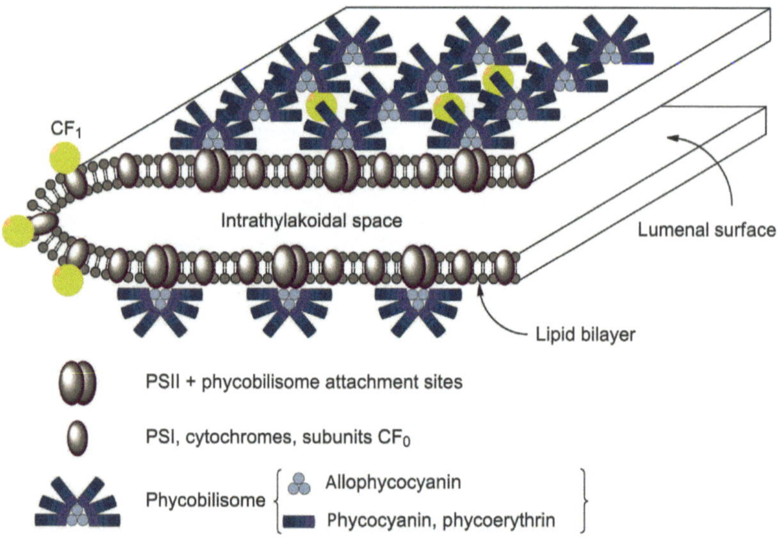

Justine Dumay *et al.*, Figure 11.1 Schematic representation of the thylakoid membrane and phycobilisomes. CF, coupling factor; PS, photosystem.

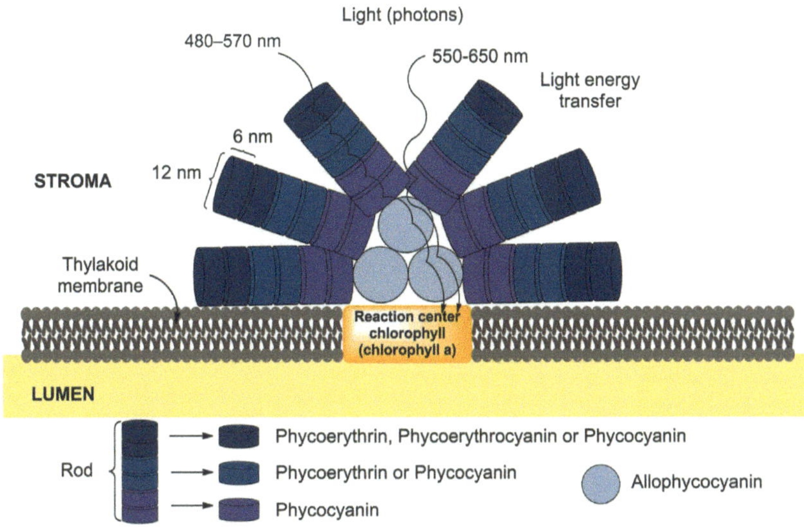

Justine Dumay *et al.*, Figure 11.3 Typical phycobilisome structure in red algae.

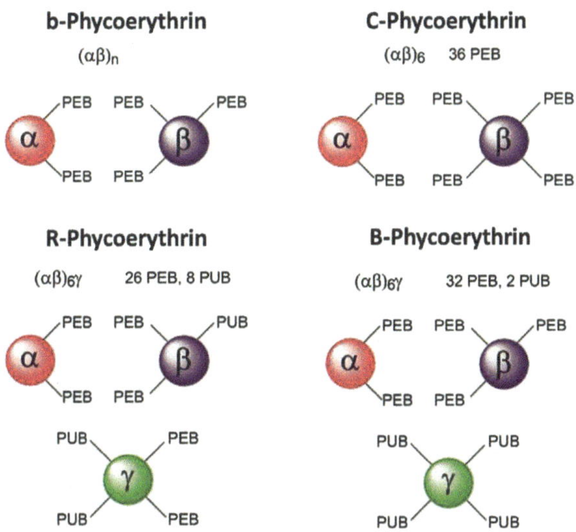

Justine Dumay *et al.*, Figure 11.7 Phycoerythrin structures.

Valérie Stiger-Pouvreau *et al.*, Figure 13.1 Illustration of the eight species of Sargassaceae from Brittany (France); up first line, from left to right and following lines: *Bifurcaria bifurcata, Cystoseira baccata, C. foeniculacea, C. humilis, C. nodicaulis, C. tamariscifolia, Halidrys siliquosa* and *Sargassum muticum. Photo credits: Klervi.Le-Lann©Lemar-iuem-ubo for pictures 1 and 8; Camille.Jegou©Lubem-ubo for pictures 2,3,4,5,6 and Erwan.Amice© CNRS-iuem-ubo for picture 7.*

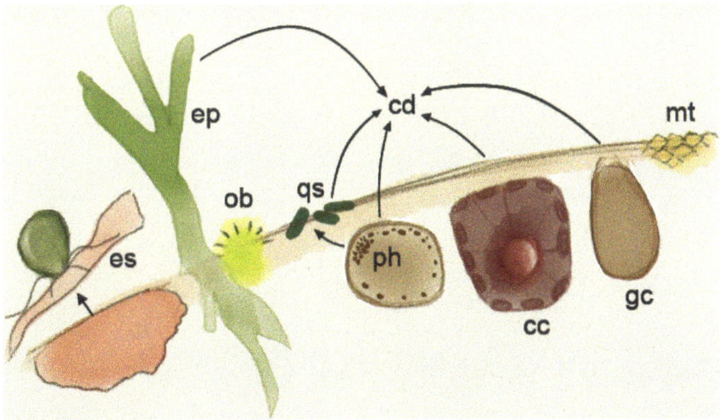

Bernardo A. P. da Gama *et al.*, Figure 14.1 The multifaceted defence system of marine macroalgae against epibionts (**ep**). Physical defences such as epithallus sloughing (**es**) of colonised surfaces (a settling *Ulva* spore is represented) has been demonstrated in red, brown and green seaweeds, while thallus microtopography (**mt**) has only recently been investigated. Cellular structures related to AF chemical defence (**cd**), in particular halogenated natural products, were identified in thalli cortex from *Laurencia* spp. (*corps en cerise* – **cc**), *Delisea pulchra, Asparagopsis armata* (gland cells – **gc**), and *Plocamium brasiliense* and are possibly still unveiled in other red algae of the class Florideophyceae. Physodes (**ph**) are known intracellular structures related to polyphenolic chemical defences in brown algae, often suggested as antifoulants. Oxidative bursts (**ob**) and bacterial quorum sensing (**qs**) inhibition are AF mechanisms recently demonstrated in a number of algae. Despite the negative effects to thalli, defended macroscopic epibionts and specific bacteria may also produce compounds that chemically defend or camouflage the basibiont alga, therefore providing an associational defence. Mechanisms such as thallus abandonment/proliferation are not represented (see text for references).

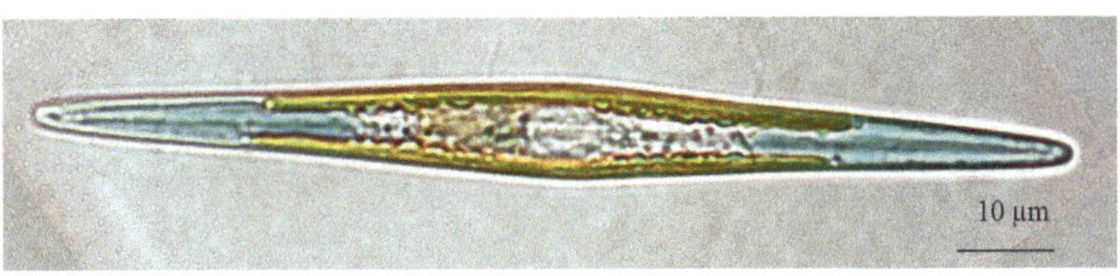

Jean-Luc Mouget *et al.*, Figure 15.1 Live cell of *Haslea ostrearia* in light microscopy (LM).

FSC
www.fsc.org
MIX
Papier | Fördert gute Waldnutzung
FSC® C083411

Kolibri 360 GmbH
Ferdinand-Jühlke-Straße 7
99095 Erfurt, Deutschland
produktsicherheit@kolibri360.de